VICKERS

INDUSTRIAL HYDRAULICS MANUAL

FIRST EDITION 1970

First Printing	April	1970
Second Printing	September	1970
Third Printing	May	1971
Fourth Printing	April	1972
Fifth Printing	March	1973
Sixth Printing	December	1974
Seventh Printing	July	1974
Eighth Printing	December	1974
Ninth Printing	November	1975
Tenth Printing	July	1976
Eleventh Printing	January	1977
Twelfth Printing	September	1977
Thirteenth Printing	June	1978
Fourteenth Printing	January	1979
Fifteenth Printing	July	1979
Sixteenth Printing	January	1980
Seventeenth Printing	October	1980
Eighteenth Printing	June	1981
Nineteenth Printing	January	1982
Twentieth Printing	October	1982
Twenty-first Printing	September	1983
Twenty-second Printing	May	1984
Twenty-third Printing	March	1985
Twenty-fourth Printing	March	1986
Twenty-fifth Printing	April	1987
Twenty-sixth Printing	March	1988
Twenty-seventh Printing	June	1989

Printed in U.S.A.

PREFACE

Like many branches of engineering, hydraulics is both ancient and modern. The use of the water wheel, for example, is so ancient that its invention precedes written history. On the other hand, the use of fluid under pressure to transmit power and to control intricate motions is relatively modern and has had its greatest development in the past two or three decades.

Power generation, the branch of hydraulics represented by the water wheel, does not concern us here. The steam engine, the internal combustion engine, the electric motor and the water turbine all have performed an admirable job in supplying motive power. However, each lacks the mechanics to direct this power to useful work. The purpose of this manual is to study the use of fluids under pressure in the transmission of power or motion under precise control.

We have often been asked the question, "Why is industrial hydraulics necessary when we have at our disposal many well known mechanical, pneumatic and electric devices?"

It is because a confined fluid is one of the most versatile means of modifying motion and transmitting power known today. It is as unyielding as steel, and yet infinitely flexible. It will change its shape to fit the body that resists its thrust, and it can be divided into parts, each part doing work according to its size, and can be reunited to work again as a whole.

It can move rapidly in one part of its length and slowly in another. No other medium combines the same degree of positiveness, accuracy, and flexibility, maintaining the ability to transmit a maximum of power in a minimum of bulk and weight.

The laws of physics governing fluids are as simple as the mechanics of solids and simpler than the laws governing electricity, vapors, or gases. The use of engineering in general, and hydraulics in particular, has been to achieve the end of extending man's physical and mental power to enable a job to be done more accurately, more quickly, and with a smaller expenditure of human energy.

Although this manual is devoted primarily to operation and maintenance of Vickers equipment, it includes general chapters covering basic hydraulics and all types of pumps, motors and controls. The Vickers equipment covered is limited to the representative series most commonly encountered in the machine tool industry.

In recent years, a trend has developed toward the establishment of standards in most phases of industry. In the field of hydraulics, probably the most significant efforts in this direction were started by the Joint Industry Conference (J. I. C.). The Joint Industry Conference was comprised of several recognized industrial associations interested in the establishment of industrial standards which would promote safety of personnel and ease of maintenance and increase the service life of equipment and tools. Since its published recommendations were received so favorably in the field of hydraulics, the efforts were continued by the American Standards Association (ASA) in cooperation with the National Fluid Power Association. The name of ASA has more recently been changed to American National Standards Institute (A. N. S. I.).

The standards established for graphical symbols and color coding of flow and pressures have been used throughout this manual. Significance of the symbols is discussed in Chapter 2 and Appendix II. The color key used in pictorials of components and in hydraulic lines is as follows:

RED	Operating or system pressure
BLUE	Exhaust flow
GREEN	Intake or drain
YELLOW	Measured (metered) flow
ORANGE	Reduced pressure, pilot pressure or charging pressure
VIOLET	Intensified pressure
BLANK	Inactive fluid

TABLE OF CONTENTS

Chapter	Title	Page
1	An Introduction to Hydraulics	1-1
2	Principles of Power Hydraulics	2-1
3	Hydraulic Fluids	3-1
4	Hydraulic Piping and Sealing	4-1
5	Reservoirs and Fluid Conditioners	5-1
6	Hydraulic Actuators	6-1
7	Directional Controls	7-1
8	Servo Valves	8-1
9	Pressure Controls	9-1
10	Volume Controls	10-1
11	Hydraulic Pumps	11-1
12	Accessories	12-1
13	Industrial Hydraulic Circuits	13-1

Appendices

I	Definitions of Technical Terms	Page 1
II	Graphical Symbols	Page 7

AN INTRODUCTION TO HYDRAULICS

The study of hydraulics deals with the use and characteristics of liquids. Since the beginning of time, man has used fluids to ease his burden. It is not hard to imagine a caveman floating down a river, astride a log with his wife--and towing his children and other belongings aboard a second log with a rope made of twisted vines.

Earliest recorded history shows that devices such as pumps and water wheels were known in very ancient times. It was not, however, until the 17th century that the branch of hydraulics with which we are to be concerned first came into use. Based upon a principle discovered by the French scientist Pascal, it relates to the use of confined fluids in transmitting power, multiplying force and modifying motions.

Pascal's Law, simply stated, says this:

Pressure applied on a confined fluid is transmitted undiminished in all directions, and acts with equal force on equal areas, and at right angles to them.

This precept explains why a full glass bottle will break if a stopper is forced into the already full chamber. The liquid is practically non-compressible and transmits the force applied at the stopper throughout the container (Fig. 1-1). The result is an exceedingly higher force on a larger area than the stopper. Thus it is possible to break out the bottom by pushing on the stopper with a moderate force.

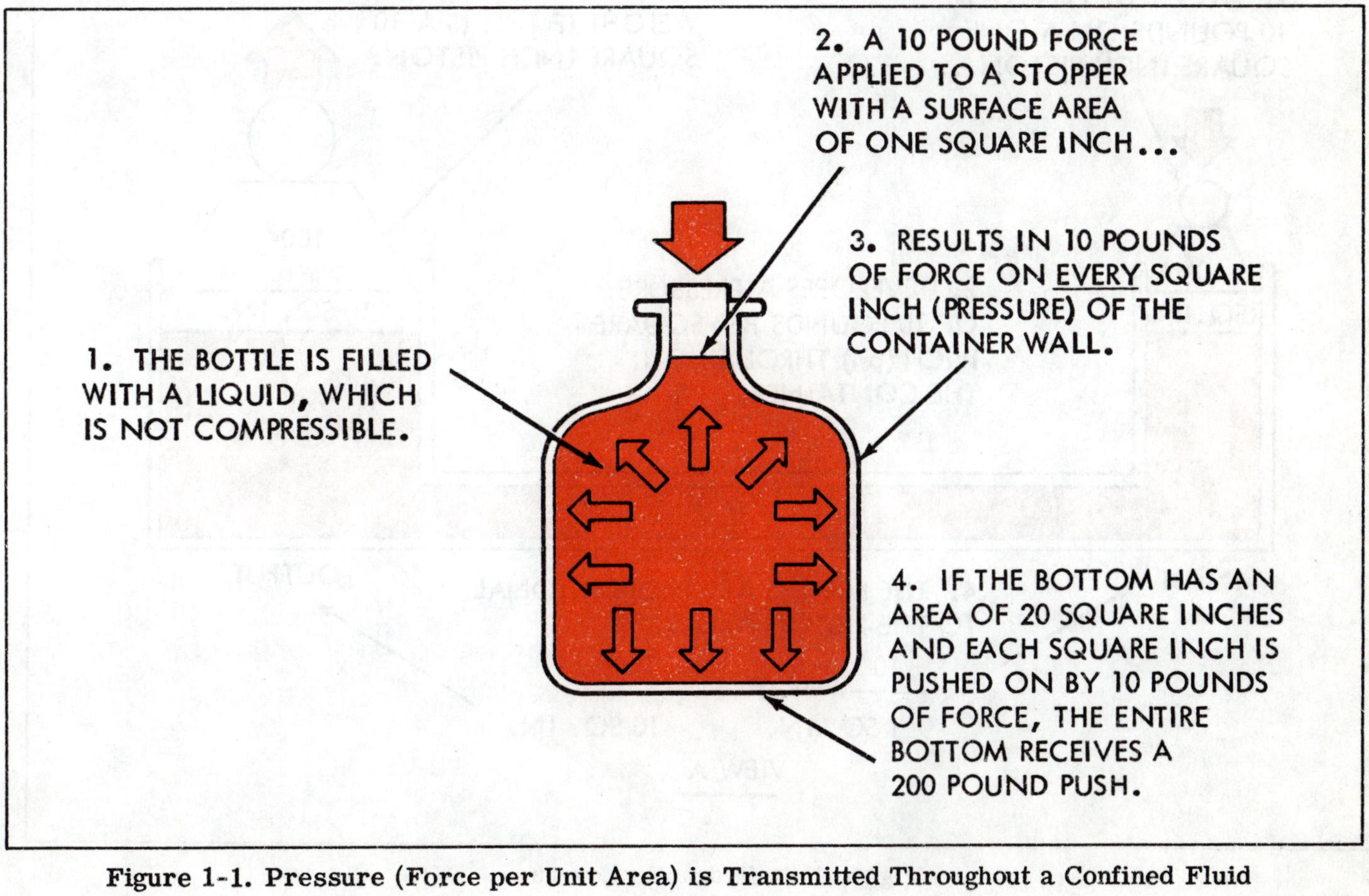

2. A 10 POUND FORCE APPLIED TO A STOPPER WITH A SURFACE AREA OF ONE SQUARE INCH...

3. RESULTS IN 10 POUNDS OF FORCE ON EVERY SQUARE INCH (PRESSURE) OF THE CONTAINER WALL.

1. THE BOTTLE IS FILLED WITH A LIQUID, WHICH IS NOT COMPRESSIBLE.

4. IF THE BOTTOM HAS AN AREA OF 20 SQUARE INCHES AND EACH SQUARE INCH IS PUSHED ON BY 10 POUNDS OF FORCE, THE ENTIRE BOTTOM RECEIVES A 200 POUND PUSH.

Figure 1-1. Pressure (Force per Unit Area) is Transmitted Throughout a Confined Fluid

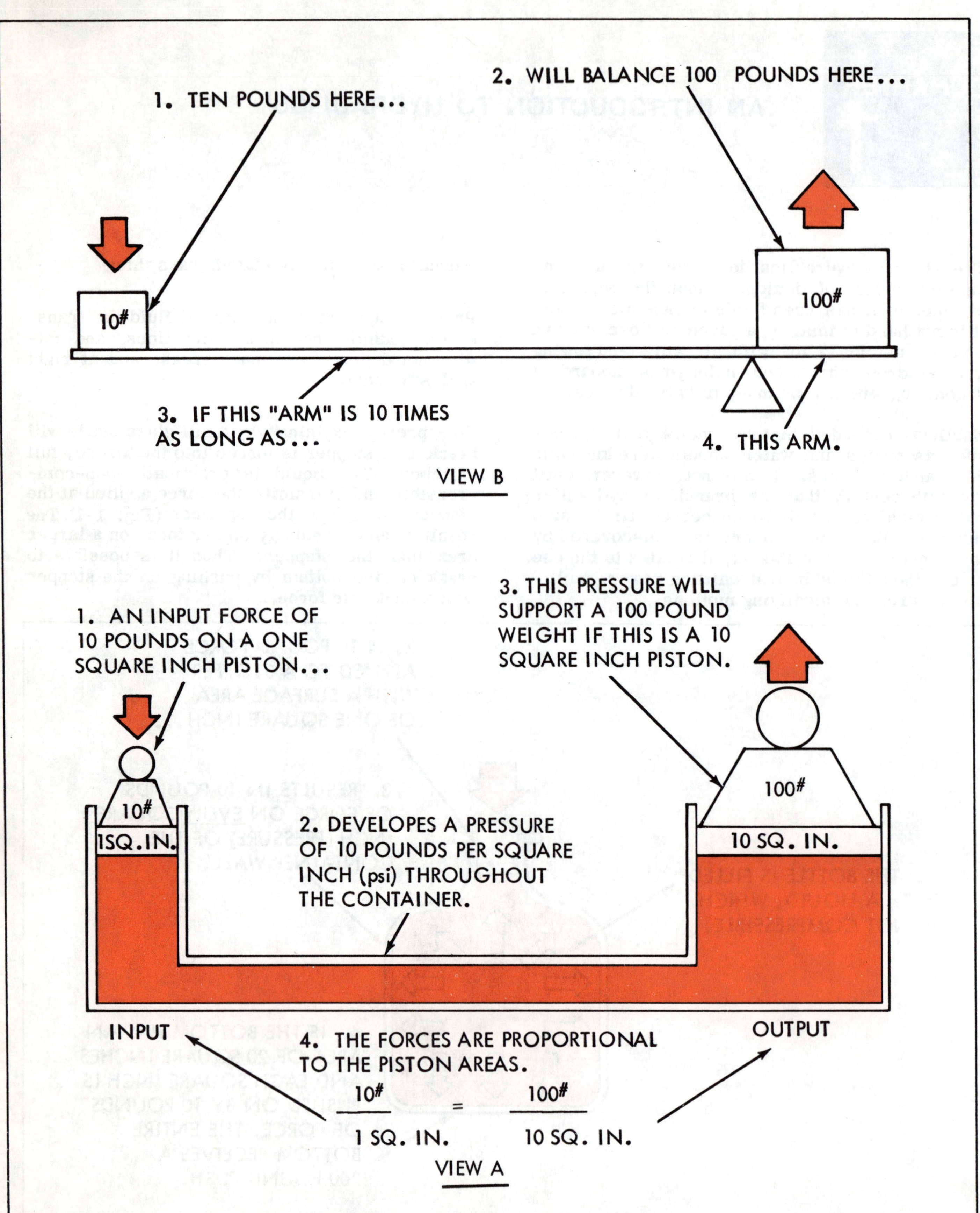

Figure 1-2. Hydraulic Leverage

Perhaps it was the very simplicity of Pascal's Law that prevented men from realizing its tremendous potential for some two centuries. Then, in the early stages of the industrial revolution, a British mechanic named Joseph Bramah utilized Pascal's discovery in developing a hydraulic press.

Bramah decided that, if a small force on a small area would create a proportionally larger force on a larger area, the only limit to the force a machine can exert is the area to which the pressure is applied.

Figure 1-2 shows how Bramah applied Pascal's principle to the hydraulic press. The applied force is the same as on the stopper in Fig. 1-1 and the small piston has the same one square inch area. The larger piston though, has an area of 10 square inches. The large piston is pushed on with 10 pounds of force per square inch, so that it can support a total weight or force of 100 pounds.

It can be seen easily that the forces or weights which will balance with this apparatus are proportional to the piston areas. Thus, if the output piston area is 200 square inches, the output force will be 2000 pounds (assuming the same 10 pounds of push on each square inch). This is the operating principle of the hydraulic jack, as well as the hydraulic press.

It is interesting to note the similarity between this simple press and a mechanical lever (view B). As Pascal had previously stated--here again force is to force as distance is to distance.

PRESSURE DEFINED

In order to determine the total force exerted on a surface, it is necessary to know the pressure or force on a unit of area. We usually express this Pressure In Pounds Per Square Inch, abbreviated psi. Knowing the pressure and the number of square inches of area on which it is being exerted, one can readily determine the total force.

(Force in Pounds = Pressure in psi x Area In Sq. In.)

CONSERVATION OF ENERGY

A fundamental law of physics states that energy can neither be created nor destroyed. The multiplication of force in Fig. 1-2 is not a matter of getting something for nothing. The large piston is moved only by the liquid displaced by the small piston making the distance each piston moves inversely proportional to its area (Fig. 1-3).

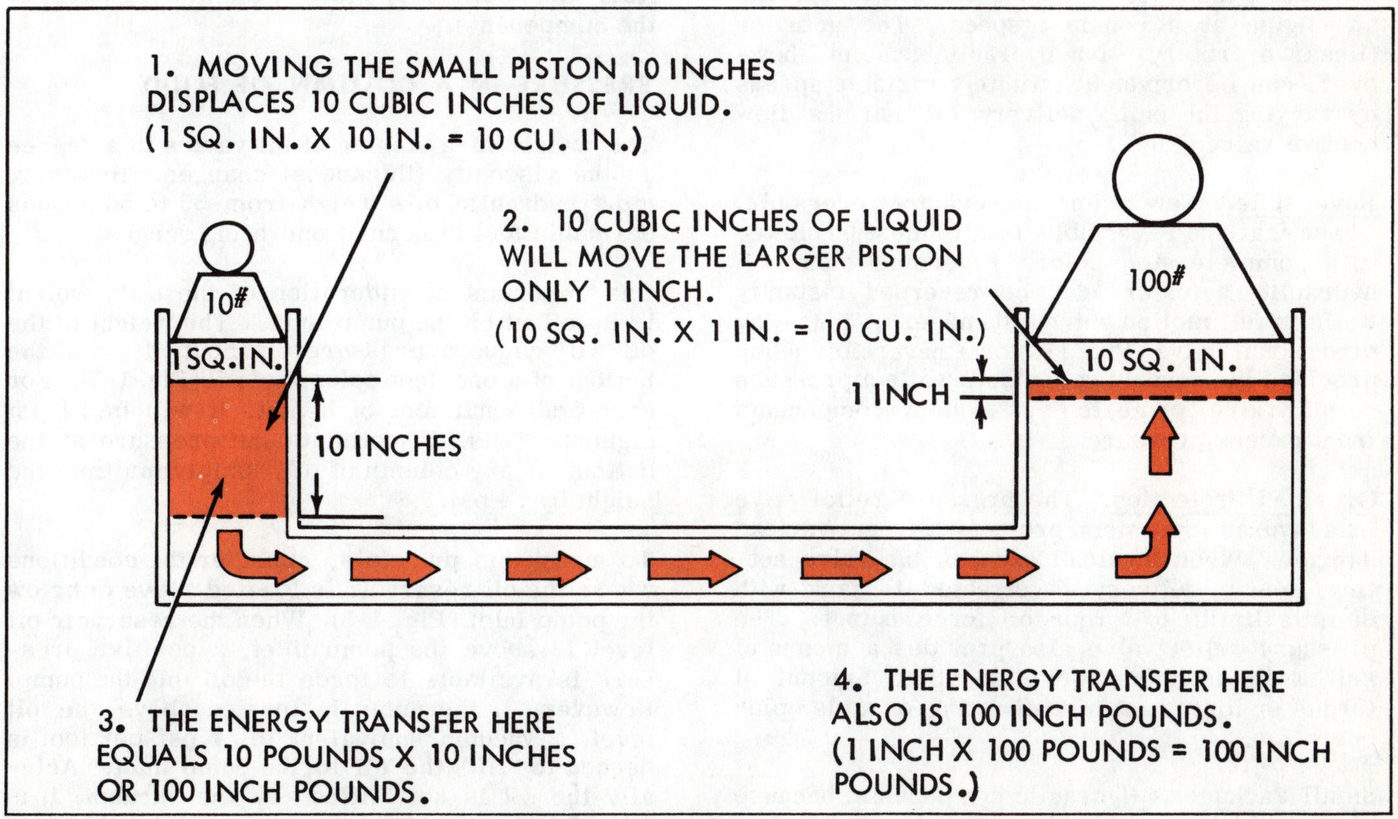

Figure 1-3. Energy Can Neither Be Created Nor Destroyed

What is gained in force must be sacrificed in distance or speed.

HYDRAULIC POWER TRANSMISSION

Hydraulics now could be defined as a means of transmitting power by pushing on a confined liquid. The input component of the system is called a pump; the output is called an actuator.

While for the sake of simplicity we have shown a single small piston, most power driven pumps would incorporate multiple pistons, vanes or gears as their pumping elements. Actuators are linear, such as the cylinder shown; or rotary, such as hydraulic motors (Fig. 1-4).

The hydraulic system is not a source of power. The power source is a prime mover such as an electric motor or an engine which drives the pump. The reader might ask, therefore, why not forget about hydraulics and couple the mechanical equipment directly to the prime mover? The answer is in the versatility of the hydraulic system, which gives it advantages over other methods of transmitting power.

ADVANTAGES OF HYDRAULICS

Variable Speed. Most electric motors run at a constant speed. It is also desirable to operate an engine at a constant speed. The actuator (linear or rotary) of a hydraulic system, however, can be driven at infinitely variable speeds by varying the pump delivery or using a flow control valve (Fig. 1-5).

Reversible. Few prime movers are reversible. Those that are reversible usually must be slowed to a complete stop before reversing them. A hydraulic actuator can be reversed instantly while in full motion without damage. A four-way directional valve (Fig. 1-6) or a reversible pump provides the reversing control, while a pressure relief valve protects the system components from excess pressure.

Overload Protection. The pressure relief valve in a hydraulic system protects it from overload damage. When the load exceeds the valve setting, pump delivery is directed to tank with definite limits to torque or force output. The pressure relief valve also provides a means of setting a machine for a specified amount of torque or force, as in a chucking or a clamping operation.

Small Packages. Hydraulic components, because of their high speed and pressure capabilities, can provide high power output with very small weight and size.

Can Be Stalled. Stalling an electric motor will cause damage or blow a fuse. Likewise, engines cannot be stalled without the necessity for restarting. A hydraulic actuator, though, can be stalled without damage when overloaded, and will start up immediately when the load is reduced. During stall, the relief valve simply diverts delivery from the pump to the tank. The only loss encountered is in wasted horsepower.

HYDRAULIC OIL

Any liquid is essentially non-compressible and therefore will transmit power instantaneously in a hydraulic system. The name hydraulics, in fact, comes from the Greek, hydor, meaning "water" and, aulos, meaning "pipe." Bramah's first hydraulic press and some presses in service today use water as the transmitting medium.

However, the most common liquid used in hydraulic systems is petroleum oil. Oil transmits power readily because it is only very slightly compressible. It will compress about 1/2 of one percent at 1000 psi pressure, a negligible amount in most systems. The most desirable property of oil is its lubricating ability. The hydraulic fluid must lubricate most of the moving parts of the components.

PRESSURE IN A COLUMN OF FLUID

The weight of a volume of oil varies to a degree as the viscosity (thickness) changes. However, most hydraulic oils weigh from 55 to 58 pounds per cubic foot in normal operating ranges.

One important consideration of the oil's weight is its effect on the pump inlet. The weight of the oil will cause a pressure of about .4 psi at the bottom of a one-foot column of oil (Fig. 1-7). For each additional foot of height, it will be .4 psi higher. Thus, to estimate the pressure at the bottom of any column of oil, simply multiply the height by .4 psi.

To apply this principle, consider the conditions where the oil reservoir is located above or below the pump inlet (Fig. 1-8). When the reservoir oil level is above the pump inlet, a positive pressure is available to force the oil into the pump. However, if the pump is located above the oil level, a vacuum equivalent to .4 psi per foot is needed to "lift" the oil to the pump inlet. Actually the oil is not "lifted" by the vacuum, it is forced by atmospheric pressure into the void

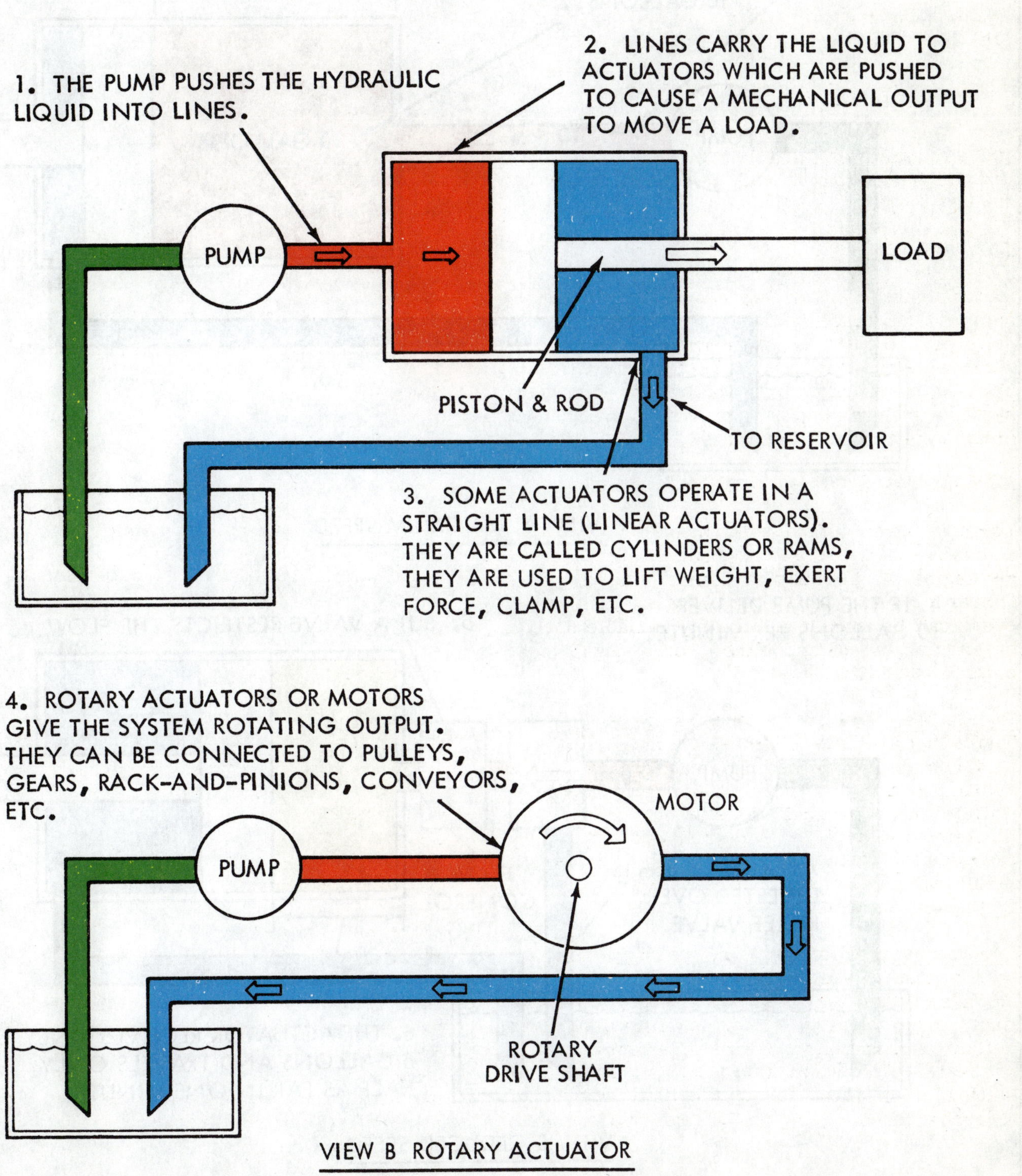

Figure 1-4. Hydraulic Power Transmission

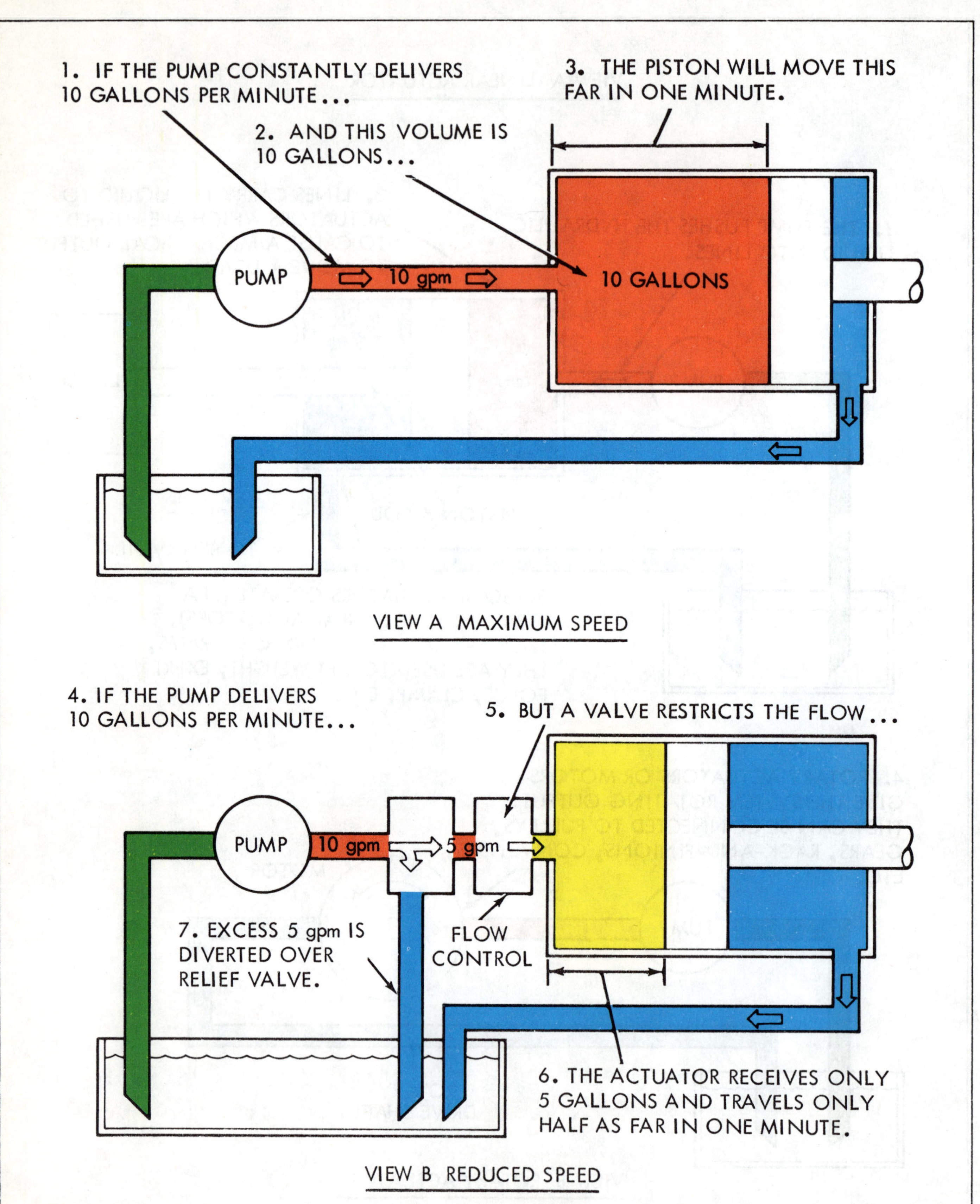

Figure 1-5. Hydraulic Drive Speed is Variable

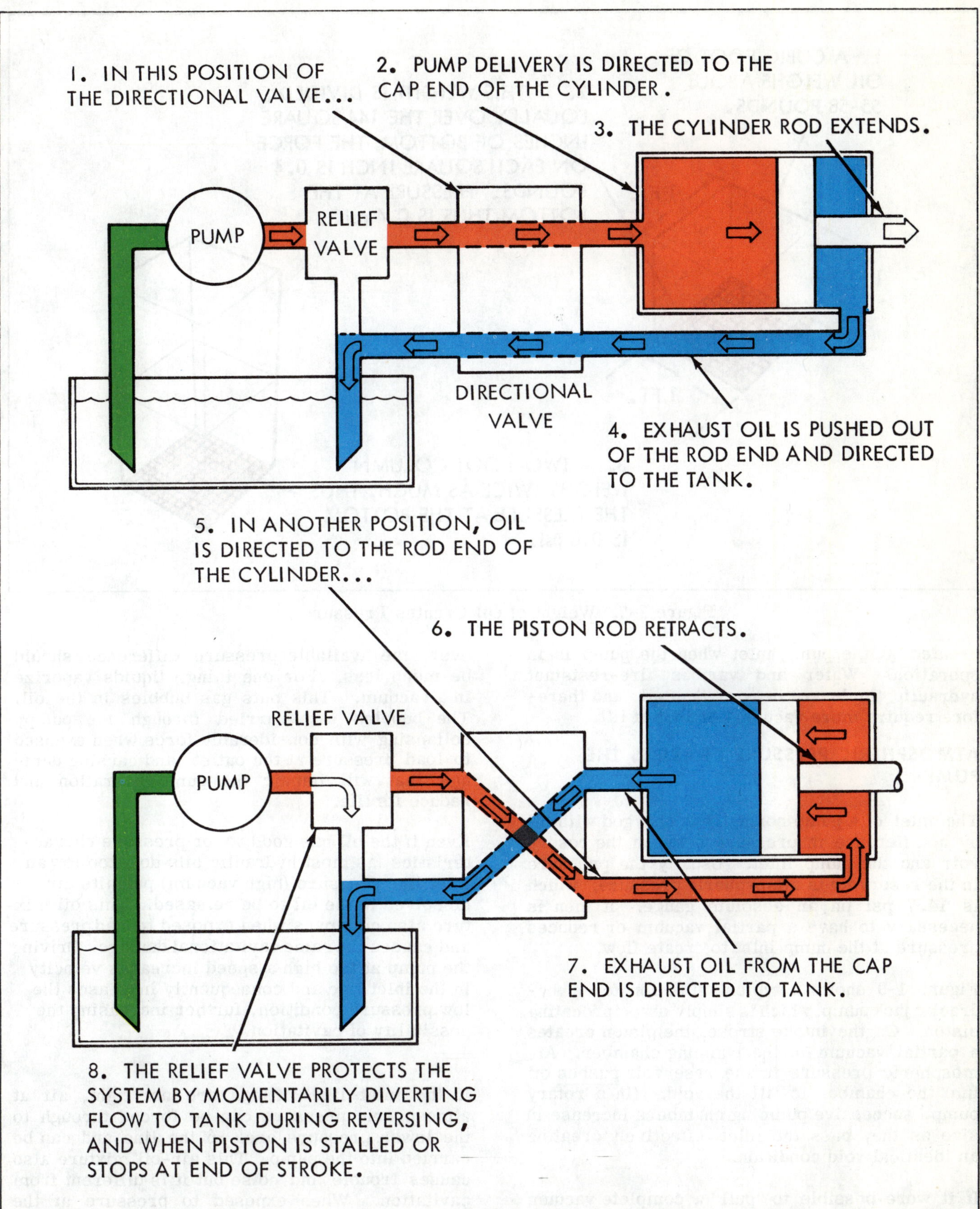

Figure 1-6. Hydraulic Drives are Reversible

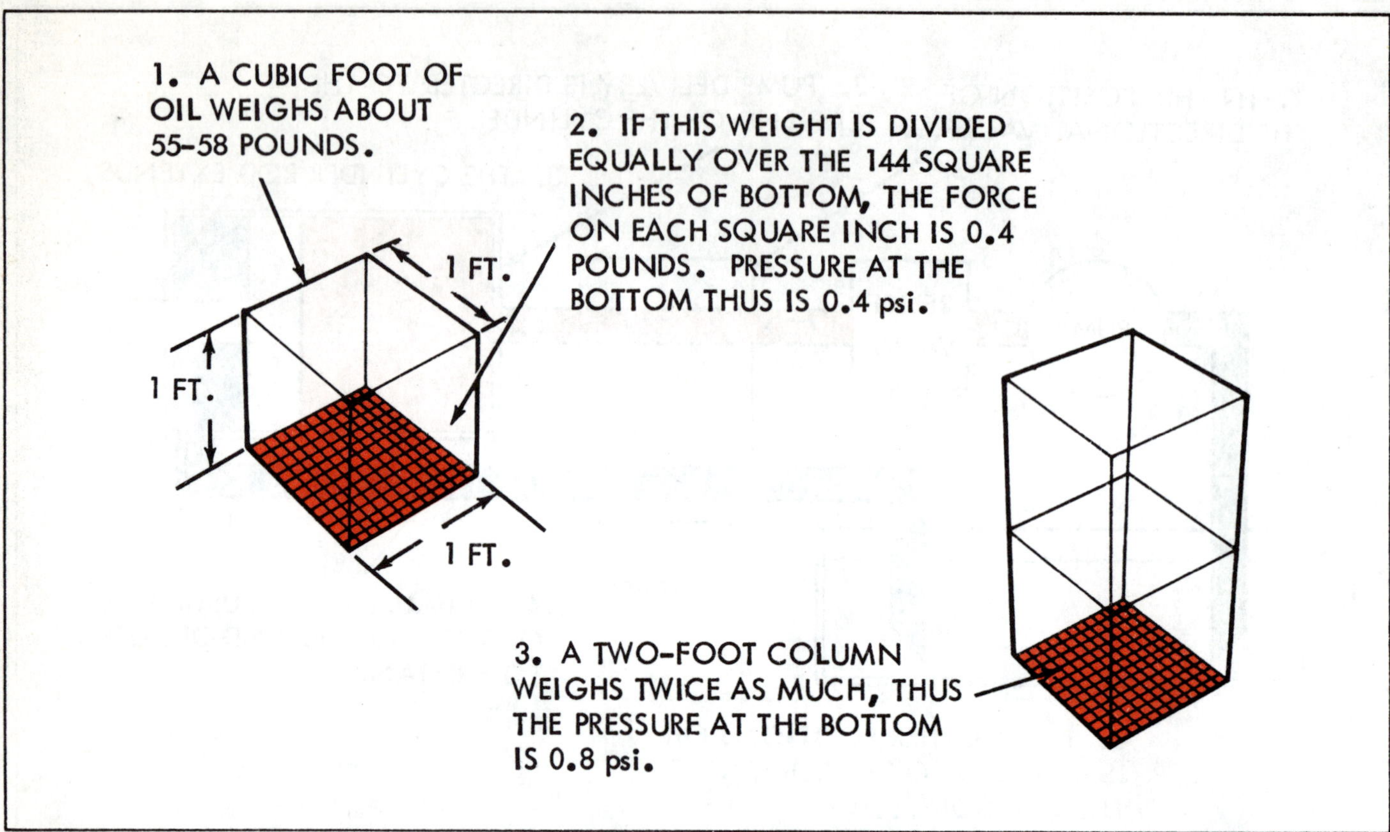

Figure 1-7. Weight of Oil Creates Pressure

created at the pump inlet when the pump is in operation. Water and various fire-resistant hydraulic fluids are heavier than oil, and therefore require more vacuum per foot of lift.

ATMOSPHERIC PRESSURE CHARGES THE PUMP

The inlet of a pump normally is charged with oil by a difference in pressure between the reservoir and the pump inlet. Usually the pressure in the reservoir is atmospheric pressure, which is 14.7 psi on an absolute gauge. It then is necessary to have a partial vacuum or reduced pressure at the pump inlet to create flow.

Figure 1-9 shows a typical situation for a hydraulic jack pump, which is simply a reciprocating piston. On the intake stroke, the piston creates a partial vacuum in the pumping chamber. Atmospheric pressure in the reservoir pushes oil into the chamber to fill the void. (In a rotary pump, successive pumping chambers increase in size as they pass the inlet, effectively creating an identical void condition.)

If it were possible to "pull" a complete vacuum at the pump inlet, there would be available some 14.7 psi to push the oil in. Practically, however, the available pressure difference should be much less. For one thing, liquids vaporize in a vacuum. This puts gas bubbles in the oil. The bubbles are carried through the pump; collapsing with considerable force when exposed to load pressure at the outlet, and causing damage that will impair the pump operation and reduce its life.

Even if the oil has good vapor pressure characteristics (as most hydraulic oils do), too low an inlet line pressure (high vacuum) permits air dissolved in the oil to be released. This oil mixture also collapses when exposed to load pressure and causes the same cavitational damage. Driving the pump at too high a speed increases velocity in the inlet line and consequently increases the low pressure condition, further increasing the possibility of cavitation.

If the inlet line fittings are not tight, air at atmospheric pressure can be forced through to the lower pressure area in the line and can be carried into the pump. This air-oil mixture also causes trouble and noise but it is different from cavitation. When exposed to pressure at the pump outlet, this additional air is compressed forming in effect a cushion, and does not col-

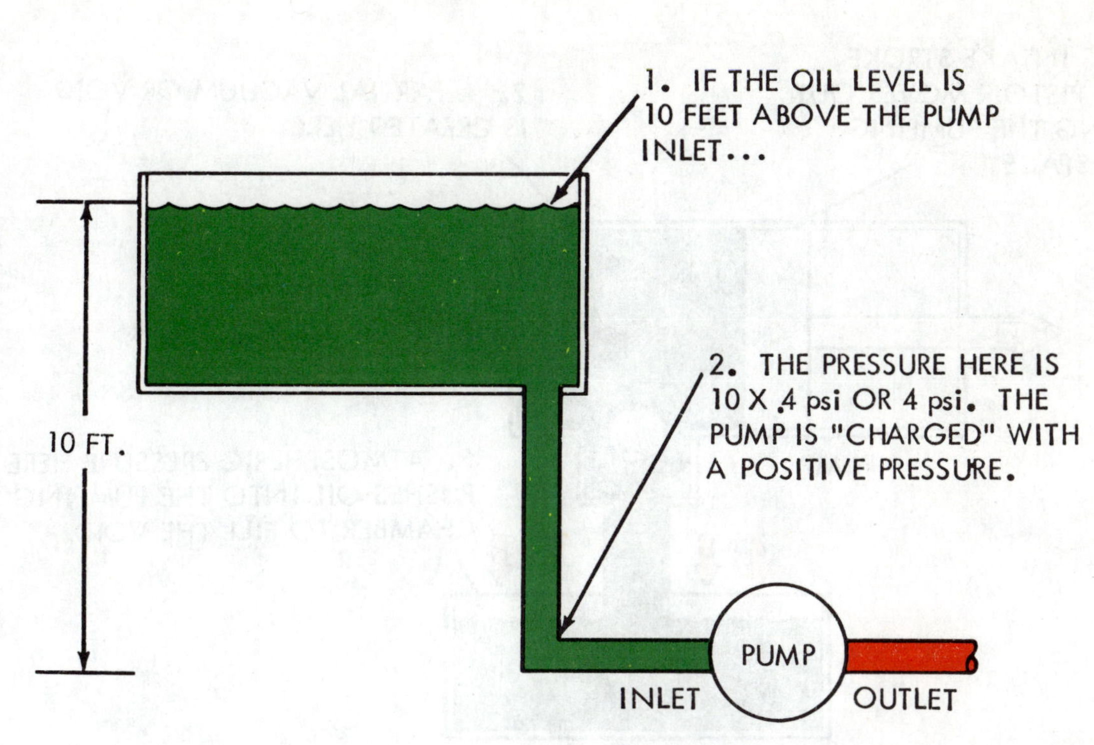

VIEW A. OIL LEVEL ABOVE PUMP CHARGES INLET

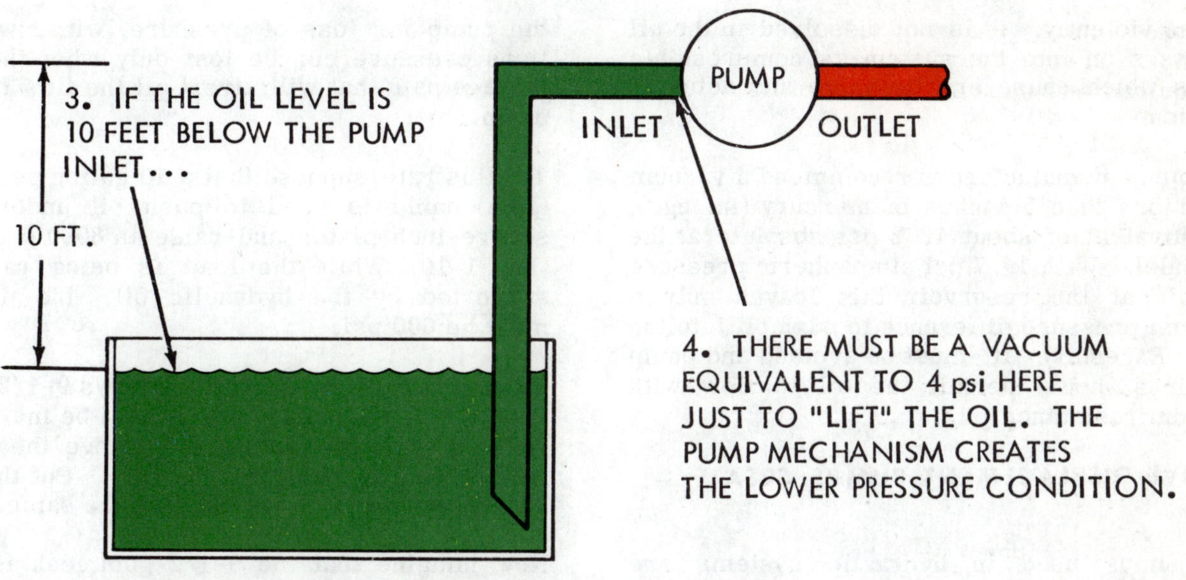

VIEW B. OIL LEVEL BELOW REQUIRES VACUUM TO "LIFT" OIL

Figure 1-8. Pump Inlet Locations

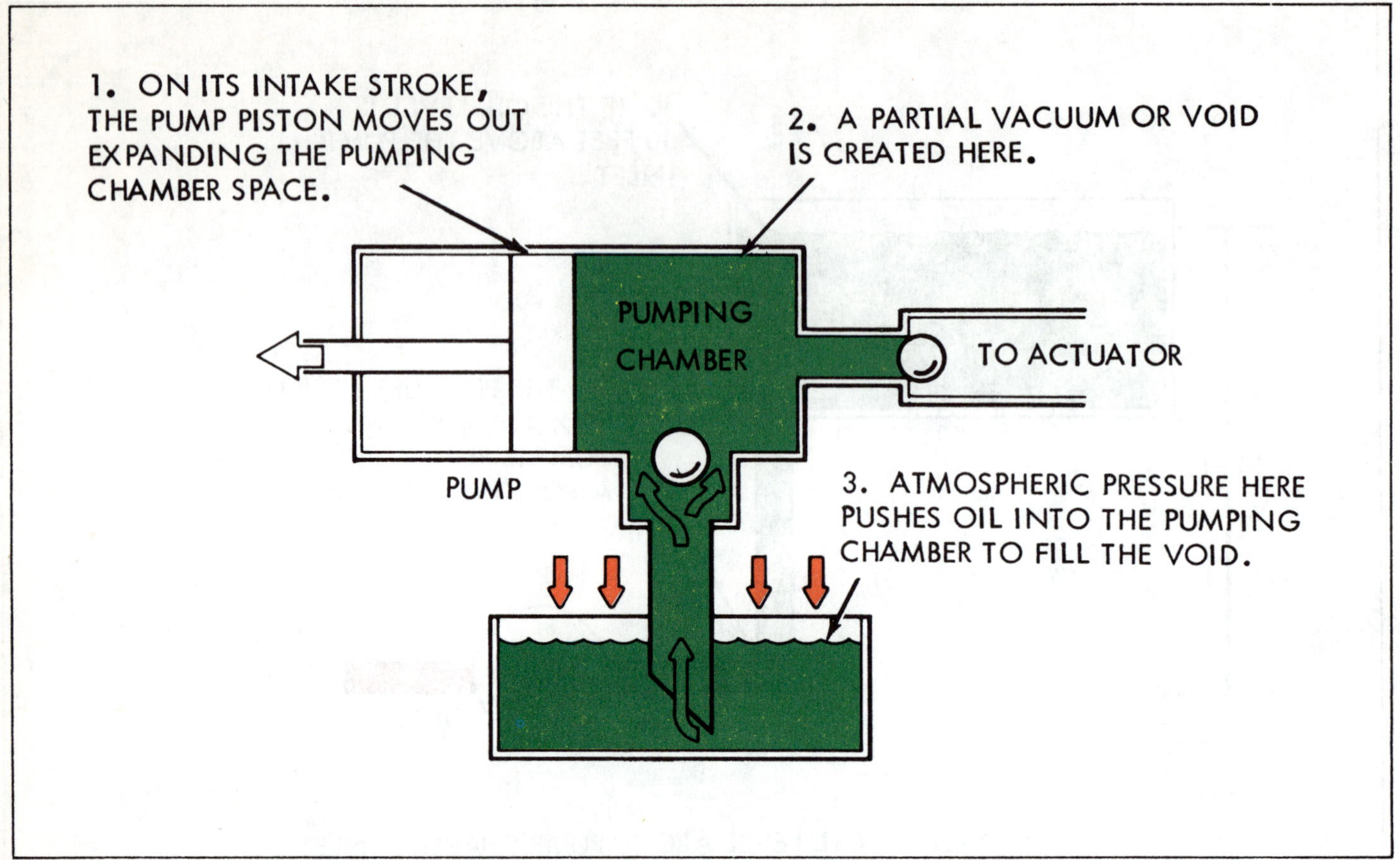

Figure 1-9. Pressure Difference Pushes Oil Into Pump

lapse as violently. It is not dissolved in the oil but passes on into the system as compressible bubbles which cause erratic valve and actuator operation.

Most pump manufacturers recommend a vacuum of no more than 5 inches of mercury (in. hg.), the equivalent of about 12.2 psi absolute at the pump inlet. With 14.7 psi atmospheric pressure available at the reservoir, this leaves only a 2-1/2 psi pressure difference to push oil into the pump. Excessive lift must be avoided and pump inlet lines should permit the oil to flow with minimum resistance.

POSITIVE DISPLACEMENT PUMPS CREATE FLOW

Most pumps used in hydraulic systems are classed as positive displacement. This means that, except for changes in efficiency, the pump output is constant regardless of pressure. The outlet is positively sealed from the inlet, so that whatever gets in is forced out the outlet port.

The sole purpose of a pump is to create flow; pressure is caused by a resistance to flow. Although there is a common tendency to blame the pump for loss of pressure, with few exceptions pressure can be lost only when there is a leakage path that will divert all the flow from the pump.

To illustrate, suppose that a 10 gallon per minute (gpm) pump is used to push oil under a 10-square-inch piston and raise an 8000 pound load (Fig. 1-10) While the load is being raised or supported by the hydraulic oil, the pressure must be 800 psi.

Even if a hole in the piston allows 9-1/2 gpm to leak at 800 psi, pressure still will be maintained. With only 1/2 gpm available to move the load, it will of course raise very slowly. But the pressure required to do so remains the same.

Now imagine that the 9-1/2 gpm leak is in the pump instead of the cylinder. There still would be 1/2 gpm moving the load and there still would be pressure. Thus, a pump can be badly worn, losing nearly all of its efficiency, and pressure still can be maintained. Maintenance of pressure alone is no indicator of a pump's condition. It is necessary to measure the flow at a given pressure to determine whether a pump is in good or bad condition.

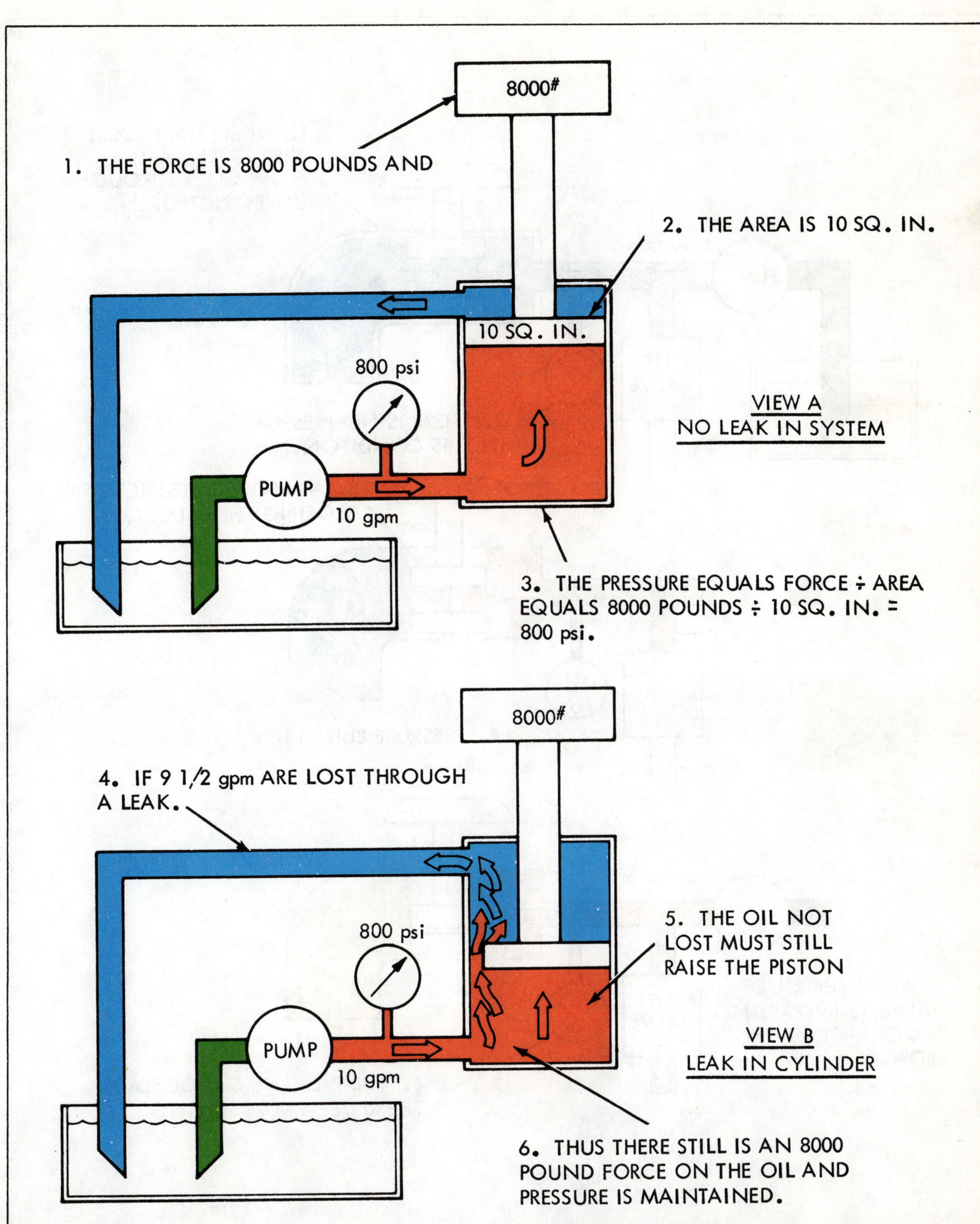

Figure 1-10. Pressure Loss Requires Full Loss of Pump Output

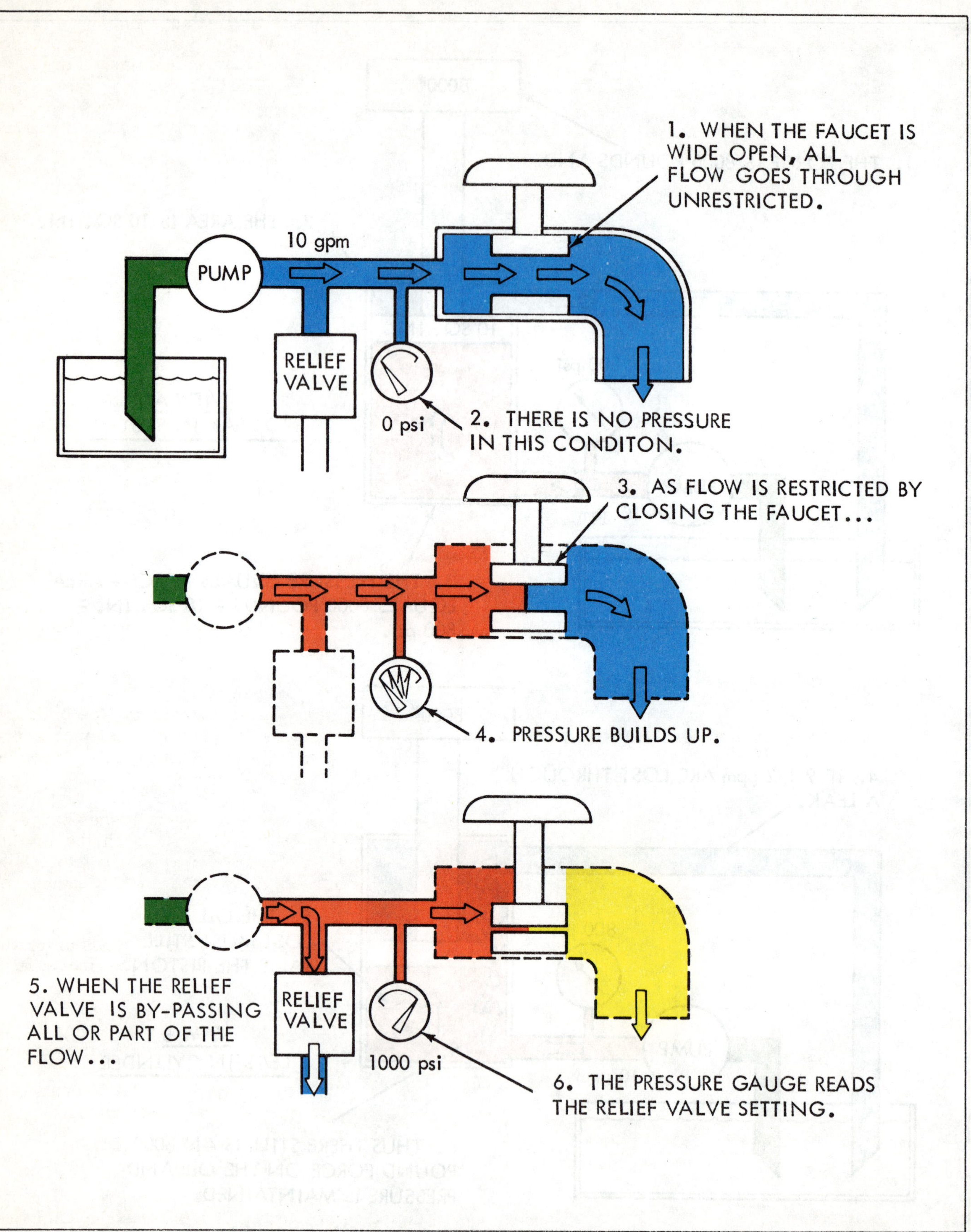

Figure 1-11. Pressure Caused by Restriction and Limited by Pressure Control Valve

HOW PRESSURE IS CREATED

Pressure results whenever the flow of a fluid is resisted. The resistance may come from (1) a load on an actuator or (2) a restriction (or orifice) in the piping.

Figure 1-10 is an example of a load on an actuator. The 8000 pound weight resists the flow of oil under the piston and creates pressure in the oil. If the weight increases, so does the pressure.

In Fig. 1-11, a 10 gpm pump has its outlet connected to a pressure relief valve set at 1000 psi and to an ordinary water faucet. If the faucet is wide open, the pump delivery flows out unrestricted and there is no reading on the pressure gauge.

Now suppose that the faucet is gradually closed. It will resist flow and cause pressure to build up on the upstream side. As the opening is restricted, it will take increasingly more pressure to push the 10 gpm through the restriction. Without the relief valve, there would theoretically be no limit to the pressure build-up. In reality, either something would break or the pump would stall the prime mover.

In our example, at the point where it takes 1000 psi to push the oil through the opening, the relief valve will begin to open. Pressure then will remain at 1000 psi. Further closing of the faucet will simply result in less oil going through it and more going over the relief valve. With the faucet completely closed, all 10 gpm will go over the relief valve at 1000 psi.

It can be seen from the above that a relief valve or some other pressure limiting device should be used in all systems using a positive displacement pump.

PARALLEL FLOW PATHS

An inherent characteristic of liquids is that they will always take the path of least resistance. Thus, when two parallel flow paths offer different resistances, the pressure will increase only to the amount required to take the easier path.

In Fig. 1-12, the oil has three possible flow paths. Since valve A opens at 100 psi, the oil will go that way and pressure will build up to only 100. Should flow be blocked beyond A, pressure would build up to 200 psi; then oil would flow through B. There would be no flow through C unless the path through valve B should also become blocked.

Similarly, when the pump outlet is directed to two actuators, the actuator which needs the lower pressure will be first to move. Since it is difficult to balance loads exactly, cylinders which must move together are often connected mechanically.

SERIES FLOW PATH

When resistances to flow are connected in series, the pressures add up. In Fig. 1-13 are shown the same valves as Fig. 1-12 but connected in series. Pressure gauges placed in the lines indicate the pressure normally required to open each valve plus back pressure from the valves downstream. The pressure at the pump is the sum of the pressures required to open individual valves.

PRESSURE DROP THROUGH AN ORIFICE

An orifice is a restricted passage in a hydraulic line or component, used to control flow or create a pressure difference (pressure drop).

In order for oil to flow through an orifice, there must be a pressure difference or pressure drop through the orifice. (The term "drop" comes from the fact that the lower pressure is always downstream.) Conversely, if there is no flow, there is no difference in pressure across the orifice.

Consider the condition surrounding the orifice in Fig. 1-14, view A. The pressure is equal on both sides; therefore, the oil is being pushed equally both ways and there is no flow.

In view B, the higher pressure pushes harder to the right and oil does flow through the orifice. In view C, there is also a pressure drop; however, the flow is less than in B because the pressure difference is lower.

An increase in pressure drop across an orifice will always be accompanied by an increase in flow.

If flow is blocked beyond an orifice (view D), the pressure will immediately equalize on both sides of the orifice in accordance with Pascal's Law. This principle is essential to the operation of many compound pressure control valves.

PRESSURE INDICATES WORK LOAD

Figure 1-10 illustrated how pressure is generated by resistance of a load. It was noted that the pressure equals the force of the load divided by the piston area.

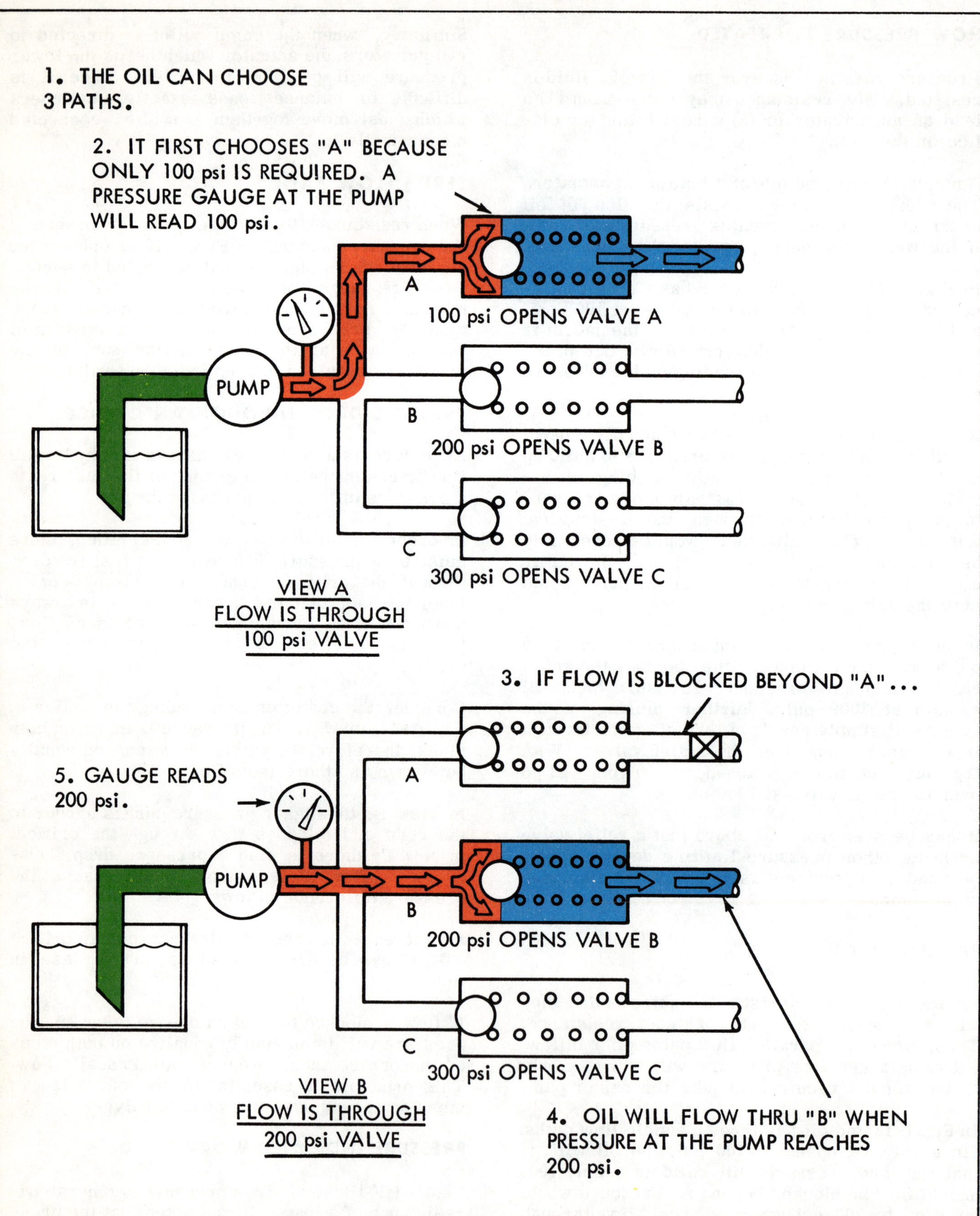

Figure 1-12. Parallel Flow Paths

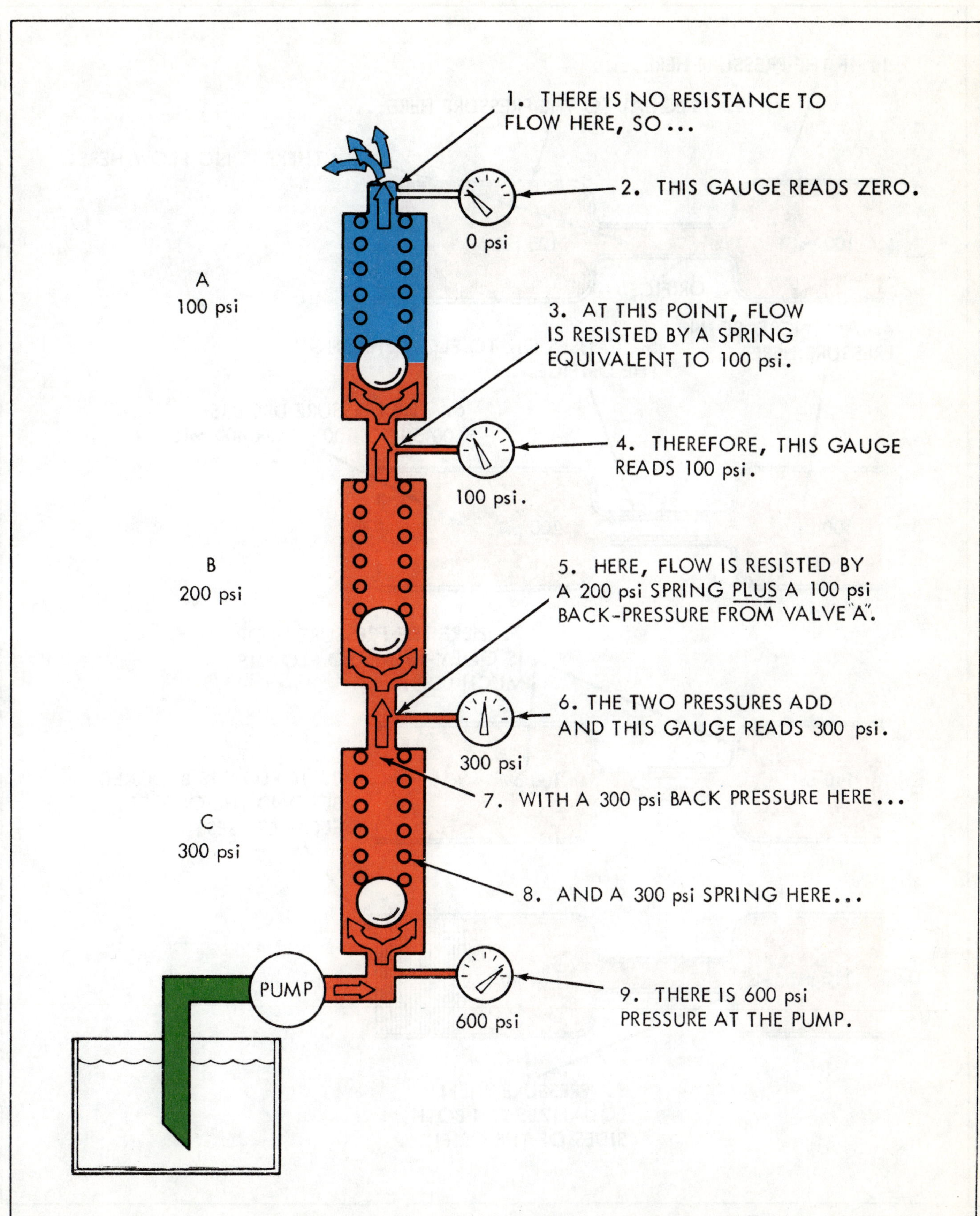

Figure 1-13. Series Resistances Add Pressure

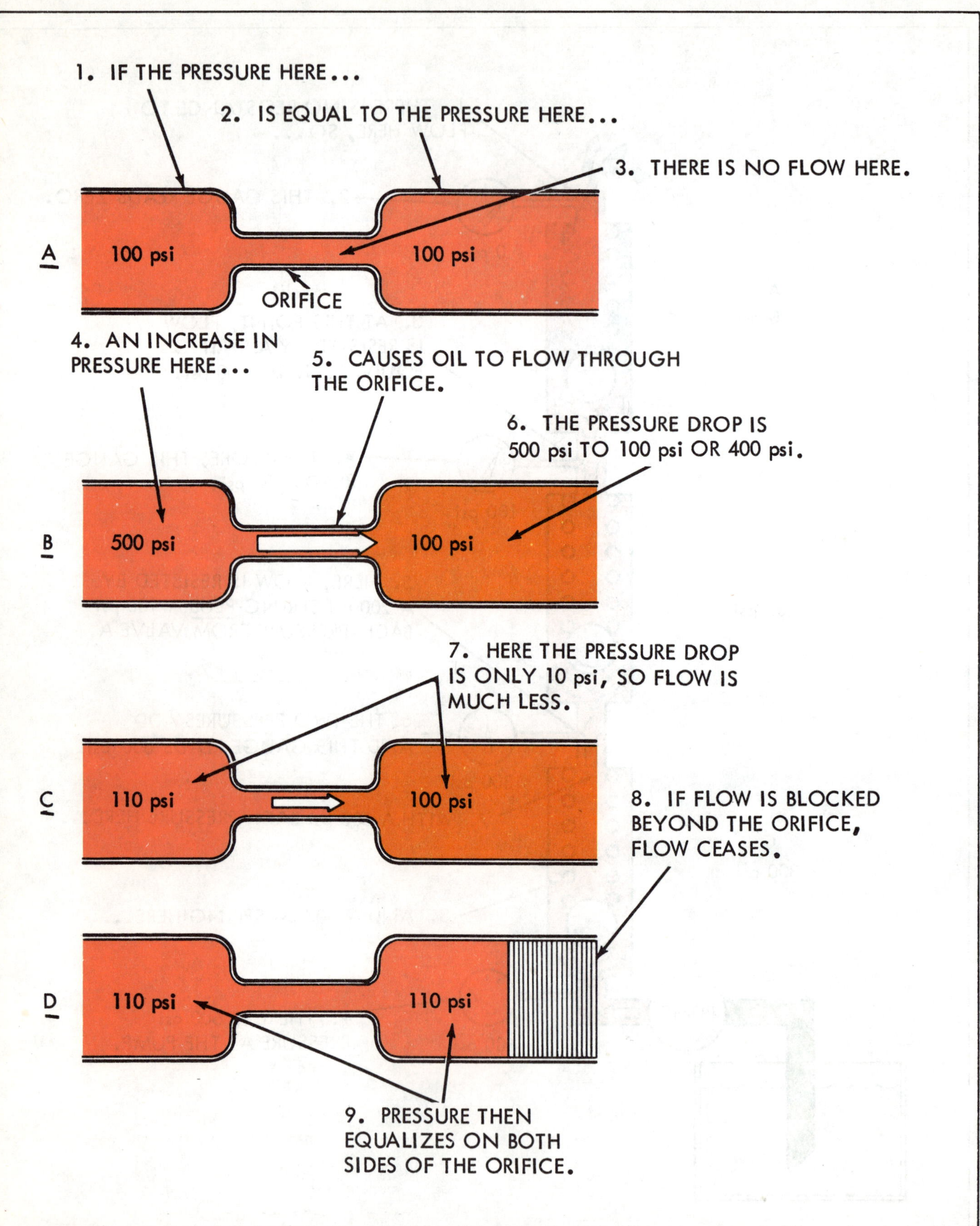

Figure 1-14. Pressure Drop and Flow Through an Orifice

We can express this relationship by the general formula:

$$P = \frac{F}{A}$$

In this relationship:

P is pressure in psi (pounds per square inch)
F is force in pounds
A is area in square inches

From this can be seen that an increase or decrease in the load will result in a like increase or decrease in the operating pressure. In other words, pressure is proportional to the load, and a pressure gauge reading indicates the work load (in psi) at any given moment.

Pressure gauge readings normally ignore atmospheric pressure. That is, a standard gauge reads zero at atmospheric pressure. An absolute gauge reads 14.7 psi at sea level atmospheric pressure. Absolute pressure is usually designated "psia".

FORCE IS PROPORTIONAL TO PRESSURE AND AREA

When a hydraulic cylinder is used to clamp or press, its output force can be computed as follows:

$$F = P \times A$$

Again:

P is pressure in psi
F is force in pounds

As an example, suppose that a hydraulic press has its pressure regulated at 2000 psi (Fig. 1-15) and this pressure is applied to a ram area of 20 square inches. The output force will then be 40,000 pounds or 20 tons.

COMPUTING PISTON AREA

The area of a piston or ram can be computed by this formula:

$$A = .7854 \times d^2$$

A is area in square inches
d is diameter of the piston in inches

The foregoing relationships are sometimes illustrated as shown to indicate the three relationships:

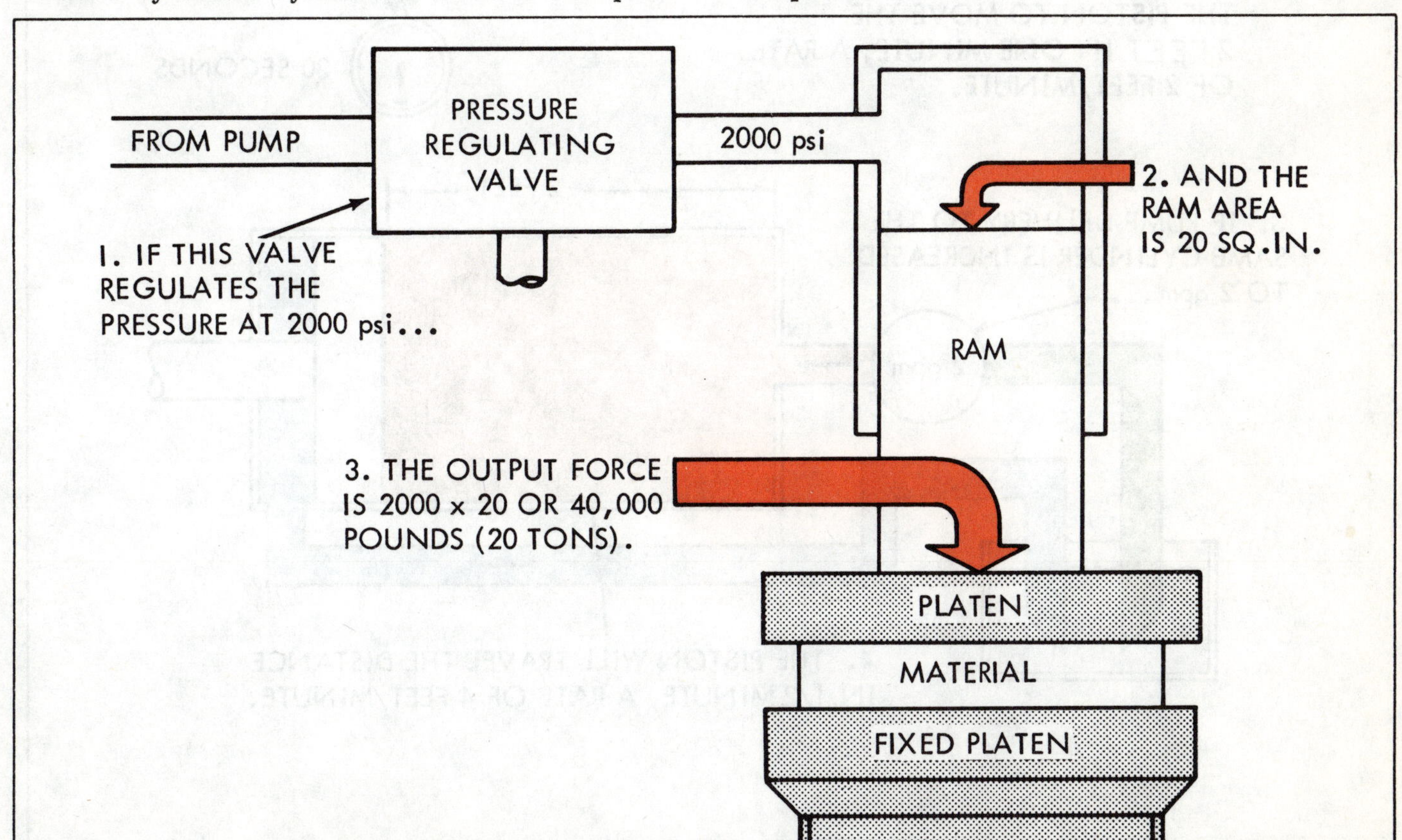

Figure 1-15. Force Equals Pressure Multiplied by Area

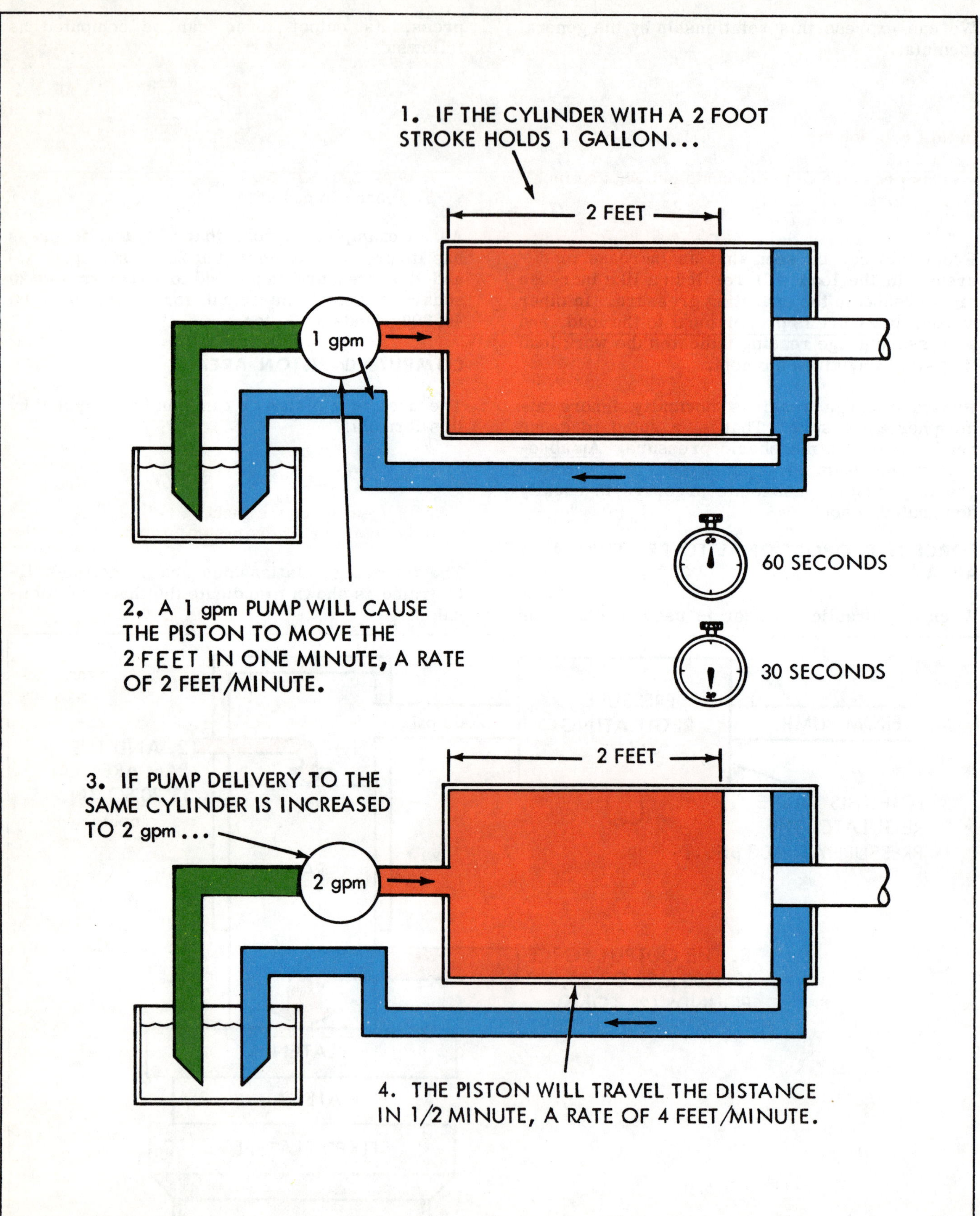

Figure 1-16. Speed Depends on Cylinder Size and Rate of Oil Flow To It

F = P x A
P = F/A
A = F/P

SPEED OF AN ACTUATOR

How fast a cylinder travels or a motor rotates depends on its size and the rate of oil flow into it. To relate flow rate to speed, consider the volume that must be filled in the actuator to effect a given amount of travel.

In Fig. 1-16, note that both cylinders have the same volume. Yet, the piston in cylinder B will travel twice as fast as cylinder A because the rate of oil flow from the pump has been doubled. If either cylinder had a smaller diameter, its rate would be faster. Or if its diameter were larger, its rate would be less, assuming of course the pump delivery remained constant.

The relationship may be expressed as follows:

$$\text{speed} = \frac{\text{vol./time}}{\text{area}}$$

$$\text{vol./time} = \text{speed x area}$$

$$\text{area} = \frac{\text{vol./time}}{\text{speed}}$$

$$\frac{v}{t} = \text{cu. in./minute}$$

a = sq. in.
s = inches/minute

From this we can conclude:(1) that the force or torque of an actuator is directly proportional to the pressure and independent of the flow, (2) that its speed or rate of travel will depend upon the amount of fluid flow without regard to pressure.

VELOCITY IN PIPES

The velocity at which the hydraulic fluid flows through the lines is an important design consideration because of the effect of velocity on friction.

Generally, the recommended velocity ranges are:

Pump Inlet Line ----2 - 4 feet per second
Working Lines -----7 - 20 feet per second

In this regard, it should be noted that:

1. The velocity of the fluid varies inversely as the square of the inside diameter.

2. Usually, friction of a liquid flowing through a line is proportional to the velocity. However, should the flow become turbulent, friction varies as the square of the velocity.

Fig. 1-17 illustrates that doubling the inside diameter of a line quadruples the cross-sectional area; thus the velocity is only one-fourth as fast in the larger line. Conversely, halving the diameter decreases the area to 1/4 and quadruples the oil velocity.

Friction creates turbulence in the oil stream and of course resists flow, resulting in an increased pressure drop through the line. Very low velocity is recommended for the pump inlet line because very little pressure drop can be tolerated there.

DETERMINING PIPE SIZE REQUIREMENTS

Two formulas are available for sizing hydraulic lines.

If the gpm and desired velocity are known, use this relationship to find the inside cross-sectional area:

$$\text{AREA} = \frac{\text{gpm x .3208}}{\text{velocity (in feet per second)}}$$

When the gpm and size of pipe are given, use this formula to find what the velocity will be:

$$\text{VELOCITY (feet per second)} = \frac{\text{gpm}}{\text{3.117 x area}}$$

In Chapter 4, you will find a nomographic chart which permits making these computations by laying a straight edge across printed scales.

SIZE RATINGS OF LINES

The nominal ratings in inches for pipes, tubes, etc. are not accurate indicators of the inside diameter.

In standard pipes, the actual inside diameter is larger than the nominal size quoted. To select pipe, you'll need a standard chart which shows actual inside diameters (see Chapter 4).

For steel and copper tubing, the quoted size is the outside diameter. To find the inside diameter, subtract twice the wall thickness (Fig. 1-18).

WORK AND POWER

Whenever a force or push is exerted through a distance, work is done.

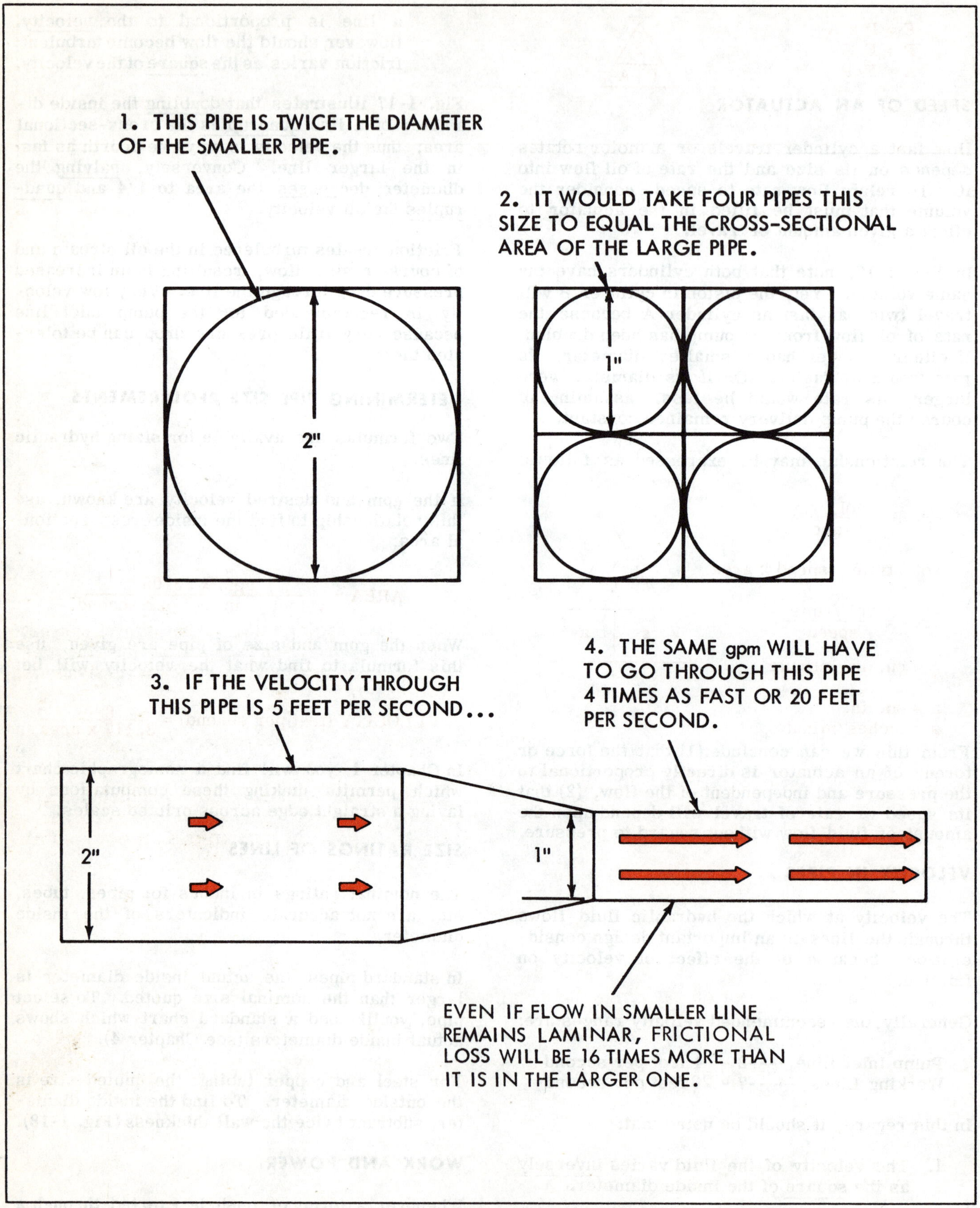

Figure 1-17. Velocity is Inversely Proportional to Pipe Cross-Section Area

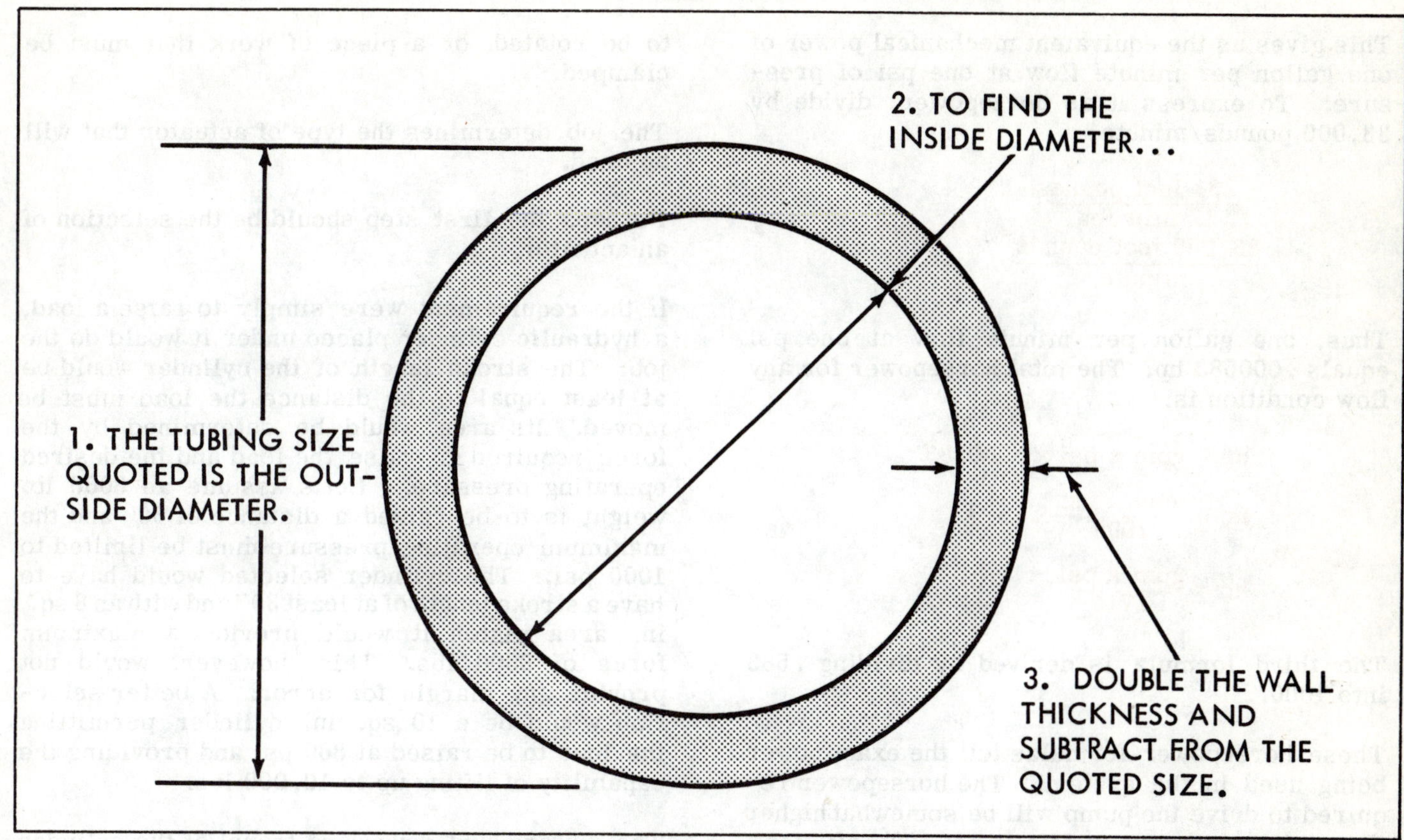

Figure 1-18. Tubing Inside Diameter

WORK = force x distance

Work is usually expressed in foot pounds. For example, if a 10 pound weight is lifted 10 feet, the work is 10 pounds x 10 feet or 100 foot pounds.

The formula above for work does not take into consideration how fast the work is done. The <u>rate</u> of doing work is called <u>power.</u>

To visualize power, think of climbing a flight of stairs. The work done is the body's weight multiplied by the height of the stairs. But it is more difficult to run up the stairs than to walk. When you run, you do the same work at a faster rate.

$$\text{POWER} = \frac{\text{force x distance}}{\text{time}} \text{ or } \frac{\text{work}}{\text{time}}$$

The standard unit of power is the <u>horsepower</u>, abbreviated hp. It is equivalent to 33,000 pounds lifted one foot in one minute. It also has equivalents in electrical power and heat.

$$1 \text{ hp} = \frac{33,000 \text{ foot pounds}}{\text{minute}} \text{ or } \frac{550 \text{ foot pounds}}{\text{second}}$$

1 hp = 746 watts (electrical power)

1 hp = 42.4 btu/minute (heat power)

Obviously, it is desirable to be able to convert hydraulic power to horsepower so that the mechanical, electrical and heat power equivalents will be known.

HORSEPOWER IN A HYDRAULIC SYSTEM

In the hydraulic system, speed and distance are indicated by the gpm flow and force by pressure. Thus, we might express hydraulic power this way:

$$\text{POWER} = \frac{\text{gallons}}{\text{minutes}} \times \frac{\text{pounds}}{\text{square inches}}$$

To change the relationship to mechanical units, we can use these equivalents:

1 gallon = 231 cubic inches (in.^{3})
12 inches = 1 foot

Thus:

$$\text{POWER} = \frac{\text{gallon}}{\text{min.}} \times \left(\frac{231 \text{ in.}^3}{\text{gallon}}\right) \times \frac{\text{pounds}}{\text{in.}^2} \times \left(\frac{1 \text{ foot}}{12 \text{ in.}}\right) =$$

$$\frac{231 \text{ foot pounds}}{12 \text{ minutes}}$$

This gives us the equivalent mechanical power of one gallon per minute flow at one psi of pressure. To express it as horsepower, divide by 33,000 pounds/minute:

$$\frac{\frac{231 \text{ foot pounds}}{12 \text{ minutes}}}{\frac{33,000 \text{ foot pounds}}{\text{minute}}} = .000583$$

Thus, one gallon per minute flow at one psi equals .000583 hp. The total horsepower for any flow condition is:

$$hp = gpm \times psi \times .000583$$

or

$$hp = \frac{gpm \times psi}{1000} \times .583$$

or

$$hp = \frac{gpm \times psi}{1714}$$

The third formula is derived by dividing .583 into 1000.

These horsepower formulas tell the exact power being used in the system. The horsepower required to drive the pump will be somewhat higher than this since the system is not 100% efficient.

If we assume an average efficiency of 80%, this relationship can be used to estimate power input requirements:

$$hp = gpm \times psi \times .0007$$

HORSEPOWER AND TORQUE

It also is often desirable to convert back and forth from horsepower to torque without computing pressure and flow.

These are general torque-power formulas for any rotating equipment:

$$torque = \frac{63025 \times hp}{rpm} \qquad hp = \frac{torque \times rpm}{63025}$$

Torque in this formula must be in pound-inches.

DESIGNING A SIMPLE HYDRAULIC SYSTEM

From the information given in this chapter, it is possible to design a simple hydraulic circuit. Following is a simple description of how the job might proceed. See figures 1-19 thru 1-21.

A Job To Be Done

All circuit design must start with the job to be done. There is a weight to be lifted, a tool head to be rotated, or a piece of work that must be clamped.

The job determines the type of actuator that will be used.

Perhaps the first step should be the selection of an actuator.

If the requirement were simply to raise a load, a hydraulic cylinder placed under it would do the job. The stroke length of the cylinder would be at least equal to the distance the load must be moved. Its area would be determined by the force required to raise the load and the desired operating pressure. Let's assume an 8000 lb. weight is to be raised a distance of 30" and the maximum operating pressure must be limited to 1000 psi. The cylinder selected would have to have a stroke length of at least 30" and with an 8 sq. in. area piston it would provide a maximum force of 8000 lbs. This, however, would not provide any margin for error. A better selection would be a 10 sq. in. cylinder permitting the load to be raised at 800 psi and providing the capability of lifting up to 10,000 lbs.

The upward and downward travel of the cylinder would be controlled by a directional valve. If the load is to be stopped at intermediate points in its travel, the directional valve should have a neutral position in which oil flow from the underside of the piston is blocked to support the weight on the cylinder. The rate at which the load must travel will determine the pump size. The 10 sq. in. piston will displace 10 cu. in. for every inch it lifts. Extending the cylinder 30" will require 300 cu. in. of fluid. If it is to move at the rate of 10" per second, it will require 100 cu. in. of fluid per second or 6000 cu. in. per minute. Since pumps are usually rated in gallons per minute, it will be necessary to divide 231 (cubic inches per gallon) into 6000 to convert the requirements into gallons per minute - 6000 ÷ 231 = 26 gpm.

The hp needed to drive the pump is a function of its delivery and the maximum pressure at which it will operate. The following formula will determine the size of the electric motor drive required:

$$hp = gpm \times psi \times .0007$$

$$hp = 26 \times 1000 \times .0007 = 18.2$$

To prevent overloading of the electric motor and to protect the pump and other components from excessive pressure due to overloads or stalling,

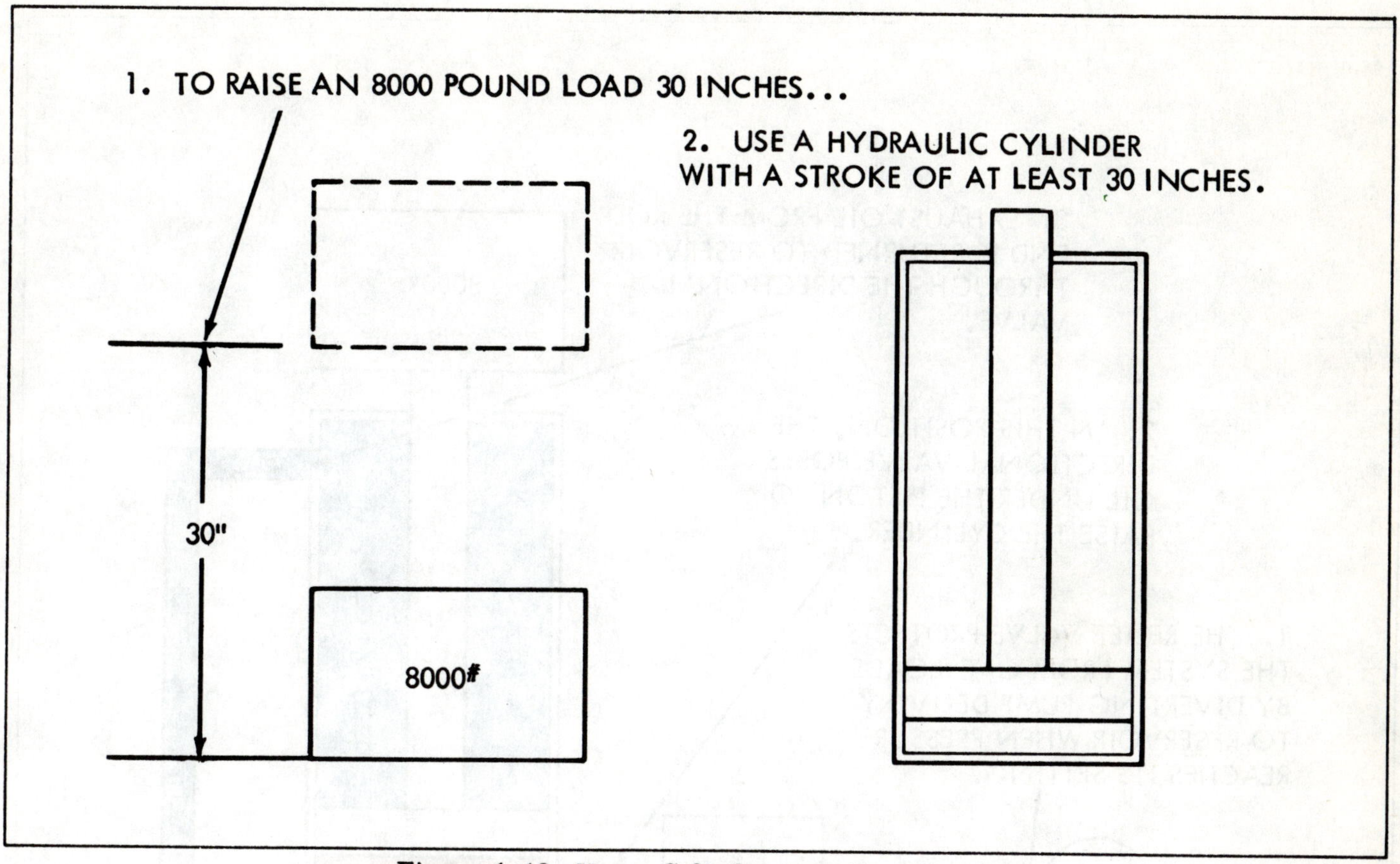

1. TO RAISE AN 8000 POUND LOAD 30 INCHES...

2. USE A HYDRAULIC CYLINDER WITH A STROKE OF AT LEAST 30 INCHES.

30"

8000#

Figure 1-19. Use a Cylinder to Raise a Load

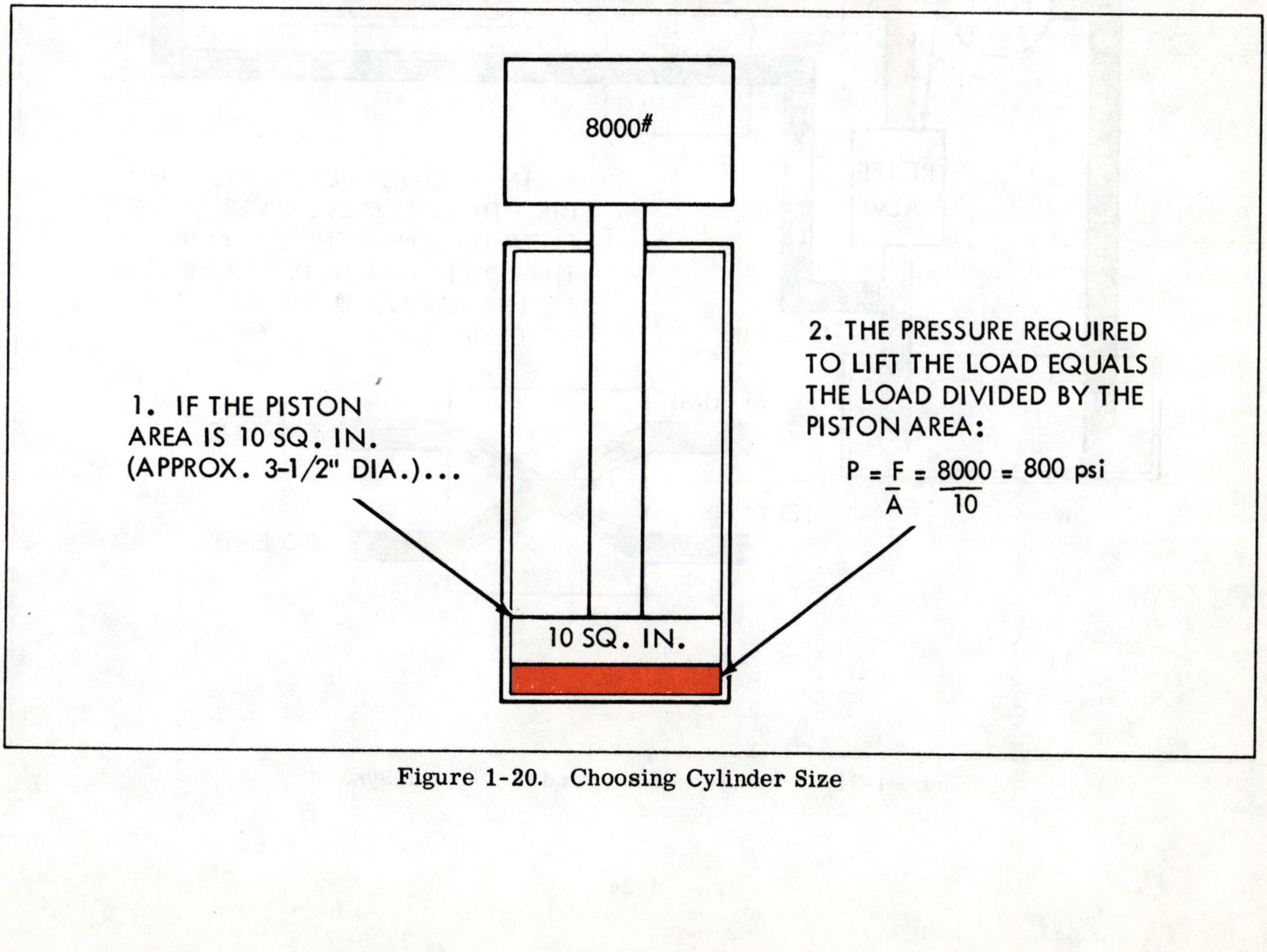

8000#

1. IF THE PISTON AREA IS 10 SQ. IN. (APPROX. 3-1/2" DIA.)...

2. THE PRESSURE REQUIRED TO LIFT THE LOAD EQUALS THE LOAD DIVIDED BY THE PISTON AREA:

$$P = \frac{F}{A} = \frac{8000}{10} = 800 \text{ psi}$$

10 SQ. IN.

Figure 1-20. Choosing Cylinder Size

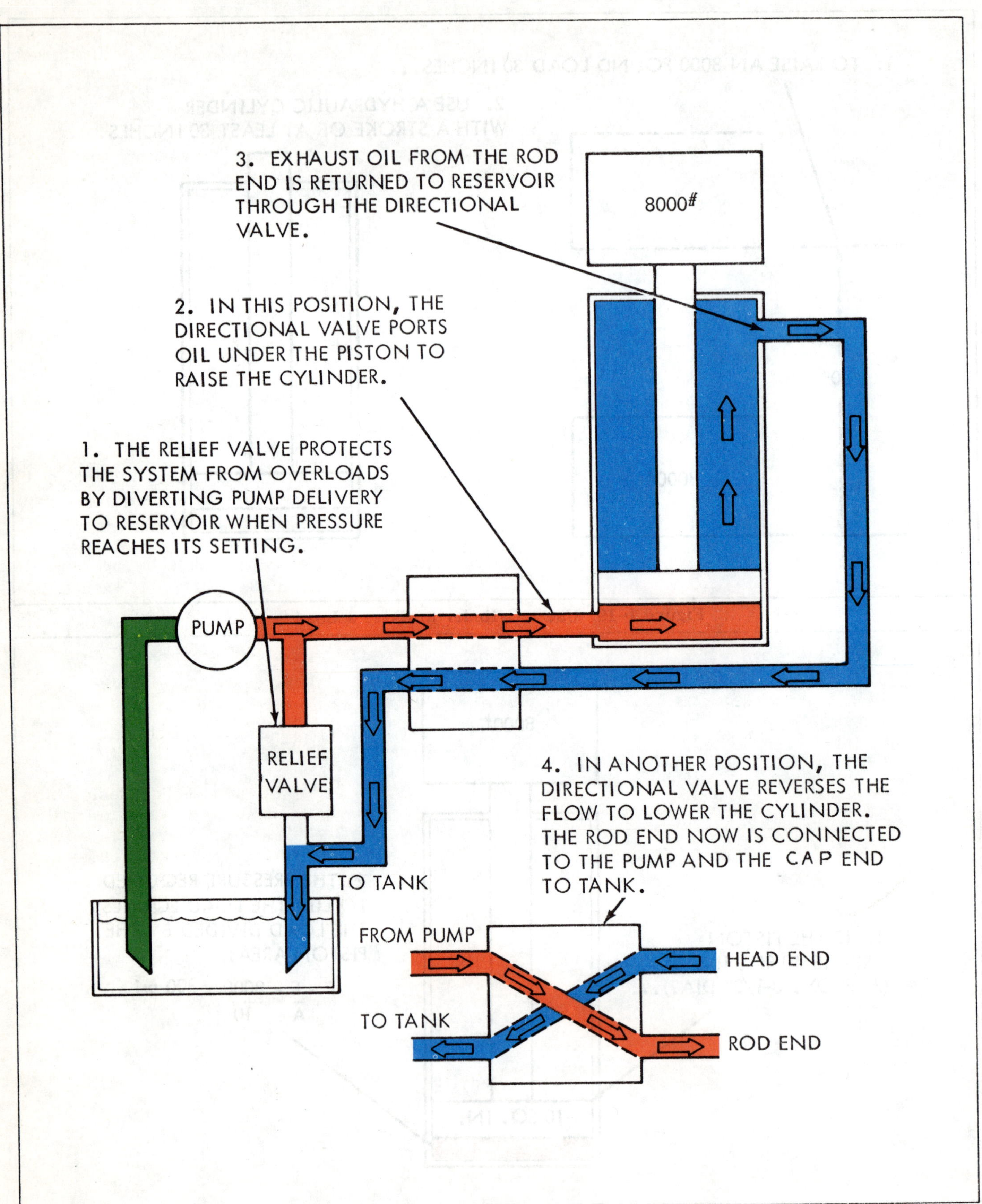

Figure 1-21 Valving to Protect and Control the System

a relief valve set to limit the maximum system pressure should be installed in the line between the pump outlet and the inlet port to the directional valve.

A reservoir sized to hold approximately two to three times the pump capacity in gallons per minute and adequate interconnecting piping would complete the system.

CONCLUSION

This chapter has presented a brief introductory overview of hydraulics to demonstrate the basic principles involved in hydraulic system operation. There are, of course, countless variations of the system presented. Many of these will be developed with a more detailed study of operating principles and components in future chapters.

QUESTIONS

1. State Pascal's Law.
2. Define pressure.
3. If a force of 1000 pounds is applied over an area of 20 square inches, what is the pressure?
4. What is meant by "conservation of energy"?
5. What is the output component of a hydraulic system named? The input component?
6. What is the prime mover?
7. Name several advantages of a hydraulic system.
8. What is the origin of the term "hydraulics"?
9. What makes petroleum oil suitable as a hydraulic fluid?
10. What is the pressure at the bottom of a 20 foot column of oil?
11. What can you say definitely about the pressures on opposite sides of an orifice when oil is flowing through it?
12. What pressure is usually available to charge the pump inlet?
13. Why should the pump inlet vacuum be minimized?
14. What is the function of the pump?
15. Why is loss of pressure usually not a symptom of pump malfunction?
16. How is pressure created?
17. If three 200 psi check valves are connected in series, how much pressure is required at the pump to push oil through all three?
18. What is the formula for pressure developed when moving a load with a cylinder?
19. What is the formula for the maximum force output of a cylinder?
20. What determines the speed of an actuator?
21. What is the relationship between fluid velocity and friction in a pipe?
22. What is work? Power?
23. How do you find the horsepower in a hydraulic system?
24. With which component does the design of a hydraulic circuit begin?
25. What determines the size pump needed in a hydraulic circuit?
26. What is the piston area of a 5-inch cylinder?
27. What does the relief valve do?
28. What does a directional valve do?

PRINCIPLES OF POWER HYDRAULICS

This chapter is divided into three sections:

* Principles of Pressure
* Principles of Flow
* Hydraulic Graphical Symbols

The first two sections will further develop the fundamentals of the physical phenomena that combine to transfer power in the hydraulic circuit. The third section, illustrating graphical symbols for circuit diagrams, will deal with the classes and functions of lines and components. All this material will serve as a background for following chapters on the equipment that makes up a hydraulic system.

PRINCIPLES OF PRESSURE

A Precise Definition

It has been noted that the term hydraulics is derived from a Greek word for water. Therefore, it might be assumed correctly that the science of hydraulics encompasses <u>any</u> device operated by water. A water wheel or turbine (Fig. 2-1) for instance, is a hydraulic device.

However, a distinction must be made between devices which utilize the impact or momentum of a moving liquid and those which are operated by pushing on a confined fluid; that is, by pressure.

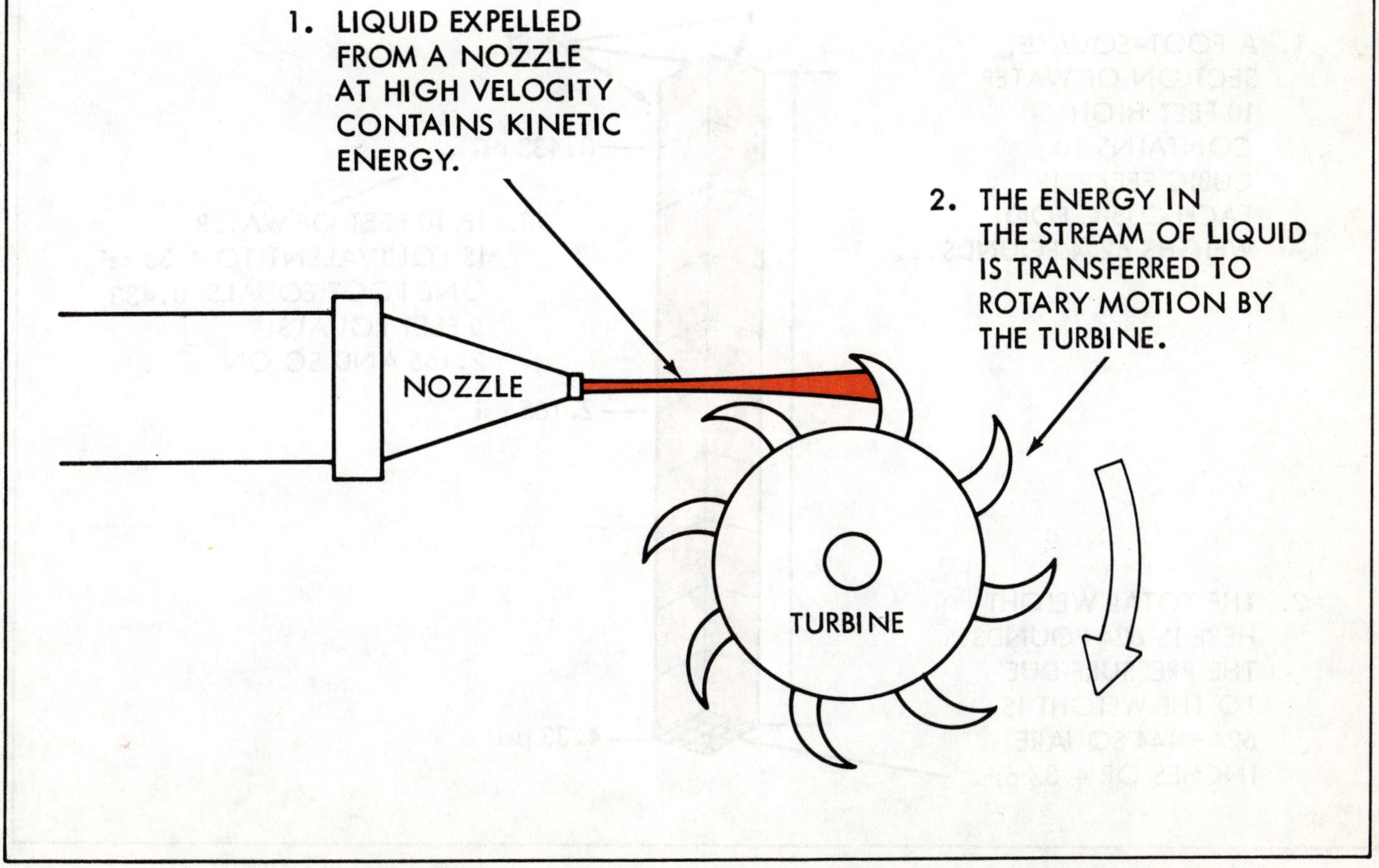

Figure 2-1. Hydrodynamic Device Uses Kinetic Energy Rather Than Pressure

Properly speaking:

* A hydraulic device which uses the impact or kinetic energy in the liquid to transmit power is called a hydrodynamic device.

* When the device is operated by a force applied to a confined liquid, it is called a hydrostatic device; pressure being the force applied distributed over the area exposed and being expressed as force per unit area (lbs./sq. in. or psi).

Of course, all the illustrations shown so far, and in fact, all the systems and equipment covered in this manual are hydrostatic. All operate by pushing on a confined liquid; that is, by transferring energy through pressure.

How Pressure is Created

Pressure results whenever there is a resistance to fluid flow or to a force which attempts to make the fluid flow. The tendency to cause flow (or the push) may be supplied by a mechanical pump or may be caused simply by the weight of the fluid.

It is well known that in a body of water, pressure increases with depth. The pressure is always equal at any particular depth due to the weight of the water above it. Around Pascal's time, an Italian scientist named Torricelli proved that if a hole is made in the bottom of a tank of water, the water runs out fastest when the tank is full and the flow rate decreases as the water level lowers. In other words, as the "head" of water above the opening lessens, so does the pressure.

Torricelli could express the pressure at the bottom of the tank only as "feet of head", or the height in feet of the column of water. Today, with the pound per square inch (psi) as a unit pressure, we can express pressure anywhere in any liquid or gas in more convenient terms. All that is required is knowing how much a cubic foot of the fluid weighs.

As shown in Fig. 2-2, a "head" of one foot of water is equivalent to .434 psi; a five foot head of water equals 2.17 psi, and so on. And as shown earlier, a head of oil is equivalent to about .4 psi per foot.

In many places, the term "head" is used to describe pressure, no matter how it is created. For instance, a boiler is said to "work up a head of steam" when pressure is created by vaporiz-

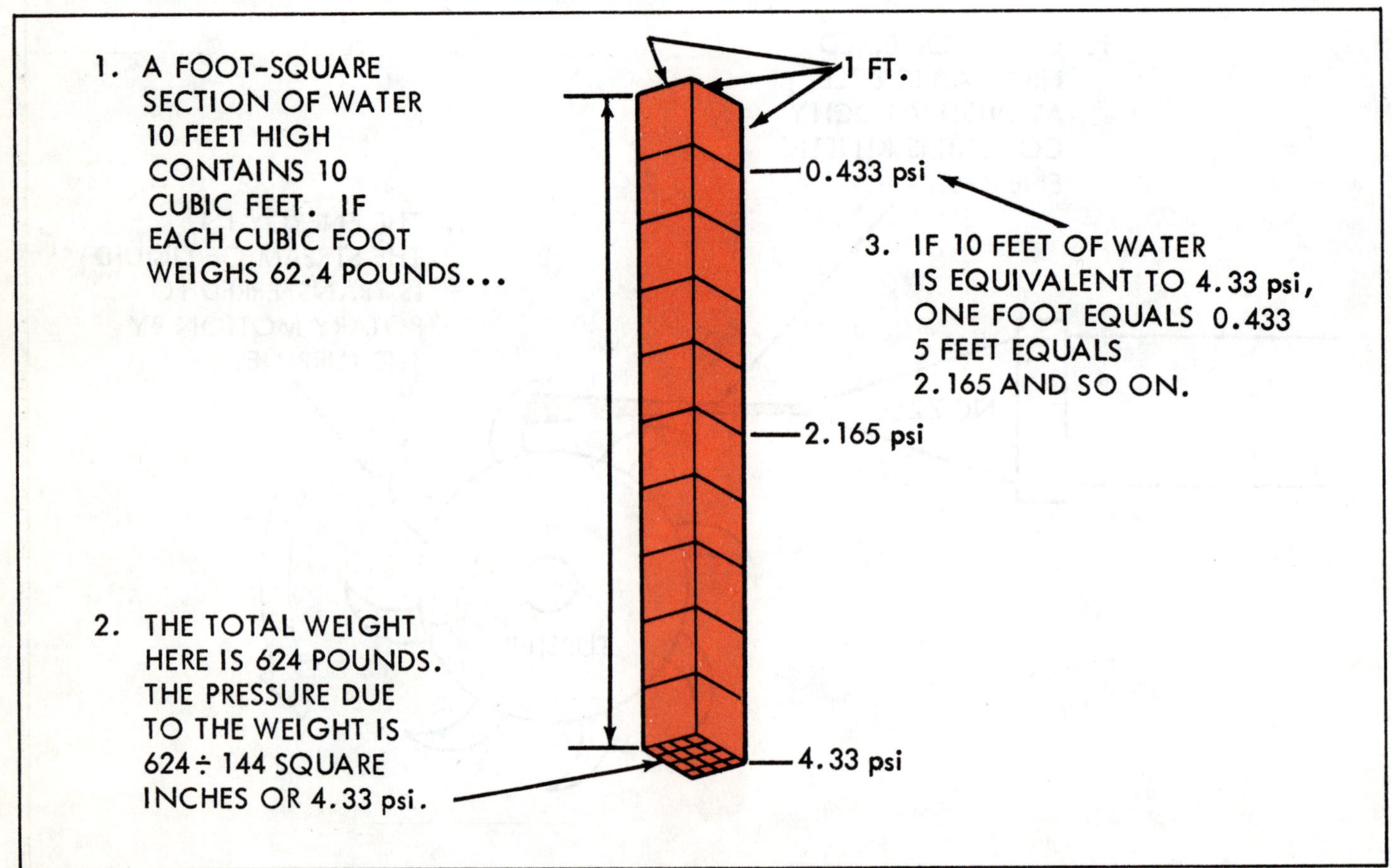

Figure 2-2. Pressure "Head" Comes From Weight of the Fluid

ing water in confinement. The terms pressure and "head" are sometimes used interchangeably.

Atmospheric Pressure

Atmospheric pressure is nothing more than pressure of the air in our atmosphere due to its weight. At sea level, a column of air one square inch in cross section and the full height of the atmosphere weighs 14.7 pounds (Fig. 2-3). Thus the pressure is 14.7 psia. At higher altitudes, of course, there is less weight in the column, so the pressure becomes less. Below sea level, atmospheric pressure is more than 14.7 psia.

Any condition where pressure is less than atmospheric pressure is called a vacuum or partial vacuum. A perfect vacuum is the complete absence of pressure or zero psia.

The Mercury Barometer

Atmospheric pressure also is measured in inches of mercury (in. Hg.) on a device known as a barometer.

The mercury barometer (Fig. 2-4), a device invented by Torricelli, is usually credited as the inspiration for Pascal's studies of pressure. Torricelli discovered that when a tube of mercury is inverted in a pan of the liquid, the column in the tube will fall only a certain distance. He reasoned that atmospheric pressure on the surface of the liquid was supporting the weight of the column of mercury with a perfect vacuum at the top of the tube.

In a normal atmosphere, the column will always be 29.92 inches high. Thus, 29.92 (usually rounded off to 30) in. Hg. becomes another equivalent of the pressure of one atmosphere.

Measuring Vacuum

Since vacuum is pressure below atmospheric, vacuum can be measured in the same units. Thus, vacuum can be expressed as psia or psi (in negative units) as well as in inches of mercury.

Most vacuum gauges, however, are calibrated in inches of mercury. A perfect vacuum, which will support a column of mercury 29.92 inches high is 29.92 in. Hg. Zero vacuum (atmospheric pressure) reads zero on a vacuum gage.

Summary of Pressure and Vacuum Scales

Since a number of ways of measuring pressure and vacuum have been discussed, it would be well to place them all together for comparison.

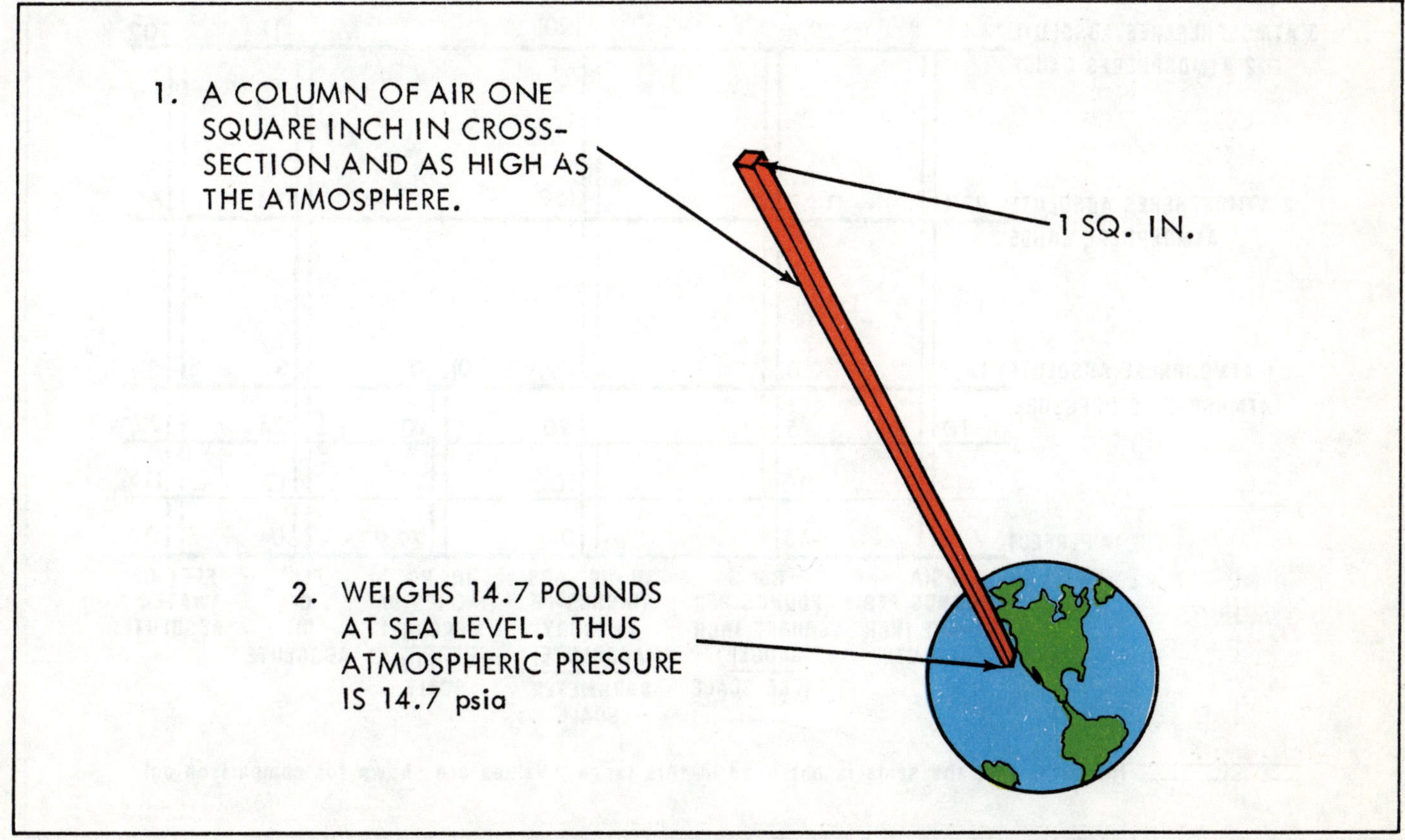

Figure 2-3. Atmospheric Pressure is a "Head" of Air

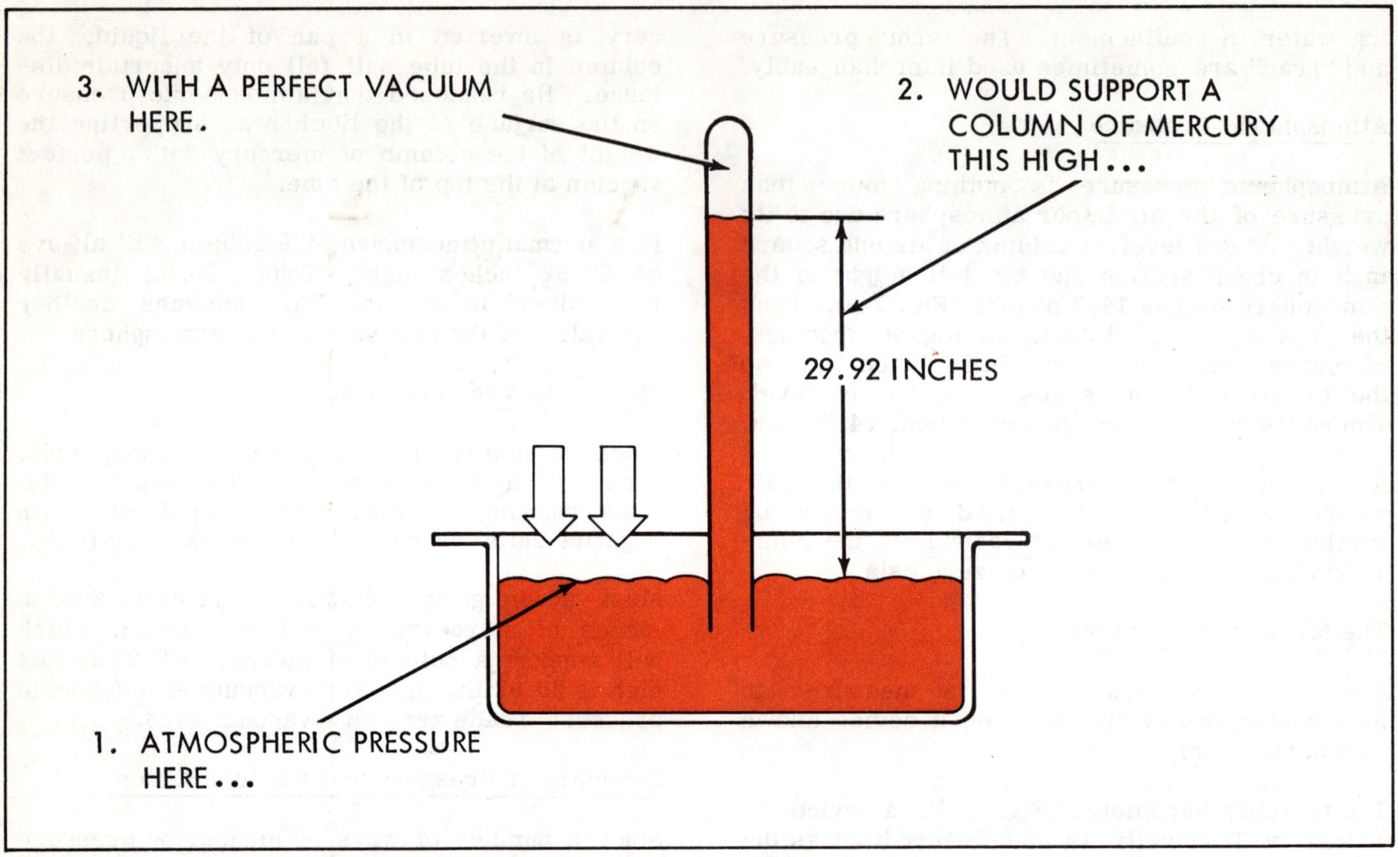

3. WITH A PERFECT VACUUM HERE.

2. WOULD SUPPORT A COLUMN OF MERCURY THIS HIGH...

29.92 INCHES

1. ATMOSPHERIC PRESSURE HERE...

Figure 2-4. The Mercury Barometer Measures Atmospheric Pressure

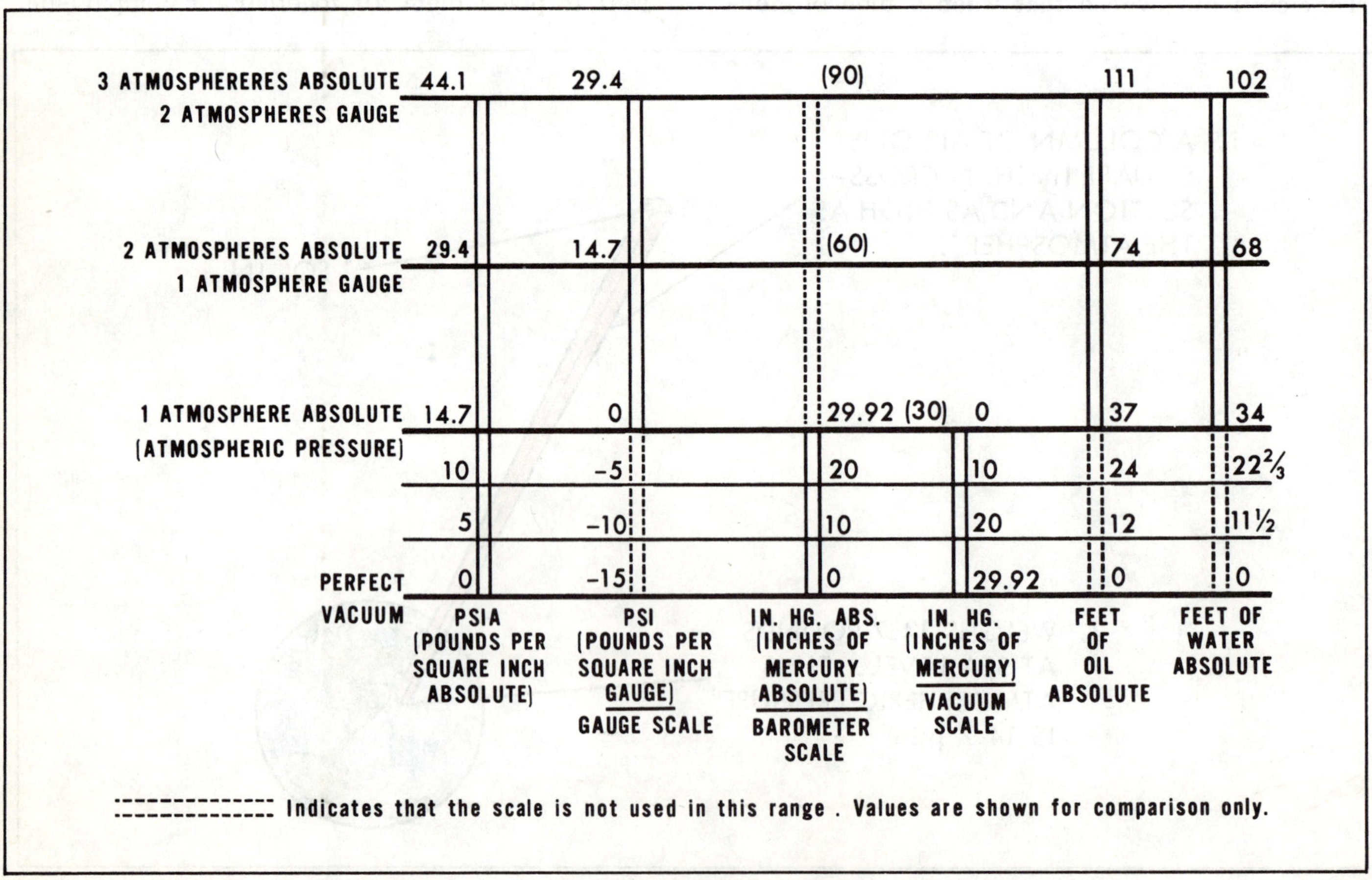

	PSIA (POUNDS PER SQUARE INCH ABSOLUTE)	PSI (POUNDS PER SQUARE INCH GAUGE) GAUGE SCALE	IN. HG. ABS. (INCHES OF MERCURY ABSOLUTE) BAROMETER SCALE	IN. HG. (INCHES OF MERCURY) VACUUM SCALE	FEET OF OIL ABSOLUTE	FEET OF WATER ABSOLUTE
3 ATMOSPHERERES ABSOLUTE 2 ATMOSPHERES GAUGE	44.1	29.4	(90)		111	102
2 ATMOSPHERES ABSOLUTE 1 ATMOSPHERE GAUGE	29.4	14.7	(60)		74	68
1 ATMOSPHERE ABSOLUTE (ATMOSPHERIC PRESSURE)	14.7	0	29.92 (30)	0	37	34
	10	-5	20	10	24	$22\frac{2}{3}$
	5	-10	10	20	12	$11\frac{1}{2}$
PERFECT VACUUM	0	-15	0	29.92	0	0

-------------- Indicates that the scale is not used in this range. Values are shown for comparison only.

Figure 2-5. Pressure and Vacuum Scale Comparison

As shown in Fig. 2-5, following is a summary of pressure and vacuum measurement:

1. An atmosphere is a pressure unit equal to 14.7 psi pressure, or 14.7 psia (the weight of a one inch square column of the air above the earth).

2. Psia (pounds per square inch absolute) is a scale which starts at a perfect vacuum (0 psia). Atmospheric pressure is 14.7 on this scale.

3. Psi (pounds per square inch gauge) is calibrated in the same units as psia but ignores atmospheric pressure. Gauge pressure may be abbreviated psig.

4. To convert from psia to psig:

Gauge Pressure + 14.7 = Absolute Pressure

Absolute Pressure - 14.7 = Gauge Pressure

5. Atmospheric pressure on the barometer scale is 29.92 in. Hg. Comparing this to the psia scale, it is evident that:

1 psi = 2 in. Hg. (approximately)

1 in. Hg. = 1/2 psi (approximately)

6. An atmosphere is equivalent to approximately 34 feet of water or 37 feet of oil.

PRINCIPLES OF FLOW

Flow is the action in the hydraulic system that gives the actuator its motion. Force can be transmitted by pressure alone, but flow is essential to cause movement. Flow in the hydraulic system is created by the pump.

How Flow is Measured

There are two ways to measure the flow of a fluid:

Velocity is the average speed of the fluid's particles past a given point or the average distance the particles travel per unit of time. It is measured in feet per second (fps) feet per minute (fpm) or inches per second (ips).

Flow rate is a measure of the volume of fluid passing a point in a given time. Large volumes are measured in gallons per minute (gpm). Small volumes may be expressed in cubic inches per minute.

Fig. 2-6 illustrates the distinction between velocity

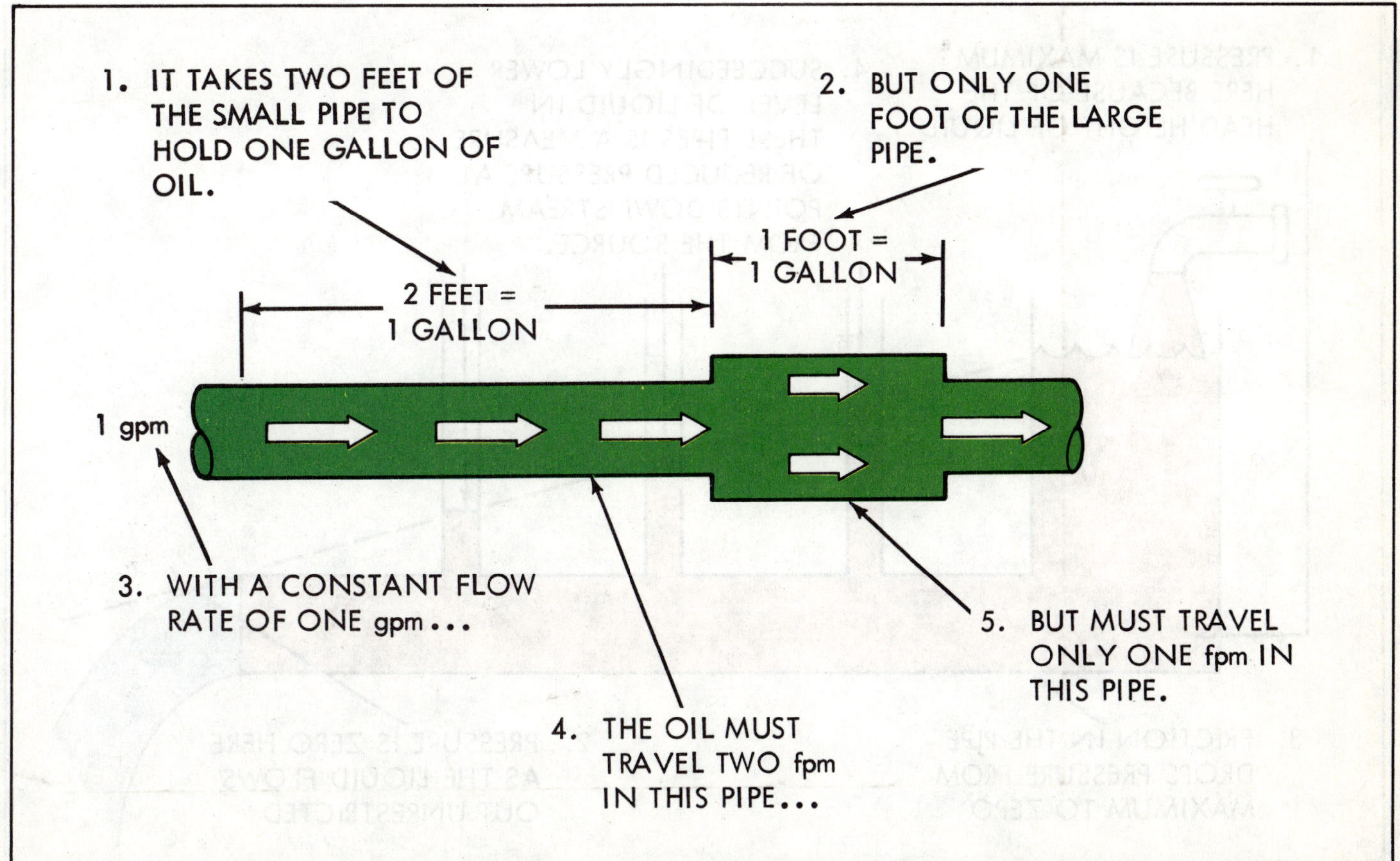

Figure 2-6. Flow is Volume Per Unit of Time; Velocity is Distance Per Unit of Time

and flow rate. A constant flow of one gallon per minute either increases or decreases in velocity when the cross section of the pipe changes size.

Flow Rate and Speed

The speed of a hydraulic actuator, as was illustrated in Chapter 1, always depends on the actuator's size and the rate of flow into it. Since the size of the actuator will be expressed in cubic inches, use these conversion factors:

1 gpm = 231 cubic inches/minute

$$\text{gpm} = \frac{\text{cubic inches/minute}}{231}$$

cubic inches/minute = gpm x 231

Flow and Pressure Drop

Whenever a liquid is flowing, there must be a condition of unbalanced force to cause motion. Therefore, when a fluid flows through a constant-diameter pipe, the pressure will always be slightly lower downstream with reference to any point upstream. The difference in pressure or pressure drop is required to overcome friction in the line.

Figure 2-7 illustrates pressure drop due to friction. The succeeding pressure drops (from maximum pressure to zero pressure) are shown as differences in head in succeeding vertical pipes.

Fluid Seeks a Level

Conversely, when there is no pressure difference on a liquid, it simply seeks a level as shown in Fig. 2-8. If the pressure changes at one point (View B) the liquid levels in the other's rise only until their weight is sufficient to make up the difference in pressure. The difference in height (head) in the case of oil is one foot per 0.4 psi. Thus it can be seen that additional pressure difference will be required to cause a liquid to flow up a pipe or to lift the fluid, since the force due to the weight of the liquid must be overcome. In circuit design, naturally, the pressure required to move the oil mass and to overcome friction must be added to the pressure needed to move the load. In most applications, good design minimizes these pressure "drops" to the point where they become almost negligible.

Laminar and Turbulent Flow

Ideally, when the particles of a fluid move

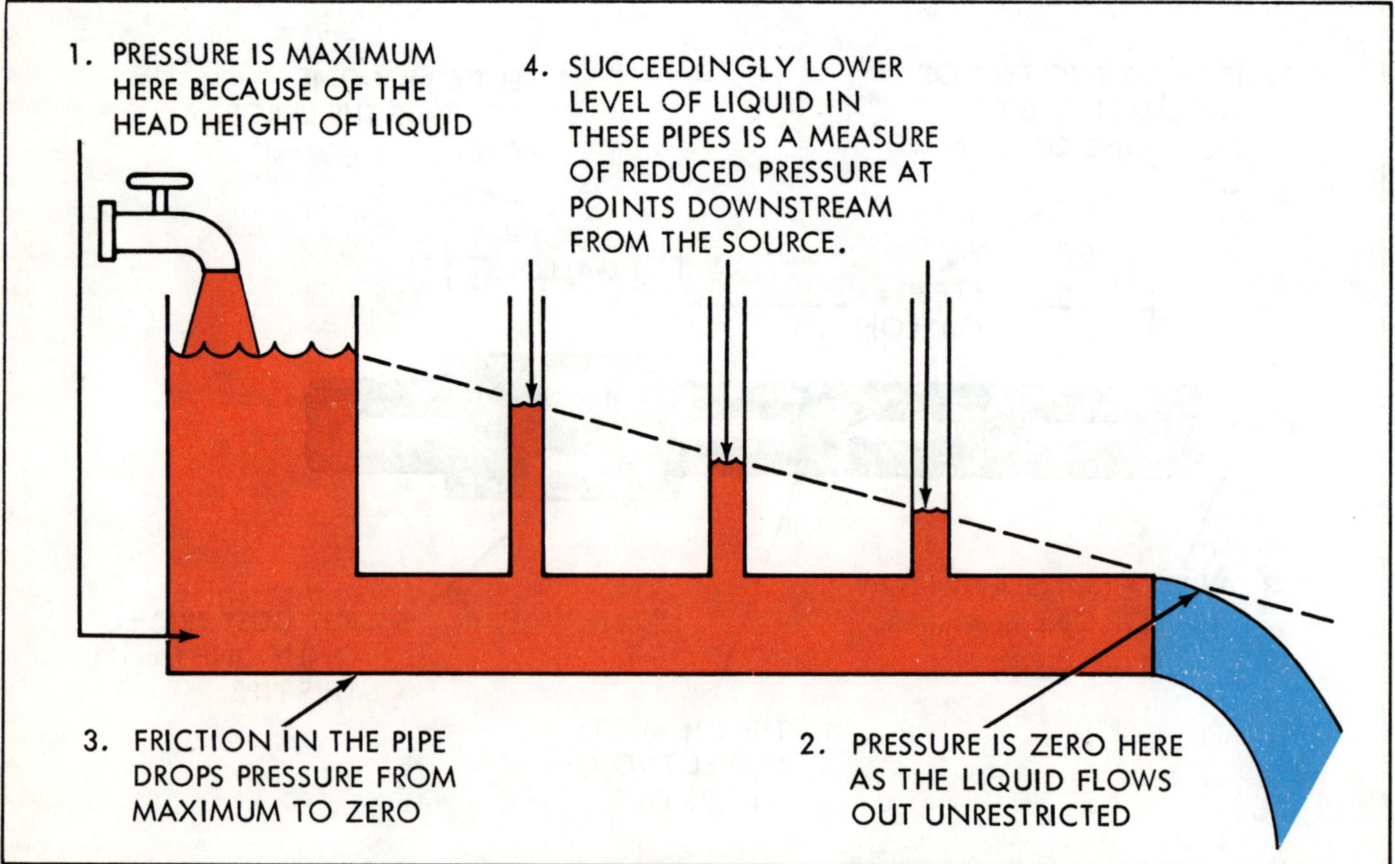

Figure 2-7. Friction in Pipes Results in a Pressure Drop

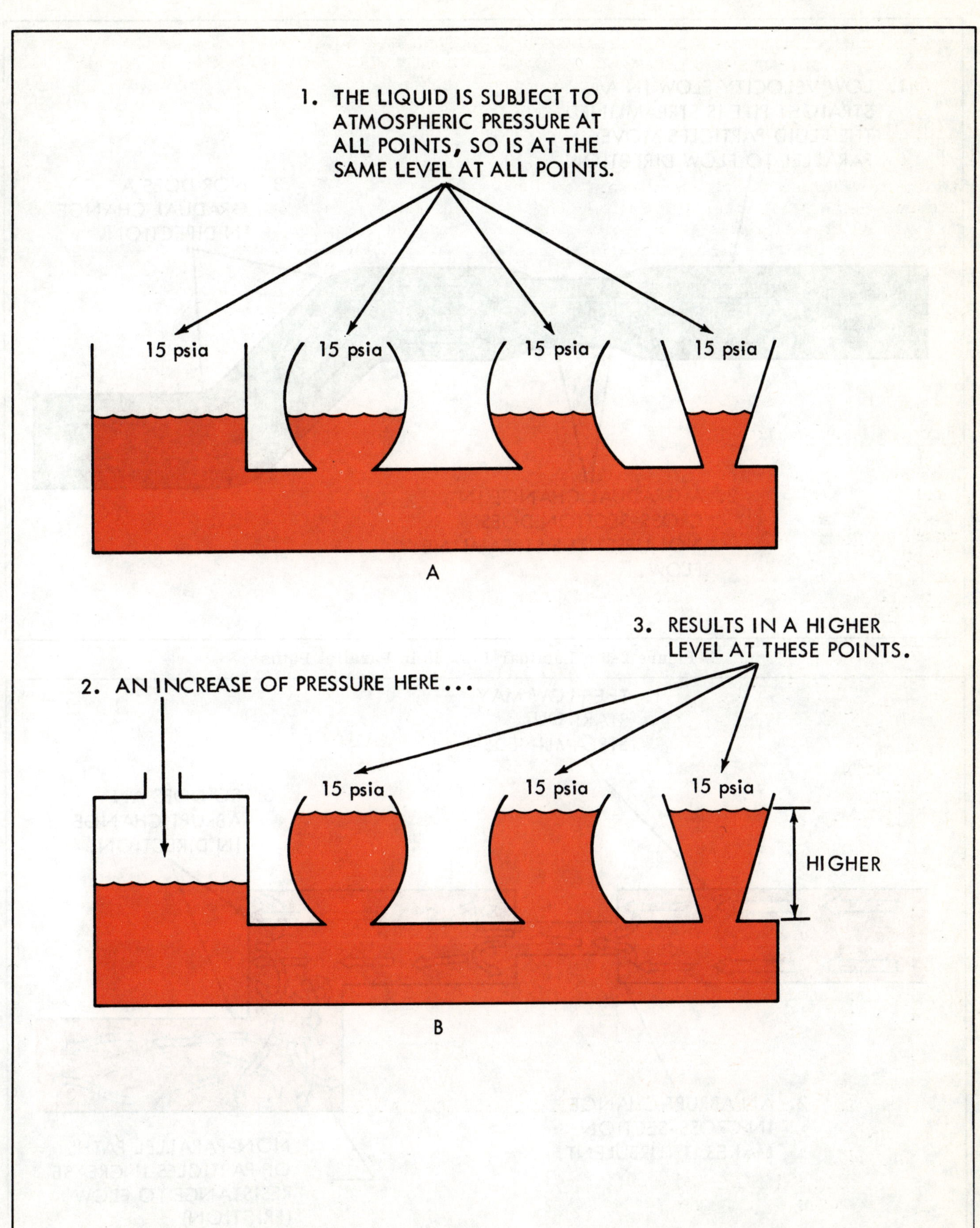

Figure 2-8. Liquid Seeks a Level or Levels Depending on the Pressure

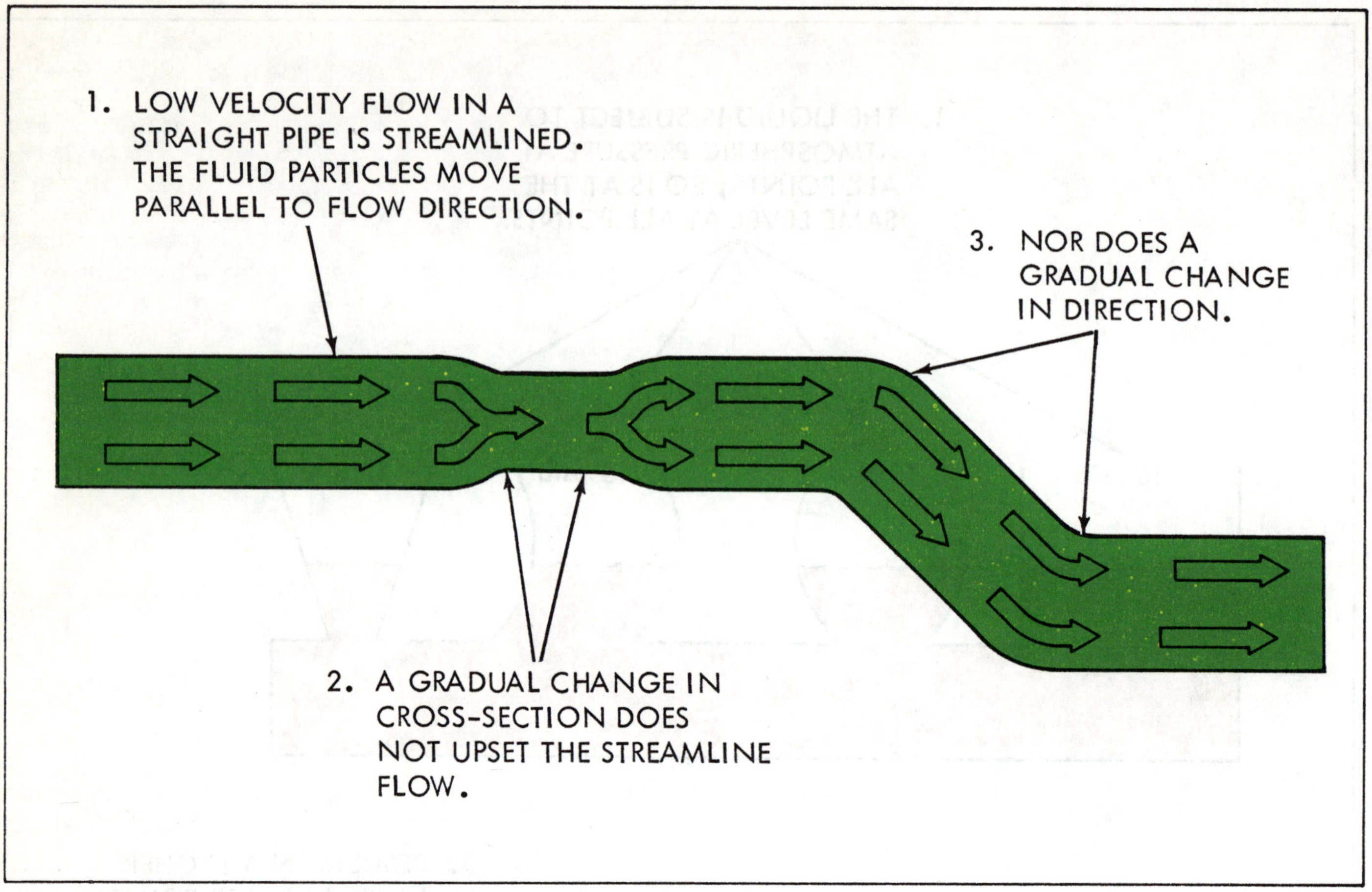

1. LOW VELOCITY FLOW IN A STRAIGHT PIPE IS STREAMLINED. THE FLUID PARTICLES MOVE PARALLEL TO FLOW DIRECTION.

2. A GRADUAL CHANGE IN CROSS-SECTION DOES NOT UPSET THE STREAMLINE FLOW.

3. NOR DOES A GRADUAL CHANGE IN DIRECTION.

Figure 2-9. Laminar Flow is in Parallel Paths

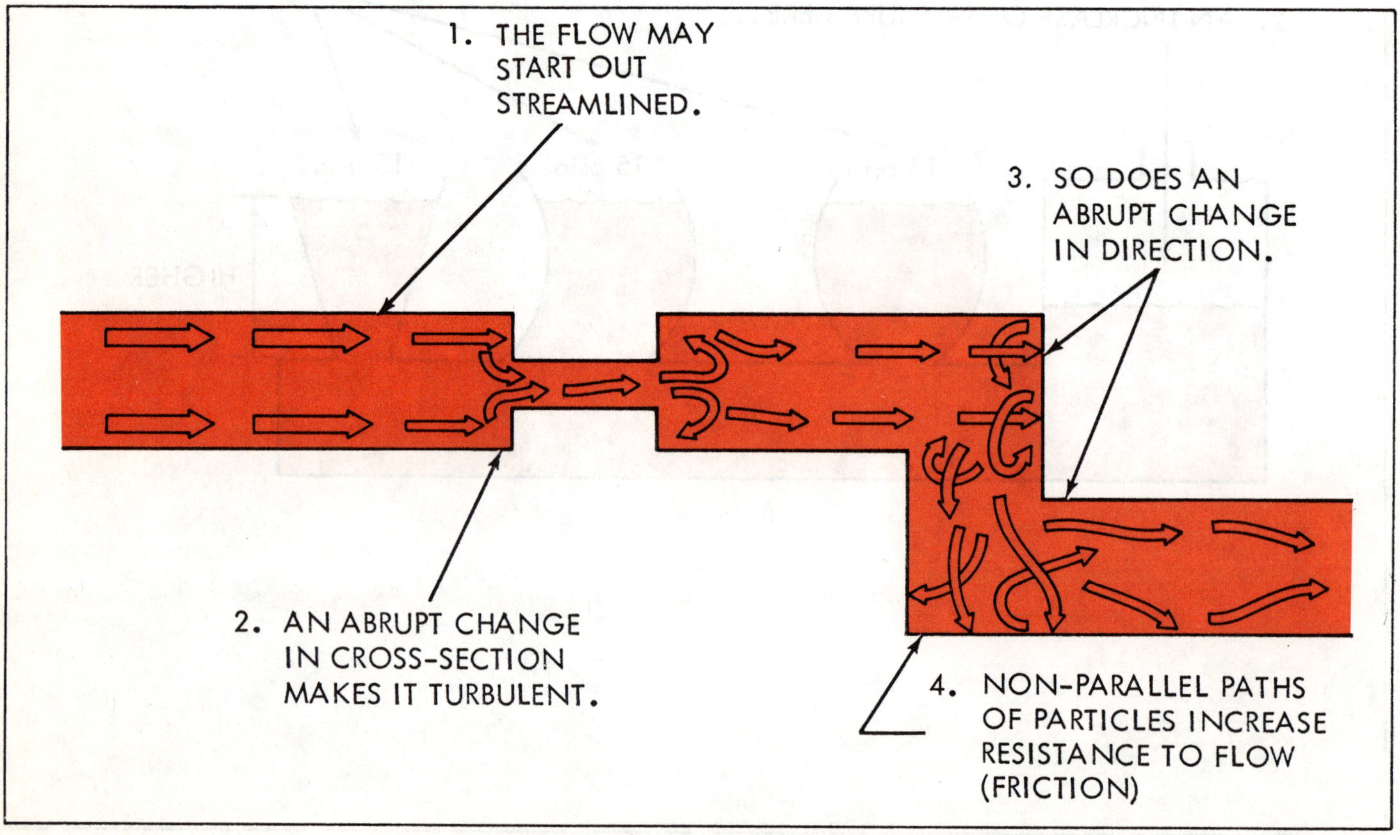

1. THE FLOW MAY START OUT STREAMLINED.

2. AN ABRUPT CHANGE IN CROSS-SECTION MAKES IT TURBULENT.

3. SO DOES AN ABRUPT CHANGE IN DIRECTION.

4. NON-PARALLEL PATHS OF PARTICLES INCREASE RESISTANCE TO FLOW (FRICTION)

Figure 2-10. Turbulence Results in Flow Resistance

through a pipe, they will move in straight, parallel flow paths (Fig. 2-9). This condition is called laminar flow and occurs at low velocity in straight piping. With laminar flow, friction is minimized.

Turbulence is the condition where the particles do not move smoothly parallel to the flow direction (Fig. 2-10). Turbulent flow is caused by abrupt changes in direction or cross section, or by too high velocity. The result is greatly increased friction, which generates heat, increases operating pressure and wastes power.

Bernoulli's Principle

Hydraulic fluid in a working system contains energy in two forms: kinetic energy by virtue of the fluid's weight and velocity and potential energy in the form of pressure.

Daniel Bernoulli, a Swiss scientist, demonstrated that in a system with a constant flow rate, energy is transformed from one form to the other each time the pipe cross-section size changes.

Bernoulli's principle says that the sums of the pressure energy and kinetic energy at various points in a system must be constant if flow rate is constant. When the pipe diameter changes (Fig. 2-11), the velocity changes. Kinetic energy thus either increases or decreases. However, energy can neither be created nor destroyed. Therefore, the change in kinetic energy must be offset by a decrease or increase in pressure.

The use of a venturi in an automobile engine carburetor (Fig. 2-12) is a familiar example of Bernoulli's principle. Air flowing through carburetor barrel is reduced in pressure as it passes through a reduced cross section of the throat. The decrease in pressure permits gasoline to flow, vaporize and mix with the air stream.

Fig. 2-13 shows the combined effects of friction and velocity changes on the pressure in a line.

HYDRAULIC SYSTEM GRAPHICAL SYMBOLS

Hydraulic circuits and their components are depicted in various ways in drawings. Depending on what the picture must convey, it may be a

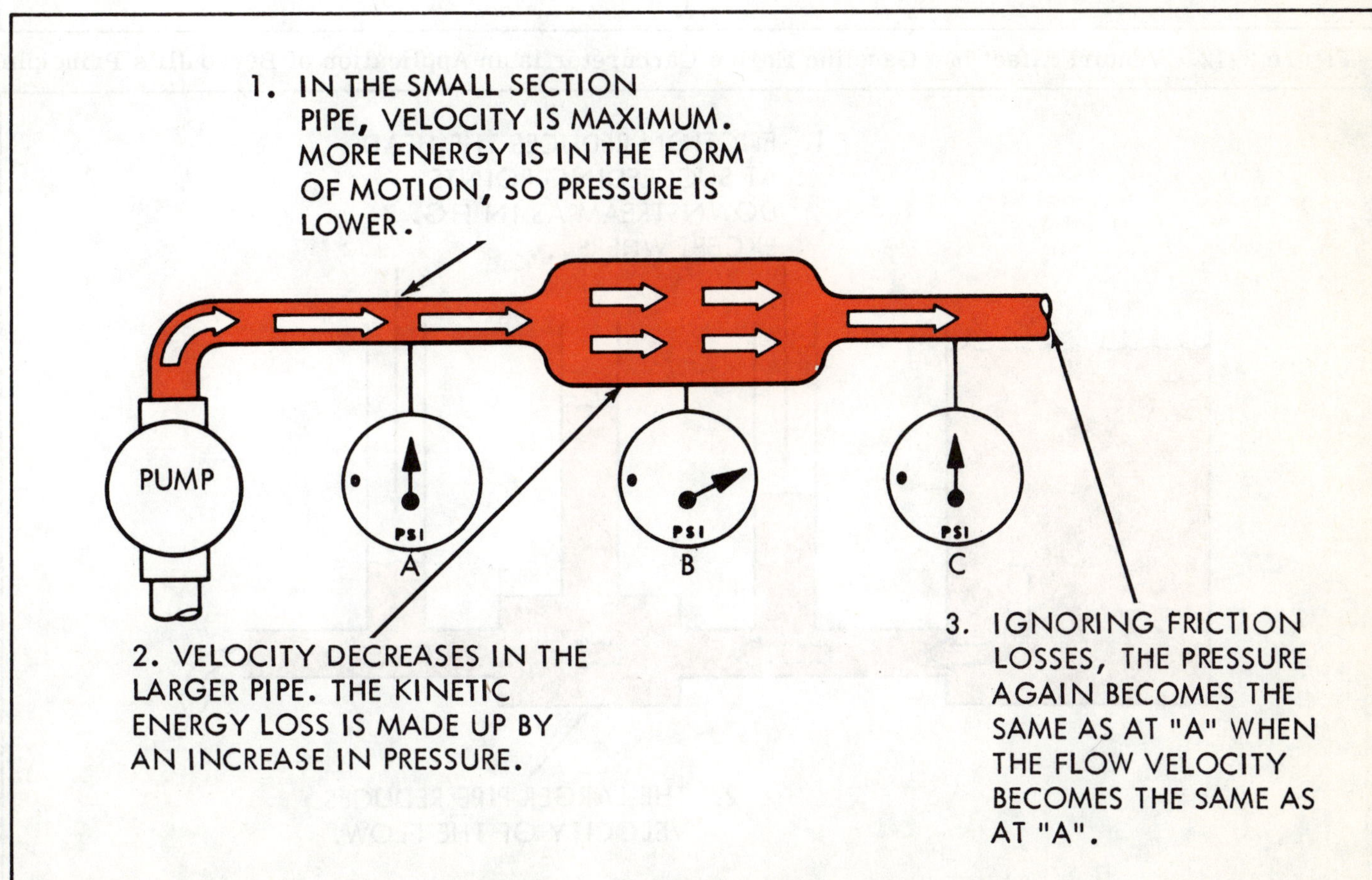

Figure 2-11. The Sum of Pressure and Kinetic Energy is Constant With a Constant Flow Rate

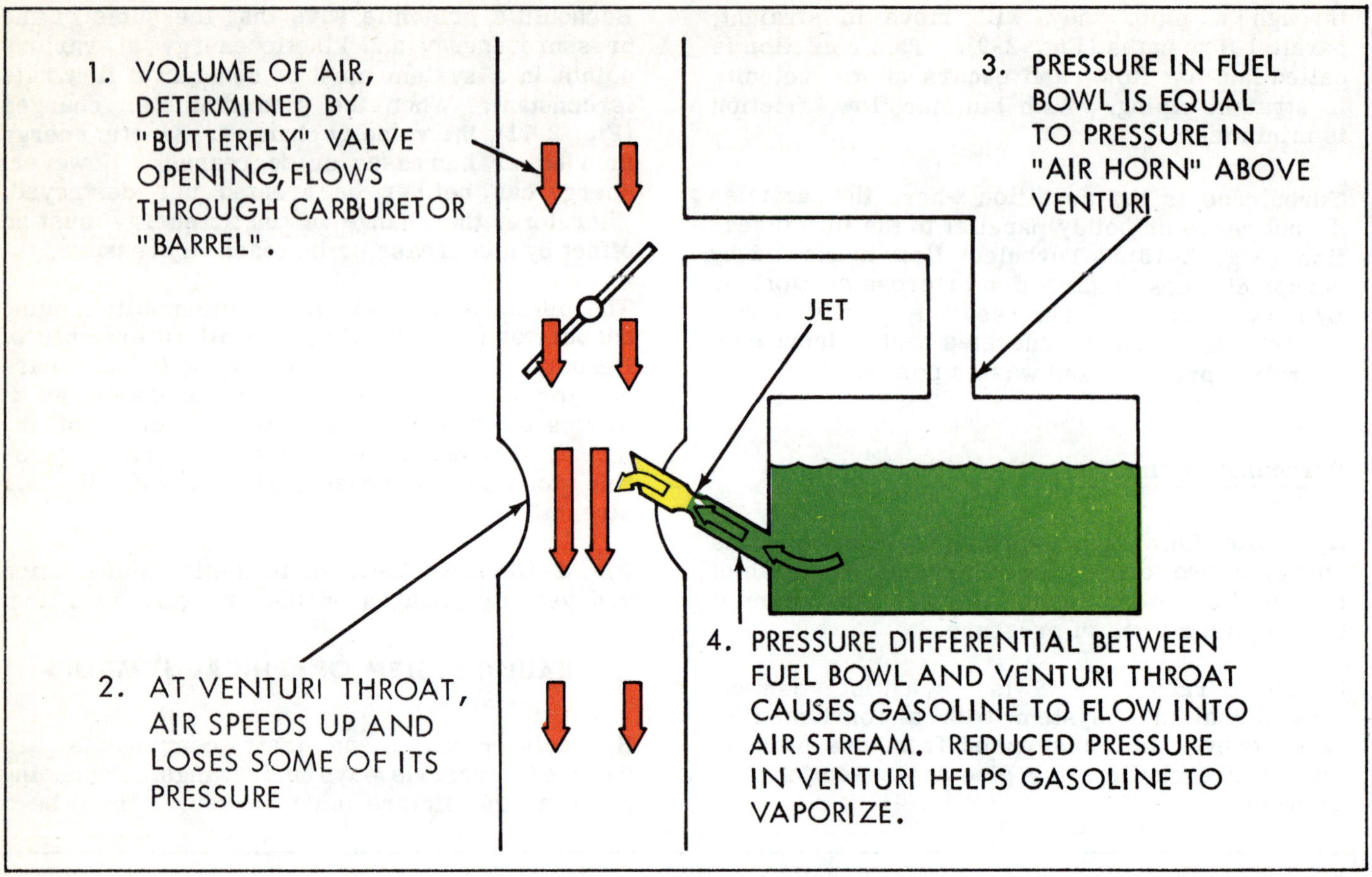

1. VOLUME OF AIR, DETERMINED BY "BUTTERFLY" VALVE OPENING, FLOWS THROUGH CARBURETOR "BARREL".

2. AT VENTURI THROAT, AIR SPEEDS UP AND LOSES SOME OF ITS PRESSURE

3. PRESSURE IN FUEL BOWL IS EQUAL TO PRESSURE IN "AIR HORN" ABOVE VENTURI

JET

4. PRESSURE DIFFERENTIAL BETWEEN FUEL BOWL AND VENTURI THROAT CAUSES GASOLINE TO FLOW INTO AIR STREAM. REDUCED PRESSURE IN VENTURI HELPS GASOLINE TO VAPORIZE.

Figure 2-12. Venturi Effect in a Gasoline Engine Carburetor is an Application of Bernoulli's Principle

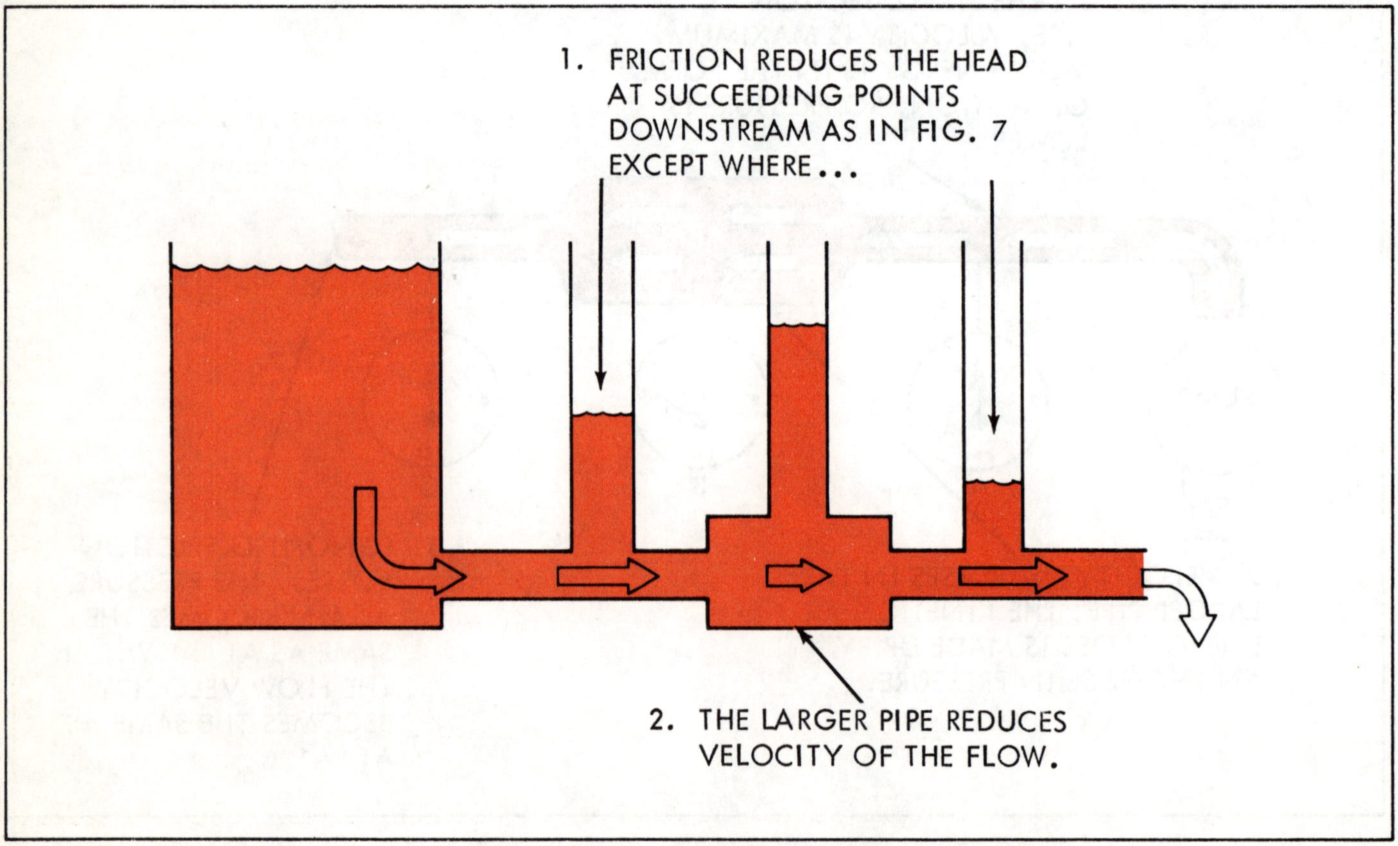

1. FRICTION REDUCES THE HEAD AT SUCCEEDING POINTS DOWNSTREAM AS IN FIG. 7 EXCEPT WHERE...

2. THE LARGER PIPE REDUCES VELOCITY OF THE FLOW.

Figure 2-13. Friction and Velocity Affect Pressure

pictorial representation of the components' exteriors; a cutaway showing internal construction; a graphical diagram which shows function; or a combination of any of the three.

All three types are of necessity used in this manual. In industry, however, the graphical symbol and diagram are most common. Graphical symbols are the "shorthand" of circuit diagrams, using simple geometric forms which show functions and inter-connections of lines and components.

The complete "standard" for graphical symbols is reproduced in Appendix of this manual. Following is a brief exposition of the most common symbols and how they are used, along with an abbreviated classification of some hydraulic lines and components.

Lines

Hydraulic pipes, tubes and fluid passages are drawn as single lines (Fig. 2-14). There are three basic classifications:

* A working line (solid) carries the main stream of flow in the system. For graphical diagram purposes, this includes the pump inlet (suction) line, pressure lines and return lines to the tank.

* A pilot line (long dashes) carries fluid that is used to control the operation of a valve or other component.

* A drain line (short dashes) carries leakage oil back to the reservoir.

Rotating Components

A circle is the basic symbol for rotating components. Energy triangles (Fig. 2-15) are placed in the symbols to show them as energy sources (pumps) or energy receivers (motors). If the component is uni-directional, the symbol has only one triangle. A reversible pump or motor is drawn with two triangles.

Cylinders

A cylinder is drawn as a rectangle (Fig. 2-16) with indications of a piston, piston rod and port con-

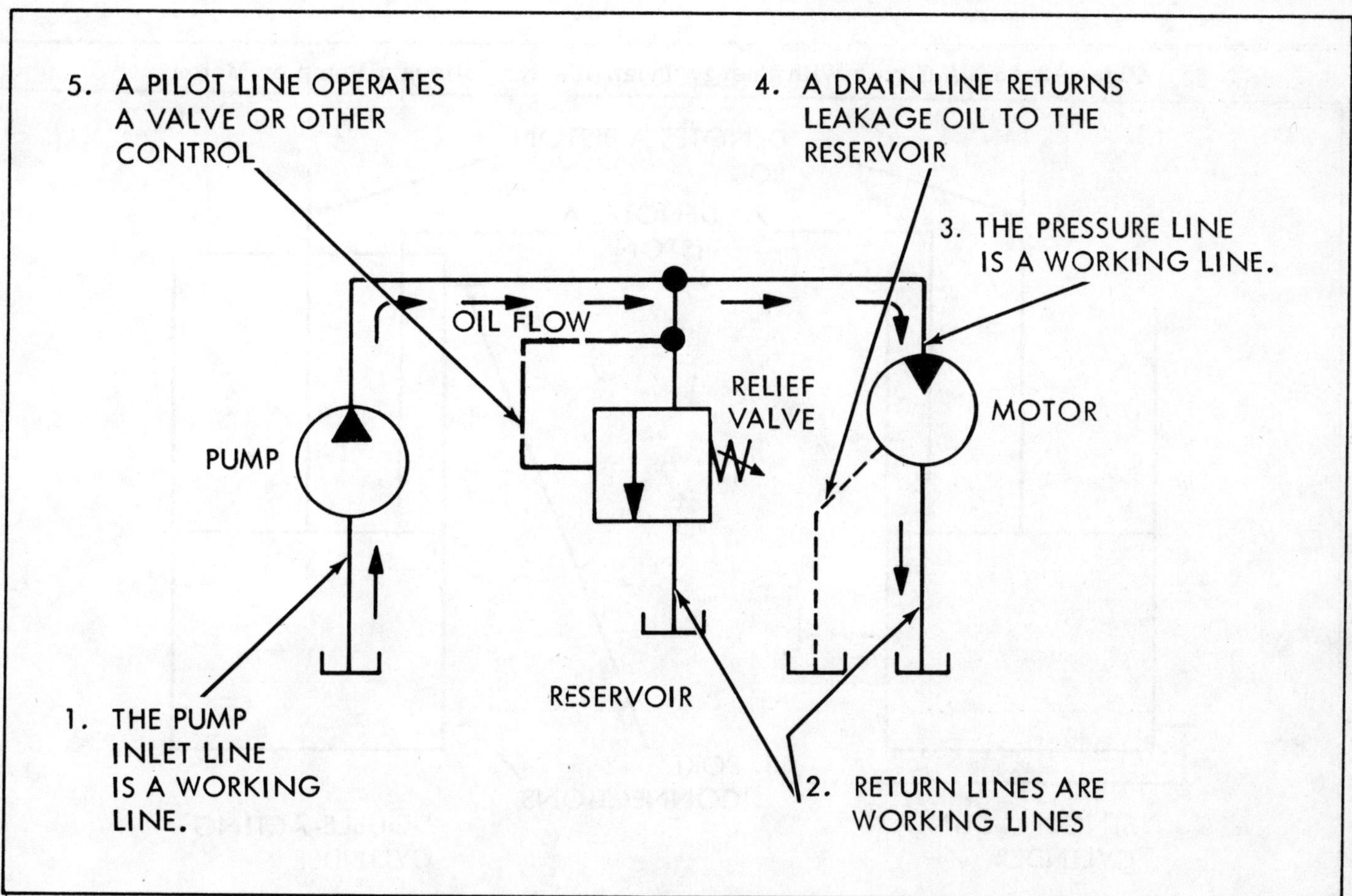

Figure 2-14. Three Classifications of Lines

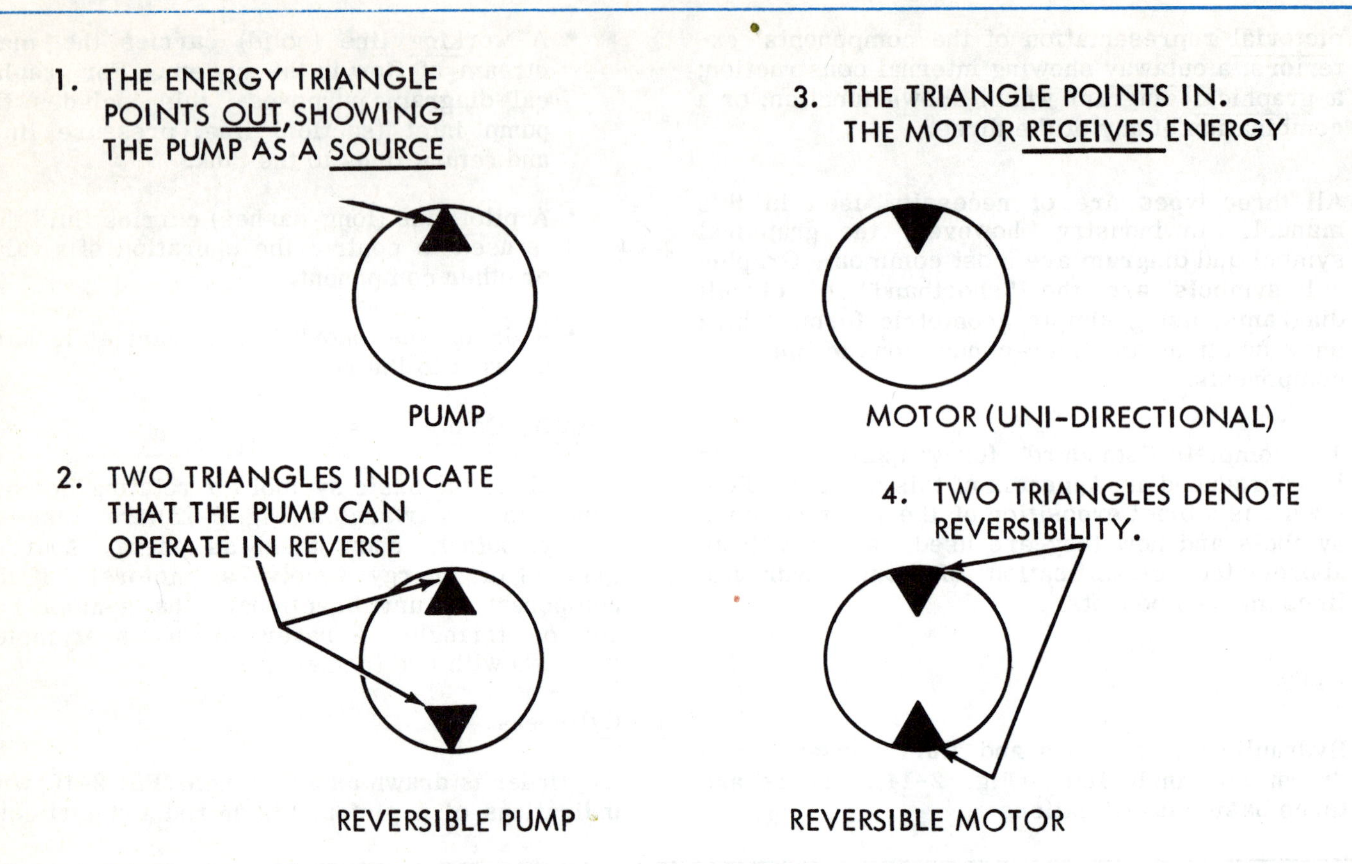

Figure 2-15. A Circle With Energy Triangles Symbolizes a Pump or Motor

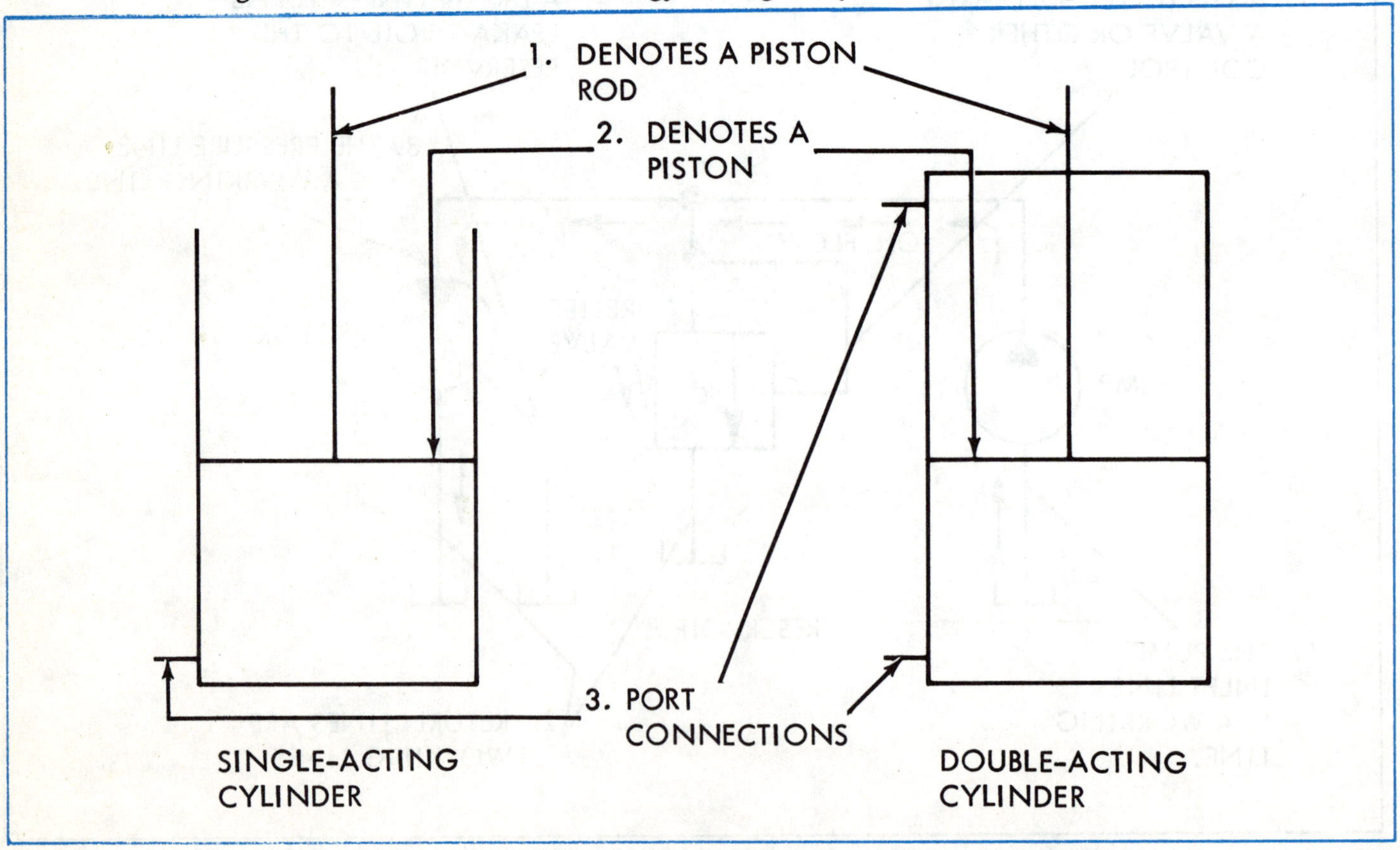

Figure 2-16. Cylinder Symbols are Single Acting or Double Acting

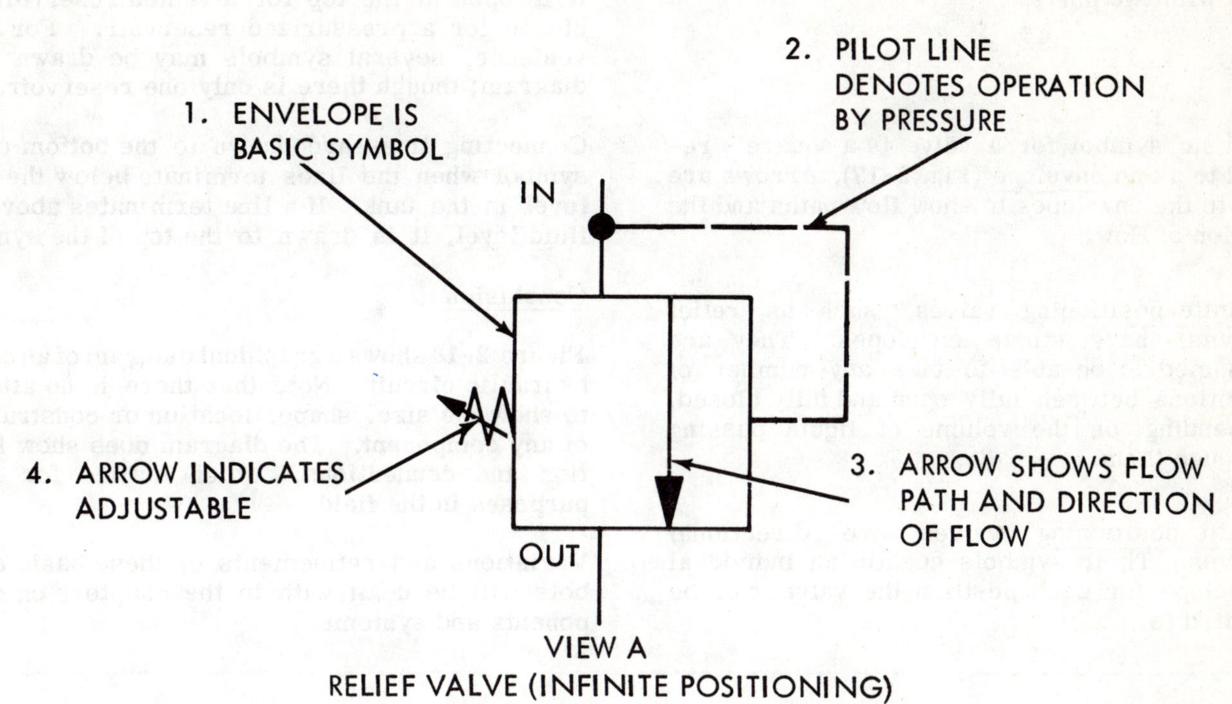

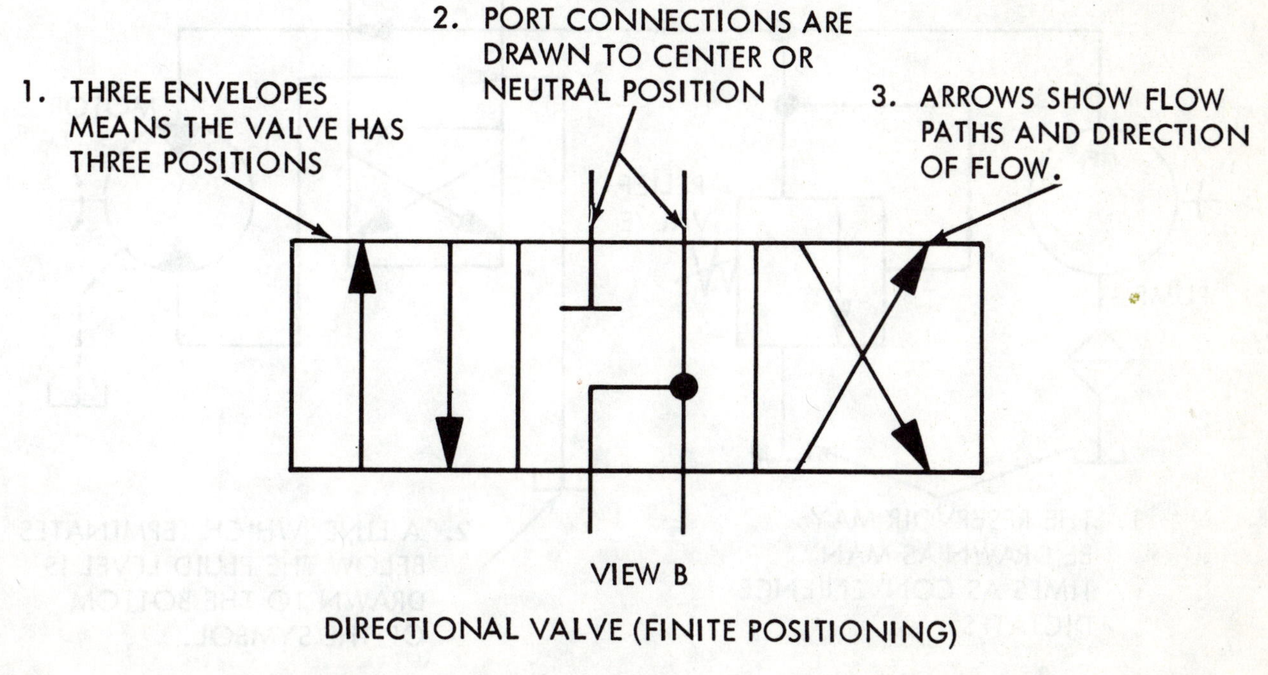

Figure 2-17. An Envelope is the Basic Valve Symbol

nection(s). A single acting cylinder is shown open at the rod end and with only a cap-end port connection. A double-acting cylinder appears closed with two ports.

Valves

The basic symbol for a valve is a square - referred to as an envelope (Fig. 2-17). Arrows are added to the envelopes to show flow paths and the direction of flow.

> Infinite-positioning valves, such as relief valves, have single envelopes. They are assumed to be able to take any number of positions between fully open and fully closed, depending on the volume of liquid passing through them.

> Finite-positioning valves are directional valves. Their symbols contain an individual envelope for each position the valve can be shifted to.

Reservoir Tank Symbol

The reservoir is drawn as a rectangle (Fig. 2-18). It is open at the top for a vented reservoir and closed for a pressurized reservoir. For convenience, several symbols may be drawn in a diagram; though there is only one reservoir.

Connecting lines are drawn to the bottom of the symbol when the lines terminate below the fluid level in the tank. If a line terminates above the fluid level, it is drawn to the top of the symbol.

Conclusion

Figure 2-18 shows a graphical diagram of an entire hydraulic circuit. Note that there is no attempt to show the size, shape, location or construction of any component. The diagram does show function and connections, which suffice for most purposes in the field.

Variations and refinements of these basic symbols will be dealt with in the chapters on components and systems.

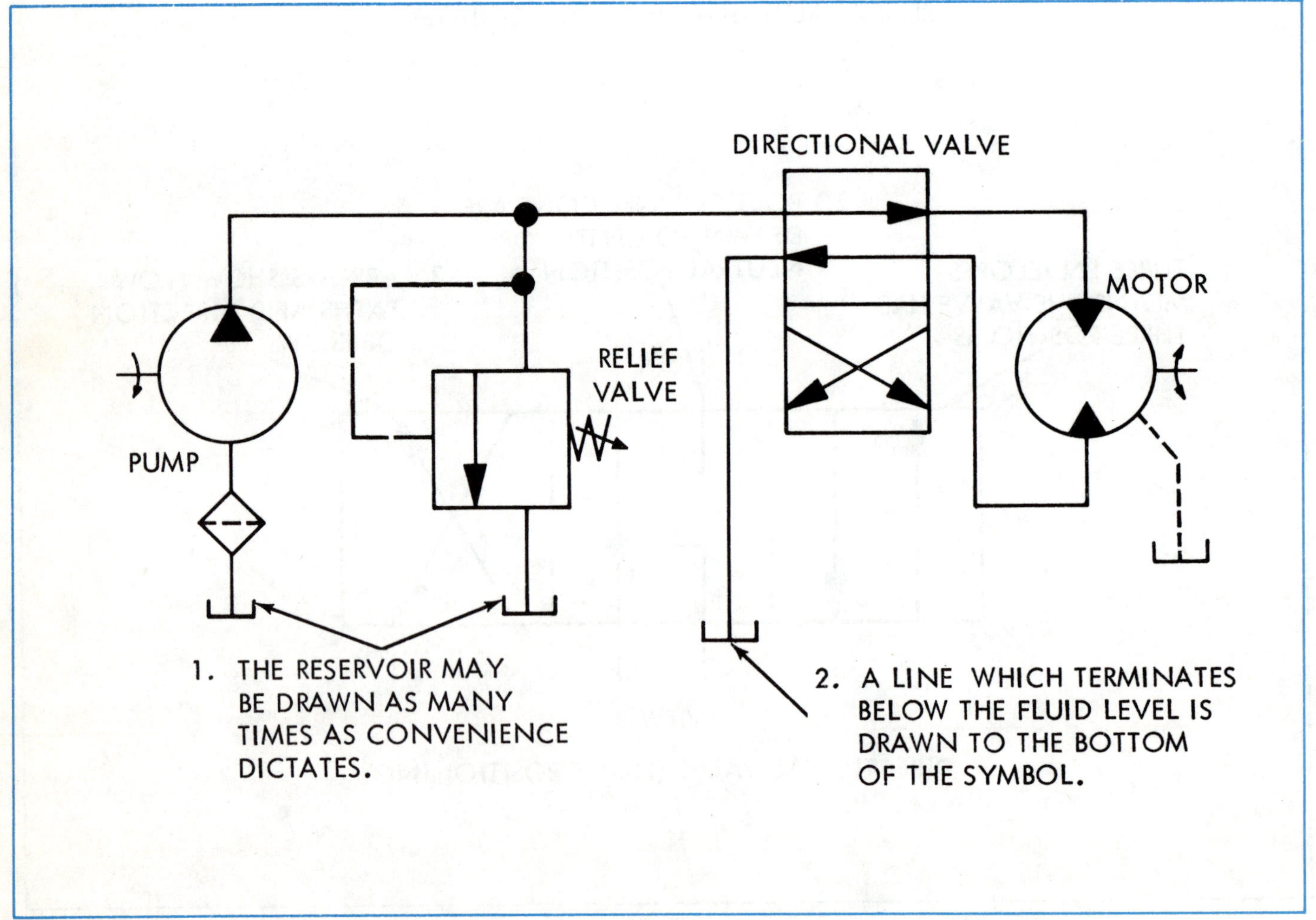

Figure 2-18. Graphical Diagram of Motor-Reversing Circuit

QUESTIONS

1. What is a hydrodynamic device?
2. How does a hydrostatic device differ?
3. Name two ways which create a tendency for a liquid to flow.
4. What is a pressure "head"?
5. How much is atmospheric pressure in psia? In psig? In inches of mercury? In feet of water?
6. How is the mercury column supported in a barometer?
7. Express 30 psig in psia.
8. What are two ways to measure flow?
9. Express 5 gpm in cubic inches per minute.
10. What happens when a liquid is subject to different pressures?
11. Pump working pressure is the sum of which individual pressures?
12. What is laminar flow?
13. What are some causes of turbulence?
14. In what two forms do we find energy in the hydraulic fluid?
15. What is Bernoulli's principle?
16. Name three kinds of working lines and tell what each does.
17. What is the basic graphical symbol for a pump or motor?
18. How many envelopes are in the symbol for a relief valve?
19. Which connecting lines are drawn to the bottom of the reservoir symbol?
20. How many positions does the directional valve in Fig. 2-18 have? The relief valve?

CHAPTER 3

HYDRAULIC FLUIDS

Selection and care of the hydraulic fluid for a machine will have an important effect on how it performs and on the life of the hydraulic components. The formulation and application of hydraulic fluids is a science of itself, far beyond the scope of this manual. In this chapter, you will find the basic factors involved in the choice of a fluid and its proper use.

A fluid has been defined in Chapter 1 as any liquid or gas. However, the term fluid has come into general use in hydraulics to refer to the liquid used as the power-transmitting medium. In this chapter, fluid will mean the hydraulic fluid, whether a specially-compounded petroleum oil or one of the special fire-resistant fluids, which may be a synthetic compound.

PURPOSES OF THE FLUID

The hydraulic fluid has four primary purposes: to transmit power, to lubricate moving parts, to seal clearances between parts, and to cool or dissipate heat.

Power Transmission

As a power transmitting medium, the fluid must flow easily through lines and component passages. Too much resistance to flow creates considerable power loss. The fluid also must be as incompressible as possible so that action is instantaneous when the pump is started or a valve shifts.

Lubrication

In most hydraulic components, internal lubrication is provided by the fluid. Pump elements and other wearing parts slide against each other on a film of fluid (Fig. 3-1). For long component life the oil must contain the necessary additives to ensure high antiwear characteristics. Not all hydraulic oils contain these additives.

Vickers recommends the new generation of industrial hydraulic oils containing adequate quantities of antiwear additives. For general hydraulic service, these oils offer superior protection against pump and motor wear and the advantage of long service life. In addition, they provide good demulsibility as well as protection against rust. These oils are generally known as antiwear type hydraulic oils.

Experience has shown that the 10W and 20-20W SAE viscosity automotive crankcase oils, having letter designation "SC", "SD", or "SE", are excellent for severe hydraulic service where there is little or no water present. The only adverse effect is that their "detergent" additives tend to hold water in a tight emulsion and prevent separation of water, even on long time standing. It should be noted that very few water problems have been experienced to date in the use of these crankcase oils in machinery hydraulic systems. Normal condensation has not been a problem.

These oils are highly recommended for mobile equipment hydraulic systems.

Sealing

In many instances, the fluid is the only seal against pressure inside a hydraulic component. In Figure 3-1, there is no seal ring between the valve spool and body to minimize leakage from the high-pressure passage to the low-pressure passages. The close mechanical fit and oil viscosity determines leakage rate.

Cooling

Circulation of the oil through lines and around the walls of the reservoir (Fig. 3-2) gives up heat that is generated in the system.

QUALITY REQUIREMENTS

In addition to these primary functions, the hydraulic fluid may have a number of other quality requirements. Some of these are:

* Prevent rust

* Prevent formation of sludge, gum and varnish

* Depress foaming

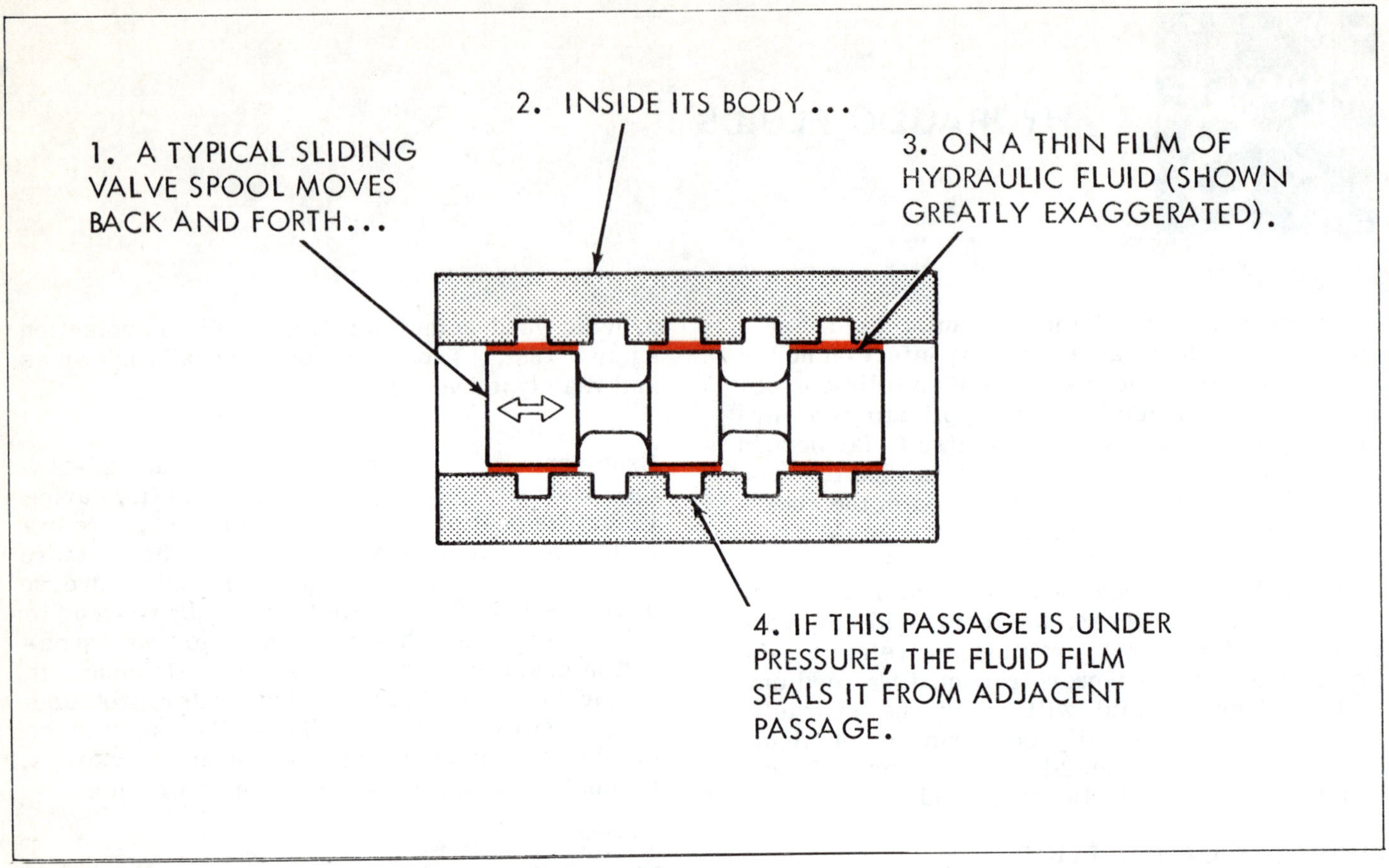

Figure 3-1. Fluid Lubricates Working Parts

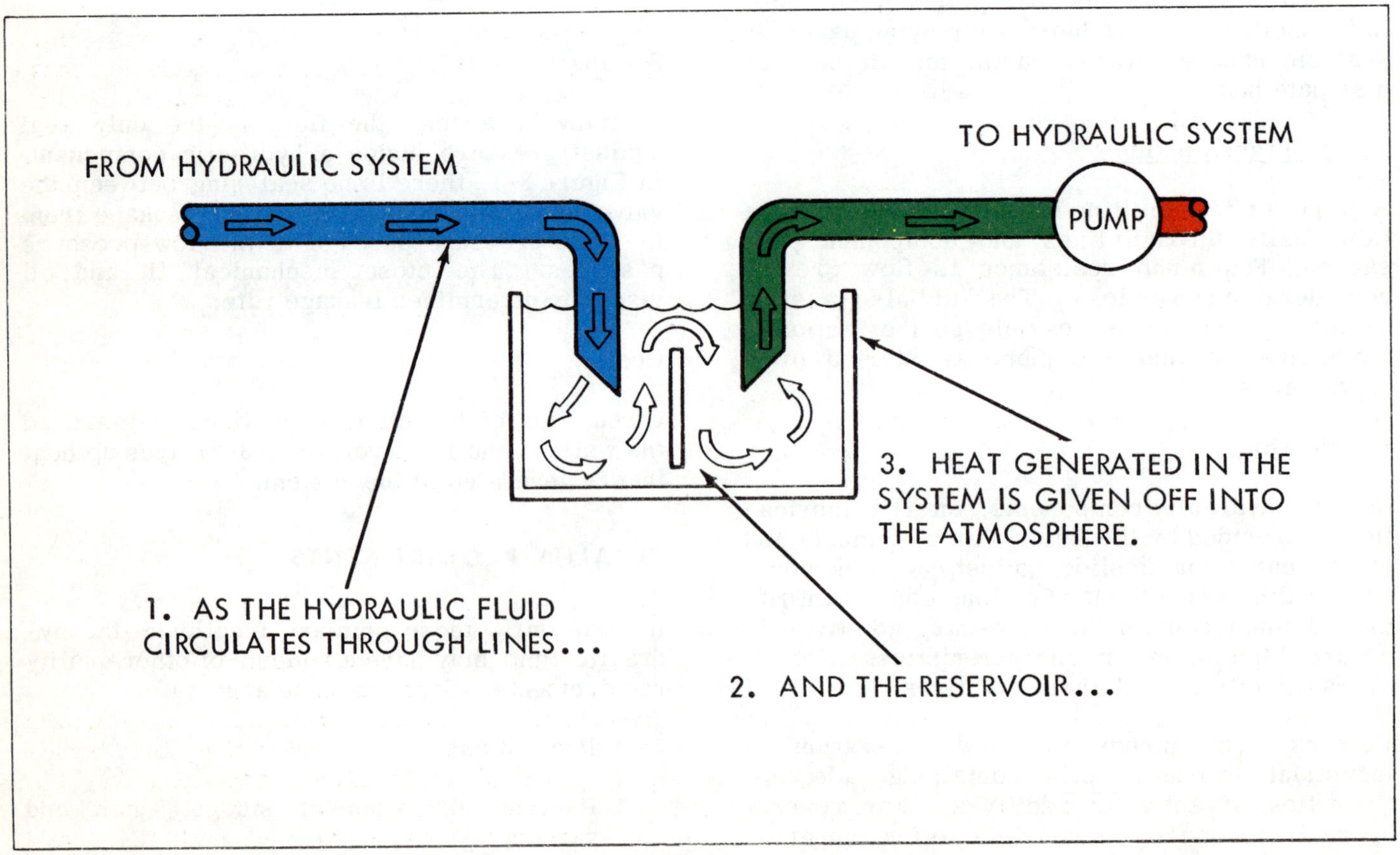

Figure 3-2. Circulation Cools the System

* Maintain its own stability and thereby reduce fluid replacement cost

* Maintain relatively stable body over a wide temperature range

* Prevent corrosion and pitting

* Separate out water

* Compatibility with seals and gaskets

These quality requirements often are the result of special compounding and may not be present in every fluid.

FLUID PROPERTIES

Let us now consider the properties of hydraulic fluids which enable it to carry out its primary functions and fulfill some or all of its quality requirements.

VISCOSITY

Viscosity is the measure of the fluid's resistance to flow; or an inverse measure of fluidity.

If a fluid flows easily, its viscosity is low. You also can say that the fluid is thin or has a low body.

A fluid that flows with difficulty has a high viscosity. It is thick or high in body.

Viscosity a Compromise

For any hydraulic machine, the actual fluid viscosity must be a compromise. A high viscosity is desirable for maintaining sealing between mating surfaces.

However, too high a viscosity increases friction, resulting in:

* High resistance to flow

* Increased power consumption due to frictional loss

* High temperature caused by friction

* Increased pressure drop because of the resistance

* Possibility of sluggish or slow operation

* Difficulty in separating air from oil in reservoir

And should the viscosity be too low:

* Internal leakage increases

* Excessive wear or even seizure under heavy load may occur due to breakdown of the oil film between moving parts.

* Pump efficiency may decrease, causing slower operation of the actuator

* Increased temperatures result from leakage losses

DEFINING VISCOSITY

Some methods of defining viscosity, in decreasing order of exactness are: Absolute (Poise) Viscosity; Kinematic Viscosity in Centistokes; Relative Viscosity in Saybolt Universal Seconds (SUS); and S. A. E. numbers. Hydraulic fluid viscosity requirements are specified in SUS as a matter of historical usage in this country.

Absolute Viscosity

Considering viscosity as the resistance when moving one layer of liquid over another is the basis for the laboratory method of measuring absolute viscosity. Poise viscosity is defined as the force per unit of area required to move one parallel surface at a speed of one centimeter-per-second past another parallel surface separated by a fluid film one centimeter thick (Fig. 3-3). (In the metric system, force is expressed in dynes; area in square centimeters). Stated another way, poise is the ratio between the shearing stress and the rate of shear of the fluid:

$$\text{ABSOLUTE VISCOSITY} = \frac{\text{Shear Stress}}{\text{Rate of Shear}}$$

$$1 \text{ POISE} = 1 \frac{\text{Dyne Second}}{\text{Square Centimeter}}$$

A smaller unit of absolute viscosity is the centipoise, which is one-hundredth of a poise:

One Centipoise = .01 Poise

Kinematic Viscosity

The concept of kinematic viscosity is the outgrowth of the use of a head of liquid to produce a flow through a capillary tube. The coefficient of absolute viscosity, when divided by the density of the liquid is called the kinematic viscosity. In the metric system, the unit of viscosity is called the Stoke and it has the units of centimeters squared per second. One one-hundredth of a Stoke is a Centistoke.

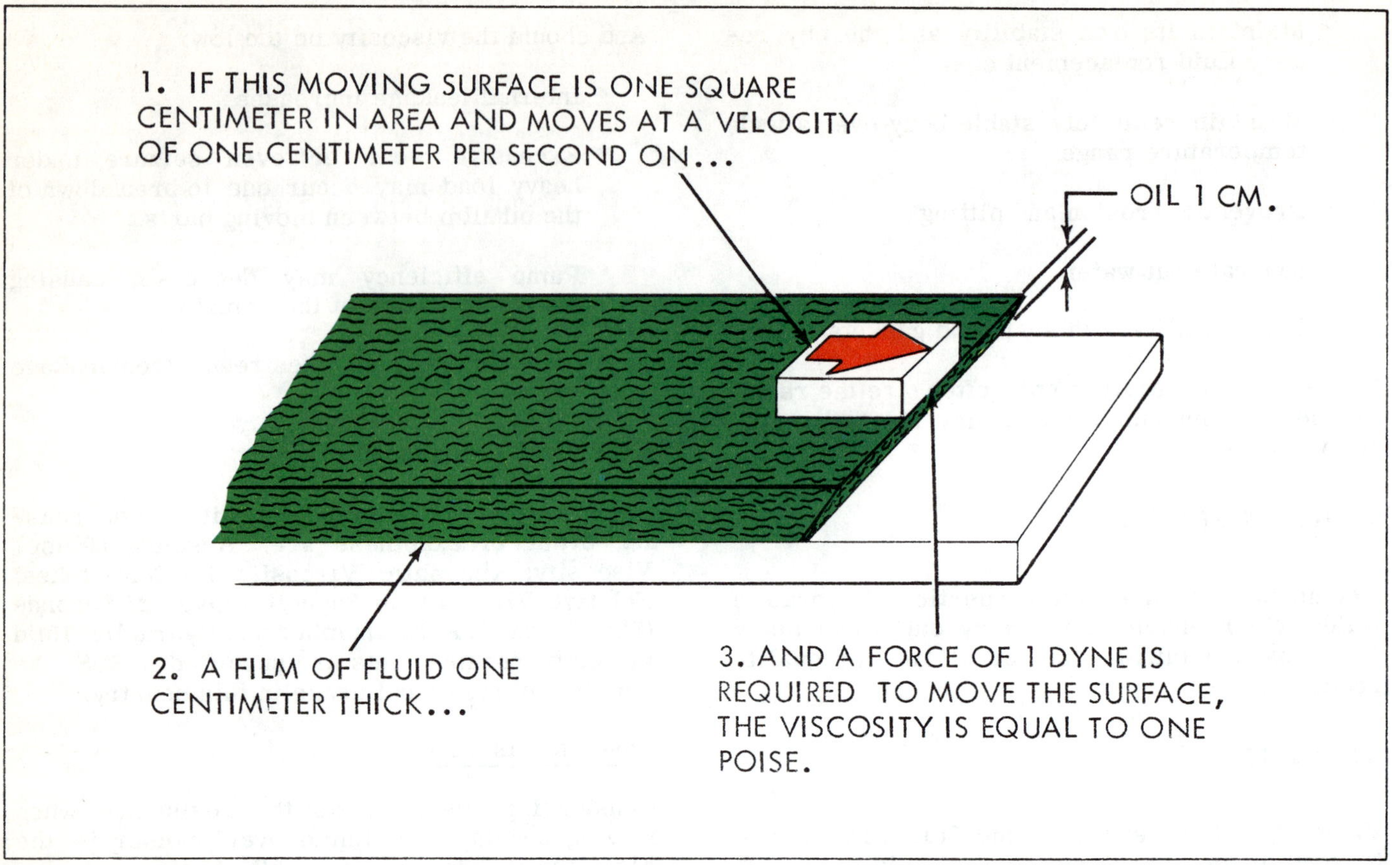

1. IF THIS MOVING SURFACE IS ONE SQUARE CENTIMETER IN AREA AND MOVES AT A VELOCITY OF ONE CENTIMETER PER SECOND ON...

OIL 1 CM.

2. A FILM OF FLUID ONE CENTIMETER THICK...

3. AND A FORCE OF 1 DYNE IS REQUIRED TO MOVE THE SURFACE, THE VISCOSITY IS EQUAL TO ONE POISE.

Figure 3-3. Measuring Absolute Viscosity

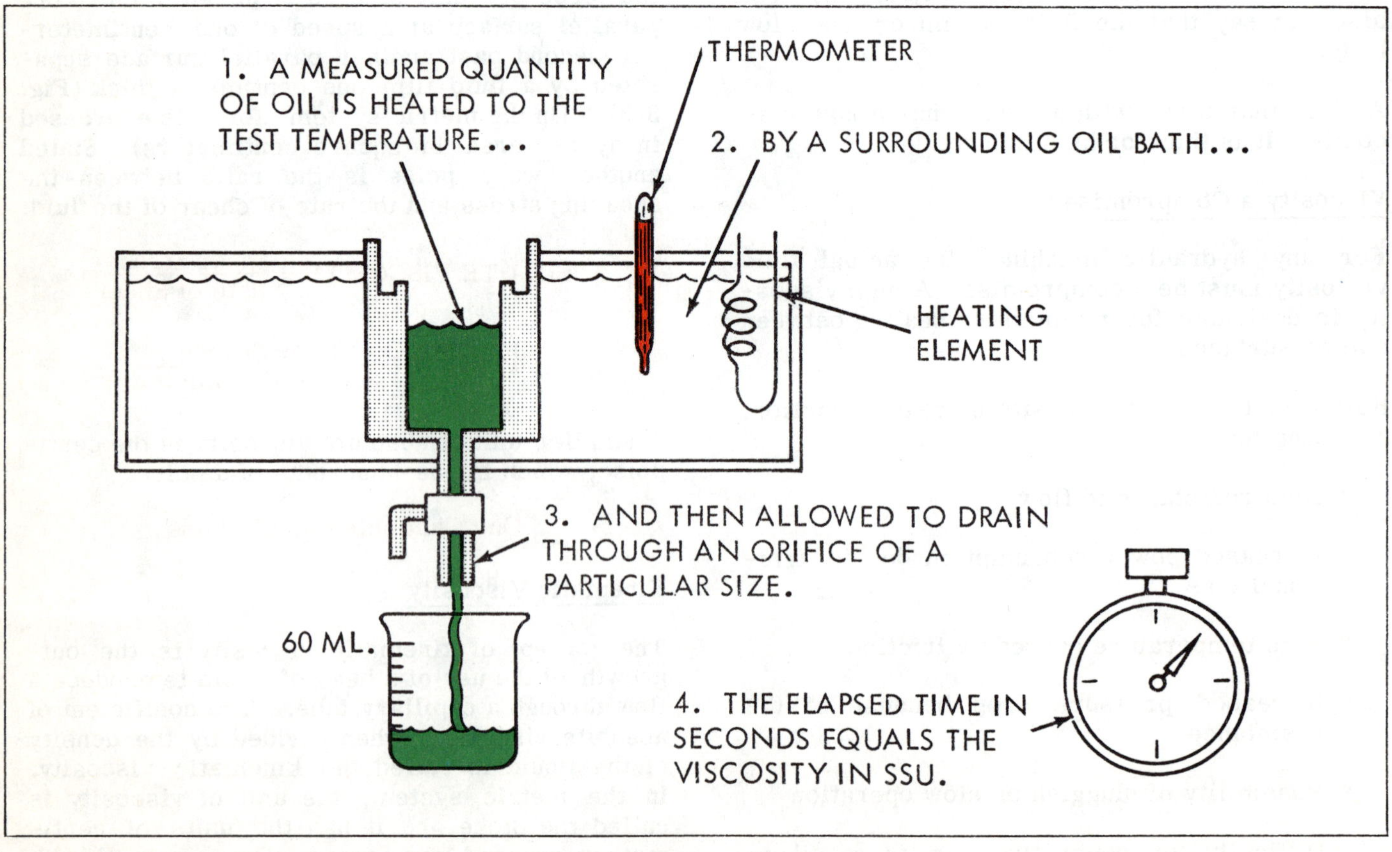

1. A MEASURED QUANTITY OF OIL IS HEATED TO THE TEST TEMPERATURE...

THERMOMETER

2. BY A SURROUNDING OIL BATH...

HEATING ELEMENT

3. AND THEN ALLOWED TO DRAIN THROUGH AN ORIFICE OF A PARTICULAR SIZE.

60 ML.

4. THE ELAPSED TIME IN SECONDS EQUALS THE VISCOSITY IN SSU.

Figure 3-4. Saybolt Viscosimeter Measures Relative Viscosity

Following are conversions between absolute and kinematic viscosity:

$$\text{Centipoise} = \text{Centistoke} \times \text{Density}$$

$$\text{Centistoke} = \frac{\text{Centipoise}}{\text{Density}}$$

SUS Viscosity

For most practical purposes, it will serve to know the relative viscosity of the fluid. Relative viscosity is determined by timing the flow of a given quantity of the fluid through a standard orifice at a given temperature. There are several methods in use. The most accepted method in this country is the Saybolt Viscosimeter (Fig. 3-4).

The time it takes for the measured quantity of liquid to flow through the orifice is measured with a stopwatch. The viscosity in Saybolt Universal Seconds (SUS) equals the elapsed time.

Obviously, a thick liquid will flow slowly, and the SUS viscosity will be higher than for a thin liquid which flows faster. Since oil becomes thicker at low temperature and thins when warmed, the viscosity must be expressed as so many SUS's at a given temperature. The tests are usually made at 100 degrees F. or 210°F.

For industrial applications, hydraulic oil viscosities usually are in the vicinity of 150 SUS at 100°F. It is a general rule that the viscosity should never go below 45 SUS or above 4000 SUS, regardless of temperature. Where temperature extremes are encountered, the fluid should have a high viscosity index (page 3-6).

SAE Numbers

SAE Numbers have been established by the Society of Automotive Engineers to specify ranges of SUS viscosities of oils at SAE test temperatures.

Winter numbers (5W, 10W, 20W) are determined by tests at 0 degrees F. Summer oil numbers (20, 30, 40, 50, etc.) designate the SUS range at 210°F. Table 1 is a chart of the temperature ranges.

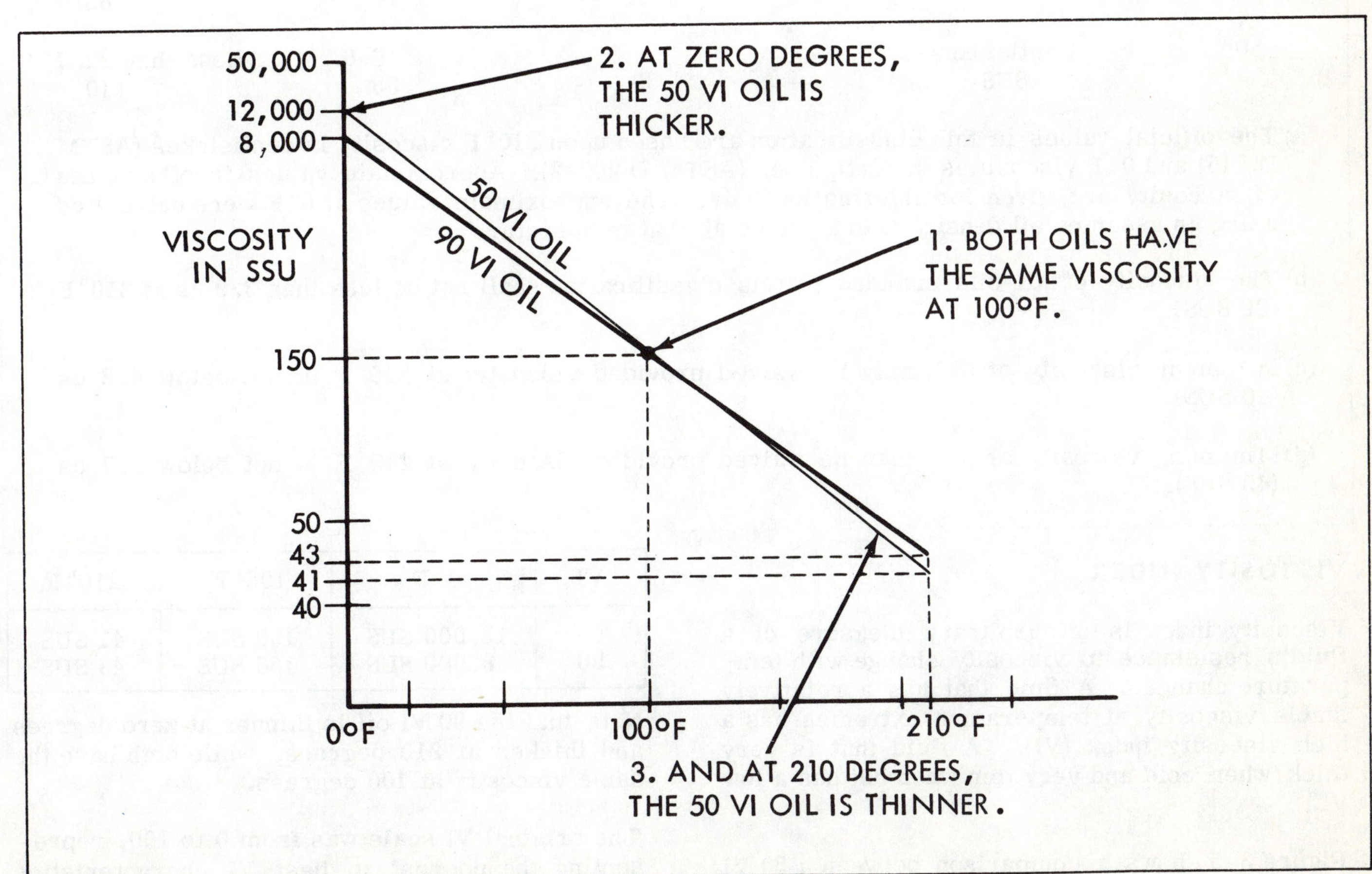

Figure 3-5. Viscosity Index (VI) is a Relative Measure of Viscosity Change with Temperature Change

TABLE 1. SAE VISCOSITY NUMBERS FOR CRANKCASE OILS

SAE VISCOSITY NUMBER	Viscosity Units (a)	Viscosity Range (b) At 0°F Minimum	At 0°F Maximum	At 210°F Minimum	At 210°F Maximum
5W	Centipoises	-	Less than 1,200	-	-
	Centistokes	-	1,300	-	-
	SUS	-	6,000	-	-
10W	Centipoises	1,200 (c)	Less than 2,400	-	-
	Centistokes	1,300	2,600	-	-
	SUS	6,000	12,000	-	-
20W	Centipoises	2,400 (d)	Less than 9,600	-	-
	Centistokes	2,600	10,500	-	-
	SUS	12,000	48,000	-	-
20	Centistokes	-	-	5.7	Less than 9.6
	SUS	-	-	45	58
30	Centistokes	-	-	9.6	Less than 12.9
	SUS	-	-	58	70
40	Centistokes	-	-	12.9	Less than 16.8
	SUS	-	-	70	85
50	Centistokes	-	-	16.8	Less than 22.7
	SUS	-	-	85	110

(a) The official values in this classification are based upon 210°F viscosity in centistokes (ASTM D 445) and 0°F viscosities in centipoises (ASTM D 260-2). Approximate values in other units of viscosity are given for information only. The approximate values at 0°F were calculated using an assumed oil density of 0.9 gm/cc at that temperature

(b) The viscosity of all oils included in this classification shall not be less than 3.9 cs at 210°F (39 SUS).

(c) Minimum viscosity at 0°F may be waived provided viscosity at 210°F is not below 4.2 cs (40 SUS).

(d) Minimum viscosity at 0°F may be waived provided viscosity at 210°F is not below 5.7 cs (45 SUS).

VISCOSITY INDEX

Viscosity index is an arbitrary measure of a fluid's resistance to viscosity change with temperature changes. A fluid that has a relatively stable viscosity at temperature extremes has a high viscosity index (VI). A fluid that is very thick when cold and very thin when hot has a low VI.

Figure 3-5 shows a comparison between a 50 VI and a 90 VI oil. Compare these actual viscosities at three temperatures:

VI	0°F	100°F	210°F
50	12,000 SUS	150 SUS	41 SUS
90	8,000 SUS	150 SUS	43 SUS

Note that the 90 VI oil is thinner at zero degrees and thicker at 210 degrees, while both have the same viscosity at 100 degrees.

The original VI scale was from 0 to 100, representing the poorest to best VI characteristics then known. Today, chemical additives and refining techniques have increased the VI of

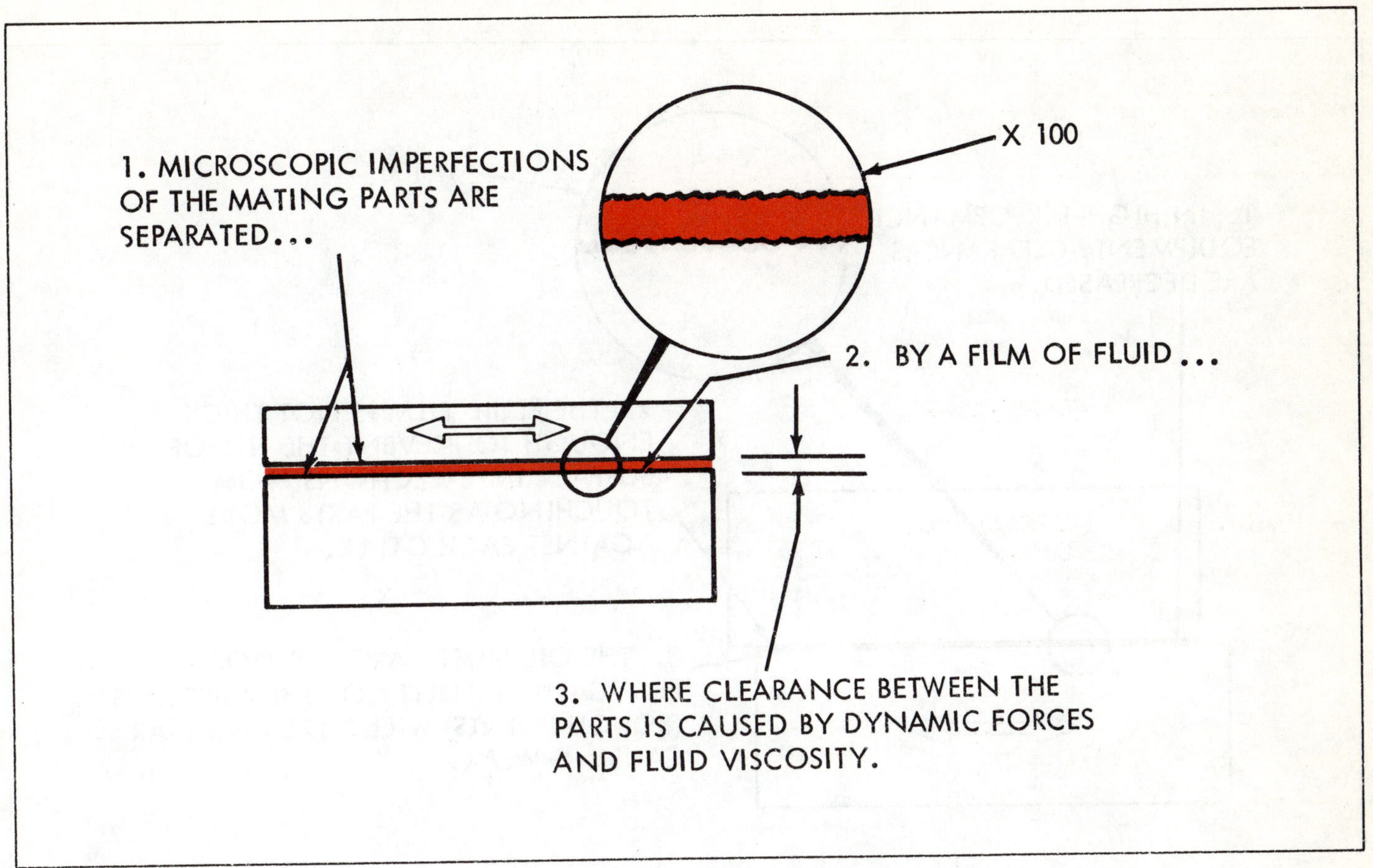

Figure 3-6. Full Film Lubrication Prevents Metal-to-Metal Contact

some oils considerably above 100. A high VI is desirable when the equipment operates in temperature extremes. However, in a machine that runs at relatively constant temperatures the viscosity index of the fluid is less critical.

POUR POINT

Pour point is the lowest temperature at which a fluid will flow. It is a very important specification if the hydraulic system will be exposed to extremely low temperature. For a thumb rule, the pour point should be 20 degrees F below the lowest temperature to be encountered.

LUBRICATING ABILITY

It is desirable for hydraulic system moving parts to have enough clearance to run together on a substantial film of fluid (Fig. 3-6). This condition is called full-film lubrication. So long as the fluid has adequate viscosity, the minute imperfections in the surfaces of the parts do not touch.

However, in certain high performance equipment, increased speeds and pressure, coupled with lower clearances, cause the film of fluid to be squeezed very thin (Fig. 3-7) and a condition called boundary lubrication occurs. Here, there may be metal-to-metal contact between the tips of the two mating part surfaces and some chemical lubricating ability is needed.

OXIDATION RESISTANCE

Oxidation, or chemical union with oxygen, is a serious reducer of the service life of a fluid. Petroleum oils are particularly susceptible to oxidation, since oxygen readily combines with both carbon and hydrogen in the oil's makeup.

Most of the oxidation products are soluble in the oil, and additional reactions take place in the products to form gum, sludge and varnish. The first stage products which stay in the oil are acid in nature and can cause corrosion throughout the system, in addition to increasing the viscosity of the oil. The insoluble gums, sludge and varnish plug orifices, increase wear and cause valves to stick.

CATALYSTS

There are always a number of oxidation catalysts or helpers in a hydraulic system. Heat, pressure, contaminants, water, metal surfaces and agitation all accelerate oxidation once it starts.

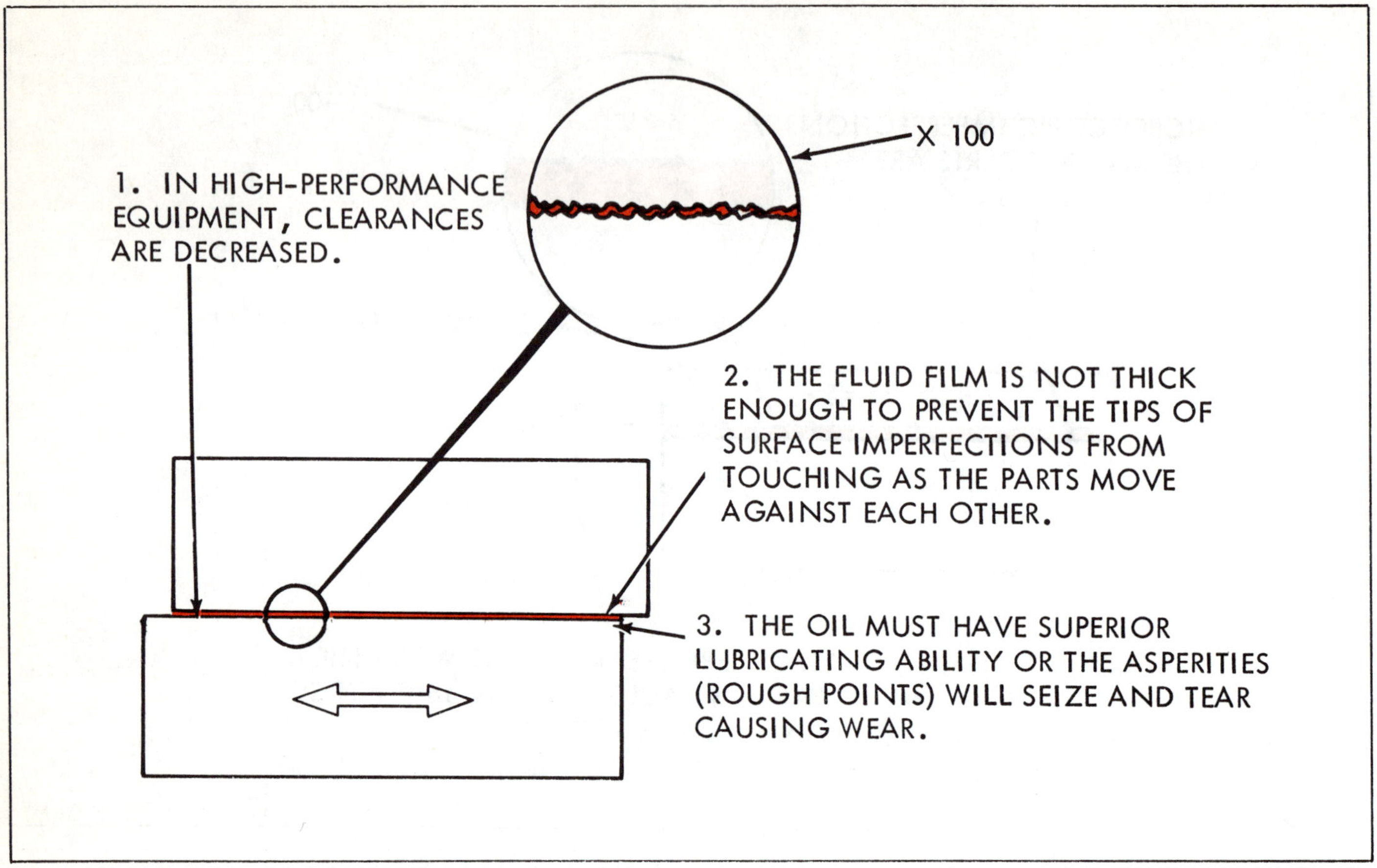

Figure 3-7. Boundary Lubrication Requires Chemical Additives

Figure 3-8. Rust Caused by Moisture in the Oil

Temperature is particularly important. Tests have shown that below 135°F, oil oxidizes very slowly. But the rate of oxidation (or any other chemical reaction) approximately doubles for every 18°F increase in temperature.

Oil refiners incorporate additives in hydraulic oils to resist oxidation, since many systems operate at considerably higher temperature. These additives either:

1. Stop oxidation from continuing immediately after it starts (chain breaker type) or ....
2. Reduce the effect of oxidation catalysts (metal deactivator type).

RUST AND CORROSION PREVENTION

Rust (Fig. 3-8) is the chemical union of iron (or steel) with oxygen. Corrosion is a chemical reaction between a metal and a chemical--usually an acid. Acids result from the chemical union of water with certain elements.

Since it is usually not possible to keep air and atmosphere-borne moisture out of the hydraulic system, there will always be opportunities for rust and corrosion to occur. During corrosion, particles of metal are dissolved and washed away (Fig. 3-9). Both rust and corrosion contaminate the system and promote wear. They also allow excessive leakage past the affected parts and may cause components to seize.

Rust and corrosion can be inhibited by incorporating additives that "plate" on the metal surfaces to prevent their being attacked chemically.

DEMULSIBILITY

Small quantities of water can be tolerated in most systems. In fact, some anti-rust compounds promote a degree of emulsification, or mixture with any water that gets into the system. This prevents the water from settling and breaking through the anti-rust film. However, very much water in the oil will promote the collection of contaminants and can cause sticky valves and accelerated wear.

With proper refining, a hydraulic oil can have a high degree of demulsibility or ability to separate out water.

USE OF ADDITIVES

Since most of the desirable properties of a fluid

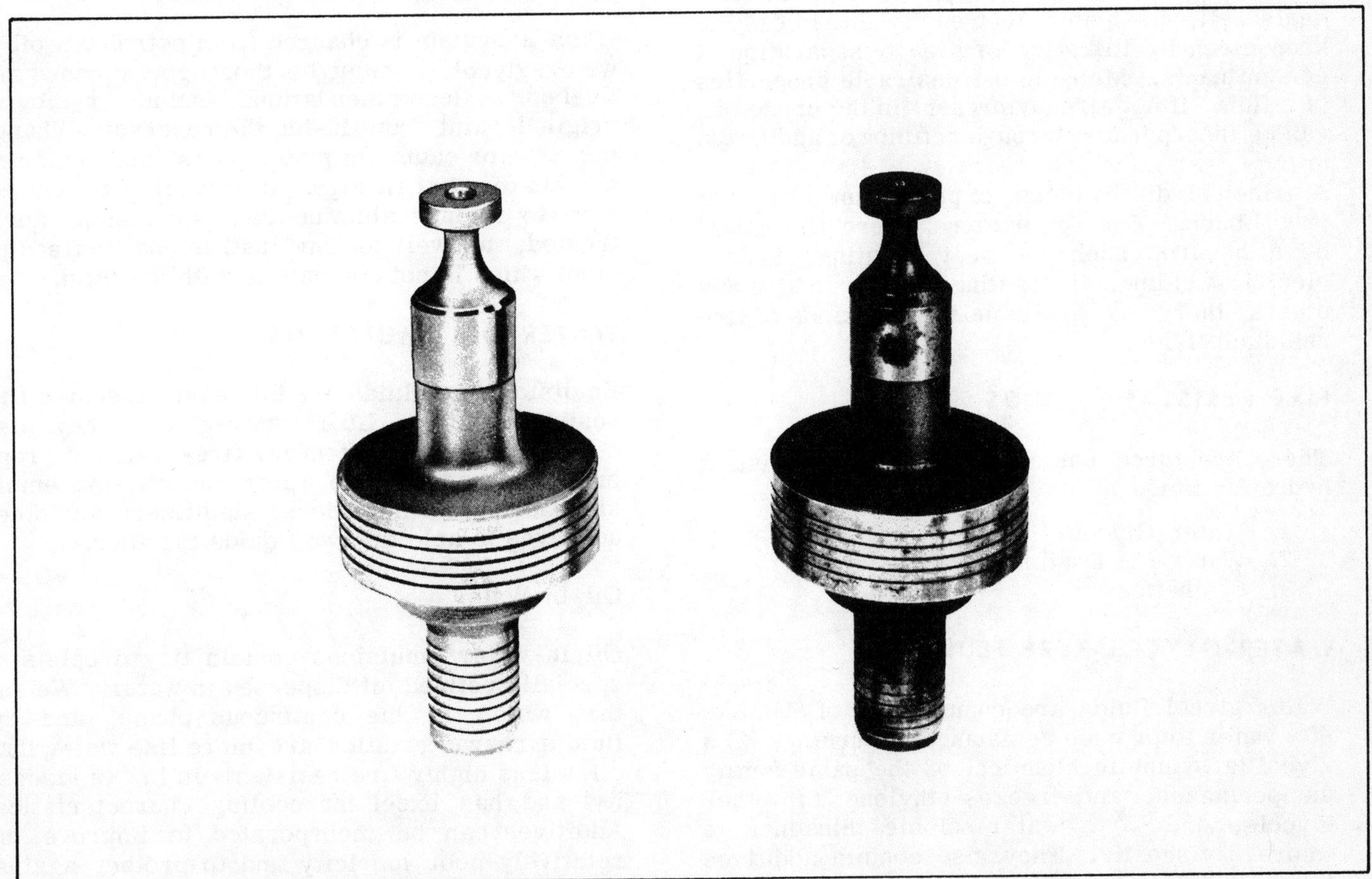

Figure 3-9. Corrosion Caused by Acid Formation in the Hydraulic Oil

are at least partly traceable to additives, it might be supposed that commercial additives could be incorporated in any oil to make it more suitable for a hydraulic system. Refiners, however, warn against this, saying that additives must be compatible with the base fluid and with each other and further that this compatibility cannot be determined in the field. Unless one has laboratory facilities for ascertaining their compatibility it is best to leave the use of additives to the discretion of the fluid manufacturer.

PETROLEUM OIL AS A HYDRAULIC FLUID

Petroleum oil is still by far the most highly used base for hydraulic fluids. The characteristics or properties of petroleum oil fluids depend on three factors:

1. The type of crude oil used
2. The degree and method of refining
3. The additives used

In general, petroleum oil has excellent lubricity. Some crude oils have better than average lubricating or anti-wear properties. Depending on their makeup, some crude oils may display higher demulsibility, more oxidation resistance at higher temperatures or higher viscosity index than others. Oil naturally protects against rust, seals well, dissipates heat easily and is easy to keep clean by filtration or gravity separation of contaminants. Most of the desirable properties of a fluid, if not already present in the crude oil, can be incorporated through refining or additives.

A principal disadvantage of petroleum oil is that it will burn. For applications where fire could be a hazard, such as heat treating, hydroelectric welding, die casting, forging and many others, there are available several kinds of fire resistant fluids.

FIRE RESISTANT FLUIDS

There are three basic types of fire resistant hydraulic fluids:

1. Water-Glycols
2. Water-Oil Emulsions
3. Synthetics

WATER-GLYCOL TYPE FLUIDS

Water glycol fluids are compounded of (1) 35 to 40% water to provide resistance to burning, (2) a glycol (a synthetic chemical of the same family as permanent anti-freeze--ethylene or other glycols, and (3) a water-soluble thickener to improve viscosity. They also contain additives to prevent foaming, rust and corrosion, and to improve lubrication.

Characteristics

Water-glycol fluids generally have good wear resistance characteristics, provided that high speeds and loads are avoided. The fluid has a high gravity (it is heavier than oil), which can create a higher vacuum at pump inlets. Certain metals such as zinc, cadmium and magnesium react with water glycol fluids and cannot be used in systems where compatible paints and enamels also must be used with these fluids.

Most of the newer synthetic seal materials are compatible with water-glycol fluid. Asbestos, leather and cork-impregnated materials should be avoided in rotating seals, since they tend to absorb water.

Some disadvantages of these fluids are: (1) it is necessary to continually measure water content and make up for evaporation to maintain required viscosity and (2) evaporation may also cause loss of certain additives, thereby reducing the life of the fluid and of the hydraulic components. Also, (3) operating temperatures must be kept low and (4) the cost (at the present time) is greater than for conventional oils.

Changing to Water-Glycol

When a system is changed from petroleum oil to water-glycol, it must be thoroughly cleaned and flushed. Recommendations include removing original paint from inside the reservoir, changing zinc or cadmium plated parts, and replacing certain die cast fittings. It may also be necessary to replace aluminum parts unless properly treated, as well as any instrumentation equipment which is not compatible with the fluid.

WATER-OIL EMULSIONS

Emulsion-type fluids are the least expensive fire resistant fluids. Like water-glycol, they also depend on water content for fire-resistant properties. In addition to water and oil, the emulsions contain emulsifiers, stabilizers and other additives to hold the two liquids together.

Oil-in-Water

Oil-in-water emulsions contain tiny droplets of specially refined oil dispersed in water. We say that water is the continuous phase, and the fluid's characteristics are more like water than oil. It is highly fire resistant, is low in viscosity and has excellent cooling characteristics. Additives can be incorporated to improve the relatively poor lubricity and to protect against rust. This fluid has in the past been used primarily with large, low-speed pumps. Now conventional hydraulic pumps are also available for use with it.

Water-in-Oil

Water-in-oil emulsions are more common in use. Tiny droplets of water are dispersed in a continuous oil phase. Like oil, these fluids have excellent lubricity and body. Additionally, the dispersed water gives the fluid a better cooling ability. Rust inhibitors are incorporated for both the water and oil phases. Anti-foam additives also are used with no difficulty.

These emulsions usually contain about 40% water as used in the system. However, some manufacturers furnish a fluid concentrate and the customer adds water when the fluid is installed. As with the water-glycol fluid, it is necessary to replenish the water to maintain proper viscosity.

Other Characteristics

Operating temperatures must be kept low with any water-oil emulsion to avoid evaporation and oxidation. The fluid must circulate and should not be repeatedly thawed and frozen or the two phases may separate. Inlet conditions should be carefully chosen because of the higher density of the fluid and its inherent high viscosity.

Emulsions seem to have a greater affinity for contamination and require extra attention to filtration, including magnetic plugs to attract iron particles.

Compatibility With Seals and Metals

Emulsion fluids are generally compatible with all metals and seals found in petroleum hydraulic systems.

Change-Over to Emulsion

When a hydraulic system is changed over to water-oil emulsion fluid, it should be completely drained, cleaned and flushed. It's essential to get out any contamination (such as water-glycol fluids) which might cause the new fluid to break down. Most seals can be left undisturbed. Butyl dynamic (moving) seals should be replaced, however. In changing from synthetic fluids, seals must be changed to those rated for petroleum oil use.

SYNTHETIC FIRE-RESISTANT FLUIDS

Synthetic fire resistant fluids are laboratory-synthesized chemicals which are themselves less flammable than petroleum oils. Typical of these are: (1) phosphate esters, (2) chlorinated (halogenated) hydro-carbons, (3) synthetic base fluids which are mixtures of 1 and 2 may contain other material as well.

Characteristics

Since the synthetics do not contain any water or other volatile material, they operate well at high temperature without loss of any essential elements. They also are suitable for high-pressure systems.

Synthetic fire resistant fluids do not operate best in low-temperature systems. Auxiliary heating may be required in cold environments.

Also, these fluids have the highest specific gravity (weight) of any type and pump inlet conditions require special care when they are used. Some vane pumps are built with special bodies to provide the improved inlet conditions needed to prevent pump cavitation when a synthetic fluid is used.

The viscosity index of synthetic fluids is generally low--ranging from 80 to as low as minus 400. Thus, they should not be used except where the operating temperature is relatively constant.

Synthetic fluids are probably the most costly hydraulic fluids being used at this time.

Seal Compatibility

Synthetic fluids are not compatible with the commonly used Nitrile (Buna) and Neoprene seals. Therefore, a changeover from petroleum, water glycol or water-oil requires dismantling all the components to replace the seals. Special seals made of compatible materials are available for replacement on all Vickers components. They can be purchased singly or in kits, or can be built into new units ordered specifically for this type fluid.

Figure 3-10 is a chart showing the types of materials that are compatible with various hydraulic fluids.

FLUID MAINTENANCE

Hydraulic fluid of any kind is not an inexpensive item. Further, changing the fluid and flushing or cleaning improperly maintained systems is time consuming and costly. Therefore, it's important to care for the fluid properly.

STORAGE AND HANDLING

Here are some simple rules to prevent contamination of the fluid during storage and handling:

		WATER-BASE FLUIDS		NON-WATER-BASE FLUIDS		
MATERIALS UNDER CONSIDERATION	PETROLEUM OILS	OIL AND WATER EMULSION	WATER-GLYCOL MIXTURE		PHOSPHATE ESTERS	
ACCEPTABLE SEAL AND PACKING MATERIALS	NEOPRENE, BUNA N	NEOPRENE, BUNA N, (NO CORK)	NEOPRENE, BUNA N, (NO CORK)		BUTYL, VITON, VYRAM, SILICONE, TEFLON FBA	
ACCEPTABLE PAINTS	CONVENTIONAL	CONVENTIONAL	AS RECOMMENDED BY SUPPLIER		"AIR CURE" EPOXY AS RECOMMENDED	
ACCEPTABLE PIPE DOPES	CONVENTIONAL	CONVENTIONAL	PIPE DOPES AS RECOMMENDED. TEFLON TAPE			
ACCEPTABLE SUCTION STRAINERS	100 MESH WIRE 1-1/2 TIMES PUMP CAPACITY	40 MESH WIRE 4 TIMES PUMP CAPACITY	50 MESH WIRE, 4 TIMES PUMP CAPACITY			
ACCEPTABLE FILTERS	CELLULOSE FIBER, 200-300 MESH WIRE, KNIFE EDGE OR PLATE TYPE	GLASS FIBER, 200-300 WIRE, KNIFE EDGE OR PLATE	CELLULOSE FIBER, 200-300 MESH WIRE, KNIFE EDGE OR PLATE	CELLULOSE FIBER, 200-300 MESH WIRE, KNIFE EDGE OR PLATE TYPE (FULLER'S EARTH OR MICRONIC TYPE MAY BE USED ON NON-ADDITIVE FLUIDS.)		
ACCEPTABLES METALS OF CONSTRUCTION	CONVENTIONAL	CONVENTIONAL	AVOID GALVANIZED METAL AND CADMIUM PLATING		CONVENTIONAL	

Figure 3-10. Compatibility of Hydraulic Fluids and Sealing Materials

1. Store drums on their sides. If possible, keep them inside or under a roof.

2. Before opening a drum, clean the top and the bung thoroughly so no dirt can get in.

3. Use only clean containers, hoses, etc. to transfer the fluid from the drum to the hydraulic reservoir. An oil transfer pump equipped with 25 micron filters is recommended.

4. Provide a 200 mesh screen in the reservoir filler pipe.

Keeping the fluid clean and free from moisture will help it last much longer and avoid contamination damage to close-fitting parts in the hydraulic components.

IN-OPERATION CARE

Proper in-operation care of hydraulic fluid includes:

1. Prevent contamination by keeping the system tight and using proper air and fluid filtration.

2. Establish fluid change intervals so the fluid will be replaced before it breaks down. If necessary, the supplier can test samples in the laboratory at intervals to help establish the frequency of change.

3. Keep the reservoir filled properly to take advantage of its heat dissipating characteristics and prevent moisture from condensing on inside walls.

4. Repair all leaks immediately.

QUESTIONS

1. Name four primary functions of the hydraulic fluid.

2. Name four quality properties of a hydraulic fluid.

3. Define viscosity. What is the common unit of viscosity?

4. How is viscosity affected by cold? By heat?

5. If viscosity is too high, what can happen to the system?

6. What is viscosity index? When is viscosity index important?

7. Which type of hydraulic fluid has the best natural lubricity?

8. Name several catalysts to oxidation of hydraulic oil.

9. How are rust and corrosion prevented?

10. What is demulsibility?

11. What are the three factors that determine the properties of a hydraulic oil?

12. What are the three basic types of fire-resistant hydraulic fluid?

13. Which type of hydraulic fluid is not compatible with Buna or Neoprene seals?

14. Which type of fire-resistant hydraulic fluid is best for high temperature operation?

15. How does the specific gravity of the fluid affect the pump inlet conditions?

16. What is the most important factor in good fluid maintenance?

CHAPTER 4

HYDRAULIC PIPING AND SEALING

This chapter is comprised of two parts. First is a description of the hydraulic system "plumbing" --the types of connecting lines and fittings used to carry fluid between the pumps, valves, actuators, etc. The second part deals with the prevention of leakage and the types of seals and seal materials required for hydraulic applications.

PIPING

Piping is a general term which embraces the various kinds of conducting lines that carry hydraulic fluid between components; plus the fittings or connectors used between the conductors. Hydraulic systems today use principally three types of conducting lines: steel pipe, steel tubing and flexible hose. At present, pipe is the least expensive of the three while tubing and hose offer more convenience in making connections and in servicing the "plumbing". The future may see plastic plumbing, which is gradually coming into use for certain applications.

Pipes

Iron and steel pipes were the first conductors used in industrial hydraulic systems and are still used widely because of their low cost. Seamless steel pipe is recommended for hydraulic systems with the pipe interior free of rust, scale and dirt.

Sizing Pipes

Pipe and pipe fittings are classified by nominal size and wall thickness. Originally, a given size pipe had only one wall thickness and the stated size was the actual inside diameter.

Later, pipes were manufactured with varying wall thicknesses: standard, extra heavy and double extra heavy (Fig. 4-1). However, the outside diameter did not change. To increase wall thickness, the inside diameter was changed. Thus, the nominal pipe size alone indicates only the thread size for connections.

Pipe Schedule

Currently, wall thickness is being expressed as a schedule number. Schedule numbers are specified by the American National Standards Institute (ANSI) from 10 to 160 (Fig. 4-2). The numbers cover ten sets of wall thickness.

For comparison, schedule 40 corresponds closely to standard. Schedule 80 essentially is extra heavy. Schedule 160 covers pipes with the greatest wall thickness under this system.

The old double, extra heavy classification is slightly thicker than schedule 160. Figures 4-1 and 4-2 show pipe sizes up to 12 inches (nominal). Larger sizes are available. Schedule 10, which appears blank in the chart in Fig. 4-2, is used only for pipes larger than 12 inches.

Pipe Sealing

Pipe threads are tapered (Fig. 4-3) as opposed to tube and some hose fittings which have straight threads. Joints are sealed by an interference fit between the male and female threads as the pipe is tightened.

This creates one of the major disadvantages of pipe. When a joint is broken, the pipe must be tightened further to reseal. Often this necessitates replacing some of the pipe with slightly longer sections. However, the difficulty has been overcome somewhat by using teflon tape or other compounds to reseal pipe joints.

Special taps and dies are required for threading hydraulic system pipes and fittings. The threads are the "dryseal" type. They differ from standard pipe threads by engaging the roots and crests before the flanks; thus avoiding spiral clearance (Fig. 4-3).

Pipe Fittings

Since pipe can have only male threads, and does not bend, various types of fittings are used to make connections and change direction (Fig. 4-4). Most fittings are female-threaded to mate with pipe although some have male threads to mate with other fittings or with the ports in hydraulic components.

The many fittings necessary in a pipe circuit present multiple opportunities for leakage, par-

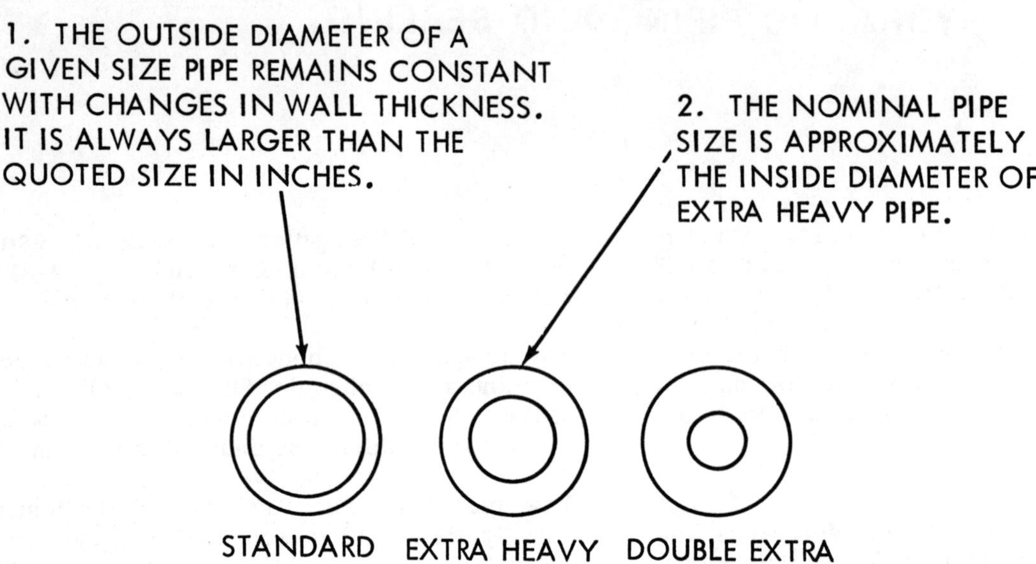

NOMINAL SIZE	PIPE O.D.	INSIDE DIAMETER		
		STANDARD	EXTRA HEAVY	DOUBLE EXTRA HEAVY
1/8	.405	.269	.215	
1/4	.540	.364	.302	
3/8	.675	.493	.423	
1/2	.840	.622	.546	.252
3/4	1.050	.824	.742	.434
1	1.315	1.049	.957	.599
1-1/4	1.660	1.380	1.278	.896
1-1/2	1.900	1.610	1.500	1.100
2	2.375	2.067	1.939	1.503
2-1/2	2.875	2.469	2.323	1.771
3	3.500	3.068	2.900	
3-1/2	4.000	3.548	3.364	
4	4.500	4.026	3.826	
5	5.563	5.047	4.813	4.063
6	6.625	6.065	5.761	
8	8.625	8.071	7.625	
10	10.750	10.192	9.750	
12	12.750	12.080	11.750	

Fig. 4-1. Early Classifications of Pipe Wall Thickness

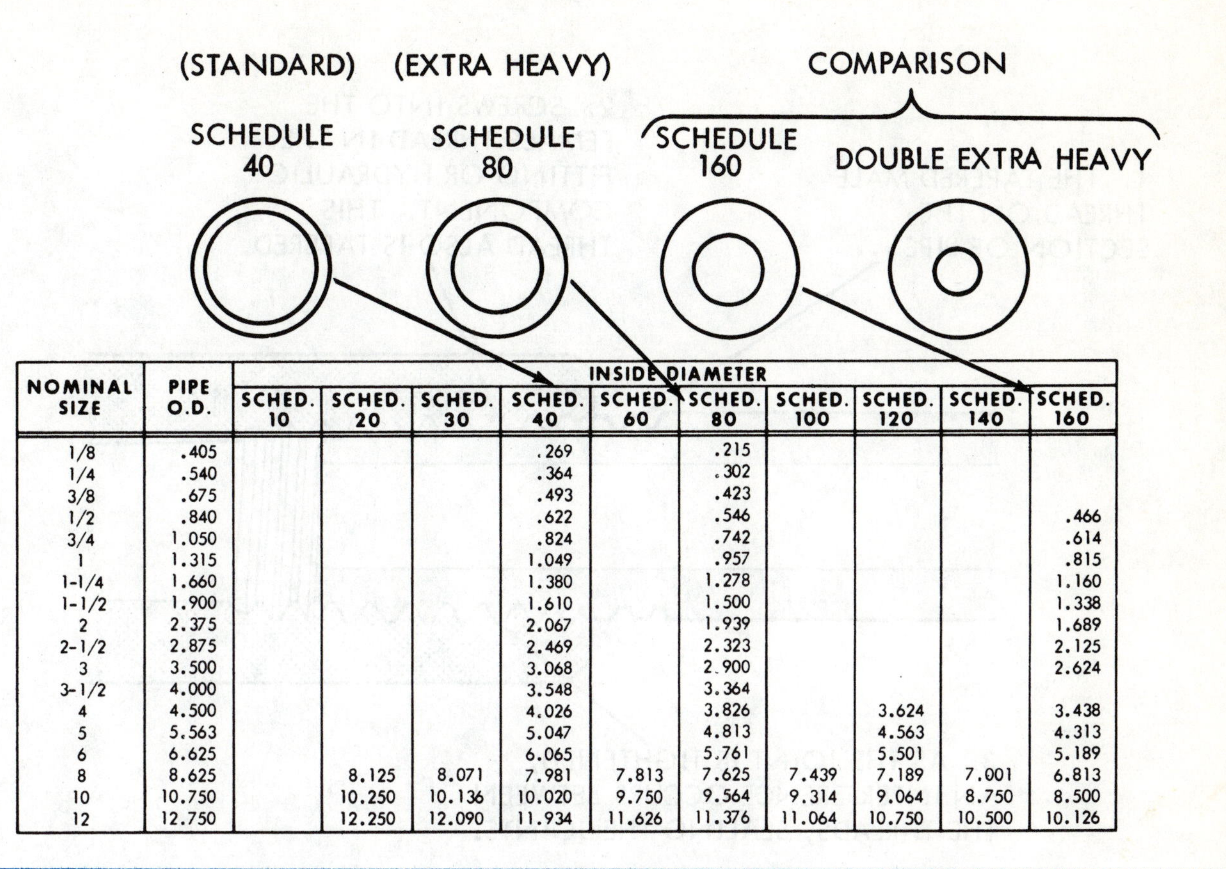

NOMINAL SIZE	PIPE O.D.	INSIDE DIAMETER									
		SCHED. 10	SCHED. 20	SCHED. 30	SCHED. 40	SCHED. 60	SCHED. 80	SCHED. 100	SCHED. 120	SCHED. 140	SCHED. 160
1/8	.405				.269		.215				
1/4	.540				.364		.302				
3/8	.675				.493		.423				
1/2	.840				.622		.546				.466
3/4	1.050				.824		.742				.614
1	1.315				1.049		.957				.815
1-1/4	1.660				1.380		1.278				1.160
1-1/2	1.900				1.610		1.500				1.338
2	2.375				2.067		1.939				1.689
2-1/2	2.875				2.469		2.323				2.125
3	3.500				3.068		2.900				2.624
3-1/2	4.000				3.548		3.364				
4	4.500				4.026		3.826		3.624		3.438
5	5.563				5.047		4.813		4.563		4.313
6	6.625				6.065		5.761		5.501		5.189
8	8.625		8.125	8.071	7.981	7.813	7.625	7.439	7.189	7.001	6.813
10	10.750		10.250	10.136	10.020	9.750	9.564	9.314	9.064	8.750	8.500
12	12.750		12.250	12.090	11.934	11.626	11.376	11.064	10.750	10.500	10.126

Fig. 4-2. Pipes Currently are Sized by Schedule Number

ticularly as pressure increases. Threaded connections are used up to 1-1/4". Where larger pipes are needed, flanges are welded to the pipe (Fig. 4-5). Flat gaskets or O-rings are used to seal flanged fittings.

TUBING

Seamless steel tubing offers significant advantages over pipe for hydraulic plumbing. Tubing can be bent into any shape, is easier to work with and can be used over and over without any sealing problems. Usually the number of joints is reduced.

In low-volume systems, tubing will handle higher pressure and flow with less bulk and weight. However, it is more expensive, as are the fittings required to make tube connections.

Tubing Sizes

A tubing size specification refers to the outside diameter. Tubing is available in 1/16 inch increments from 1/8 inch to one inch O. D. and in 1/4 inch increments beyond one inch. Various wall thicknesses are available for each size. The inside diameter, as previously noted, equals the outside diameter less twice the wall thickness.

Tube Fittings

Tubing is never sealed by threads, but by various kinds of fittings (Fig. 4-6). Some of these fittings seal by metal-to-metal contact. They are known as compression fittings and may be either the flared or flareless type. Others use O-rings or comparable seals. In addition to threaded fittings, flanged fittings also are available to be welded to larger sized tubing.

1. Flare Fitting. The 37-degree flare fitting is the most common fitting for tubing that can be flared. The fittings shown in Fig. 4-6-A-B seal by squeezing the flared end of the tube against a seal as the compression nut is tightened. A sleeve or extension of the nut supports the tube to damp out vibration. The standard 45-degree flare fitting is used for automotive and refrigeration tubing. It also is made in an inverted design with male threads on the compression nut.

2. Sleeve or O-Ring Compression Fittings. For tubing that can't be flared, or to simply avoid the need of flaring, there are various sleeve or ferrule compression fittings (views D-F), and O-ring compression fittings (view E). The O-ring fitting allows considerable variation in the length and squareness of the tube cut.

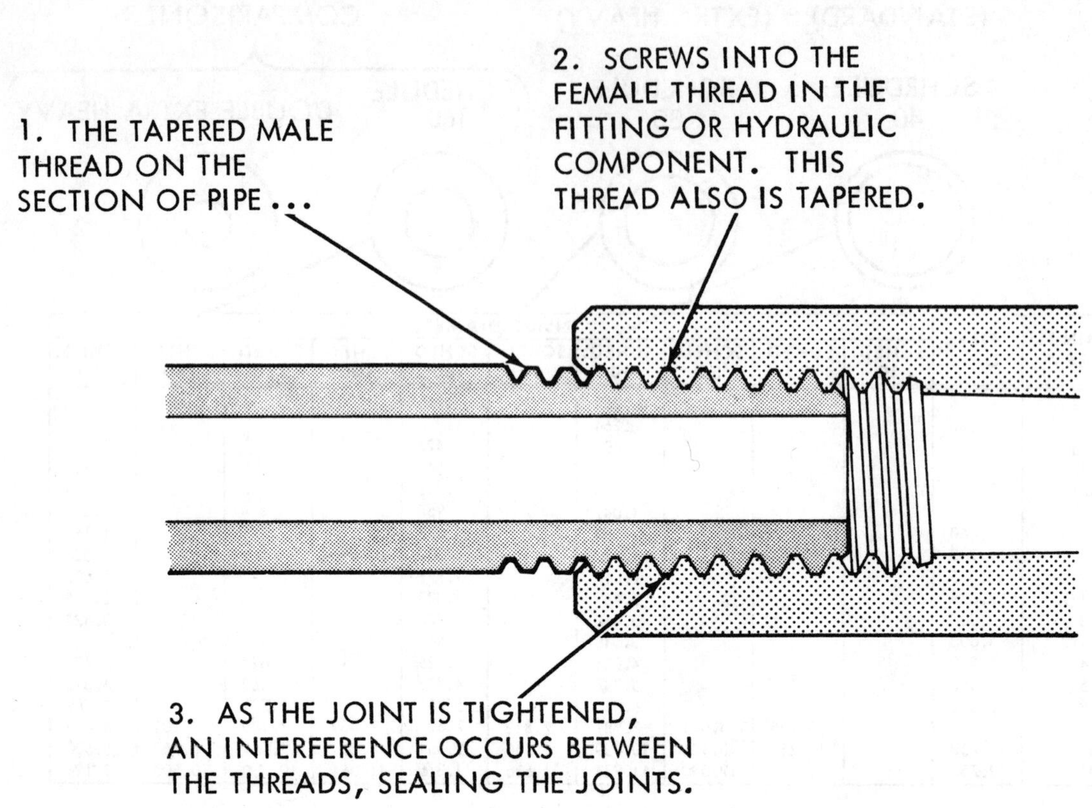

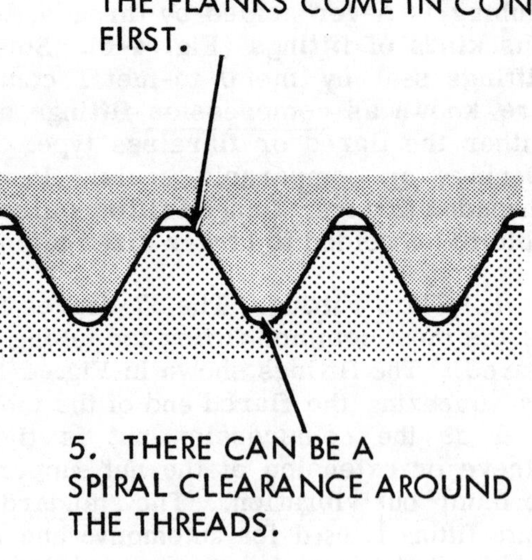

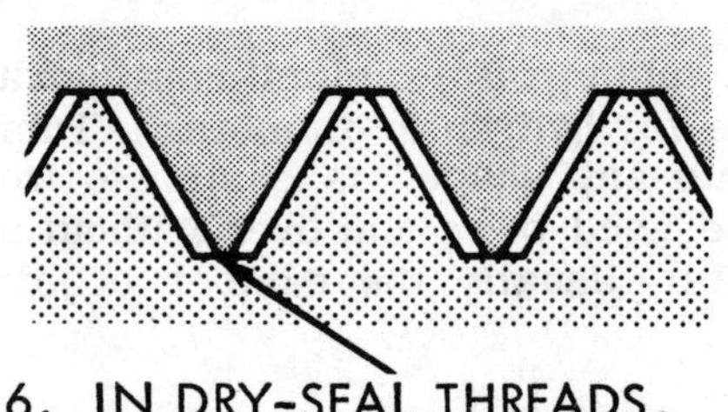

Fig. 4-3. Hydraulic Pipe Threads are Dry-Seal Tapered Type

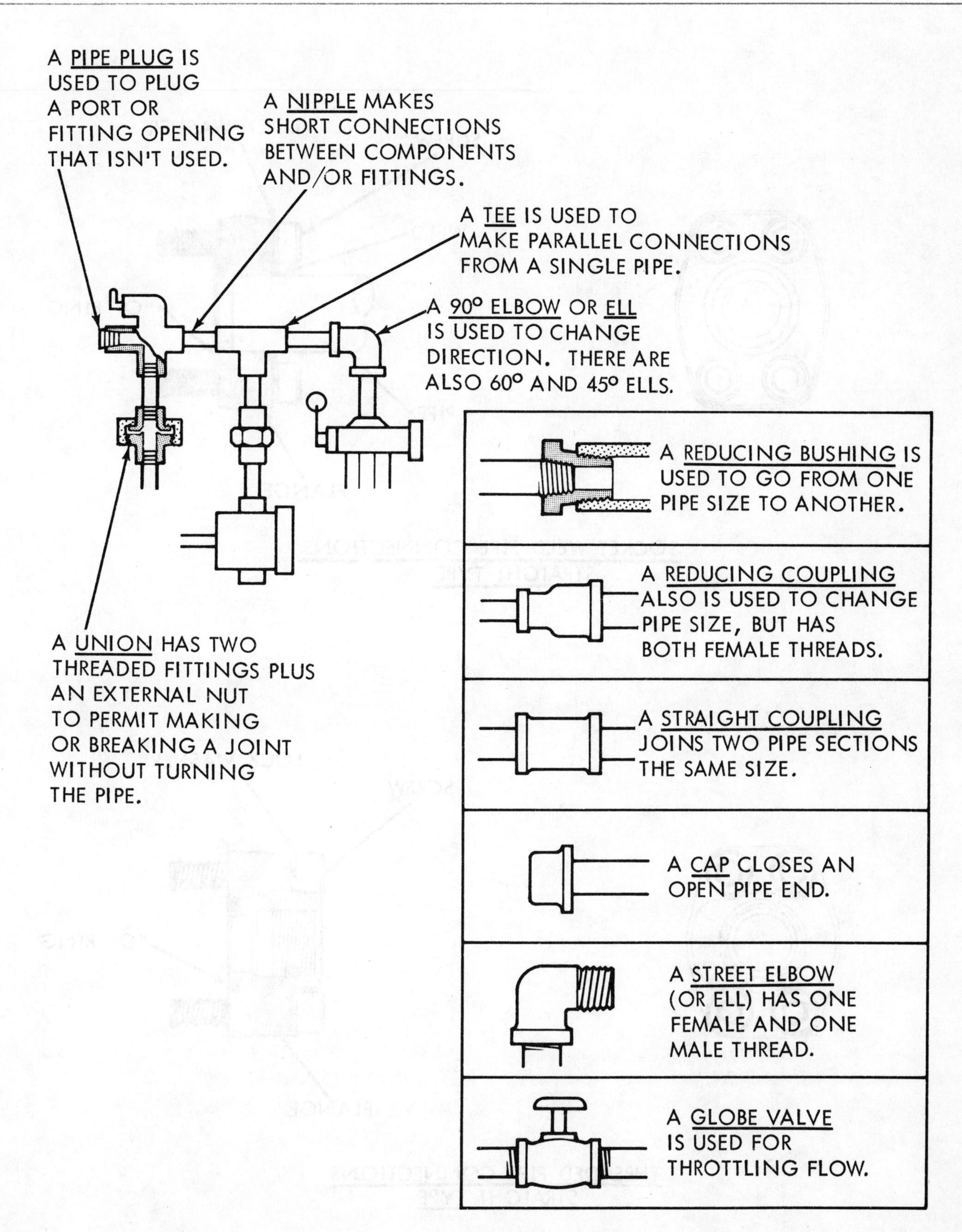

Fig. 4-4. Fittings Make the Connections Between Pipes and Components

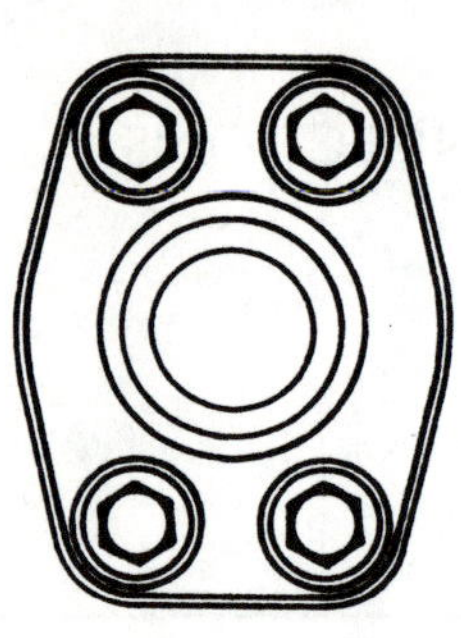

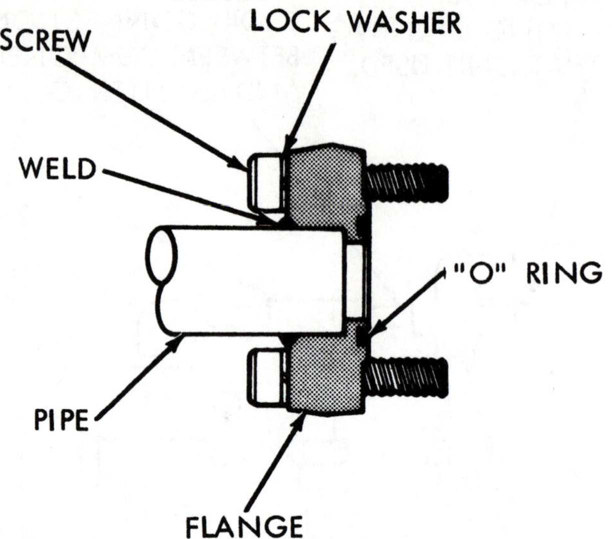

SOCKET WELD PIPE CONNECTIONS
STRAIGHT TYPE

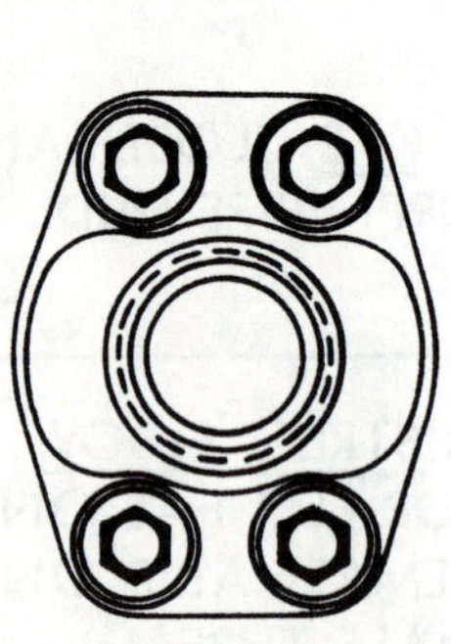

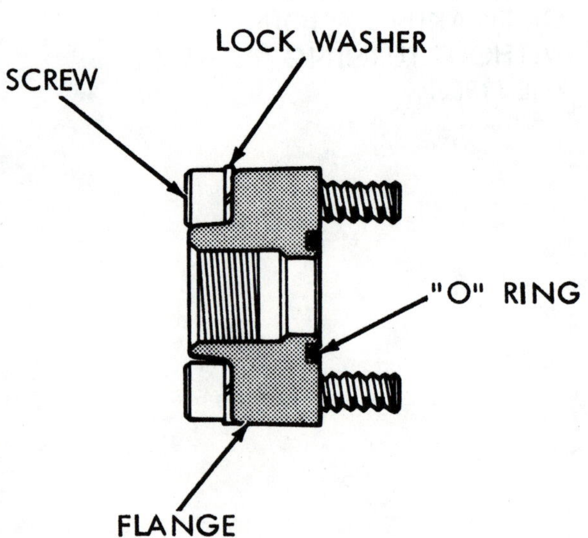

THREADED PIPE CONNECTIONS
STRAIGHT TYPE

Fig. 4-5. Flanged Connections for Large Pipe

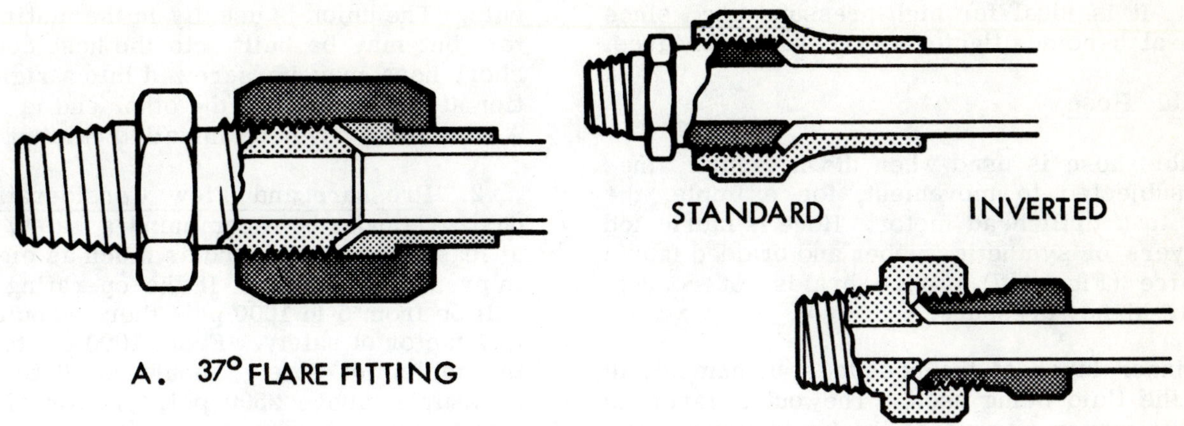

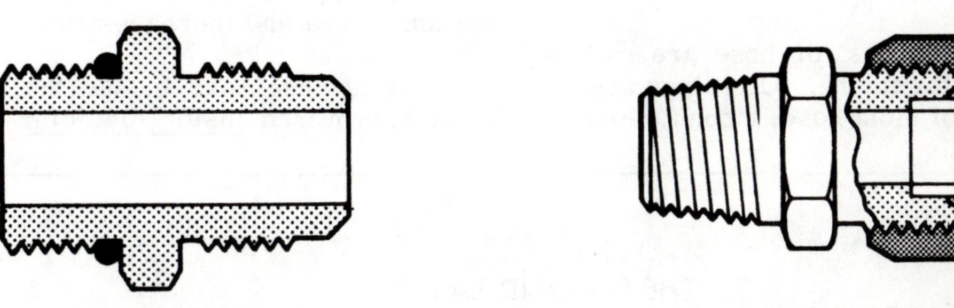

C. STRAIGHT THREAD "O" RING CONNECTOR

D. FERRULE COMPRESSION FITTING

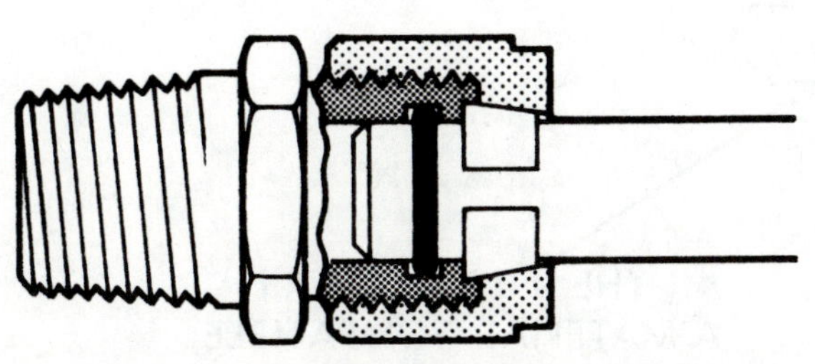

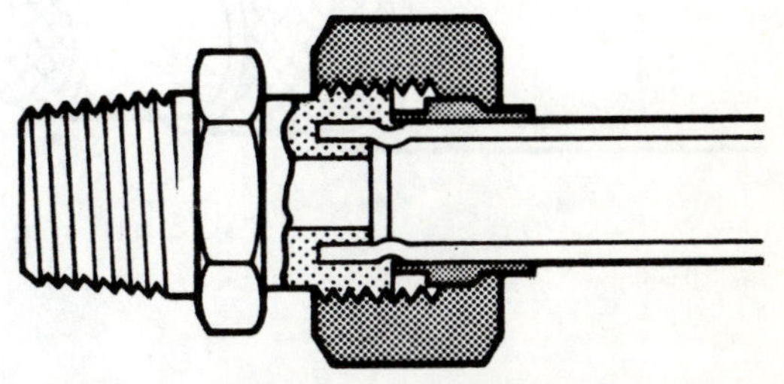

E. "O" RING COMPRESSION FITTING

F. SLEEVE COMPRESSION FITTING

Fig. 4-6. Threaded Fittings and Connectors Used With Tubing

3. Straight Thread O-Ring Connector. When the hydraulic component is equipped with straight thread ports, fittings as shown in view 4-6, C can be used. It is ideal for high-pressure use, since the seal becomes tighter as pressure increases.

Flexible Hose

Flexible hose is used when the hydraulic lines are subjected to movement, for example, the lines to a drill head motor. Hose is fabricated in layers of synthetic rubber and braided fabric or wire (Fig. 4-7). Wire braids, of course, permit higher pressure.

The inner layer of the hose must be compatible with the fluid being used. The outer layer is usually rubber to protect the braid layer. The hose may have as few as three layers, one being braid, or may have multiple layers depending on the operating pressure. When there are multiple wire layers, they may alternate with rubber layers, or the wire layers may be placed directly over one another.

1. Hose Fittings. Fittings for hose are essentially the same as for tubing. Couplings are fabricated on the ends of most hose, though there are reusable screw-on or clamp-on connectors. It is usually desirable to connect the hose ends with union-type fittings which have free-turning nuts. The union is usually in the mating connector but may be built into the hose coupling. A short hose may be screwed into a rigid connection at one end before the other end is connected. A hose must never be installed twisted.

2. Pressure and Flow Considerations. Industry standards recommend a safety factor of at least four to one, and as much as eight to one, in pressure capacity. If the operating pressure will be from 0 to 1000 psi, there should be an 8 to 1 factor of safety. From 1000 psi to 2500 psi, the factor of safety should be 6 to 1; and at pressures above 2500 psi, a factor of safety of 4 to 1 is recommended.

$$\text{Factor of Safety (FS)} = \frac{\text{(BP) Burst Pressure}}{\text{(WP) Working Pressure}}$$

In any nominal size pipe, the greater the schedule number the thicker the walls and the higher the burst strength. This decreases inside cross sectional area and increases flow velocity.

Thus, it is necessary to see that the conductor has the required inside diameter to handle the

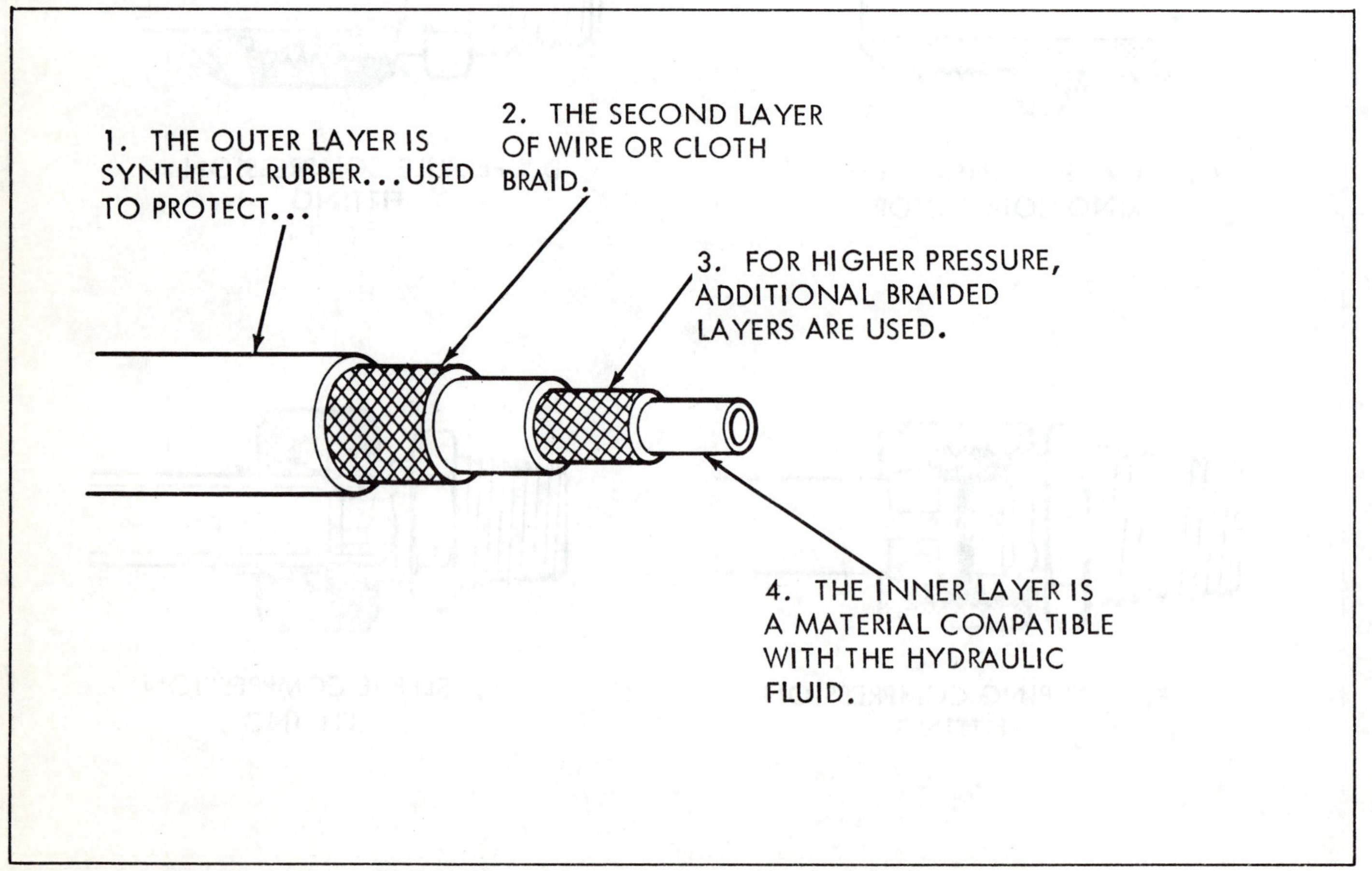

Fig. 4-7. Flexible Hose is Constructed in Layers

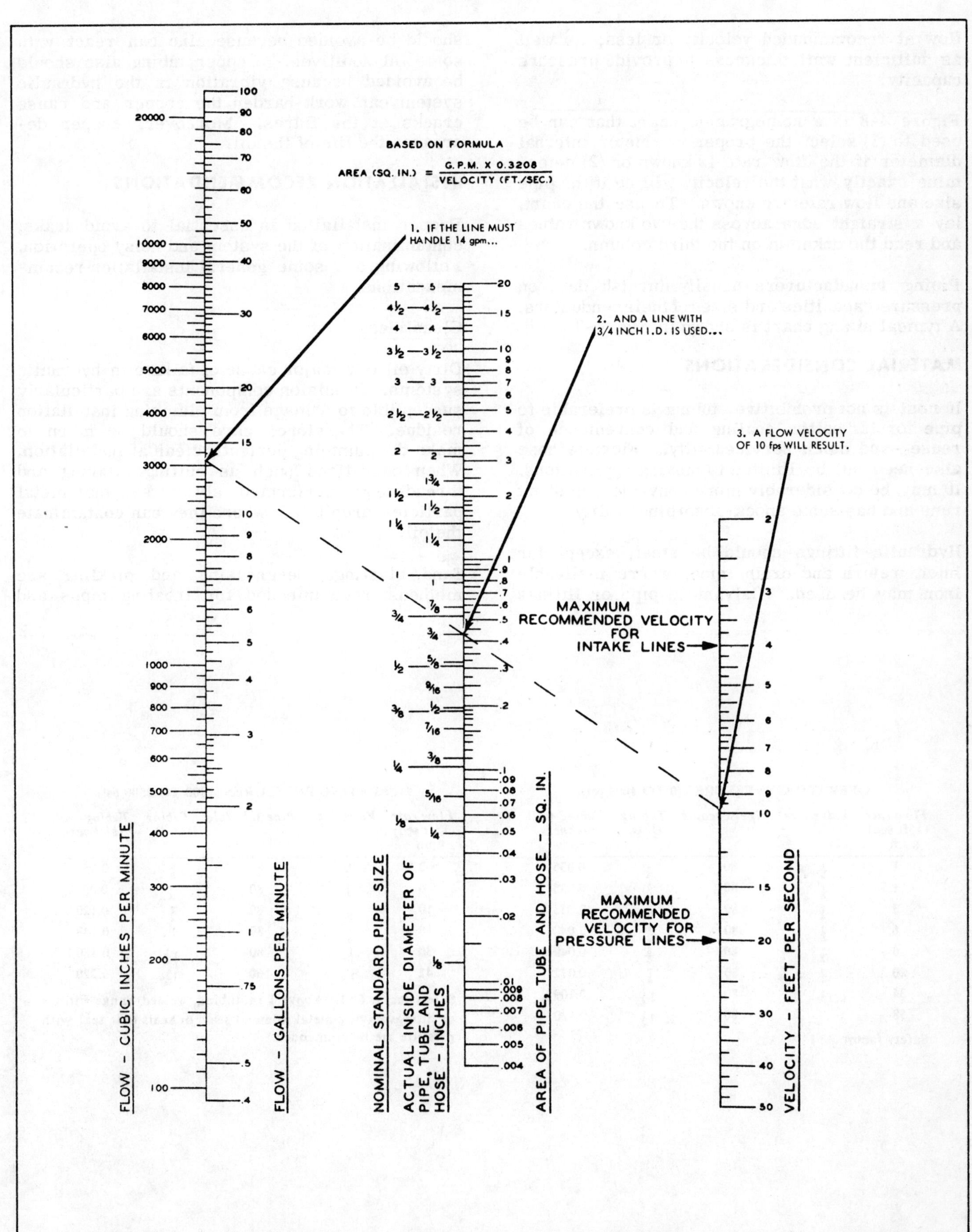

Fig. 4-8. Conductor I. D. Selection Chart

flow at recommended velocity or less, as well as sufficient wall thickness to provide pressure capacity.

Figure 4-8 is a nomographic chart that can be used to (1) select the proper conductor internal diameter if the flow rate is known or (2) determine exactly what the velocity will be if the pipe size and flow rate are known. To use the chart, lay a straight edge across the two known values and read the unknown on the third column.

Piping manufacturers usually furnish data on pressure capacities and sizes of their conductors. A typical sizing chart is shown in Fig. 4-9.

MATERIAL CONSIDERATIONS

If cost is not prohibitive, tubing is preferable to pipe for its better sealing and convenience of reuse, and quick serviceability. Flexible hose also need not be limited to moving applications. It may be considerably more convenient in short runs and has some shock-absorbing ability.

Hydraulic fittings should be steel, except for inlet, return and drain lines, where malleable iron may be used. Galvanized pipe or fittings should be avoided because zinc can react with some oil additives. Copper tubing also should be avoided because vibration in the hydraulic system can work-harden the copper and cause cracks at the flares. Moreover, copper decreases the life of the oil.

INSTALLATION RECOMMENDATIONS

Proper installation is essential to avoid leaks, contamination of the system and noisy operation. Following are some general installation recommendations.

Cleanliness

Dirty oil is a major cause of failure in hydraulic systems. Precision components are particularly susceptible to damage from plumbing installation residue. Therefore, care should be taken to make the plumbing perfectly clean at installation. When operations such as cutting, flaring and threading are performed, always see that metal particles aren't left where they can contaminate the oil.

Sand blasting, de-greasing and pickling are methods recommended for treating pipes and

OPERATING PRESSURES (0 TO 1000 psi)

Flow rate (15ft sec) gpm	*Valve size*	*Pipe schedule*	*Tubing O. D.*	*Tubing-wall thickness*
1	$\frac{1}{8}$	80	$\frac{1}{4}$	0.035
1.5	$\frac{1}{8}$	80	$\frac{5}{16}$	0.035
3	$\frac{1}{4}$	80	$\frac{3}{8}$	0.035
6	$\frac{3}{8}$	80	$\frac{1}{2}$	0.042
10	$\frac{1}{2}$	80	$\frac{5}{8}$	0.049
20	$\frac{3}{4}$	80	$\frac{7}{8}$	0.072
34	1	80	$1\frac{1}{4}$	0.109
58	$1\frac{1}{4}$	80	$1\frac{1}{2}$	0.120

Safety factor 8 : 1

OPERATING PRESSURES (1000 TO 2500 psi)

Flow rate (15ft sec) gpm	*Valve size*	*Pipe schedule*	*Tubing O. D.*	*Tubing-wall thickness*
2.5	$\frac{1}{4}$	80	$\frac{3}{8}$	0.058
6	$\frac{3}{8}$	80	$\frac{5}{8}$	0.095
10	$\frac{1}{2}$	80	$\frac{3}{4}$	0.120
18	$\frac{3}{4}$	80	1	0.148
30	1	80	$1\frac{1}{4}$	0.180
42	$1\frac{1}{4}$	160	$1\frac{1}{2}$	0.220

Safety factor 6 : 1. Above $\frac{1}{2}$ in. tubing, welded flange fittings or fittings having metal to metal seals or seals that seal with pressure are recommended.

Fig. 4-9. Pipe and Tube Sizing Chart

tubing before they are installed. Additional information on these processes can be obtained from component manufacturers (Fig. 4-10) and from distributors of commercial cleaning equipment.

Supports

Long hydraulic lines are susceptible to vibration or shock when the fluid flowing through them is suddenly stopped or reversed. Leakage can be caused by loosening or work-hardening of joints. Therefore, the lines should be supported at intervals with clamps or brackets. It is usually best to keep these supports away from the fittings for ease of assembly and disassembly. Soft materials such as wood and plastic are best for this purpose.

Consider Function of Lines

There are a number of special considerations relating to the function of the lines that should be mentioned.

1. The pump inlet port is usually larger than the outlet to accommodate a larger intake line. It is good practice to maintain this size throughout the entire length of the pump inlet. Keep the line as large as specified and as short as possible. Also avoid bends and keep the number of fittings in the inlet line to a minimum.

2. Since there usually is a vacuum at the pump inlet, inlet line connections must be tight. Otherwise air can enter the system.

3. In return lines, restrictions cause pressure to build up resulting in wasted power. Adequate line sizes should be used to assure low flow rates. Here too, fittings and bends should be held to a minimum.

4. Loose return lines also can let air into the system by aspiration. The lines must be tight and must empty below the oil level to prevent splashing and aeration.

5. Lines between actuators and speed control valves should be short and rigid for precise flow control.

Hose Installation

Flexible hose should be installed so there is no kinking during machine operation. Some slack should always be present to relieve strain and permit absorption of pressure surges.

Twisting the hose and unusually long loops also are undesirable. Clamps may be required to avoid chafing or tangling with moving parts. Hose subject to rubbing should be encased in a protective sleeve or guard.

SEALS AND LEAKAGE

Excessive leakage anywhere in a hydraulic circuit reduces efficiency and results in power loss or creates a housekeeping problem, or both.

Internal Leakage

Most hydraulic system components are built with operating clearances which allow a certain amount of internal leakage. Moving parts, of course, must be lubricated and leakage paths may be designed in solely for this purpose. In addition, some hydraulic controls have internal leakage paths built in to prevent "hunting" or oscillation of valve spools and pistons.

Internal leakage, of course, is not loss of fluid. The fluid eventually is returned to the reservoir either through an external drain line or by way of an internal passage in the component.

Additional internal leakage occurs as a component begins to wear and clearances between parts increase. This increase in internal leakage can reduce the efficiency of a system by slowing down the work and generating heat.

Finally, if the internal leakage path becomes large enough, all the pump's output may be bypassed and the machine will not operate at all.

External Leakage

External leakage is unsightly and can be very hazardous. It is expensive because the oil that leaks out seldom can be returned to the system. The principal cause of external leakage is improper installation. Joints may leak because they weren't put together properly or because vibration or shock in the line loosened them. Failure to connect drain lines, excessive operating pressure and contamination in the fluid all are common reasons for seals becoming damaged.

Sealing

Sealing is required to maintain pressure, to prevent fluid loss, and to keep out contamination. There are various methods of sealing hydraulic components, depending on whether the seal must be positive or non-positive, whether the sealing application is static or dynamic, how much pressure must be contained, and other factors.

A positive seal prevents even a minute amount of fluid from getting past.

VICKERS

Industrial and Mobile Machinery

PREPARATION OF PIPES, TUBES, AND FITTINGS BEFORE INSTALLATION IN A HYDRAULIC SYSTEM

General requirements. When installing the various iron and steel pipes, tubes, and fittings of a hydraulic system, it is necessary that they be absolutely clean, free from scale, and all kinds of foreign matter. To attain this end, the following steps should be taken.

1. Tubing, pipes, and fittings should be brushed with boiler tube wire brush or cleaned with commercial pipe cleaning apparatus. The inside edge of tubing and pipe should be reamed after cutting to remove burrs.

2. Short pieces of pipe and tubing and steel fittings are sandblasted to remove rust and scale. Sandblasting is a sure and efficient method for short straight pieces and fittings. Sandblasting is not used, however, if there is the slightest possibility that particles of sand will remain in blind holes or pockets in the work after flushing.

3. In the case of longer pieces of pipe or short pieces bent to complex shapes where it is not practical to sandblast, the parts are pickled in a suitable solution until all rust and scale is removed. Preparation for pickling requires thorough degreasing in TRI-CHLOR-ETHYLENE or other commercial degreasing solution.

4. Neutralize pickling solution.

5. Rinse parts and prepare for storage.

6. Tubing must not be welded, brazed, or silver soldered after assembly as proper cleaning is impossible in such cases. It must be accurately bent and fitted so that it will not be necessary to spring it into place.

7. If flange connections are used, flanges must fit squarely on the mounting faces and be secured with screws of the correct length. Screws or stud-nuts must be drawn up evenly to avoid distortion in the valve or pump body.

8. Be sure that all openings into the hydraulic system are properly covered to keep out dirt and metal slivers when work such as drilling, tapping, welding, or brazing is being done on or near the unit.

9. Threaded fittings should be inspected to prevent metal slivers from the threads getting into the hydraulic system.

10. Before filling the system with hydraulic oil, be sure that the hydraulic fluid is as specified and that it is clean. Do not use cloth strainers or fluid that has been stored in contaminated containers.

11. Use a #120 mesh screen when filling the reservoir. Operate the system for a short time to eliminate air in the lines. Add hydraulic fluid if necessary.

12. Safety precautions. Dangerous chemicals are used in the cleaning and pickling operations to be described. They should be kept only in the proper containers and handled with extreme care.

PICKLING PROCESS

1. Thoroughly degrease parts in degreaser, using tri-chlor Ethylene or other commercial degreasing solution.

2. Tank No. 1
   Solution. Use a commercially available derusting compound in solution as recommended by the manufacturer. The solution should not be used at a temperature exceeding that recommended by the manufacturer, otherwise the inhibitor will evaporate and leave a straight acid solution. The length of time the part will be immersed in this solution will depend upon the temperature of the solution and the amount of rust or scale which must be removed. The operator must use his judgment on this point.

3. After pickling, rinse parts in cold running water and immerse in tank No. 2. The solution in this tank should be a neutralizer mixed with water in a proportion recommended by the manufacturer. This solution should be used at recommended temperatures and the parts should remain immersed in the solution for the period of time recommended by the manufacturer.

4. Rinse parts in hot water.

5. Place in tank No. 3. The solution in this tank should contain antirust compounds as recommended by the manufacturer. Usually the parts being treated should be left to dry with antirust solution remaining on them.

   If pieces are stored for any period of time ends of the pipes should be plugged to prevent the entrance of foreign matter. Do not use rags or waste as they will deposit lint on the inside of the tube or pipe. Immediately before using pipes, tubes and fittings they should be thoroughly flushed with suitable degreasing solution.

Figure 4-10. Preparing Pipes, Tubes, and Fittings.

A non-positive seal allows a small amount of internal leakage; such as the clearance of a spool in its bore to provide a lubricating film.

Static Seals

A seal that is compressed between two rigidly connected parts is classified as a static seal. The seal itself may move somewhat as pressure is alternately applied and released, but the mating parts do not move in relation to each other.

Some examples of static seals are mounting gaskets, pipe thread connections, flange joint seals (Fig. 4-11), compression fitting ferrules (Fig. 4-6) and O-rings. Static sealing applications are relatively simple. They are essentially "non-wearing" and usually are trouble-free if assembled properly.

Dynamic Seals

Dynamic seals are installed between parts which do move relative to one another. Thus, at least one of the parts must rub against the seal and therefore dynamic seals are subject to wear. This naturally makes their design and application more difficult.

O-Ring Seals

Probably the most common seal in use in modern hydraulic equipment is the O-ring (Fig. 4-12). An O-ring is a molded, synthetic rubber seal which has a round cross-section in the free state.

The O-ring is installed in an annular groove machined into one of the mating parts. At installation, it is compressed at both the inside and outside diameters. However, it is a pressure-actuated seal as well as a compression seal. Pressure forces the O-ring against one side of its groove and outward at both diameters. It thus seals positively against two annular surfaces and one flat surface. Increased pressure results in a higher force against the sealing surfaces. The O-ring, therefore, is capable of containing extremely high pressure.

O-rings are used principally in static applications. However, they are also found in dynamic applications where there is a short reciprocating motion between the parts. They are not generally suitable for sealing rotating parts or for applications where vibration is a problem.

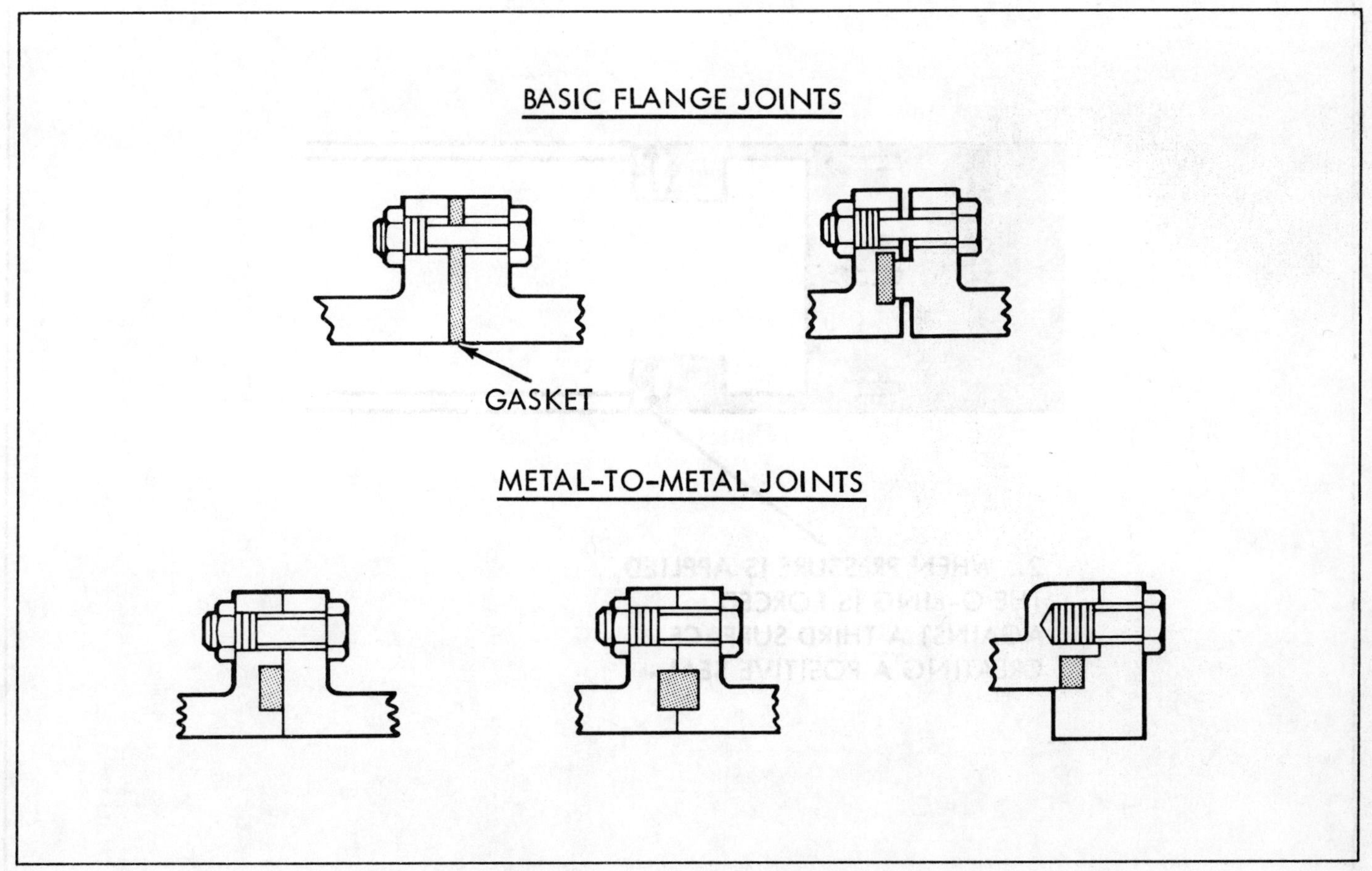

Fig. 4-11. Flange Gaskets and Seals are Typical Static Applications

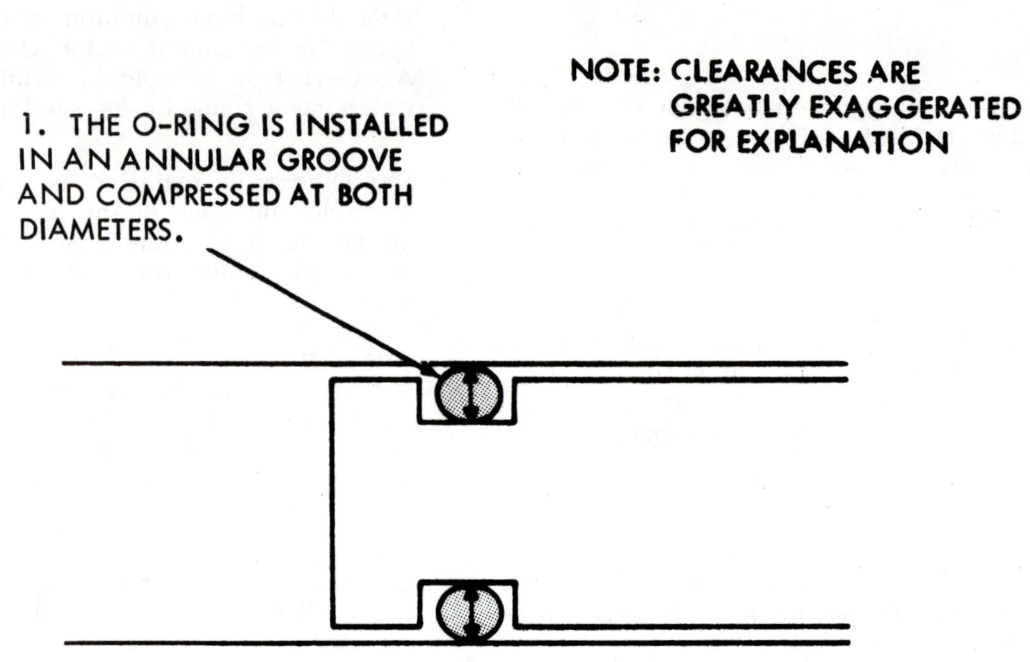

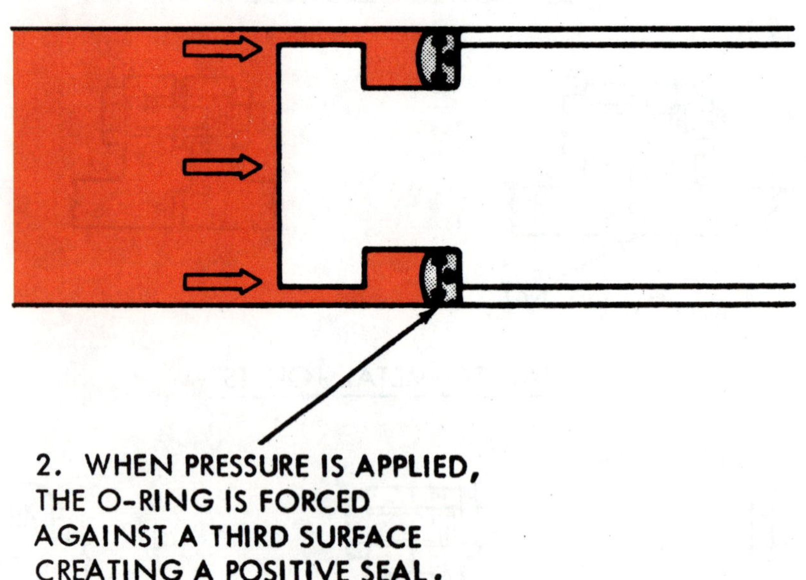

Fig. 4-12. An O-Ring is a Positive Seal

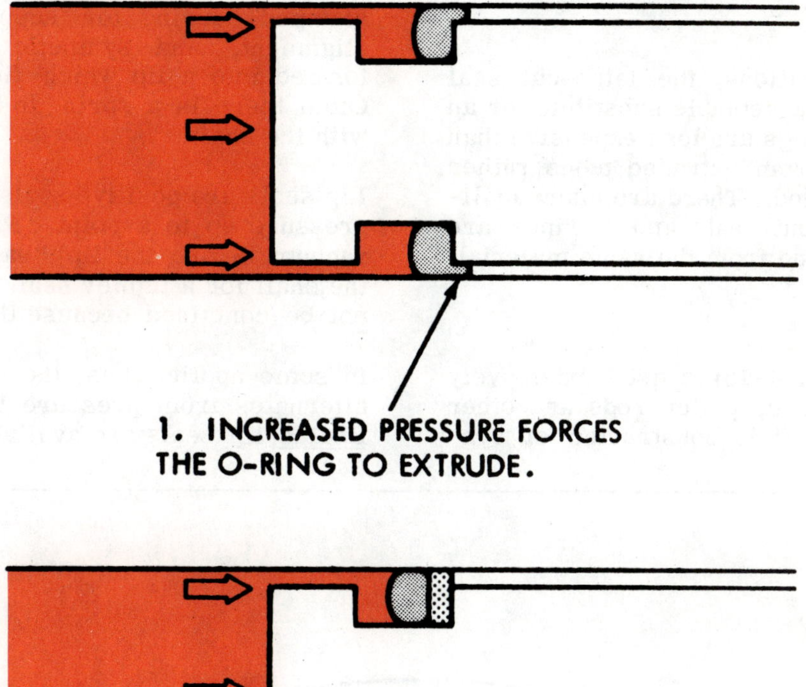

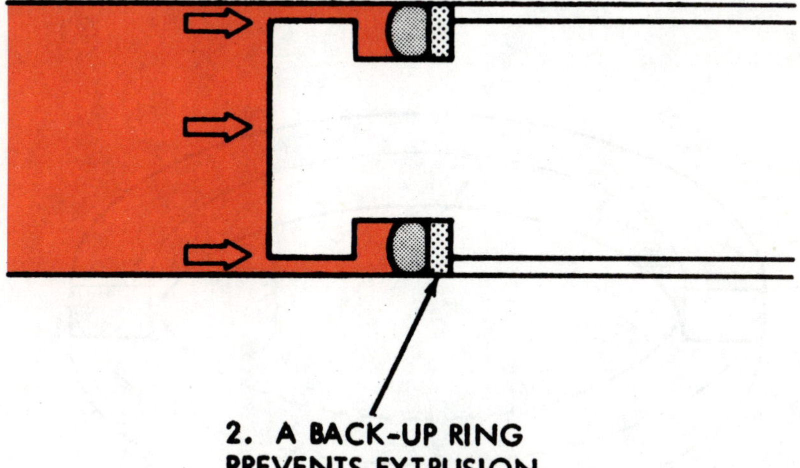

Fig. 4-13. A Back-up Ring is a Non-Extrusion Ring

Back-Up (Non-Extrusion) Rings

At high pressure, the O-ring has a tendency to extrude into the clearance space between the mating parts (Fig. 4-13). This may not be objectionable in a static application. But this extrusion can cause accelerated wear in a dynamic application. It is prevented by installing a stiff back-up ring in the O-ring groove opposite the pressure source. If the pressure alternates, back-up rings can be used on both sides of the O-ring.

Lathe-Cut Rings

In many static applications, the lathe-cut seal (Fig. 4-14) makes an acceptable substitute for an O-ring. Lathe-cut rings are less expensive than O-rings, being cut from extruded tubes rather than individually molded. There are many applications where lathe-cut seals and O-rings are interchangeable if made from the same material.

T-Ring Seals

The T-ring seal (Fig. 4-15) is used extensively to seal cylinder pistons, piston rods and other reciprocating parts. It is constructed of synthetic rubber molded in the shape of a "T", and reinforced by back-up rings on either side. The sealing edge is rounded and seals very much like an O-ring. Obviously, this seal will not have the O-ring's tendency to roll. The T-ring is not limited to short-stroke applications.

Lip Seals

Lip seals are low-pressure dynamic seals, used principally to seal rotating shafts.

A typical lip seal (Fig. 4-16) is constructed of a stamped housing for support and installation alignment, and synthetic rubber or leather formed into a lip which fits around the shaft. Often there is a spring to hold the lip in contact with the shaft.

Lip seals are positive seals. Sealing is aided by pressure up to a point. Pressure on the lip (or vacuum behind the lip) "balloons" it out against the shaft for a tighter seal. High pressure cannot be contained because the lip has no back-up.

In some applications, the chamber being sealed alternates from pressure to vacuum condition. Double lip seals are available for these applica-

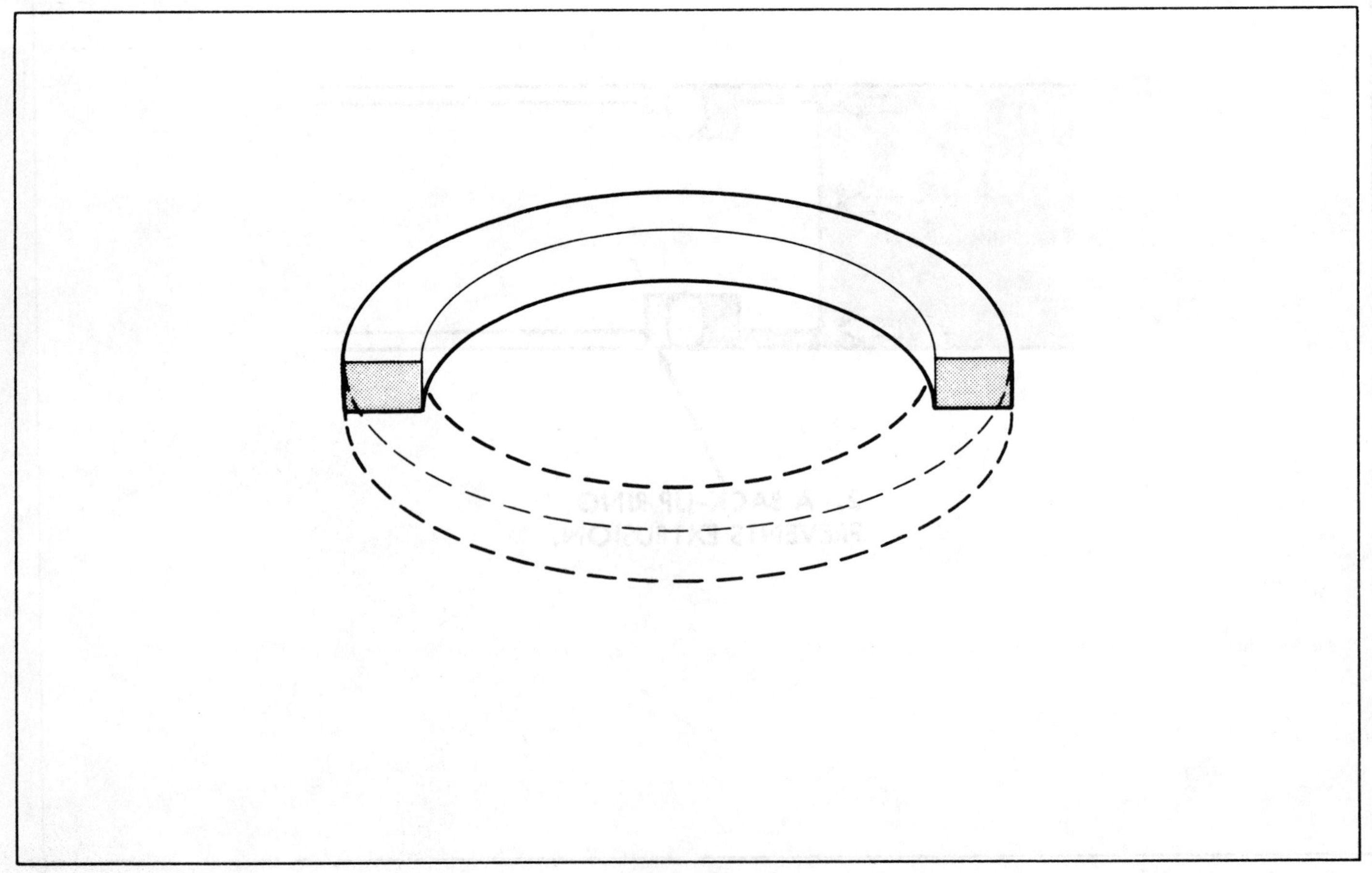

Fig. 4-14. Lathe-Cut Seal is Rectangular in Section

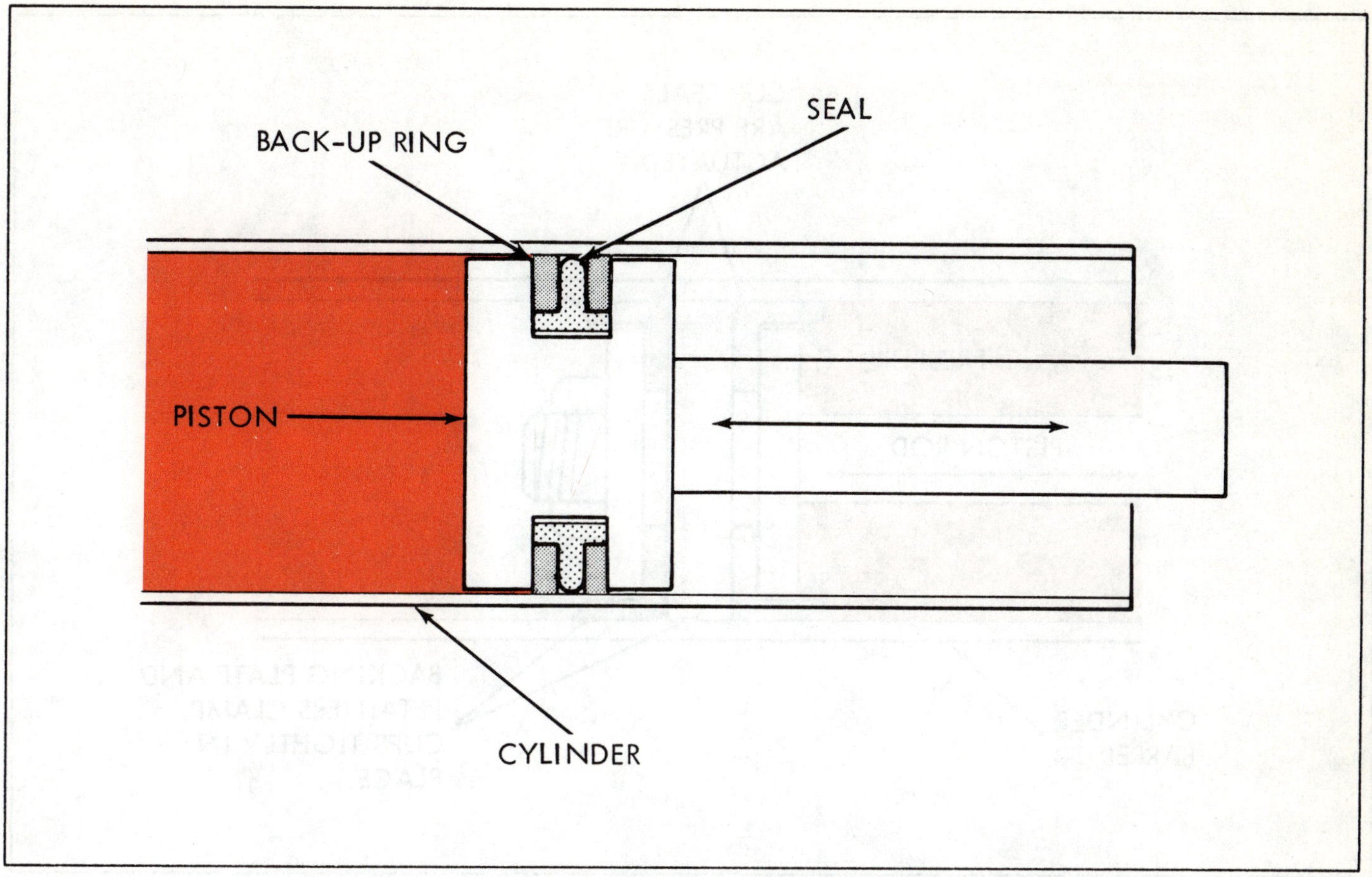

Fig. 4-15. T-Ring is a Dynamic Seal for Reciprocating Parts

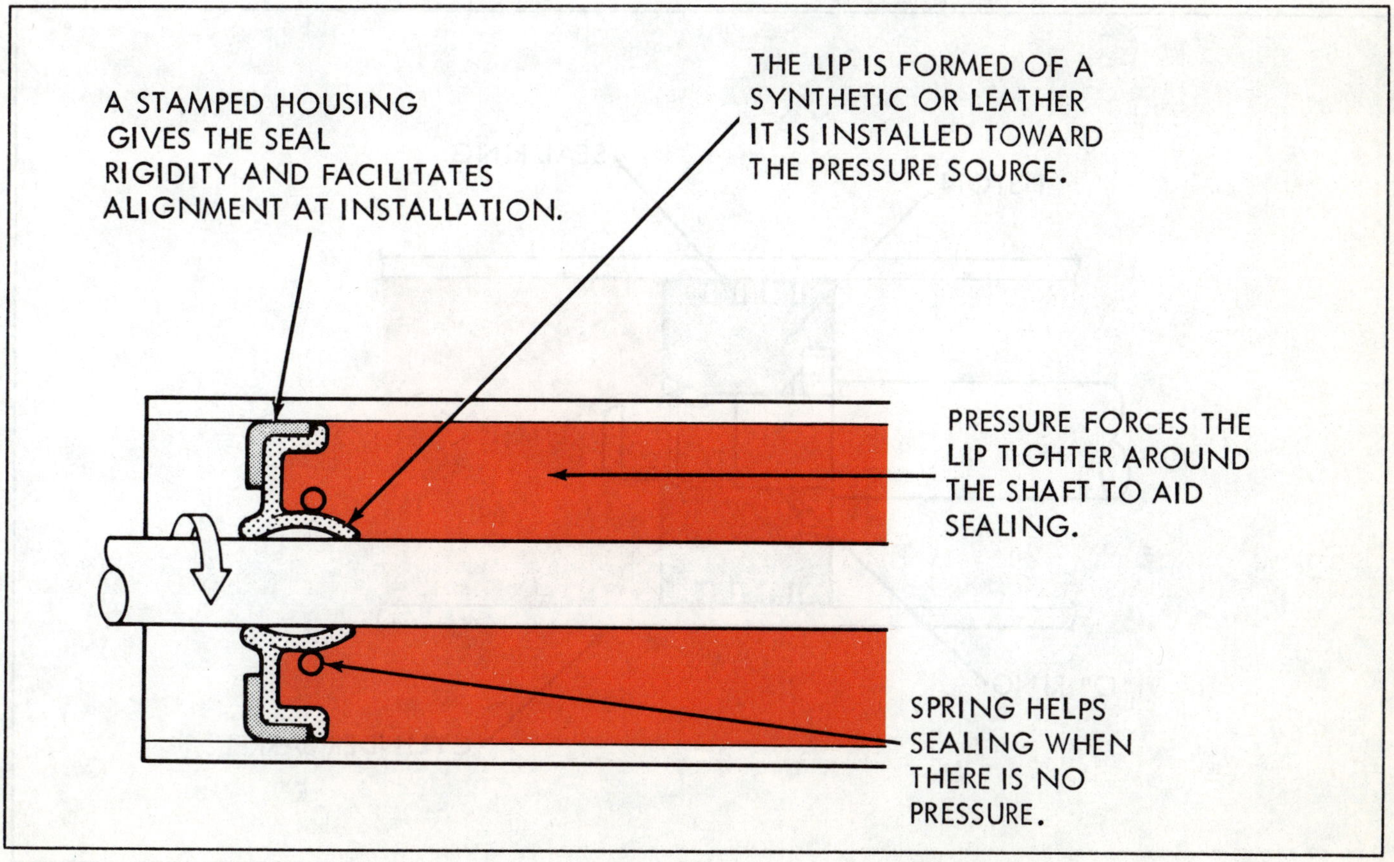

Fig. 4-16. Lip Seals are Used on Rotating Shafts

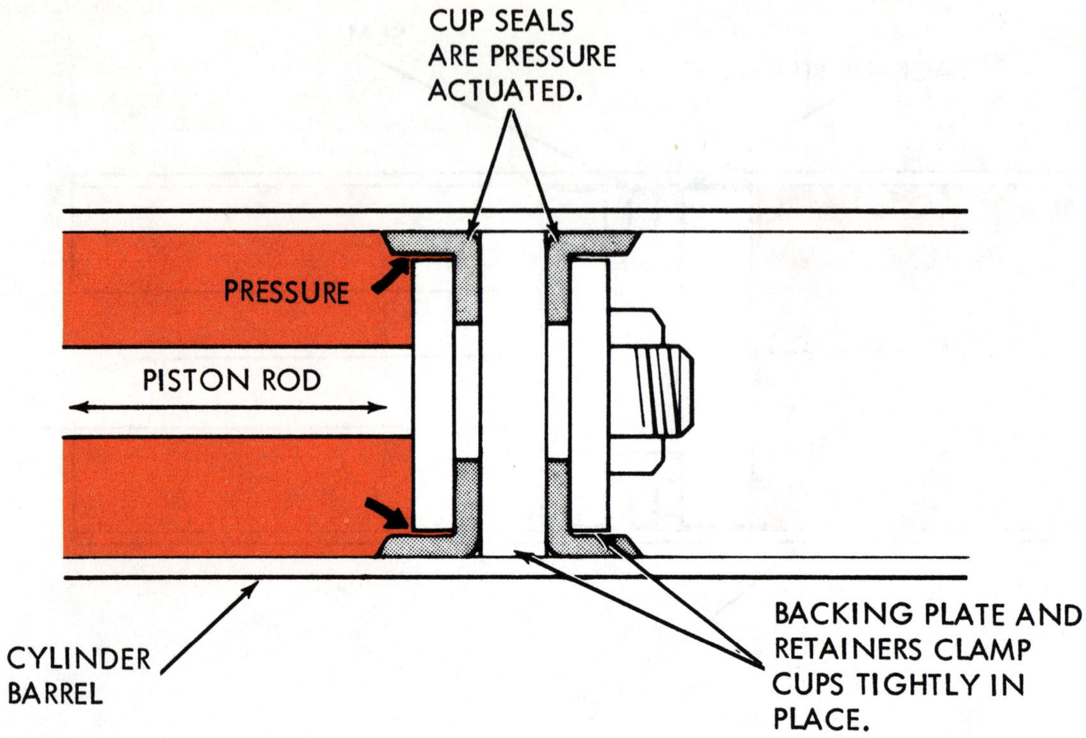

Fig. 4-17. Cup Seals are Used on Cylinder Pistons

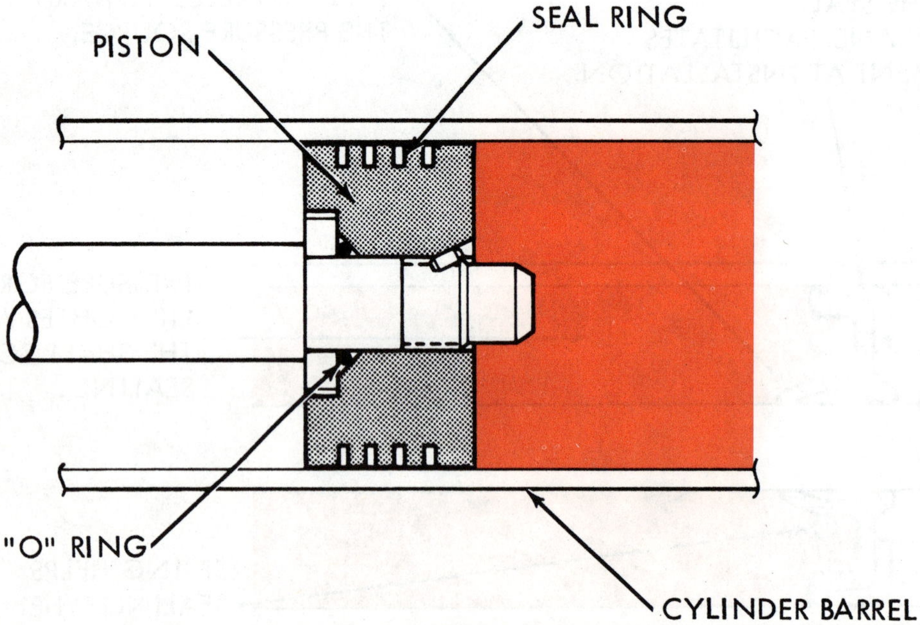

Fig. 4-18. Piston Rings are Used for Cylinder Pistons

tions to prevent air or dirt from getting in and oil from getting out.

Cup Seals

A cup seal (Fig. 4-17) is a positive seal used on many cylinder pistons. It is pressure actuated in both directions. Sealing is accomplished by forcing the cup lip outward against the cylinder barrel. This type of seal is backed up and will handle very high pressures.

Cup seals must be clamped tightly in place. The cylinder piston actually is nothing more than the backing plate and retainers that hold the cup seals.

Piston Rings

Piston rings (Fig. 4-18) are fabricated from cast iron or steel, highly polished and sometimes plated. They offer considerably less resistance to motion than leather or synthetic seals. They are most often found on cylinder pistons.

One piston ring does not necessarily form a positive seal. Sealing becomes more positive when several rings are placed side-by-side. Very high pressures can be handled.

Compression Packings

Compression packings (Fig. 4-19) were among the earliest sealing devices used in hydraulic systems and are found in both static and dynamic applications. Packings are being replaced in most static applications by O-rings or lathe-cut seals.

Most packings in use today are molded or formed into "U" or "V" shapes, and multiple packings are used for more effective sealing. The packings are compressed by tightening a flanged follower ring against them. Proper adjustment is critical, because excessive tightening will accelerate wear. In some applications, the packing ring is spring-loaded to maintain the correct force and take up wear.

Face Seal

A face seal (Fig. 4-20) is used in applications where a high pressure seal is required around a rotating shaft. Sealing is accomplished by constant contact between two flat surfaces, often carbon and steel. The stationary sealing member is attached to the body of the component. The other is attached to the shaft and turns against the

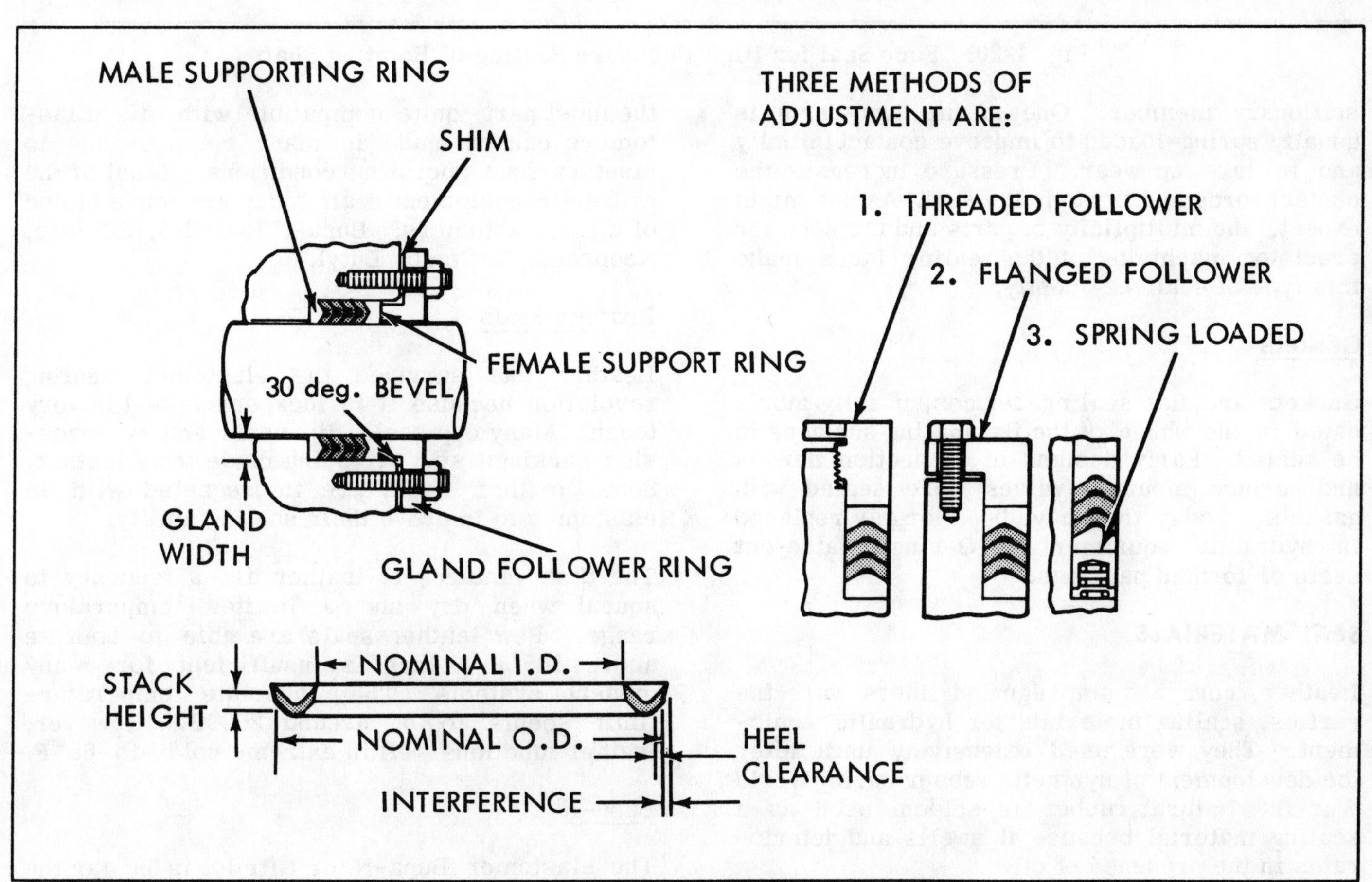

Fig. 4-19. Compression Packings

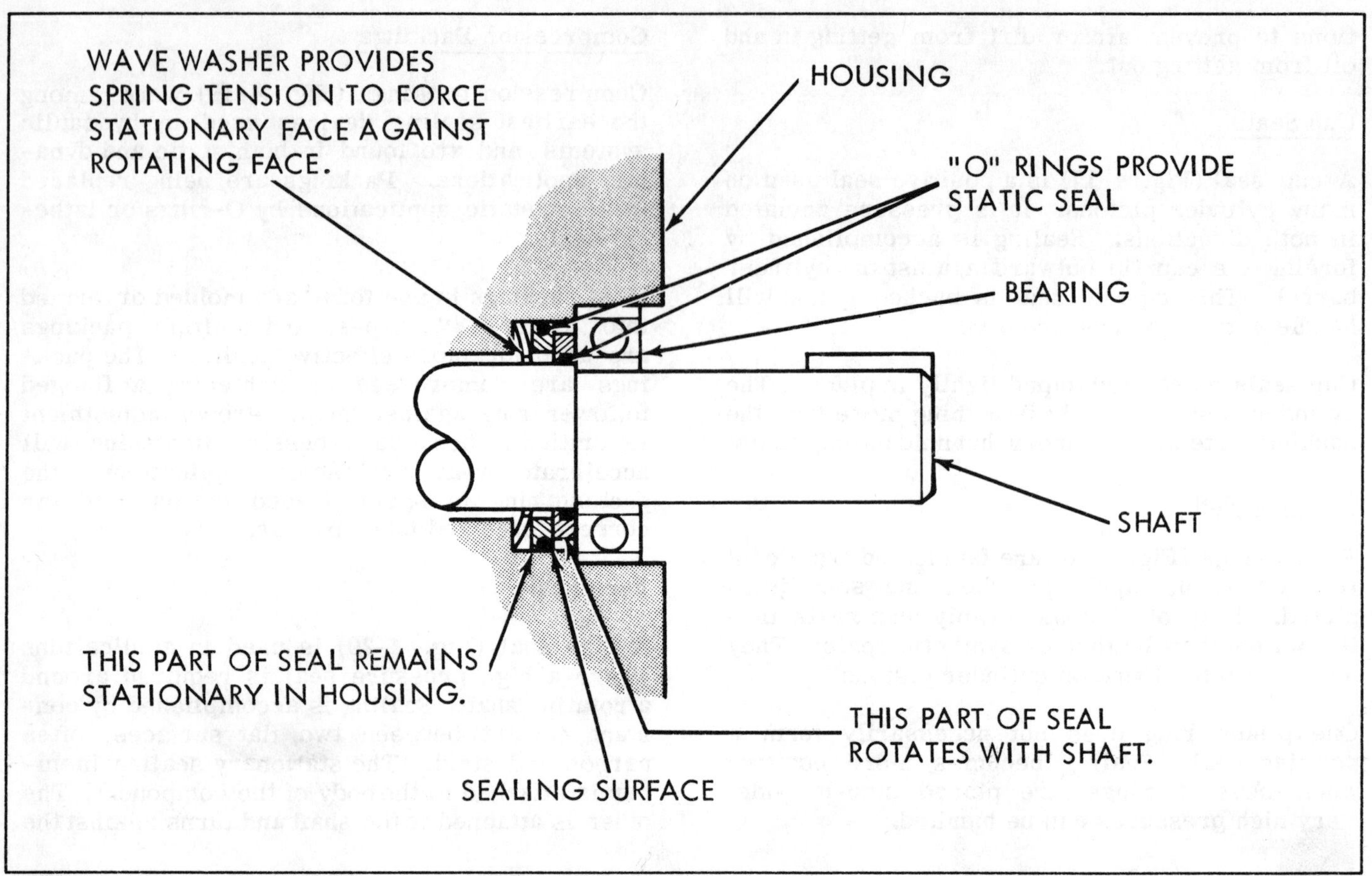

Fig. 4-20. Face Seal for High Pressure Sealing of Rotating Shaft

stationary member. One of the two parts is usually spring-loaded to improve contact initially and to take up wear. Pressure increases the contact force and tightens the seal. As one might expect, the multiplicity of parts and the need for precision machining of the sealing faces make this type of seal very costly.

Gaskets

Gaskets are flat sealing devices, usually fabricated in the shape of the flat mating surfaces to be sealed. Early designs of connection flanges and surface mounted valves were sealed with gaskets. Today they have been largely replaced in hydraulic equipment by O-rings, lathe-cut seals or formed packings.

SEAL MATERIALS

Leather, cork and impregnated fibers were the earliest sealing materials for hydraulic equipment. They were used extensively until after the development of synthetic rubber during World War II. Natural rubber is seldom used as a sealing material because it swells and deteriorates in the presence of oil.

Synthetic rubbers (elastomers), however, are for the most part quite compatible with oil. Elastomers can be made in many compositions to meet various operating conditions. Most of the hydraulic equipment seals today are made of one of these elastomers: Buna-N (Nitrile), Silicone, Neoprene, Teflon or Butyl.

Leather Seals

Leather has survived the elastomer sealing revolution because it is inexpensive and is very tough. Many cup seals, lip seals and compression packings still are being made from leather. Some leather seals are impregnated with an elastomer to improve their sealing ability.

The disadvantages of leather are a tendency to squeal when dry and a limited temperature range. Few leather seals are able to operate above 165°F, which is insufficient for many modern systems. Their absolute temperature limit seems to be around 200°F. However, leather functions well in extreme cold--to -65°F.

Buna-N

The elastomer Buna-N (or Nitrile) is by far the most widely used sealing material in modern hydraulic systems. It is moderately tough, wears well and is inexpensive. There are a number of compositions compatible with petroleum oil--most of them easily molded into any required seal shape.

Buna-N has a reasonably wide temperature range, retaining its sealing properties from -40° to 230°F. At moderately high temperatures, it retains its shape in most petroleum oils where other materials tend to swell. It does swell, however, in some synthetic fluids.

Silicone

Silicone is an elastomer with a much wider temperature range than Buna-N, and is therefore a popular material for rotating shaft seals and static seals in systems that run from very cold to very hot. It retains its shape and sealing ability to -60°F and is generally satisfactory up to 400 or 500°F.

At high temperature, silicone tends to absorb oil and swell. This however, is no particular disadvantage in static applications. Silicone is not used for reciprocating seals, because it tears and abrades too easily. Silicone seals are compatible with most fluids; even more so with fire-resistant fluids than petroleum.

Neoprene

One of the earliest elastomers used in hydraulic system sealing was neoprene. A tough material, it still is in limited use in low temperature systems using petroleum fluids. Above 150°F neoprene is unsuitable as a sealing material because of a tendency to vulcanize or "cook".

Plastics, Fluoro-Plastics and Fluoro-Elastomers

Several sealing materials are synthesized by combining fluorine with an elastomer or plastic. They include Kel-F, Viton A and Teflon. Nylon is another synthetic material with similar properties. It is often used in combination with the elastomers to give them reinforcement. Both nylon and teflon are used for back-up rings as well as sealing materials. Teflon, of course, is used in a tape form for sealing pipe joints. All have exceptionally high heat resistance (to 500°F) and are compatible with most fluids.

PREVENTING LEAKAGE

The three general considerations in preventing leakage are:

1. Design to minimize the possibility. (Back, gasket or sub-plate mounting)

2. Proper installation.

3. Control of operating conditions.

Let us explore each briefly.

Anti-Leakage Designs

We have already noted that designs using straight thread connectors and welded flanges are less susceptible to leakage than pipe connections. Back-mounting of valves with all pipe connections made permanently to a mounting plate has made a great difference in preventing leakage and in making it easier to service a valve (Fig. 4-21). Most valves being built today are the back-mounted design. (The term gasket-mounted was originally applied to this design because gaskets were used on the first back-mounted valves. Gasket-mounted or sub-plate mounted is still used to refer to back-mounted valves sealed by O-rings or lathe-cut seals.)

A further advance from back-mounting is the use of manifolds (Fig. 4-22). Some are drilled and some combine mounting plates with passage plates (sandwiched and brazed together), providing interconnections between valves and eliminating a good deal of external plumbing.

Proper Installation

Installation recommendations were covered earlier in this chapter. Careful installation, with attention to avoiding pinching or cocking a seal, usually assures a leak-proof connection. Manufacturers often recommend a special driver for inserting lip type shaft seals to be certain they are installed correctly. Vibration and undue stress at joints, which are common causes of external leakage, also are avoided by good installation practice.

Operating Conditions

Control over operating conditions can be very important to seal life. These are the operating factors that can help prevent leakage:

1. Avoid Contamination. An atmosphere contaminated with moisture, dirt or any abrasive material shortens the life of shaft seals and piston rod seals exposed to the air. Protective devices should be used in contaminated atmospheres. Equally important is clean fluid to avoid damage to internal seals.

2. Fluid Compatibility. Some fire-resistant fluids attack and disintegrate certain elastomer seals. Few seals, in fact, are compatible with all fluids. The fluid supplier should always be

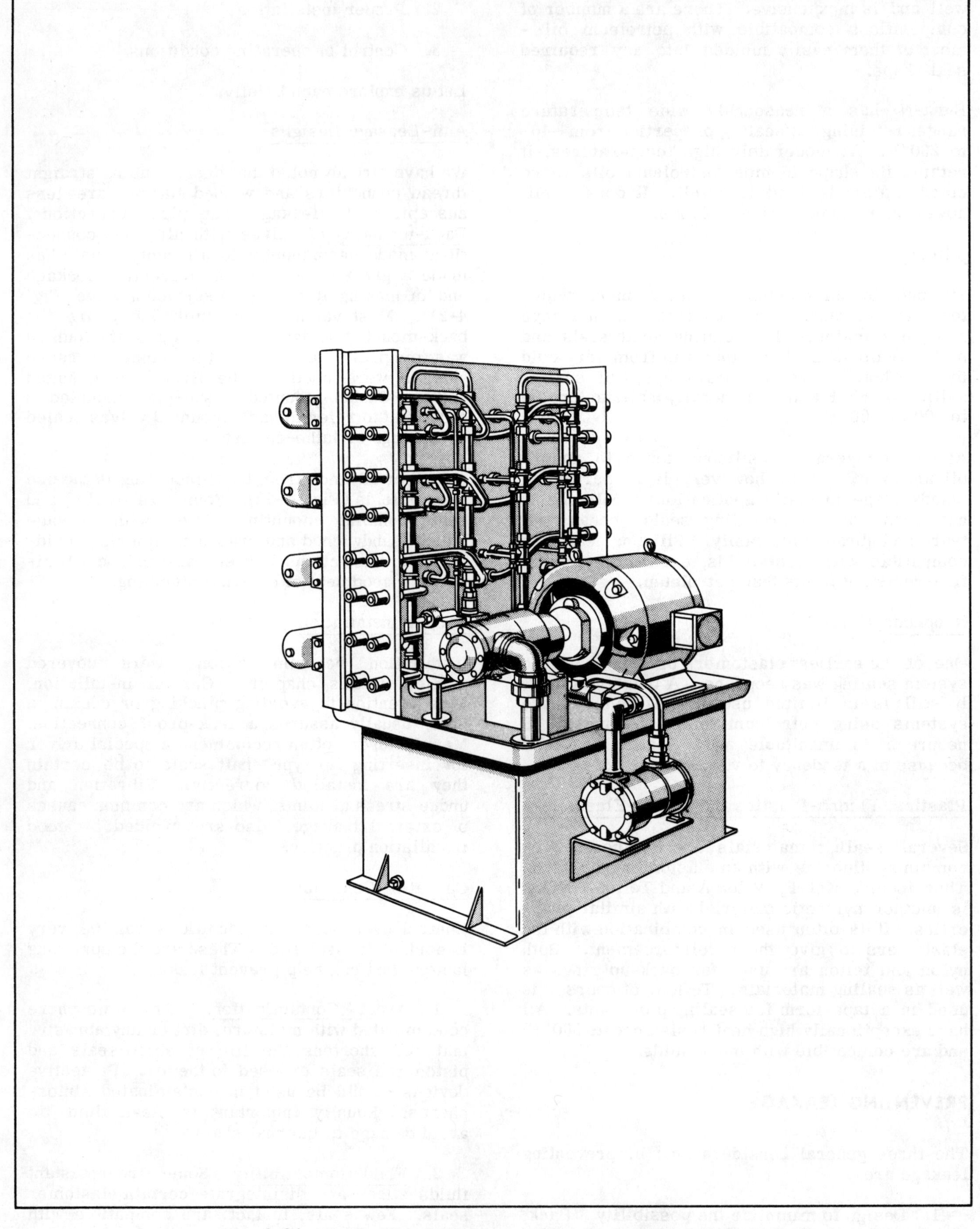

Fig. 4-21. Back-Mounting Leaves Pipe Connections Undisturbed

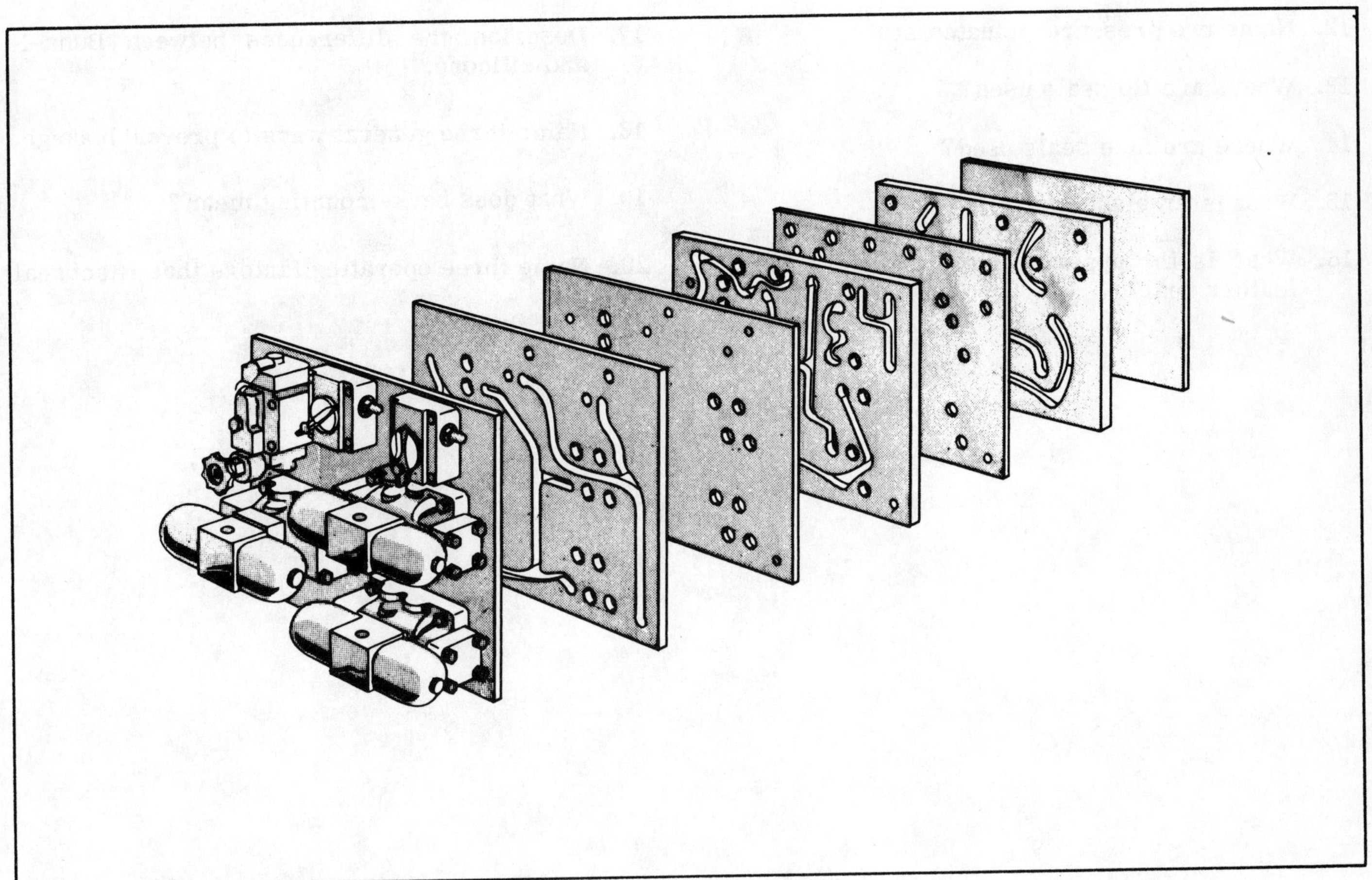

Fig. 4-22. Manifold Contains Interconnecting Passages to Eliminate Piping Between Valves

consulted when in doubt whether to change seals when a change is made in the type of fluid. (See Chapter 3.) Fluid additives (added by the machine user) also may attack seals and should be used only at the recommendation of the fluid supplier.

3. Temperature. At extremely low temperatures, a seal may become too brittle to be effective. At too high a temperature, a seal may harden, soften or swell. The operating temperature should always be kept well within the temperature range of the seals being used.

4. Pressure. Excess fluid pressure from overloads puts an additional strain on oil seals and may "blow" a seal causing a leak.

5. Lubrication. No seal should ever be installed or operated dry. All must be lubricated or the seal will wear quickly and leak. Leather seals should be soaked in fluid before installation. Elastomer seals are not as absorptive as leather but should be coated with fluid before installation.

QUESTIONS

1. How is a pipe size specified?
2. What is the schedule number of standard pipe?
3. How does a pipe thread seal?
4. What advantages does tubing have over pipe?
5. To what does the specified size of tubing refer?
6. How are tubing connections sealed?
7. How does a flexible hose contain pressure?
8. Name some methods for cleaning hydraulic pipes.
9. Give two reasons for pipe supports.
10. What is a positive seal?
11. What is a static sealing application?

12. Name two pressure-actuated seals.

13. Where are lip seals used?

14. Where are face seals used?

15. What is an elastomer?

16. What is the upper temperature limit for a leather seal?

17. Describe the differences between Buna-N and silicone.

18. Name three general ways to prevent leakage.

19. What does back-mounting mean?

20. Name three operating factors that affect seal life.

RESERVOIRS AND FLUID CONDITIONERS

This chapter deals with conditioning the fluid; that is, providing storage space for all the fluid required in the system plus a reserve, keeping the fluid clean, and maintaining the proper operating temperature.

The storage space for the fluid, of course, is the oil reservoir. The fluid is kept clean by using strainers, filters and magnetic plugs to the degree required by the conditions.

The design of the circuit has considerable effect upon the fluid temperature. Heat exchangers, however, are sometimes required, particularly where operating temperatures are critical or the system cannot dissipate all the heat that is generated.

RESERVOIRS

The designer of industrial hydraulic systems has an advantage over his counterpart in aerospace or mobile equipment. This advantage is in a good deal of flexibility in reservoir design.

With almost no location or sizing problems, the reservoir for a piece of shop equipment can usually be designed to perform a number of functions. It is first a storehouse for the fluid until called for by the system. The reservoir also should provide a place for air to separate out of the fluid and should permit contaminants to settle out as well. In addition, a well-designed reservoir will help dissipate any heat that is generated in the system.

Reservoir Construction

A typical industrial reservoir, conforming to industry standards, is shown in Fig. 5-1. The tank is constructed of welded steel plate with extensions of the end plates supporting the unit on the floor. The entire inside of the tank is painted with a sealer to reduce rust which can result from condensed moisture. This sealer must be chosen for compatibility with the fluid being used.

The reservoir is designed for easy fluid maintenance. The bottom of the tank is dished and has a drain plug at the lowest point so the tank can be drained completely. Easily removable covers as shown are desirable for access for cleaning. A sight glass for checking the fluid also is highly desirable. (It is far more likely that periodic checks will be made through a sight glass than with a dipstick or cover which must be removed.)

The filler hole is provided with a fine mesh screen to keep out contamination when the fluid is replenished.

Breather. A vented breather cap is used on most reservoirs and should also contain an air filtering screen. In dirty atmospheres, an oil bath air filter may be better. The filter or breather must be large enough to handle the air flow required to maintain atmospheric pressure whether the tank is empty or filled. In general, the higher the flow rate, the larger the breather required. On a pressurized reservoir, of course, a breather is not used. It is replaced by an air valve to regulate the pressure in the tank between preset limits.

Baffle Plate. A baffle plate (Fig. 5-2) extends lengthwise through the center of the tank; it is usually about 2/3 the height of the oil level and is used to separate the pump inlet line from the return line so that the same fluid cannot recirculate continuously, but must take a circuitous route through the tank.

Thus, the baffle (1) prevents local turbulence in the tank, (2) allows foreign material to settle to the bottom, (3) gives the fluid an opportunity to get rid of entrapped air, and (4) helps increase heat dissipation through the tank walls.

Line Connections and Fittings

Most lines to the reservoir terminate below the oil level. The line connections at the tank cover are often "packed" (sealed) slip-joint type flanges. This design prevents dirt from entering through these openings, and makes it easy to remove inlet line strainers for cleaning.

Pump inlet and return lines must be well below the fluid level; otherwise, the oil may become aerated and foam. Drain lines, however, may terminate above the fluid level if necessary to

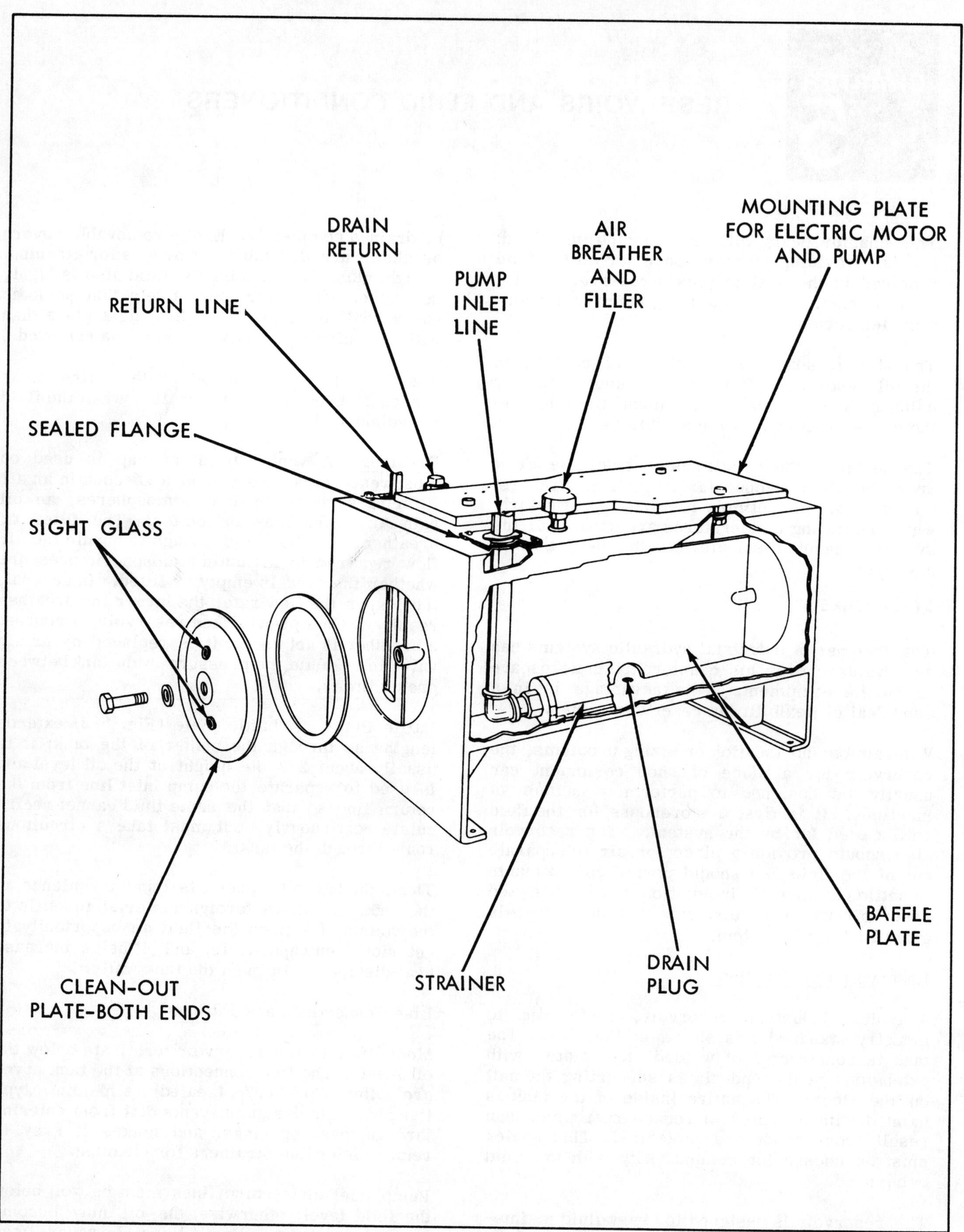

Figure 5-1. Reservoir is Designed for Easy Maintenance

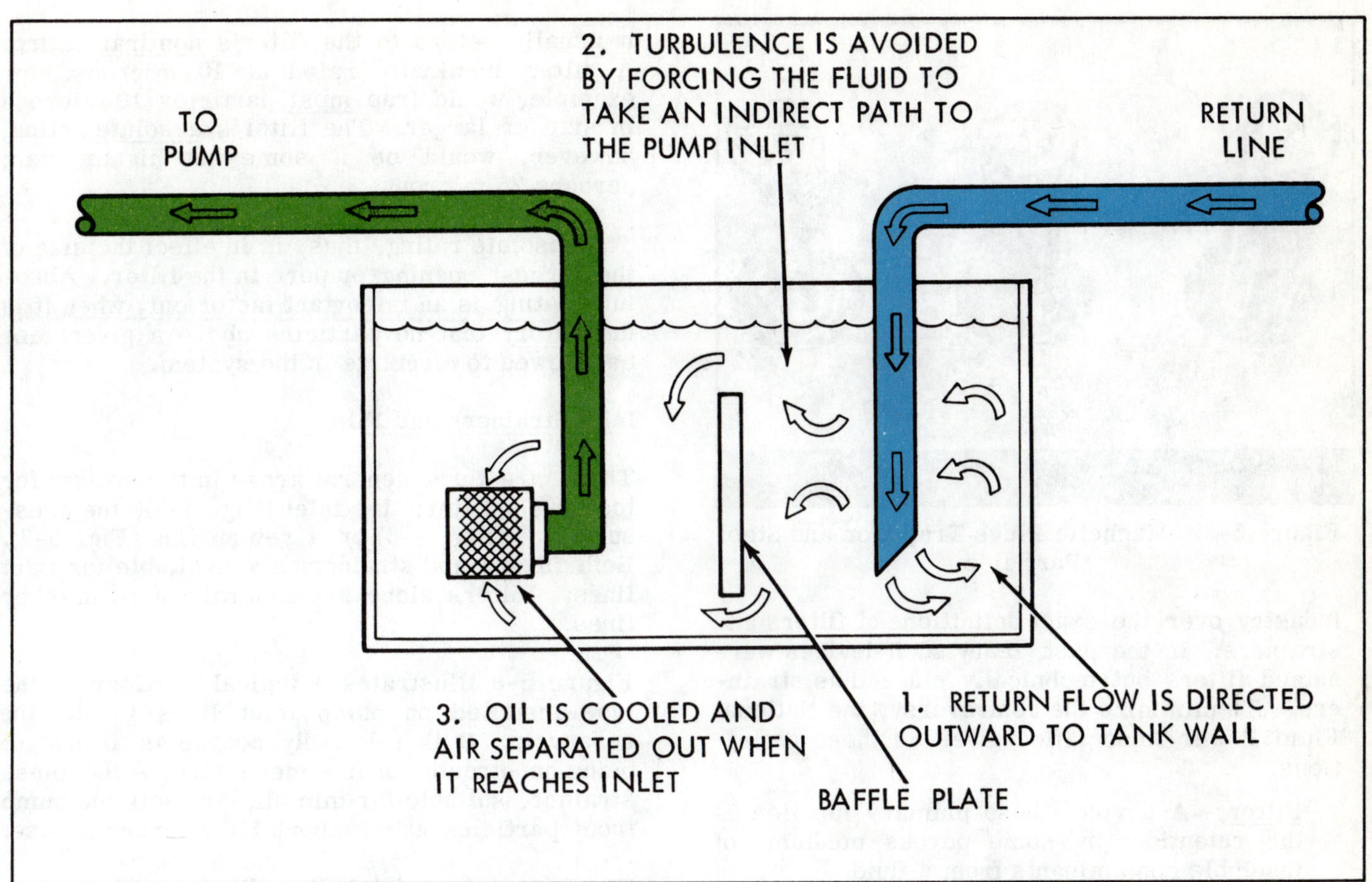

Figure 5-2. Baffle Plate Controls Direction of Flow in Tank

avoid pressure build-up in drain passages or siphoning oil through them. Connections above the fluid level must be tightly sealed to prevent the entry of air into the system. Connections below the fluid level need only be tightened sufficiently to remain connected.

Lines that terminate near the tank bottom and are not equipped with strainers should be cut at a 45-degree angle. This prevents the line opening from "bottoming" in the tank and cutting off flow. On a return line, the angled opening is often pointed so that flow is directed at the tank walls and away from the pump inlet line.

Reservoir Sizing

A large tank is always desirable to promote cooling and separation of contaminants. At a minimum, the tank must store all the fluid the system will require and maintain the level high enough to prevent a "whirlpool" effect at the pump inlet line opening. If this occurs, air will be taken in with the fluid.

Heat expansion of the fluid, changes in fluid level due to system operation, inside tank area exposed to water condensation and the amount of heat generated in the system all are factors to consider. In industrial equipment, it's customary to provide a reservoir that holds two or three gallons of liquid for each gallon per minute (gpm) of pump delivery.

Sizing Thumb Rule:

In mobile and aerospace systems, the benefits of a large reservoir may have to be sacrificed because of space limitations.

FILTERS AND STRAINERS

Hydraulic fluids are kept clean in the system principally by devices such as filters and strainers. Magnetic plugs (Fig. 5-3) also are used in some tanks to trap iron and steel particles carried by fluid. Recent studies have indicated that even particles as small as 1 - 5 microns have a degrading effect, causing failures in servo systems and hastening oil deterioration in many cases.

Filter or Strainer

There will probably always be controversy in the

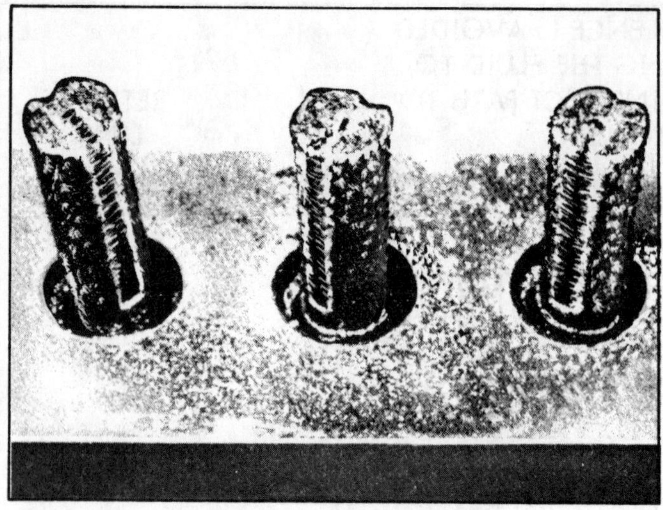

Figure 5-3. Magnetic Plugs Trap Iron and Steel Particles

industry over the exact definitions of filters and strainers. In the past, many such devices were named filters but technically classed as strainers. To minimize the controversy, the National Fluid Power Association gives us these definitions:

Filter - A device whose primary function is the retention, by some porous medium, of insoluble contaminants from a fluid.

Strainer - A coarse filter.

To put it simply, whether the device is a filter or strainer, its function is to trap contaminants from fluid flowing through it. "Porous medium" simply refers to a screen or filtering material that allows fluid flow through it but stops various other materials.

"Mesh" and Micron Ratings

A simple screen or a wire strainer is rated for filtering "fineness" by a mesh number or its near equivalent, standard sieve number. The higher the mesh or sieve number, the finer the screen.

Filters, which may be made of many materials other than wire screen, are rated by micron size. A micron is one-millionth of a meter or 39-millionths of an inch. For comparison, a grain of salt is about 70 microns across. The smallest particle a sharp eye can see is about 40 microns.

Figure 5-4 compares various micron sizes with mesh and standard sieve sizes.

Nominal and Absolute Ratings

When a filter is specified as so many microns, it usually refers to the filter's nominal rating. A filter nominally rated at 10 microns, for example, would trap most particles 10 microns in size or larger. The filter's absolute rating, however, would be a somewhat higher size; perhaps 25 microns.

The absolute rating, thus, is in effect the size of the largest opening or pore in the filter. Absolute rating is an important factor only when it is mandatory that no particles above a given size be allowed to circulate in the system.

Inlet Strainers and Filters

There are three general areas in the system for locating a filter: the inlet (Fig. 5-5), the pressure line (Fig. 5-6) or a return line (Fig. 5-7). Both filters and strainers are available for inlet lines. Filters alone are generally used in other lines.

Figure 5-8 illustrates a typical strainer of the type installed on pump inlet lines inside the reservoir. It is relatively coarse as filters go, being constructed of fine mesh wire. A 100-mesh strainer, suitable for thin oil, protects the pump from particles above about 150 microns in size.

There also are inlet line filters. These are usually mounted outside the reservoir near the pump inlet. They, too, must be relatively coarse. A fine filter (unless it's very large) creates more pressure drop than can be tolerated in an inlet line.

Pressure Line Filters

A number of filters are designed for installation right in the pressure line (Fig. 5-6) and can trap much smaller particles than inlet line filters. Such a filter might be used where system components such as valves are less dirt-tolerant than the pump. The filter thus would trap this fine contamination from the fluid as it leaves the pump.

Pressure line filters, of course, must be able to withstand the operating pressure of the system.

Return Line Filters

Return line filters (Fig. 5-7) also can trap very small particles before the fluid returns to the reservoir. They are particularly useful in systems which do not have a large reservoir to allow contaminants to settle out of the fluid. A return line filter is nearly a must in a system with a high-performance pump, which has very close clearances and usually cannot be sufficiently protected by an inlet line filter.

RELATIVE SIZE OF MICRONIC PARTICLES

MAGNIFICATION 500 TIMES

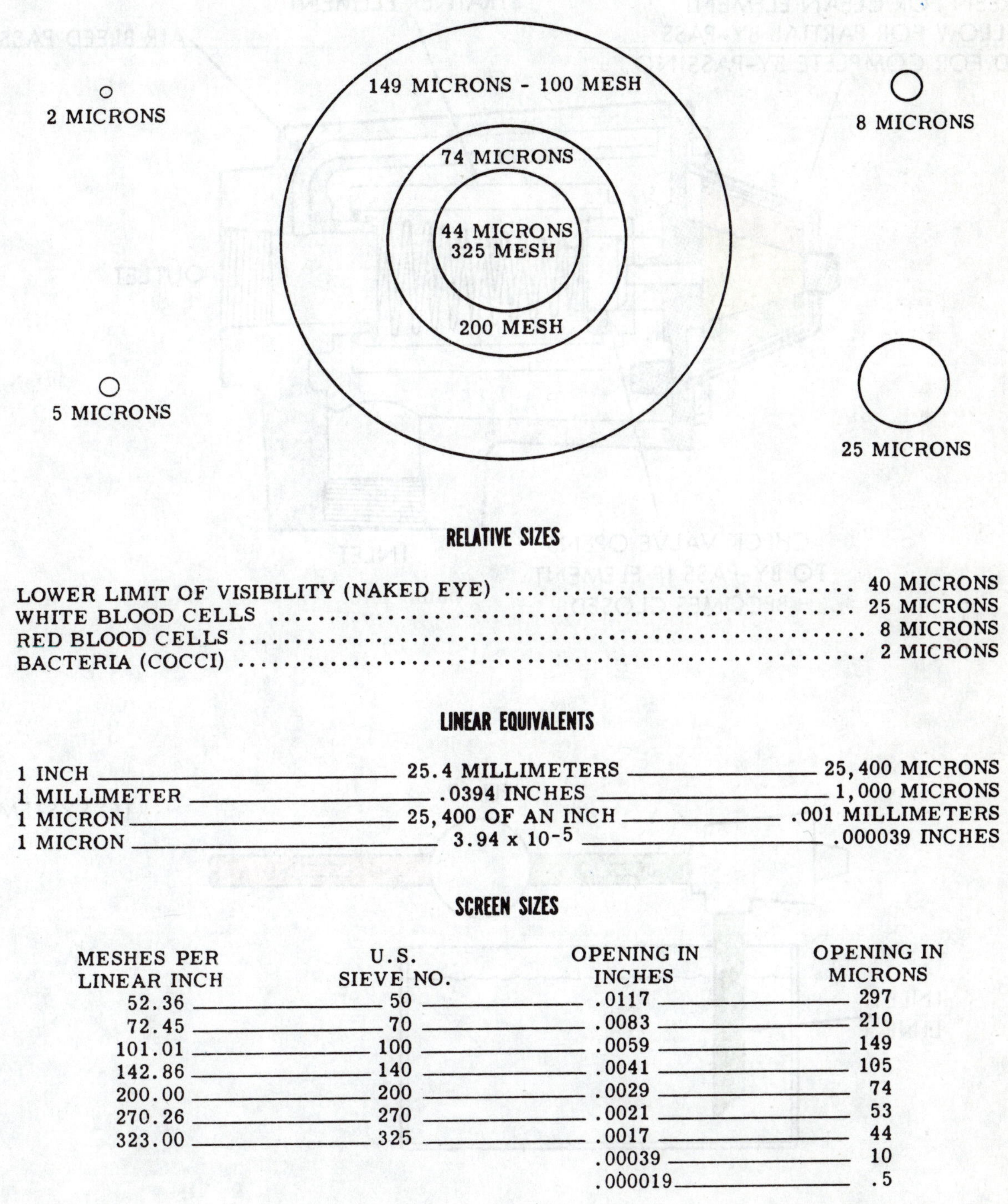

RELATIVE SIZES

LOWER LIMIT OF VISIBILITY (NAKED EYE)	40 MICRONS
WHITE BLOOD CELLS	25 MICRONS
RED BLOOD CELLS	8 MICRONS
BACTERIA (COCCI)	2 MICRONS

LINEAR EQUIVALENTS

1 INCH	25.4 MILLIMETERS	25,400 MICRONS
1 MILLIMETER	.0394 INCHES	1,000 MICRONS
1 MICRON	25,400 OF AN INCH	.001 MILLIMETERS
1 MICRON	3.94×10^{-5}	.000039 INCHES

SCREEN SIZES

MESHES PER LINEAR INCH	U.S. SIEVE NO.	OPENING IN INCHES	OPENING IN MICRONS
52.36	50	.0117	297
72.45	70	.0083	210
101.01	100	.0059	149
142.86	140	.0041	105
200.00	200	.0029	74
270.26	270	.0021	53
323.00	325	.0017	44
		.00039	10
		.000019	.5

Figure 5-4. A Micron is 39 Millionths of an Inch

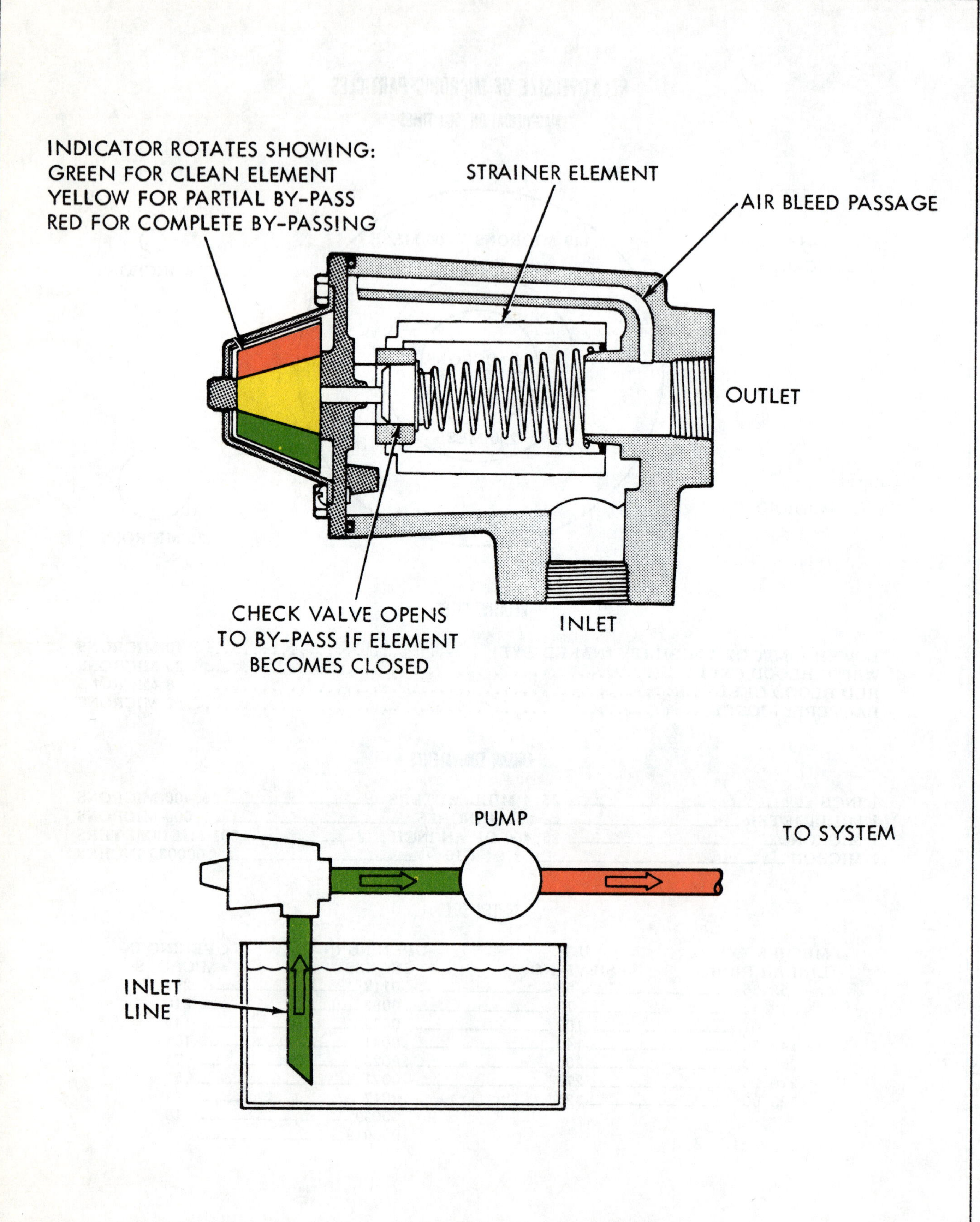

Figure 5-5. Inlet Line Filter Protects Pump

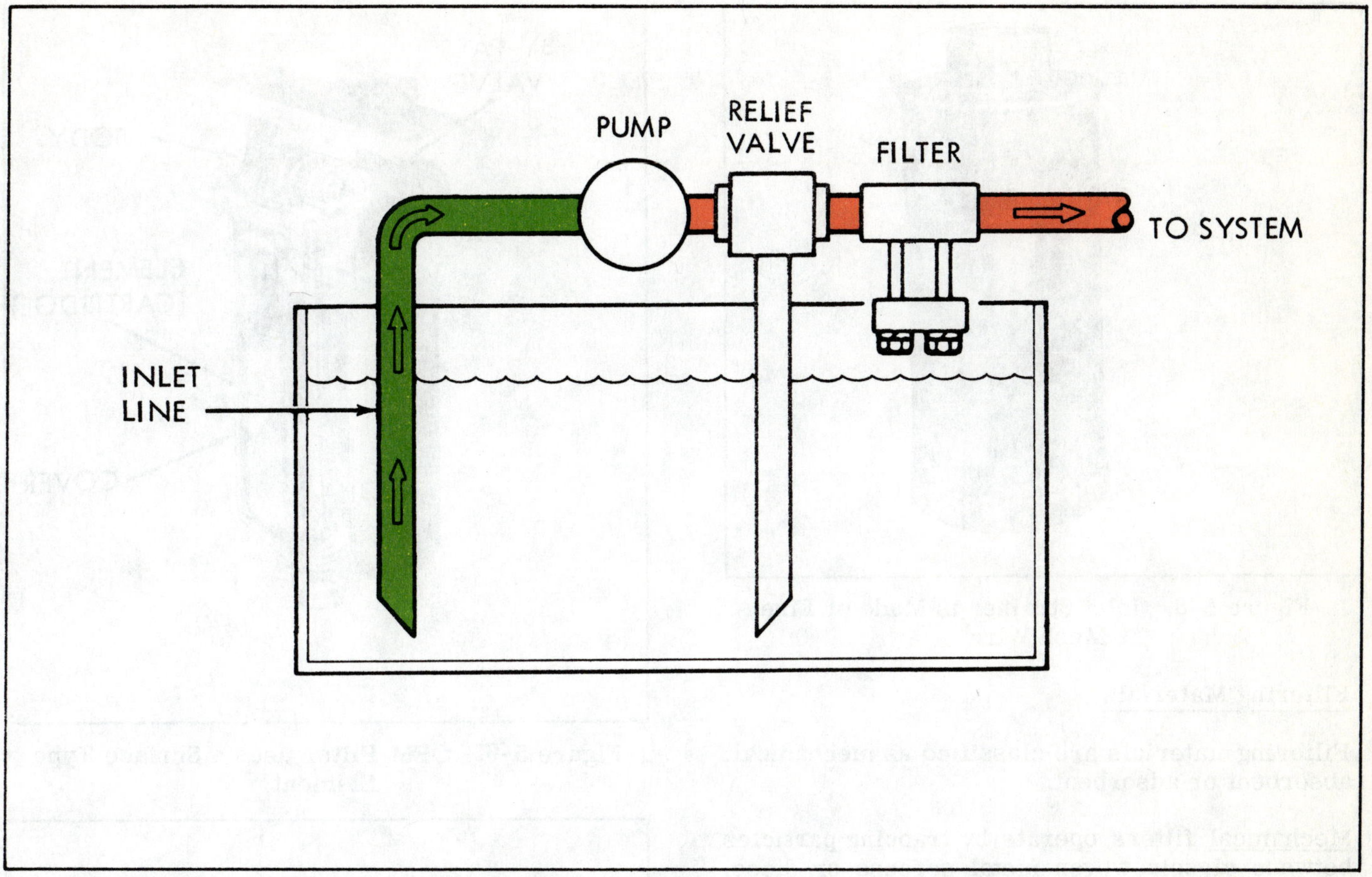

Figure 5-6. Pressure Line Filter is Downstream from Pump

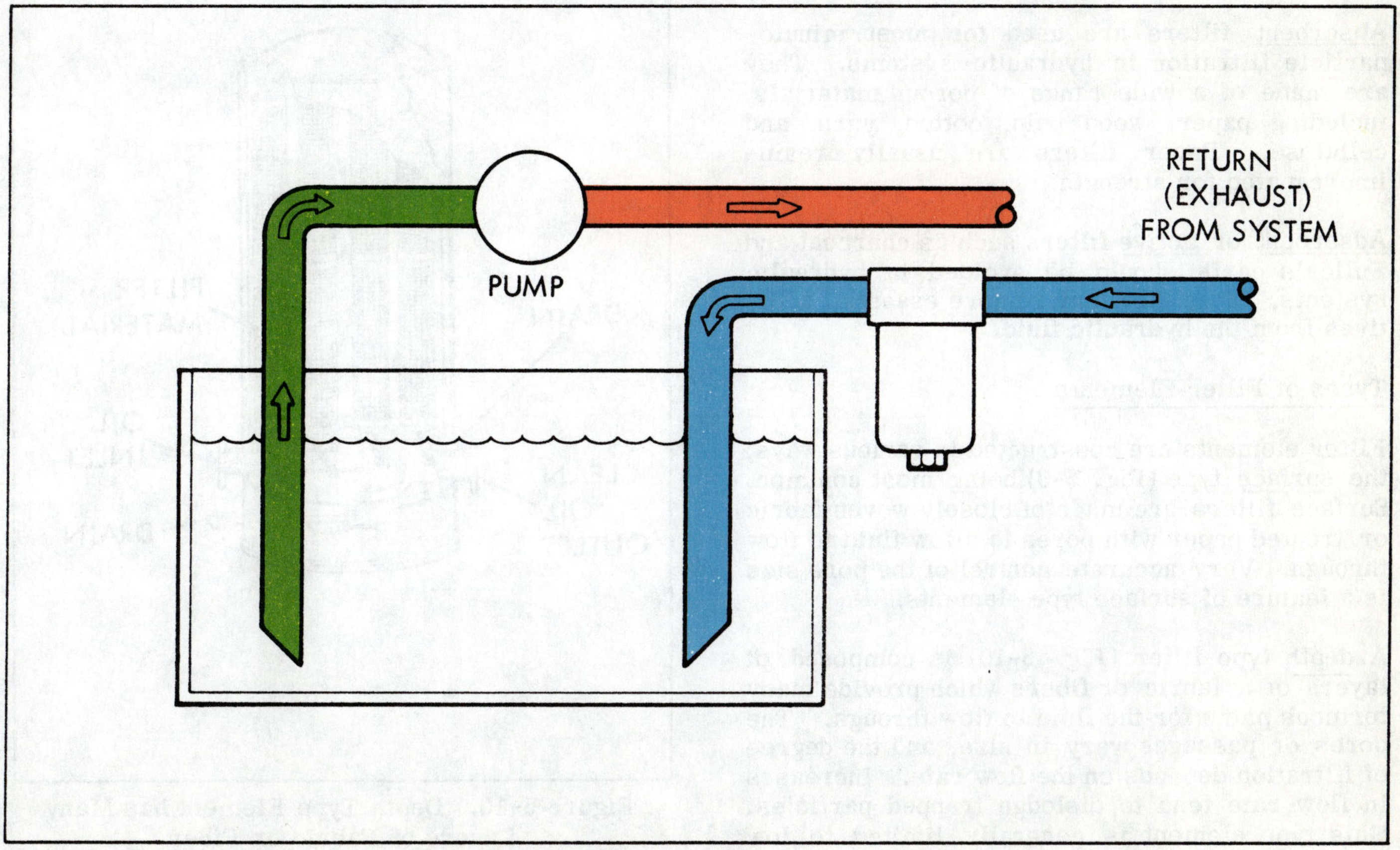

Figure 5-7. Return Line Filter Keeps Contamination From Reservoir

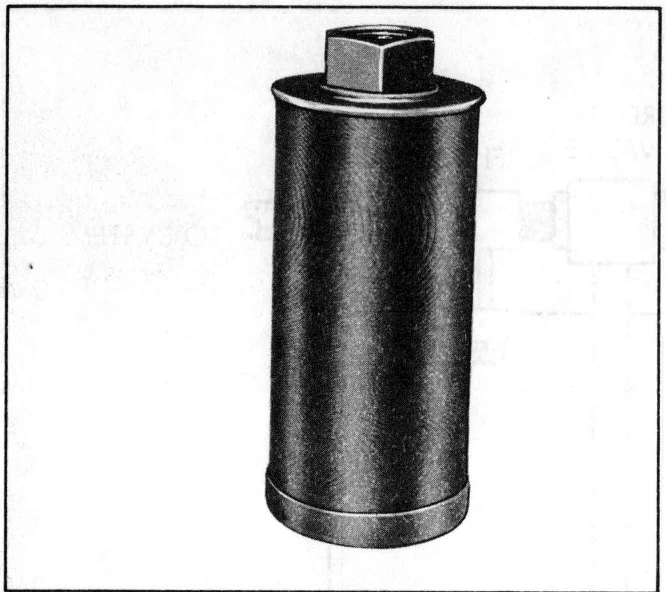

Figure 5-8. Inlet Strainer is Made of Fine Mesh Wire

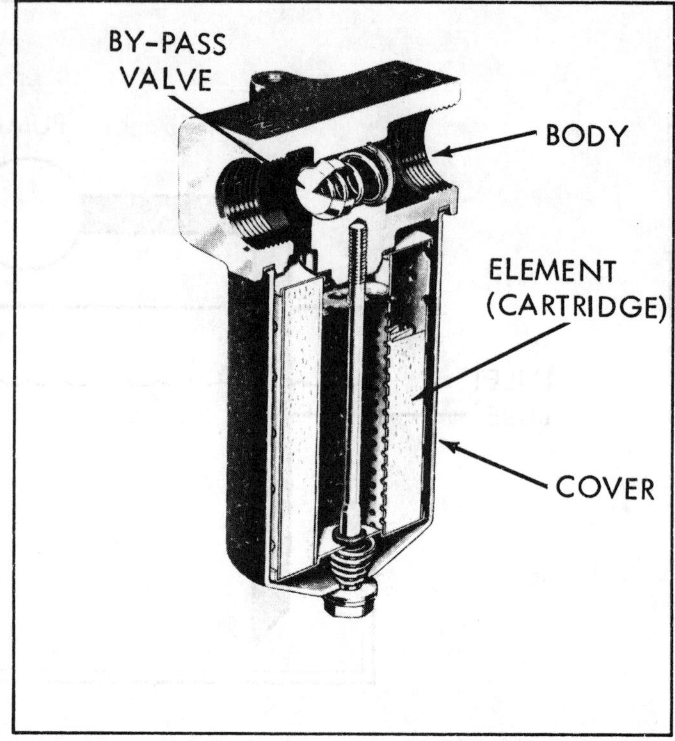

Figure 5-9. OFM Filter uses a Surface Type Element

Filtering Materials

Filtering materials are classified as mechanical, absorbent or adsorbent.

Mechanical filters operate by trapping particles between closely woven metal screens or discs. Most mechanical filters are relatively coarse.

Absorbent filters are used for most minute-particle filtration in hydraulic systems. They are made of a wide range of porous materials, including paper, wood pulp, cotton, yarn, and cellulose. Paper filters are usually resin-impregnated for strength.

Adsorbent or active filters such as charcoal and Fuller's earth should be avoided in hydraulic systems, since they may remove essential additives from the hydraulic fluid.

Types of Filter Elements

Filter elements are constructed in various ways; the surface type (Fig. 5-9) being most common. Surface filters are made of closely woven fabric or treated paper with pores to allow fluid to flow through. Very accurate control of the pore size is a feature of surface type elements.

A depth type filter (Fig. 5-10) is composed of layers of a fabric or fibers which provide many tortuous paths for the fluid to flow through. The pores or passages vary in size, and the degree of filtration depends on the flow rate. Increases in flow rate tend to dislodge trapped particles. This type element is generally limited to low flow, low pressure-drop conditions.

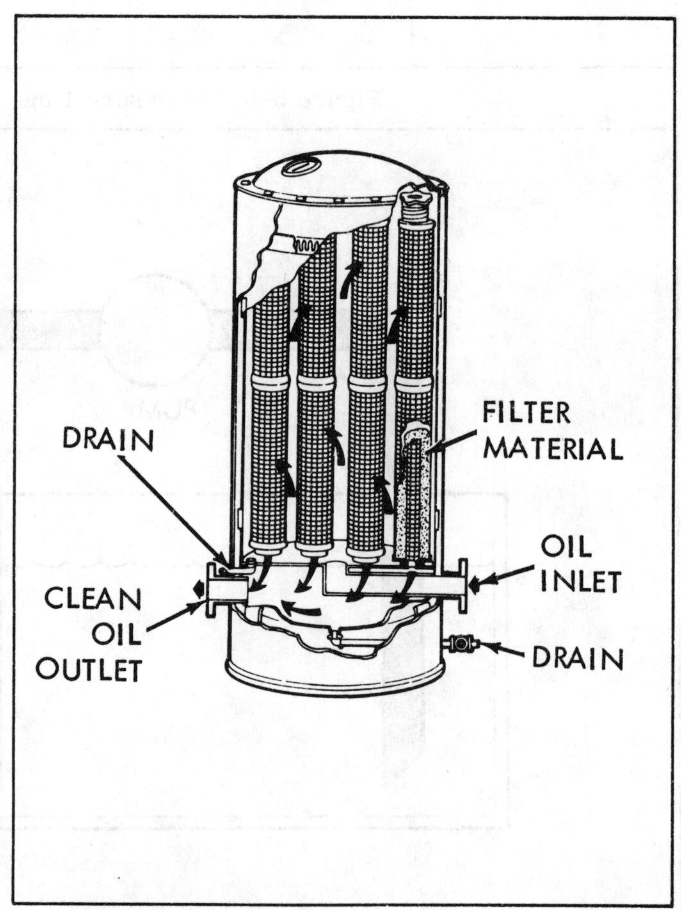

Figure 5-10. Depth Type Element has Many Layers of Fabric or Fiber

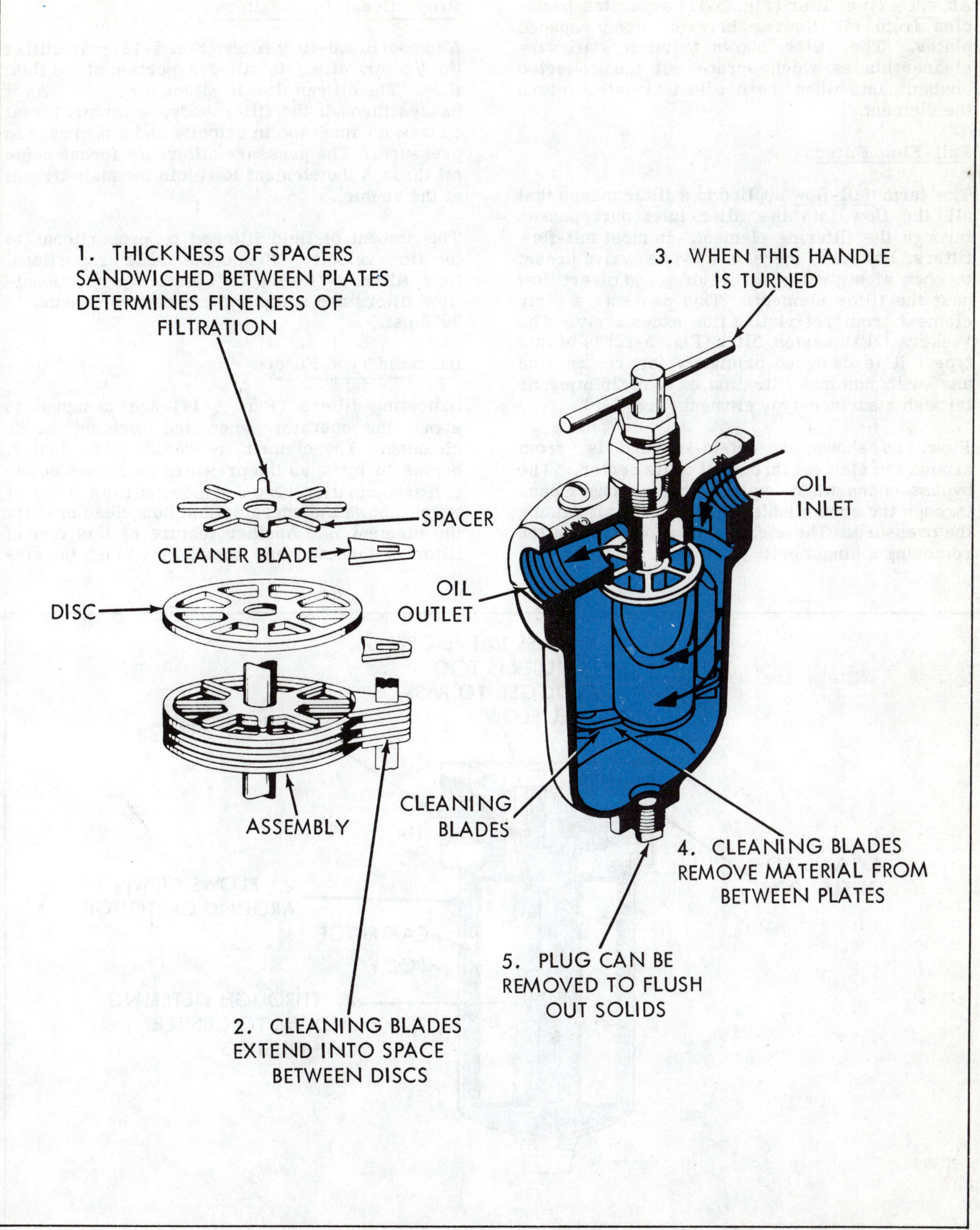

Figure 5-11. Edge-Type Filter Traps Particles Between Finely Spaced Plates

An edge type filter (Fig. 5-11) separates particles from oil flowing between finely spaced plates. The filter shown features stationary cleaner blades which scrape out the collected contaminants when the handle is twisted to turn the element.

Full-Flow Filters

The term full-flow applied to a filter means that all the flow into the filter inlet port passes through the filtering element. In most full-flow filters, however, there is a bypass valve preset to open at a given pressure drop and divert flow past the filter element. This prevents a dirty element from restricting flow excessively. The Vickers OFM series filter (Fig. 5-12) is of this type. It is designed primarily for return line use with nominal filtration to 10 or 25 microns through a surface-type element (Fig. 5-9).

Flow, as shown, is out-to-in; that is, from around the element through it to its center. The bypass opens when total flow can no longer pass through the contaminated element without raising the pressure. The element is replaceable after removing a single bolt.

Proportional-Flow Filters

A proportional-flow filter (Fig. 5-13) may utilize the Venturi effect to filter a portion of the fluid flow. The oil can flow in either direction. As it passes through the filter body, a venturi throat causes an increase in velocity and a decrease in pressure. The pressure difference forces some oil through the element to rejoin the main stream at the venturi.

The amount of fluid filtered is proportional to the flow velocity. Hence the name proportional flow filter. Vickers OF1 series proportional-flow filters are suitable for pressure-line use to 3000 psi.

Indicator Type Filters

Indicating filters (Fig. 5-14) are designed to signal the operator when the element needs cleaning. The element is designed so that it begins to move as the pressure increases due to dirt accumulation. One end is linked to an indicator which shows the operator just how clean or dirty the element is. Another feature of this type of filter is the ease and speed with which the ele-

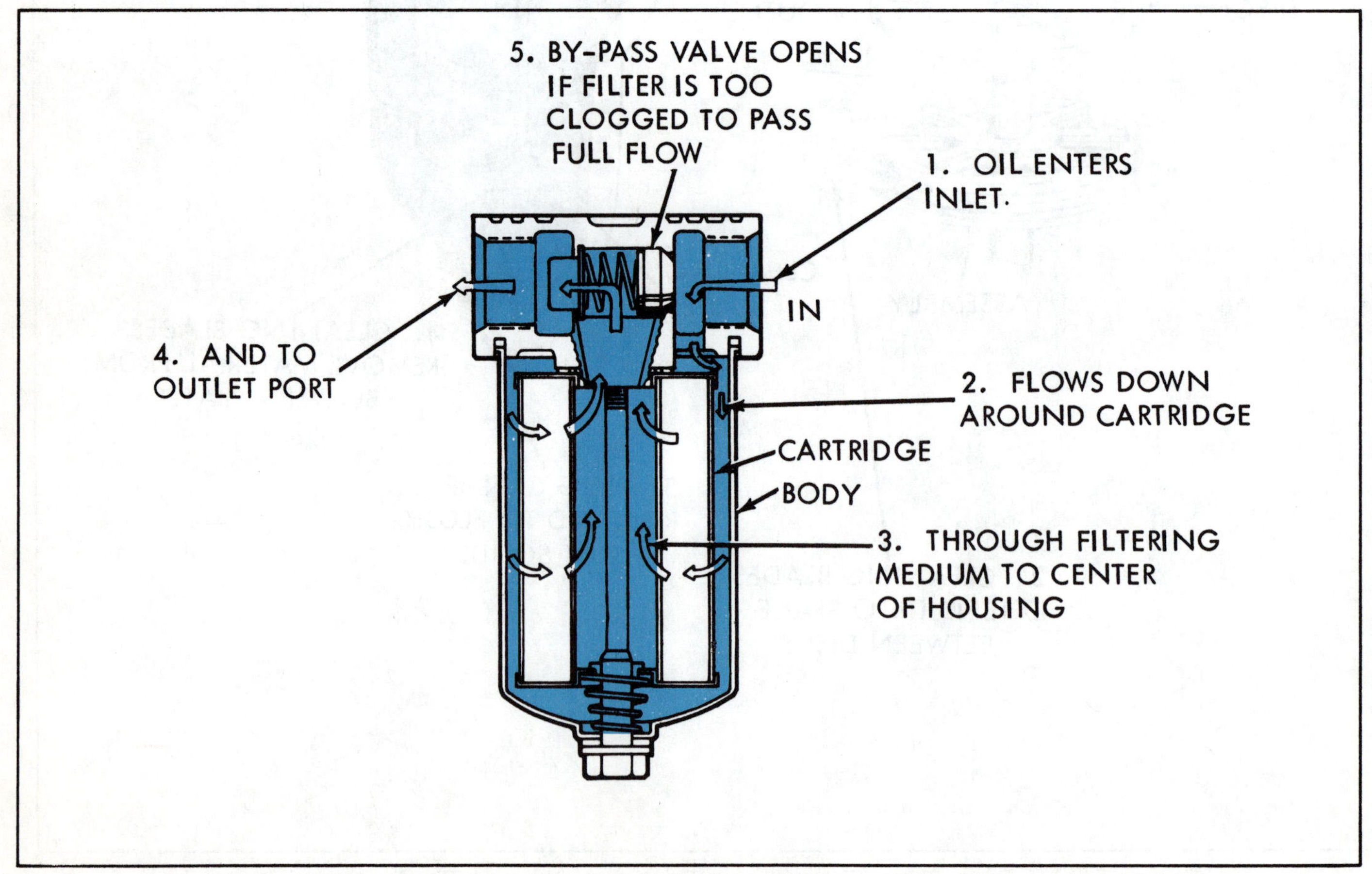

Figure 5-12. OFM Filter Handles Full Flow

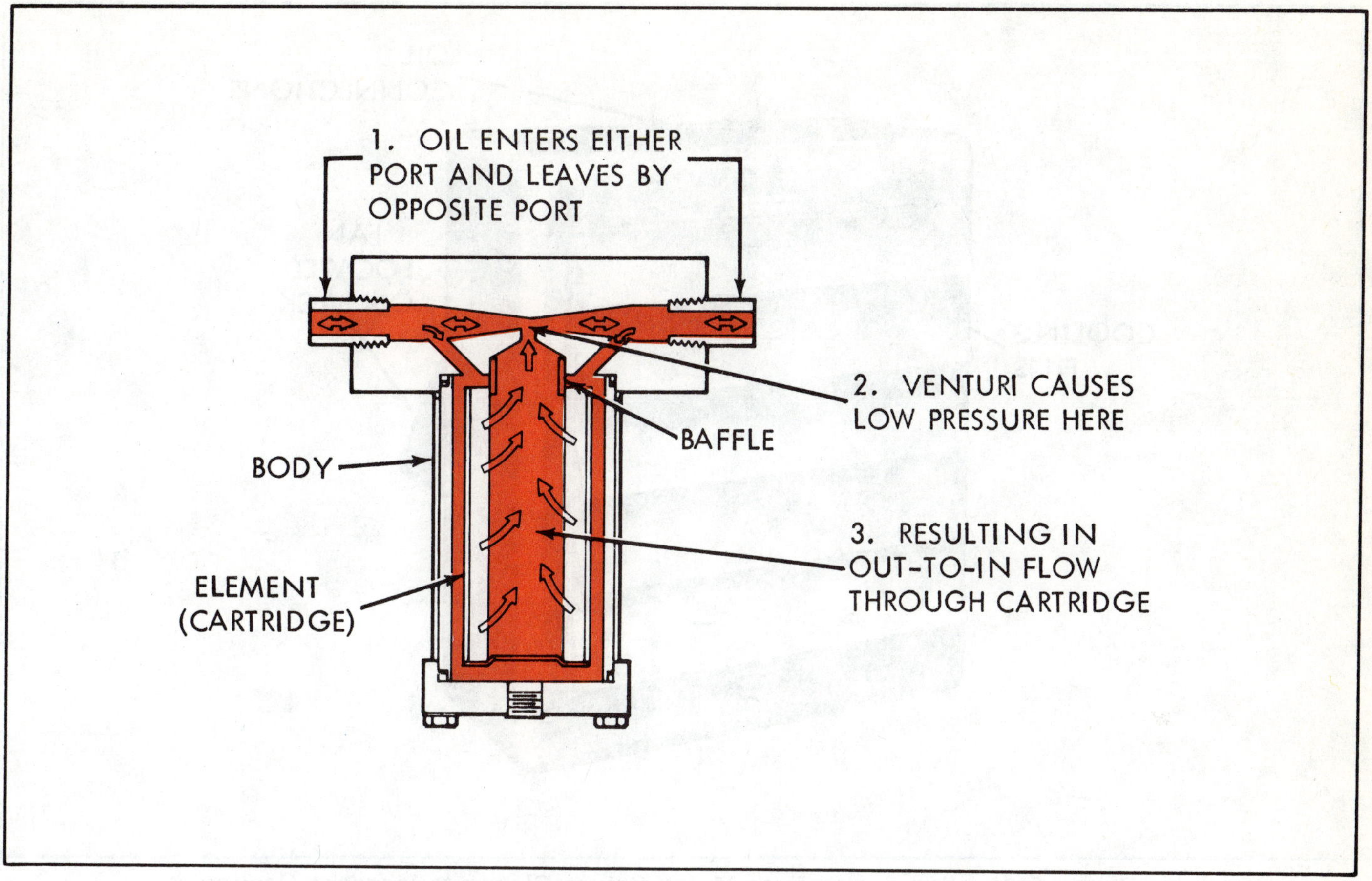

Figure 5-13. Proportional Filter Operates on Venturi Principle

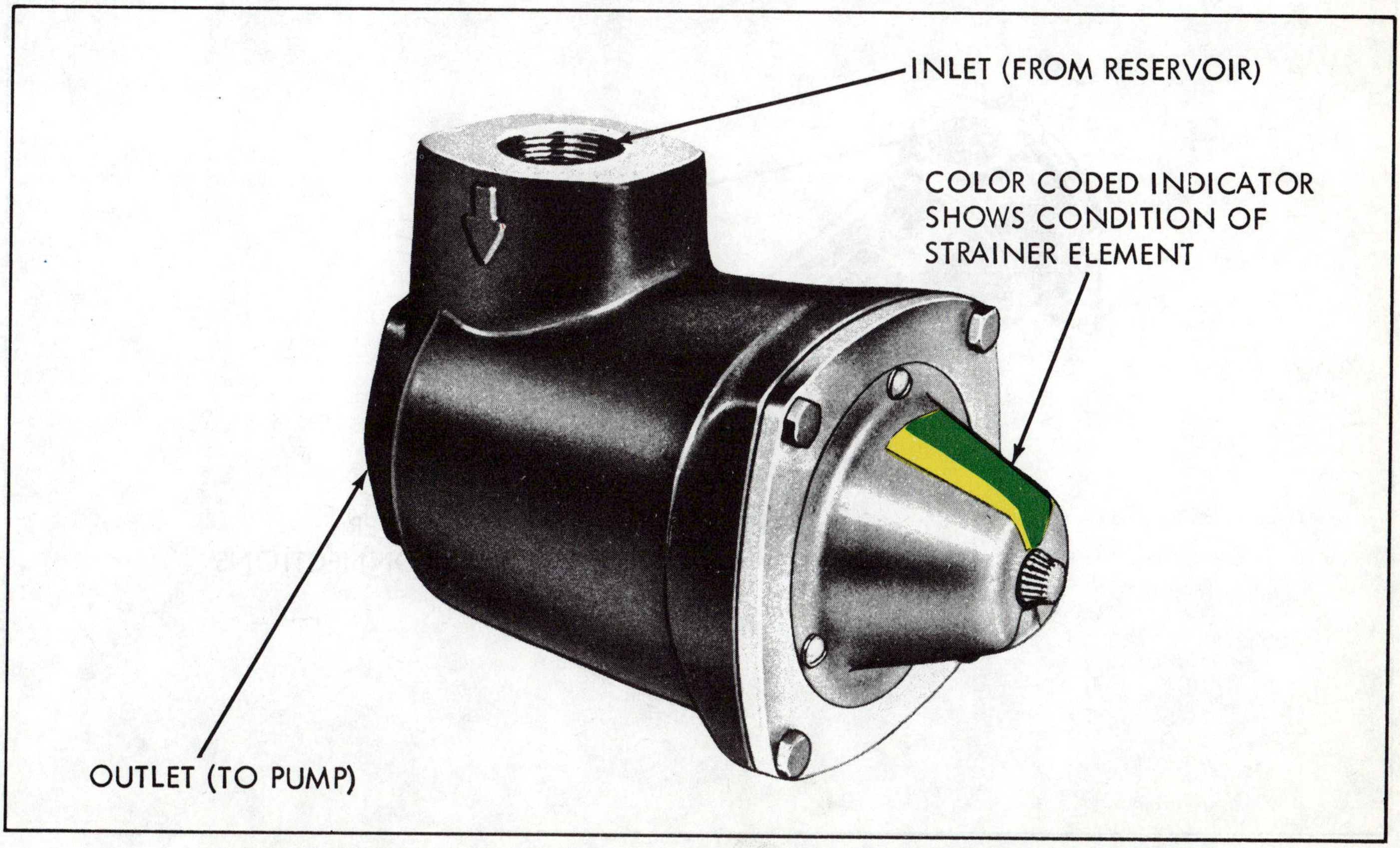

Figure 5-14. Indicating Filter Signals Operator when Cleaning is Required

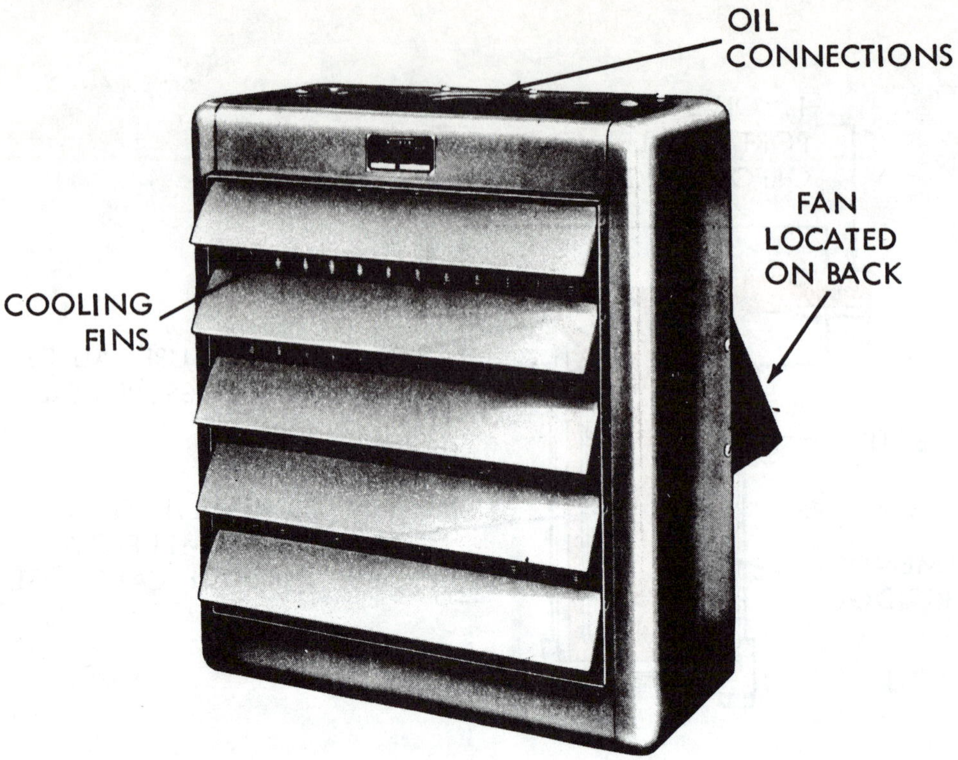

OIL
CONNECTIONS

FAN
LOCATED
ON BACK

COOLING
FINS

Figure 5-15. Air Cooler Uses Motor Driven Blower to Increase Cooling

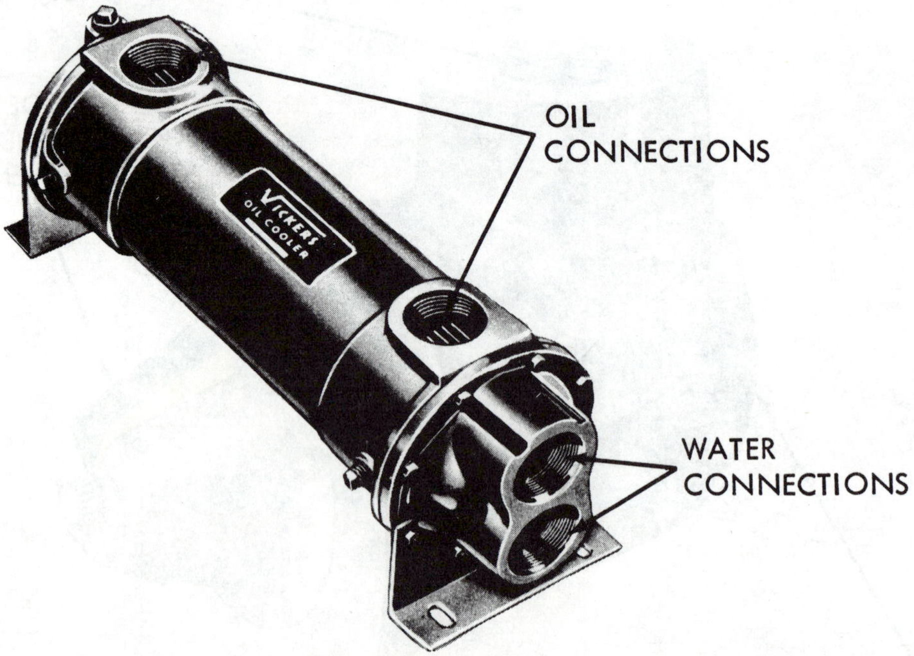

OIL
CONNECTIONS

WATER
CONNECTIONS

Figure 5-16. Shell-and-Tube Heat Exchanger Uses Water to Cool or Warm Oil

ment can be removed and replaced. Most filters of this kind are designed for inlet line installation.

HEAT EXCHANGERS

Since no system is 100 percent efficient or ever can be, heat is a common problem. For this reason, we customarily think of cooling when the fluid must be temperature conditioned. In fact, we will call the two heat exchangers illustrated here coolers. They are designed principally to cool the fluid.

However, there are some applications where the fluid must be heated. For example, some fluids with low viscosity index will not flow readily when cold and must be warmed and kept warm by heaters.

Air Coolers

An air cooler (Fig. 5-15) is used where water for cooling is not readily available. The fluid is pumped through tubes bonded to fins. The fins are aluminum or some other metal which transfers heat easily from the tube to the outside air. The cooler may incorporate a blower to increase the heat transfer.

Water Coolers

In a typical water cooler (Fig. 5-16), hydraulic fluid is circulated through the unit and around the tubes containing the water. The water carries away heat from the hydraulic fluid and can be regulated thermostatically to maintain a desired temperature. The unit may be used as a heater by circulating hot rather than cold water through it.

QUESTIONS

1. Name three functions of the reservoir.
2. Where should the reservoir drain plug be located?
3. What is the most desirable method of checking fluid level in the reservoir?
4. What is the purpose of the reservoir breather?
5. What does a reservoir baffle plate accomplish?
6. Why is a return line often cut at a 45-degree angle?
7. What would probably be an adequate size reservoir for a system with a 5 gpm pump ?
8. What is a filter ? A strainer ?
9. What is the micron size of a 170 sieve screen?
10. How large is a micron?
11. What is meant by absolute micron rating?
12. Name three possible locations for a filter.
13. What type of filter element provides precise control of pore size?
14. What does full-flow mean?
15. What is the purpose of an indicator type filter?

HYDRAULIC ACTUATORS

In this chapter we will consider the output member or actuator, where design of the system actually begins. The type of job done and the power requirements determine what type and size motor or cylinder will be used. Only after the actuator is chosen and sized can the remaining circuit components be selected to complete the system.

CYLINDERS

Cylinders are linear actuators. By linear, we mean simply that the output of a cylinder is straight-line motion and/or force.

TYPES OF CYLINDERS

Cylinders are classified as single- or double-acting and as differential or non-differential. Variations include ram or piston and rod design; and solid or telescoping rods. The differences are illustrated in Figures 6-1 through 6-6, with the graphical symbol for each type.

Ram Type Cylinder (Fig. 6-1). Perhaps the simplest actuator is the ram type. It has only one fluid chamber and exerts force in only one direction. Most are mounted vertically and retract by the force of gravity on the load. Practical for long strokes, ram type cylinders are used in elevators, jacks and automobile hoists.

Telescoping Cylinder (Fig. 6-2). A telescoping cylinder is used where the collapsed length must be shorter than could be obtained with a standard cylinder. Up to 4 and 5 sleeves can be used;

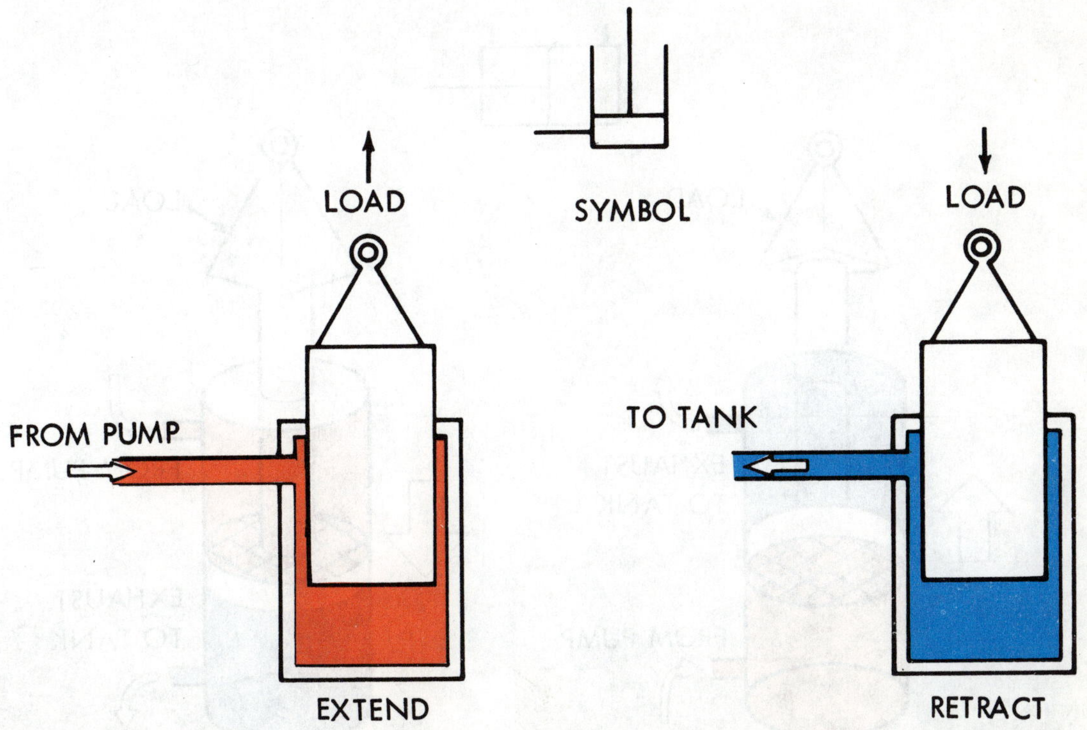

LOAD
SYMBOL
LOAD
FROM PUMP
TO TANK
EXTEND
RETRACT

Fig. 6-1. Ram Type Cylinder is Single Acting

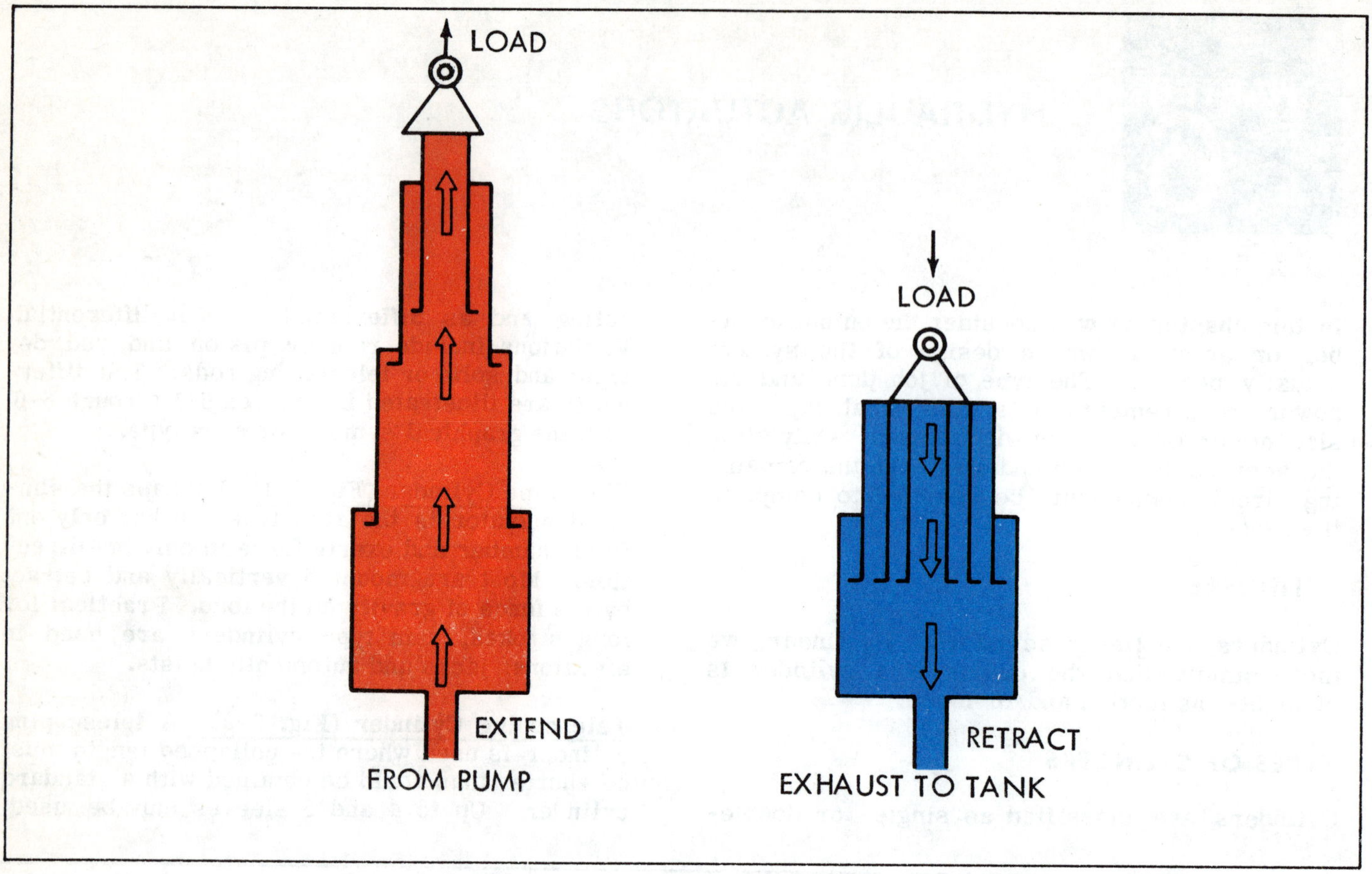

Fig. 6-2. Telescoping Rod Increases Stroke Length

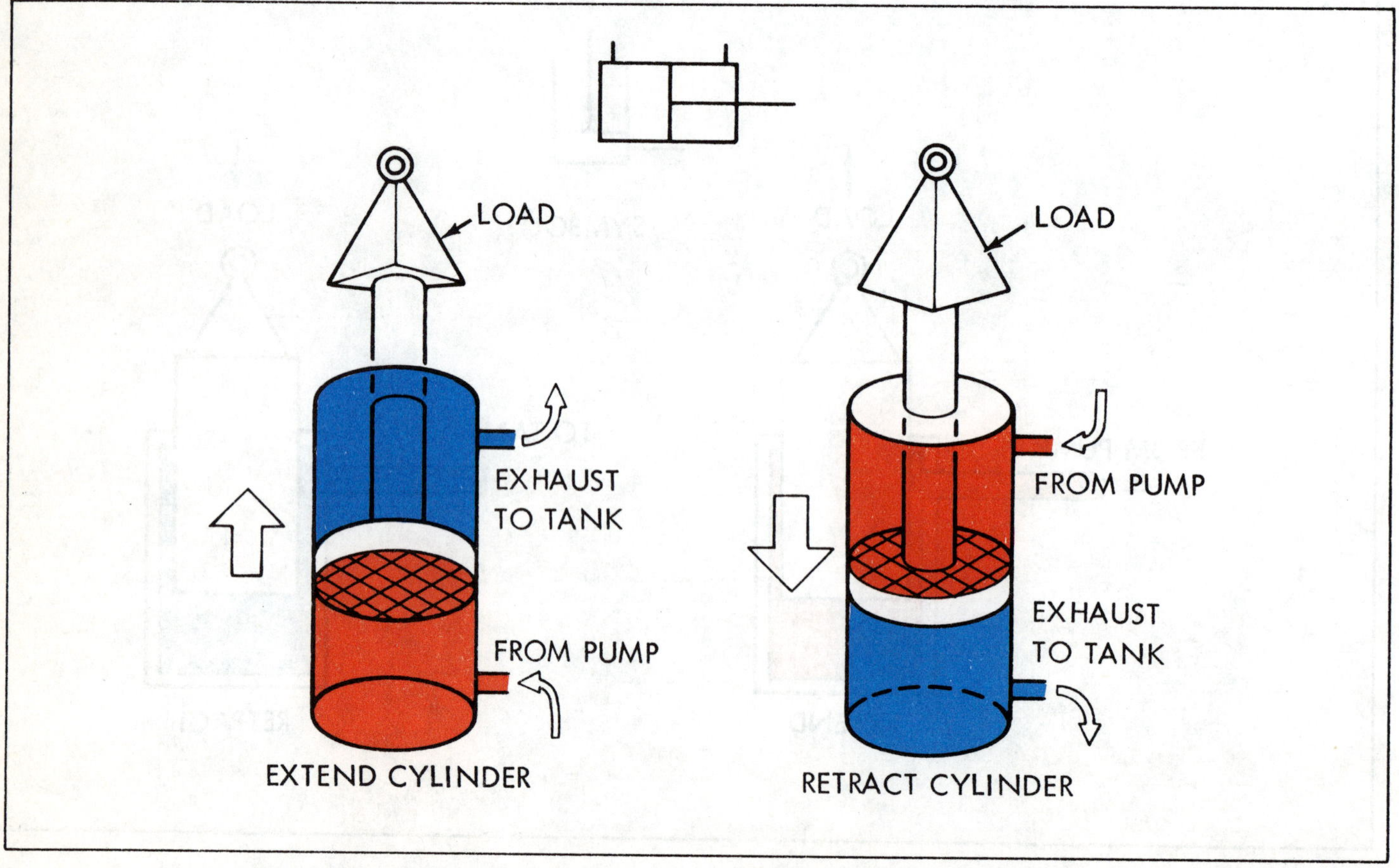

Fig. 6-3. Standard Double-Acting Cylinder has Two Power Strokes

while most are single-acting, double-acting units are available.

Standard Double-Acting (Fig. 6-3). The double-acting cylinder is so named because it is operated by hydraulic fluid in both directions. This means it is capable of a power stroke either way. The standard double-acting cylinder is classed as a differential cylinder because there are unequal areas exposed to pressure during the forward and return movements. The difference being a function of the cross-sectional area of the rod. The extending stroke is slower, but capable of exerting a greater force than can be obtained when the piston and rod are being retracted.

Double-Rod Cylinder (Fig. 6-4). Double-rod cylinders are used where it is advantageous to couple a load to each end, or where equal displacement is needed on each end. They too are double-acting cylinders but are classified as non-differential. With identical areas on either side of the piston, they can provide equal speeds and/or equal forces in either direction. Any double-acting cylinder may be used as a single-acting unit by draining the inactive end to tank.

CYLINDER CONSTRUCTION

The essential parts of a cylinder (Fig. 6-5) are a barrel; a piston and rod; end caps and suitable seals. Barrels usually are seamless steel tubing, honed to a fine finish on the inside. The piston, usually cast iron or steel, incorporates seals to reduce leakage between it and the cylinder barrel. Step cut automotive type piston rings are used where some leakage can be tolerated. For supporting loads or very low feed rates, a T-ring or "O" ring with 2 heavy duty back-up rings is often used. The ports of the cylinder are in the end caps, which may be attached directly to each end of the barrel, or secured by tie bolts. The rod packing is a cartridge type including both the seal and wiper for easy replacement.

CYLINDER MOUNTINGS

Various cylinder mountings (Fig. 6-6) provide flexibility in anchoring the cylinder. Rod ends are usually threaded for attachment directly to the load or to accept a clevis, yoke or similar coupling device.

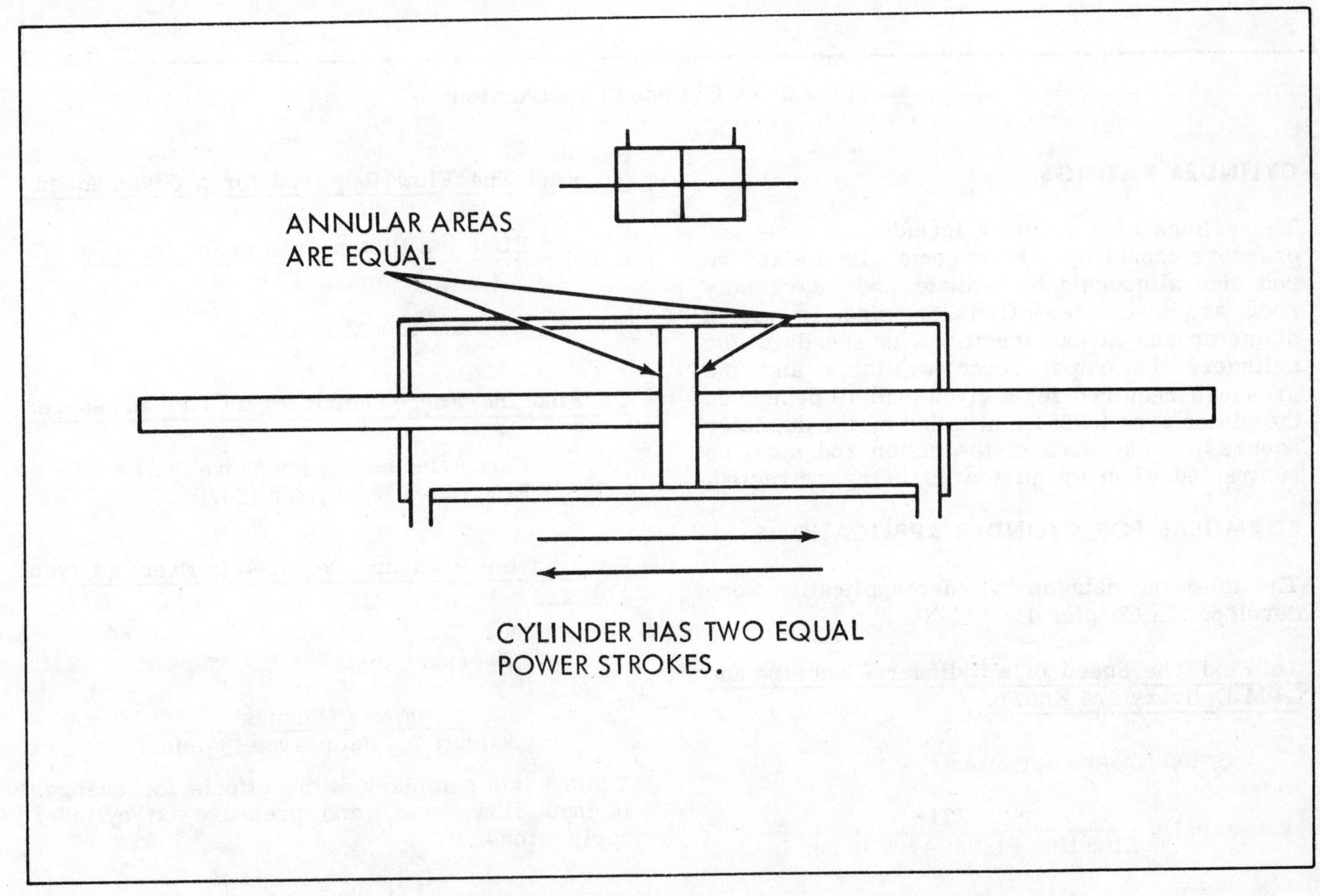

Fig. 6-4. Double-Rod Cylinder is Double-Acting but Non-Differential

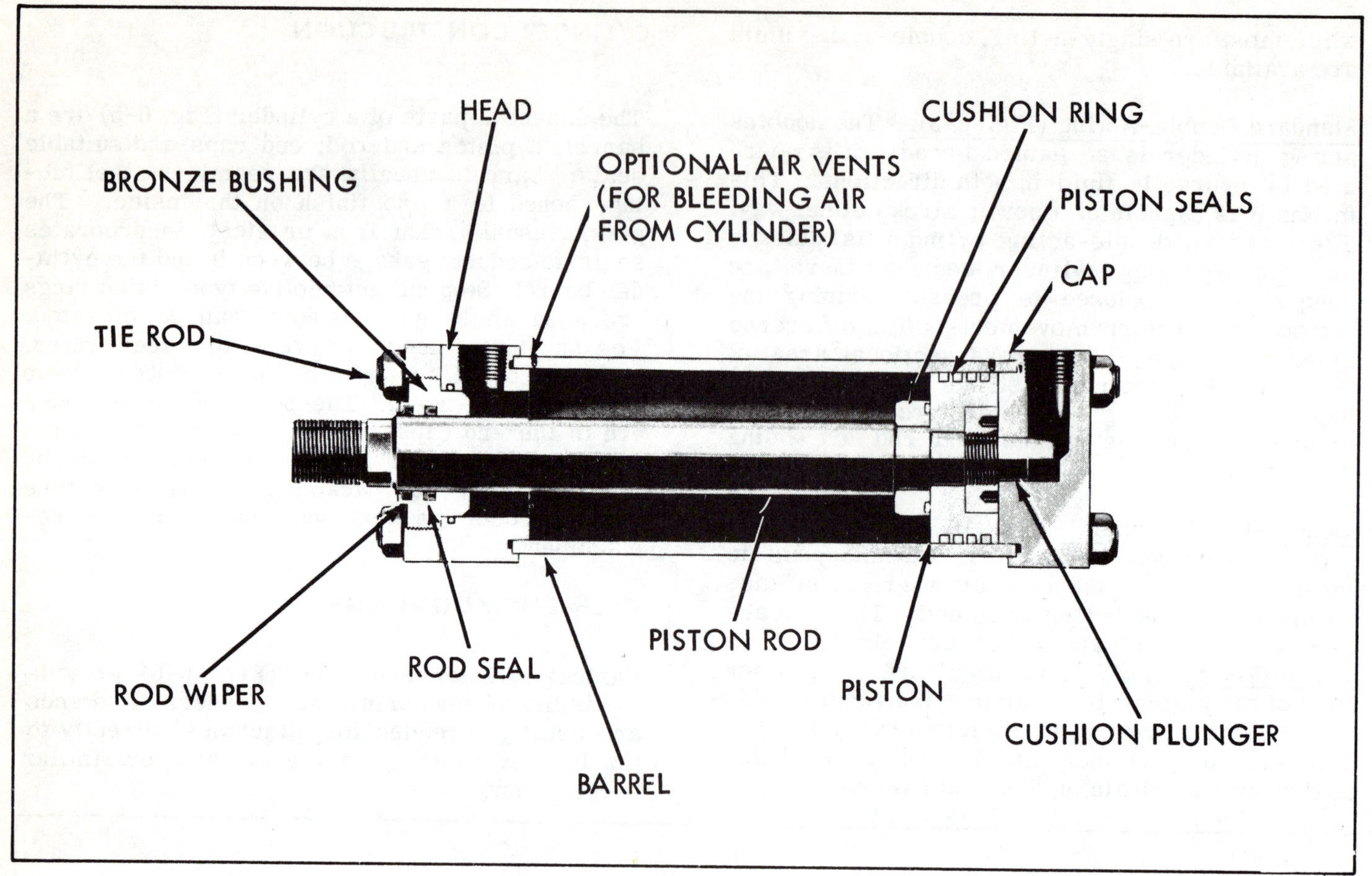

Fig. 6-5. Cylinder Construction

CYLINDER RATINGS

The ratings of a cylinder include its size and pressure capability. Most come with a standard rod size although intermediate and heavy duty rods are available. Cylinder size is piston diameter and stroke length. The speed of the cylinder, the output force available and the pressure required for a given load all depend on the piston area (.7854 multiplied by the diameter squared). The area of the piston rod must be subtracted when the piston is being retracted.

FORMULAS FOR CYLINDER APPLICATIONS

The following data on cylinder application were developed in Chapter 1:

To Find The Speed of a Cylinder When Size and GPM Delivery are Known:

$$\text{Speed (Inches per Minute)} = \text{GPM} \times \frac{231}{\text{Effective Piston Area in Sq. In.}}$$

To Find The Flow Required for a Given Speed:

$$\text{GPM} = \frac{\text{Effective Piston Area in Sq. In.} \times \text{Speed*}}{231}$$

* Inches Per Minute

To Find the Force Output for a Given Pressure:

Force (Pounds) = Pressure (psi) x Effective Piston Area (Sq. In.)

To Find the Pressure Required to Exert a Given Force:

$$\text{Pressure (psi)} = \frac{\text{Force (Pounds)}}{\text{Effective Piston Area (Sq. In.)}}$$

Table 1 is a summary of the effects for changes in input flow, size, and pressure on cylinder applications.

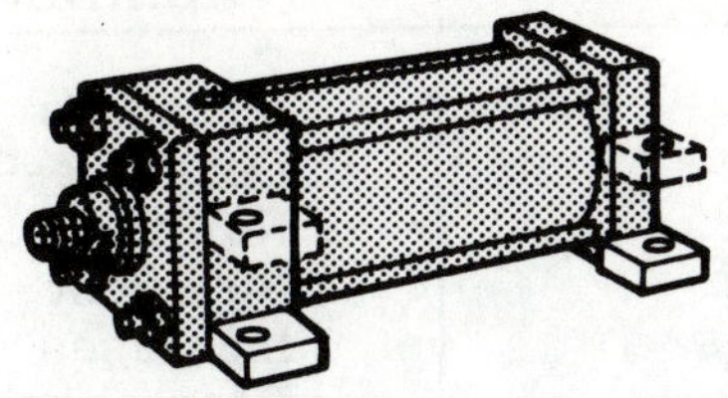

FOOT AND CENTERLINE LUG MOUNTS

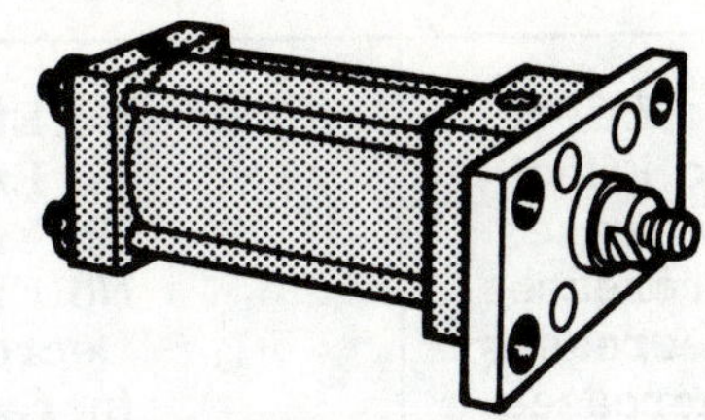

RECTANGULAR FLANGE MOUNT

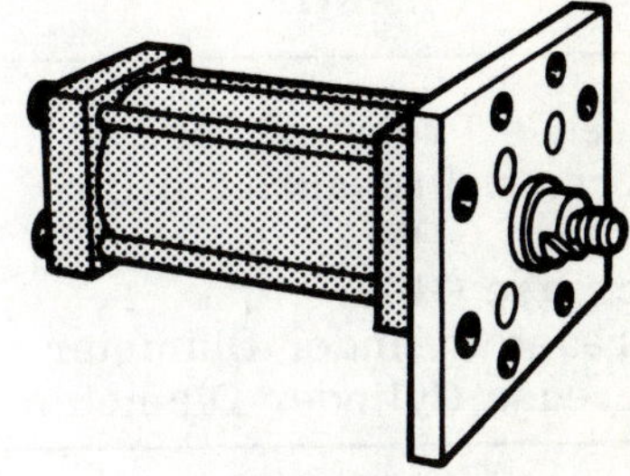

SQUARE FLANGE MOUNT

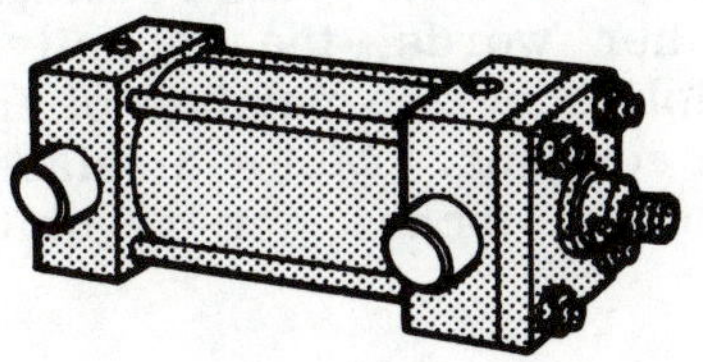

TRUNNION MOUNT

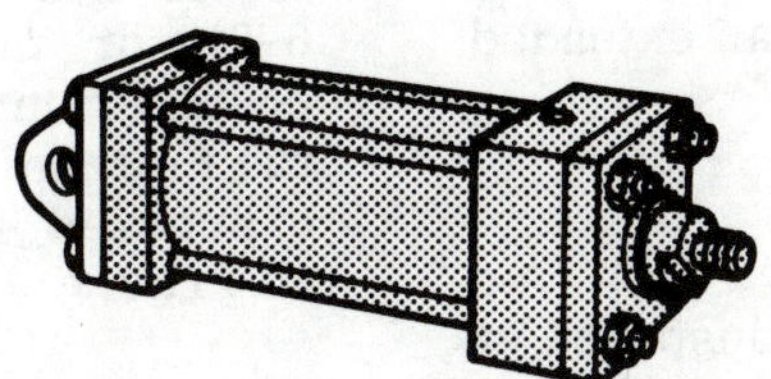

CLEVIS MOUNT

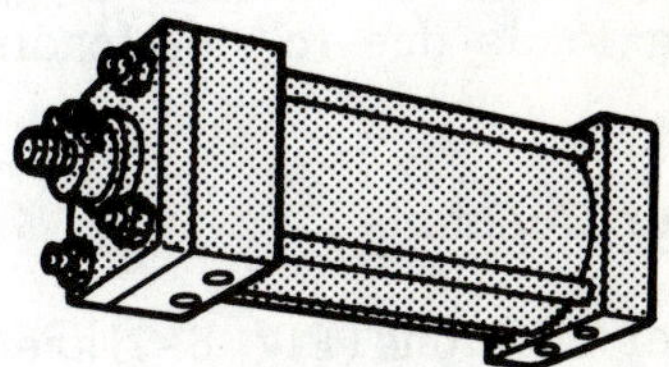

FLUSH SIDE MOUNT

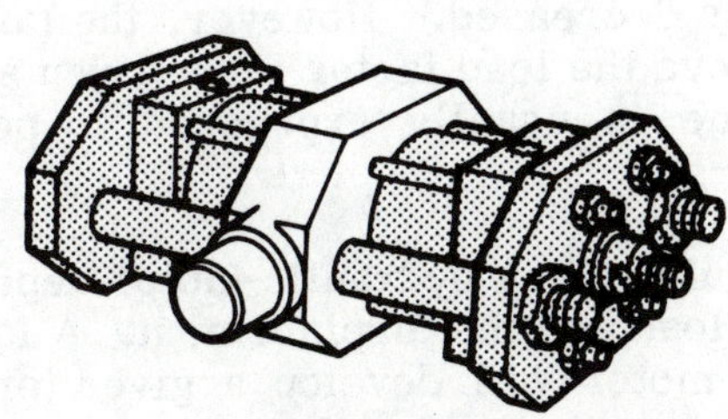

INTERMEDIATE TRUNNION MOUNT

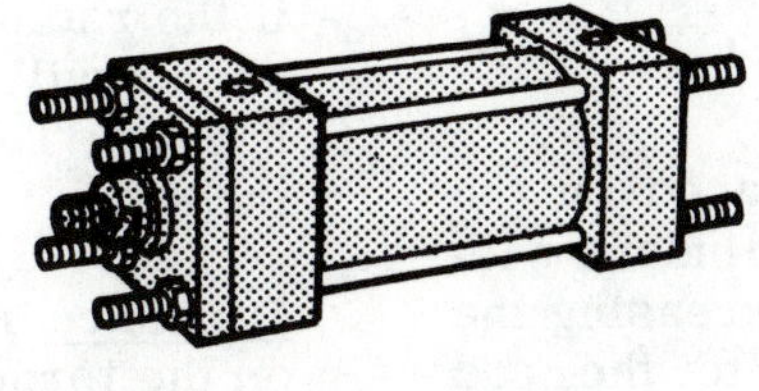

EXTENDED TIE ROD

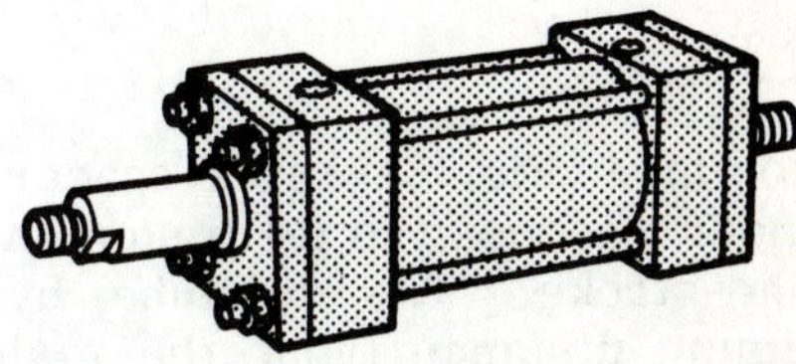

DOUBLE ROD END

Fig. 6-6. Cylinder Mountings

TABLE 1

CHANGE	SPEED	EFFECT ON OPERATING PRESSURE	OUTPUT FORCE AVAILABLE
Increase Pressure Setting	No Effect	No Effect	Increases
Decrease Pressure Setting	No Effect	No Effect	Decreases
Increase GPM	Increases	No Effect	No Effect
Decrease GPM	Decreases	No Effect	No Effect
Increase Cylinder Diameter	Decreases	Decreases	Increases
Decrease Cylinder Diameter	Increases	Increases	Decreases

Above table assumes a constant work load.

Table 2 lists piston areas, output forces and speeds for cylinders of various sizes.

CYLINDER OPTIONS

Optional equipment available includes piston ring seals for rapid-cycling operations, cylinder cushions to decelerate the load near the end of the stroke and stop tubes to prevent excessive bearing loads due to side loading on an extended rod.

Cylinder Cushions

Cylinder cushions (Fig. 6-7) are often installed at either or both ends of a cylinder to slow it down near the end of the stroke and prevent the piston from hammering against the end cap.

Deceleration begins when the tapered cushion ring or plunger enters the cap and begins to restrict exhaust flow from the barrel to the port. During the final fraction of the stroke, the exhaust oil must discharge through an adjustable orifice. The cushion feature also includes a check valve to bypass the orifice on the return stroke.

Stop Tubes

A stop tube (Fig. 6-8) is a spacer placed on the cylinder rod next to the piston on cylinders with a long stroke. The stop tube, by increasing the minimum distance from the piston to the rod bushing, provides more support for side loading on the rod, thus minimizing chances of rod bearing failure.

HYDRAULIC MOTORS

Motor is the name usually given to a rotary hydraulic actuator. Motors very closely resemble pumps in construction. Instead of pushing on the fluid as the pump does, as output members in the hydraulic system, they are pushed by the fluid and develop torque and continuous rotating motion. Since both inlet and outlet ports may at times be pressurized, most hydraulic motors are externally drained.

MOTOR RATINGS

Hydraulic motors are rated according to displacement (size), torque capacity and maximum pressure limitations.

Displacement is the amount of fluid which the motor will accept in turning one revolution (Fig. 6-9); or in other words, the capacity of one chamber multiplied by the number of chambers the mechanism contains. Motor displacement is expressed in cubic inches per revolution (cu. in./rev.)

Torque is the force component of the motor's output. It is defined as a turning or twisting effort. Motion is not required to have torque, but motion will result if the torque is sufficient to overcome friction and resistance of the load.

Figure 6-10 illustrates typical torque requirements for raising a load with a pulley. Note that the torque is always present at the driveshaft, but is equal to the load multiplied by the radius. A given load will impose less torque on the shaft if the radius is decreased. However, the larger radius will move the load faster for a given shaft speed. Torque is usually expressed in pound inches.

Pressure required in a hydraulic motor depends on the torque load and the displacement. A large displacement motor will develop a given torque with less pressure than a smaller unit. The size or torque rating of a motor usually is expressed in pound inches of torque per 100 psi of pressure.

$$\frac{\text{lb. in.}}{\text{(100 psi)}}$$

Formulas for Motor Applications

Following are the formulas for applying hydraulic motors and determining flow and pressure requirements.

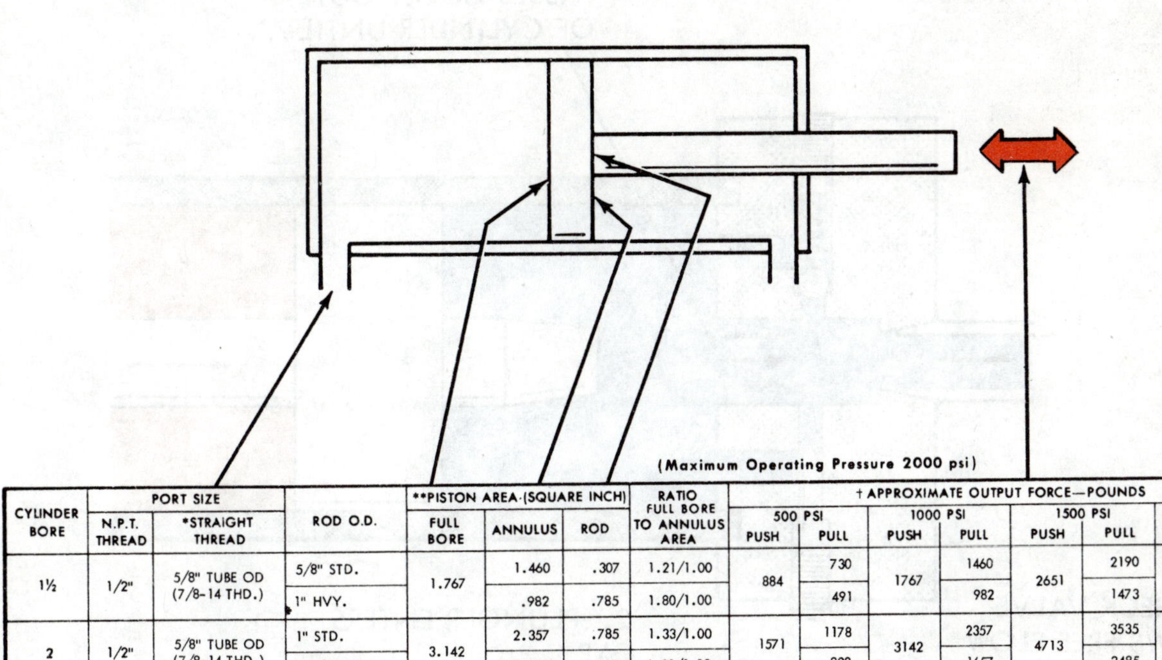

(Maximum Operating Pressure 2000 psi)

CYLINDER BORE	PORT SIZE N.P.T. THREAD	PORT SIZE *STRAIGHT THREAD	ROD O.D.	**PISTON AREA (SQUARE INCH) FULL BORE	ANNULUS	ROD	RATIO FULL BORE TO ANNULUS AREA	† APPROXIMATE OUTPUT FORCE—POUNDS 500 PSI PUSH	500 PSI PULL	1000 PSI PUSH	1000 PSI PULL	1500 PSI PUSH	1500 PSI PULL	2000 PSI PUSH	2000 PSI PULL
1½	1/2"	5/8" TUBE OD (7/8-14 THD.)	5/8" STD.	1.767	1.460	.307	1.21/1.00	884	730	1767	1460	2651	2190	3534	2920
			1" HVY.		.982	.785	1.80/1.00		491		982		1473		1964
2	1/2"	5/8" TUBE OD (7/8-14 THD.)	1" STD.	3.142	2.357	.785	1.33/1.00	1571	1178	3142	2357	4713	3535	6284	4714
			1-3/8" HVY.		1.657	1.485	1.90/1.00		828		1657		2485		3314
2½	1/2"	3/4" TUBE OD (1-1/16-12 THD.)	1" STD.	4.909	4.124	.785	1.19/1.00	2455	2062	4909	4124	7364	6186	9818	8248
			1-3/8" INT'MED.		3.424	1.485	1.43/1.00		1712		3424		5136		6848
			1-3/4" HVY.		2.504	2.405	1.96/1.00		1252		2504		3756		5008
3¼	3/4"	3/4" TUBE OD (1-1/16-12 THD.)	1-3/8" STD.	8.296	6.811	1.485	1.22/1.00	4148	3405	8296	6811	12444	10216	16592	13622
			1-3/4" INT'MED.		5.891	2.405	1.41/1.00		2945		5891		8836		11782
			2" HVY.		5.154	3.142	1.61/1.00		2577		5154		7731		10308
4	3/4"	3/4" TUBE OD (1-1/16-12 THD.)	1-3/4" STD.	12.566	10.161	2.405	1.24/1.00	6283	5080	12566	10161	18849	15241	25132	20322
			2" INT'MED.		9.424	3.142	1.33/1.00		4712		9424		14136		18848
			2-1/2" HVY.		7.666	4.900	1.64/1.00		3833		7666		11500		15332
5	3/4"	1" TUBE OD (1-5/16-12 THD.)	2" STD.	19.635	16.493	3.142	1.19/1.00	9818	8246	19635	16493	29453	24739	39270	32986
			2-1/2" INT'MED.		14.735	4.900	1.33/1.00		7367		14735		22102		29470
			3-1/2" HVY.		10.014	9.621	1.96/1.00		5007		10014		15021		20028
6	1"	1" TUBE OD (1-5/16-12 THD.)	2-1/2" STD.	28.274	23.374	4.900	1.21/1.00	14137	11687	28274	23374	42411	35061	56548	46748
			3-1/2" INT'MED.		18.653	9.621	1.52/1.00		9326		18653		27979		37306
			4" HVY.		15.708	12.566	1.80/1.00		7854		15708		23562		31416
7	1-1/4"	1-1/2" TUBE OD (1-7/8-12 THD.)	3" STD.	38.485	31.416	7.069	1.23/1.00	19242	15708	38485	31416	57728	47124	76970	62832
			4" INT'MED.		25.9,19	12.566	1.48/1.00		12959		25919		38878		51838
			5" HVY.		18.850	19.635	2.04/1.00		9425		18850		28275		37700
8	1-1/2"	1-1/2" TUBE OD (1-7/8-12 THD.)	3-1/2" STD.	50.265	40.644	9.621	1.24/1.00	25133	20332	50265	40644	75398	60966	100530	81288
			4-1/2" INT'MED.		34.361	15.904	1.46/1.00		17180		34361		51541		68722
			5-1/2" HVY.		26.507	23.758	1.90/1.00		13253		26507		39760		53014

*Straight thread connections available upon request.

† "Pull" force values apply in both directions for cylinders with double-ended piston rods.

**Fluid displacement per inch of stroke is the same value (in cubic inches) as piston area (in square inches).

Table 2

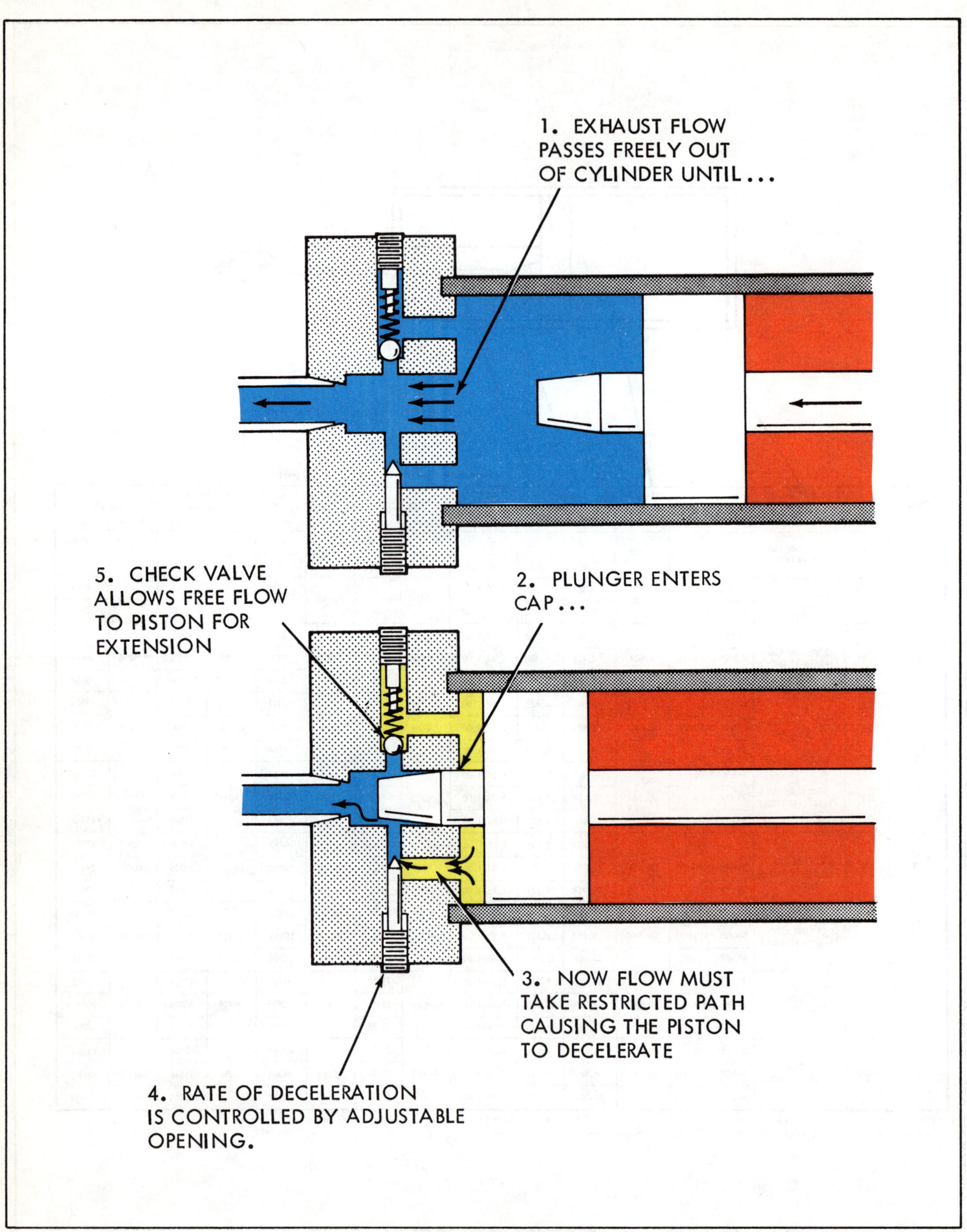

Fig. 6-7. Cylinder Cushions

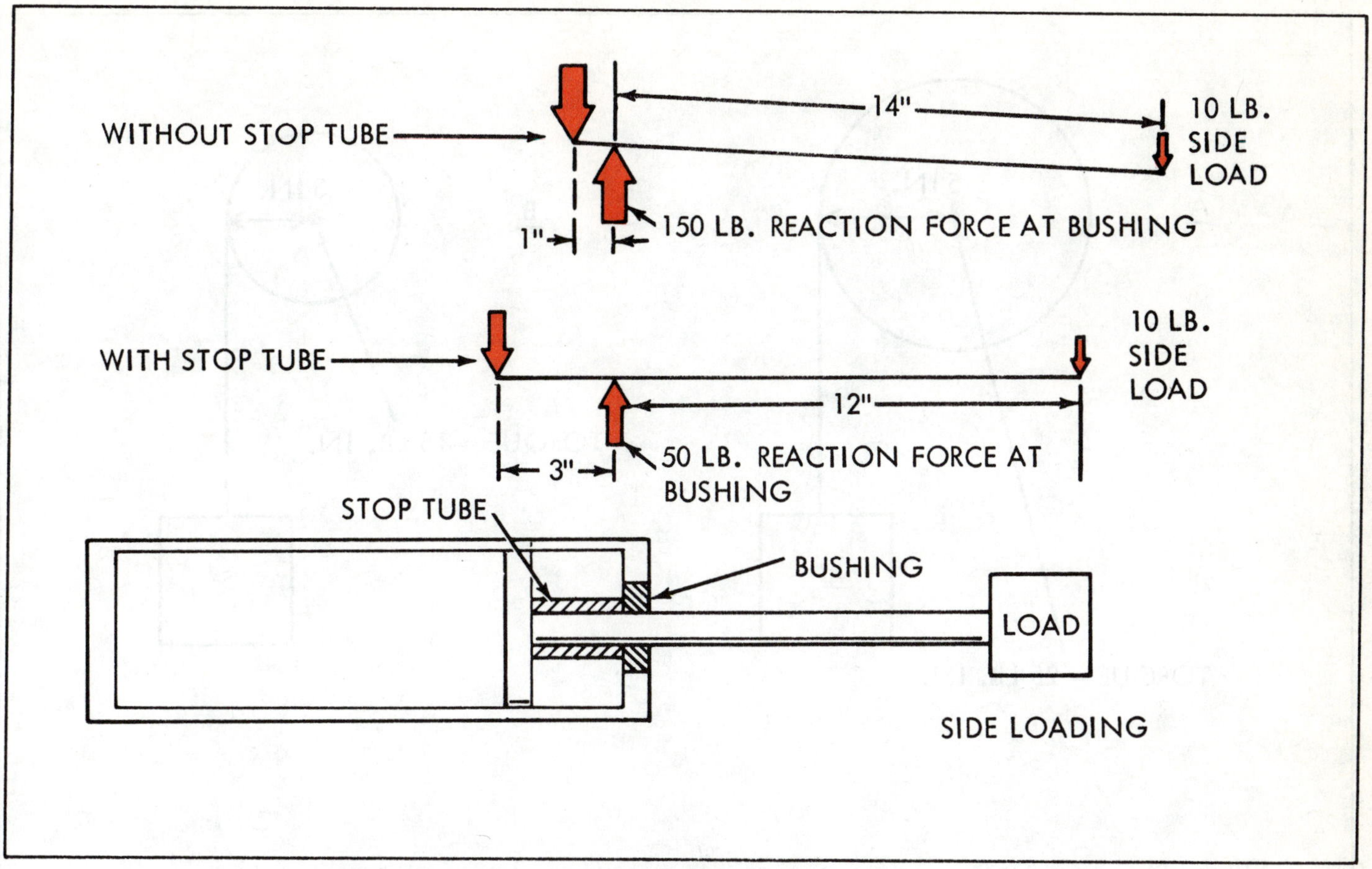

WITHOUT STOP TUBE
14"
10 LB. SIDE LOAD
1"
150 LB. REACTION FORCE AT BUSHING
WITH STOP TUBE
10 LB. SIDE LOAD
12"
3"
50 LB. REACTION FORCE AT BUSHING
STOP TUBE
BUSHING
LOAD
SIDE LOADING

Fig. 6-8. Stop Tube Limits Piston Travel

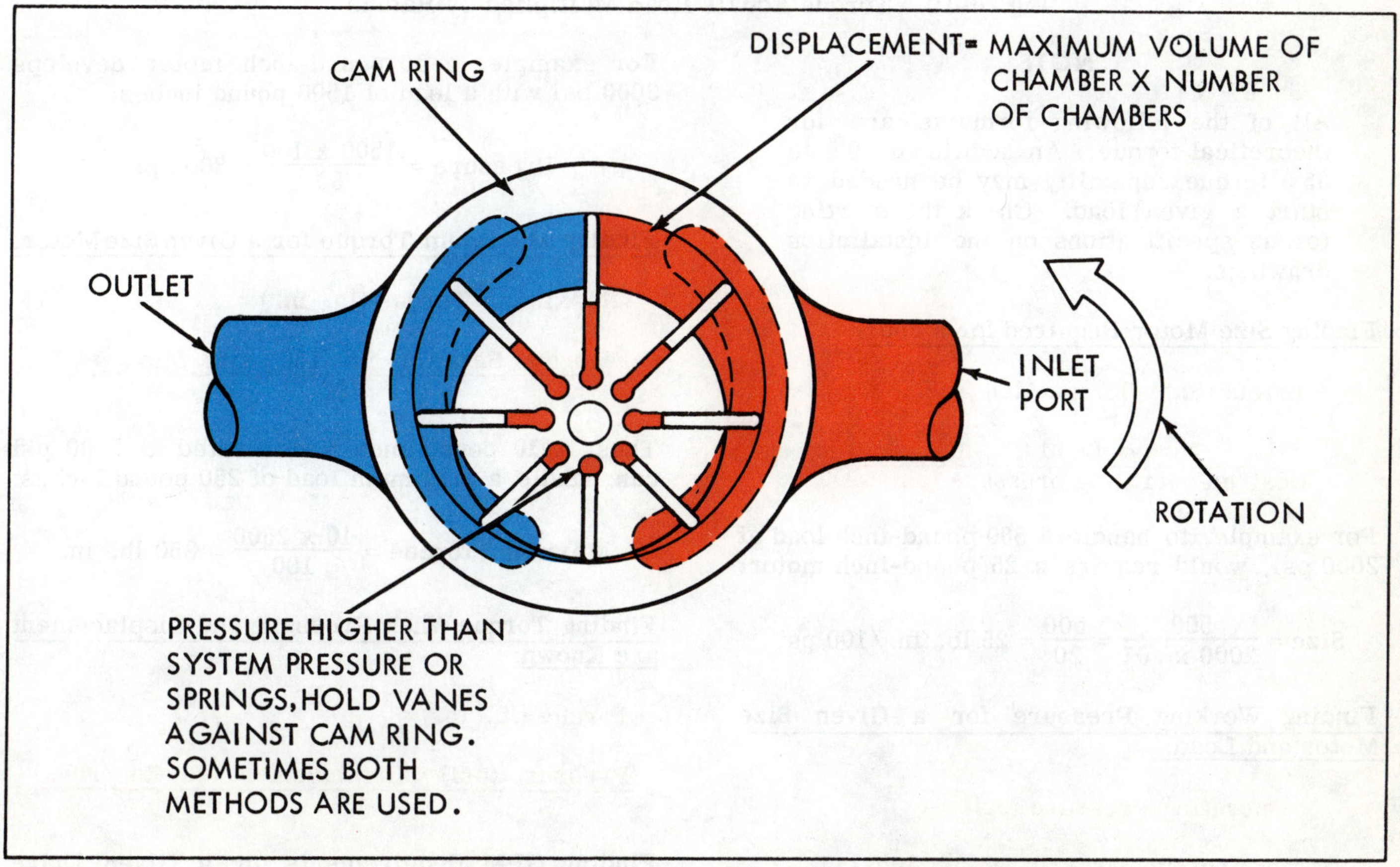

CAM RING
DISPLACEMENT= MAXIMUM VOLUME OF CHAMBER X NUMBER OF CHAMBERS
OUTLET
INLET PORT
ROTATION
PRESSURE HIGHER THAN SYSTEM PRESSURE OR SPRINGS, HOLD VANES AGAINST CAM RING. SOMETIMES BOTH METHODS ARE USED.

Fig. 6-9. Motor Displacement is Capacity Per Revolution

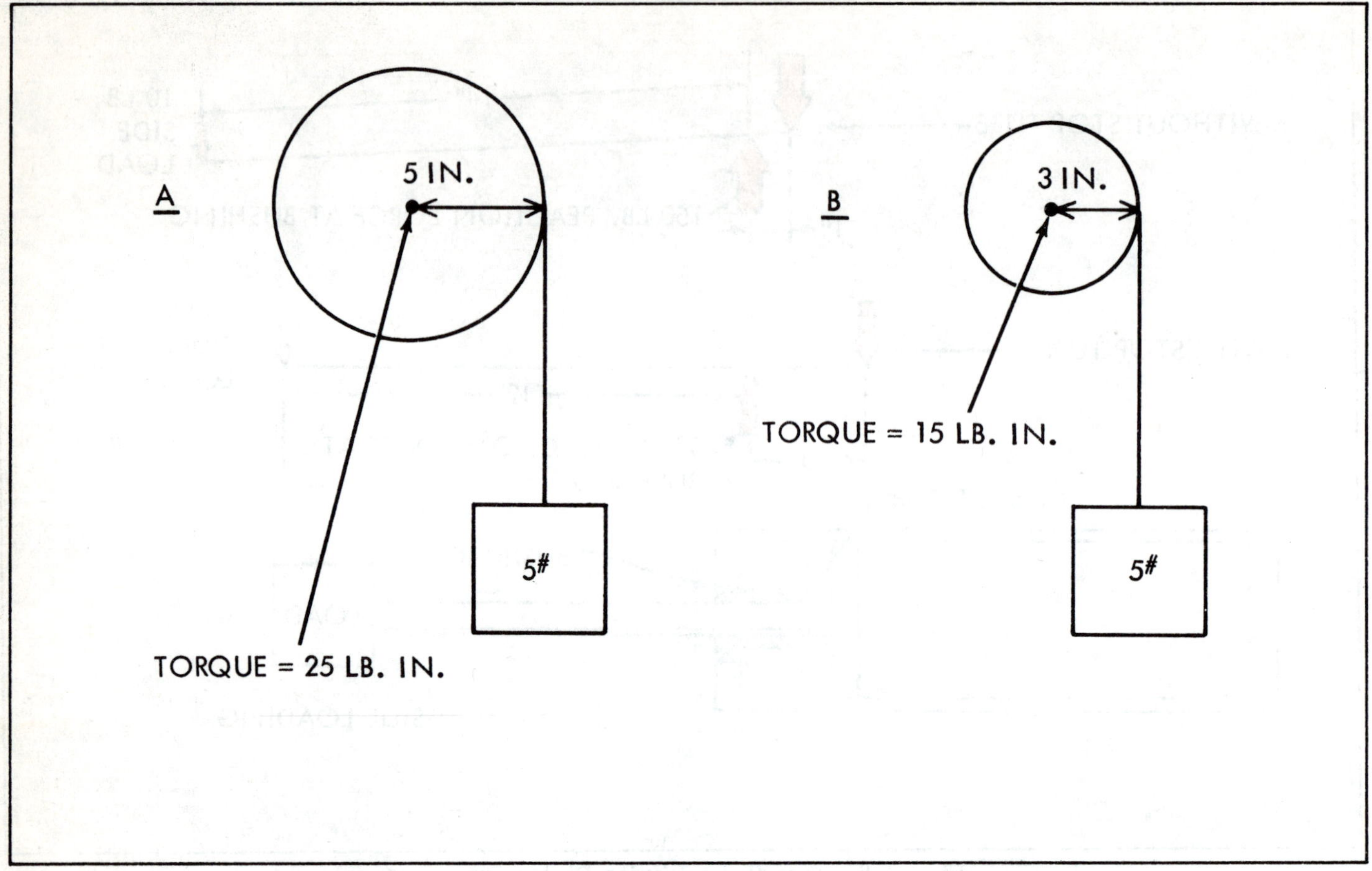

Fig. 6-10. Torque Equals Load Multiplied by Radius

NOTE

All of the following formulas are for theoretical torque. An additional 10% to 35% torque capability may be needed to start a given load. Check the starting torque specifications on the installation drawings.

Finding Size Motor Required for a Job:

$$\text{Torque Rate (lb. in./100 psi)} = \frac{\text{Torque Load (lb. in.)}}{\text{Desired operating pressure (psi)} \times .01}$$

For example, to handle a 500 pound-inch load at 2000 psi, would require a 25 pound-inch motor:

$$\text{Size} = \frac{500}{2000 \times .01} = \frac{500}{20} = 25 \text{ lb. in./100 psi}$$

Finding Working Pressure for a Given Size Motor and Load:

$$\text{Operating Pressure (psi)} = \frac{\text{Torque Load (lb. in.)} \times 100}{\text{Motor Torque Rate (lb. in./100 psi)}}$$

For example, a 50 pound-inch motor develops 3000 psi with a load of 1500 pound inches:

$$\text{Pressure} = \frac{1500 \times 100}{50} = 3000 \text{ psi}$$

Finding Maximum Torque for a Given Size Motor:

$$\text{Maximum Torque (lb. in.)} = \frac{\text{Torque Rate (lb. in./100 psi)} \times \text{Max. psi}}{100}$$

Thus, a 10 pound-inch motor rated at 2500 psi can handle a maximum load of 250 pound inches:

$$\text{Maximum torque} = \frac{10 \times 2500}{100} = 250 \text{ lb. in.}$$

Finding Torque When Pressure and Displacement are Known:

$$\text{Torque (lb. in.)} = \frac{\text{Pressure (psi)} \times \text{Displacement (cu. in./rev.)}}{2\pi}$$

Finding GPM Requirements for a Given Drive Speed:

TABLE 3

Change	Speed	Effect On Operating Pressure	Torque Available
Increase Pressure Setting	No Effect	No Effect	Increases
Decrease Pressure Setting	No Effect	No Effect	Decreases
Increase GPM	Increases	No Effect	No Effect
Decrease GPM	Decreases	No Effect	No Effect
Increase Displacement (Size)	Decreases	Decreases	Increases
Decrease Displacement (Size)	Increases	Increases	Decreases

Above table assumes a constant load.

$$\text{GPM} = \frac{\text{Speed (rpm) x Displacement (cu. in./rev.)}}{231}$$

A motor with a displacement of 10 cubic inches per revolution would require just over 43 gpm to run at 1000 rpm.

$$\text{GPM} = \frac{1000 \times 10}{231} = 43.2 \text{ gpm}$$

Finding Drive Speed When Displacement and GPM are Known:

$$\text{RPM} = \frac{\text{GPM x 231}}{\text{Displacement (cu. in./rev.)}}$$

Table 3 summarizes the effects on speed, pressure and torque capacity for changes in motor applications. Note that the basic principles are identical to the cylinder chart on page 6-6.

GEAR MOTORS

A gear motor (Fig. 6-11) develops torque through pressure on the surfaces of gear teeth. The two gears mesh and rotate together, with only one gear coupled to the drive shaft. The motor is reversible by reversing flow. The displacement of a gear motor is fixed and is roughly equal to the volume between two teeth multiplied by the number of teeth.

It is evident in Figure 6-11 that the gears are not in balance with respect to pressure loads. High pressure at the inlet and low pressure at the outlet result in high side loads on the shaft and gears as well as the bearings which support them. It is possible to balance this side loading by internal passages and ports, which place corresponding pressure conditions 180° degrees apart. This type of balancing is more often found in vane motors, however. See Figure 6-12.

Gear motors of this type are frequently limited to about 2000 psi operating pressure and the 2400 rpm range. Principal advantages have been their simplicity and a rather high dirt tolerance. These are offset, however, by somewhat lower efficiency. With current emphasis on higher performance and more sophisticated filtering equipment, the trend is toward piston type motors in many machinery and mobile equipment applications.

VANE MOTORS

In a vane motor, torque is developed by pressure on exposed surfaces of rectangular vanes which slide in and out of slots in a rotor splined to the driveshaft (Fig. 6-12). As the rotor turns, the vanes follow the surface of a cam ring, forming sealed chambers which carry the fluid from the inlet to the outlet. In the balanced design shown in Fig. 6-12 pressure build up at either port is directed to two interconnected chambers within the motor located 180° apart. Any side loads which are generated oppose and cancel each other.

Figure 6-13 shows this configuration in the "square" design, reversible vane motor. Note that the rotor turns inside the cam ring and between the body and pressure plate. Pivoted rocker arms are attached to the rotor and force the vanes outward against the elliptical cam ring. In operation, pressure under the vanes also holds them in contact with the ring.

Shuttle Valves in Pressure Plate

The pressure plate (Fig. 6-14) is designed to hold the rotating unit tightly sealed through pressure on its outer surface. Two shuttle valves in the pressure plate interconnect passages to maintain this pressure no matter which port is pressurized. The motor rotation is reversed by reversing fluid flow to and from the ports.

A special modification of this motor (Fig. 6-15) permits operation in either direction without rocker arms or shuttle valves. Oil under pressure from an external source is directed to the pressure plate and under the vanes to hold them against the ring.

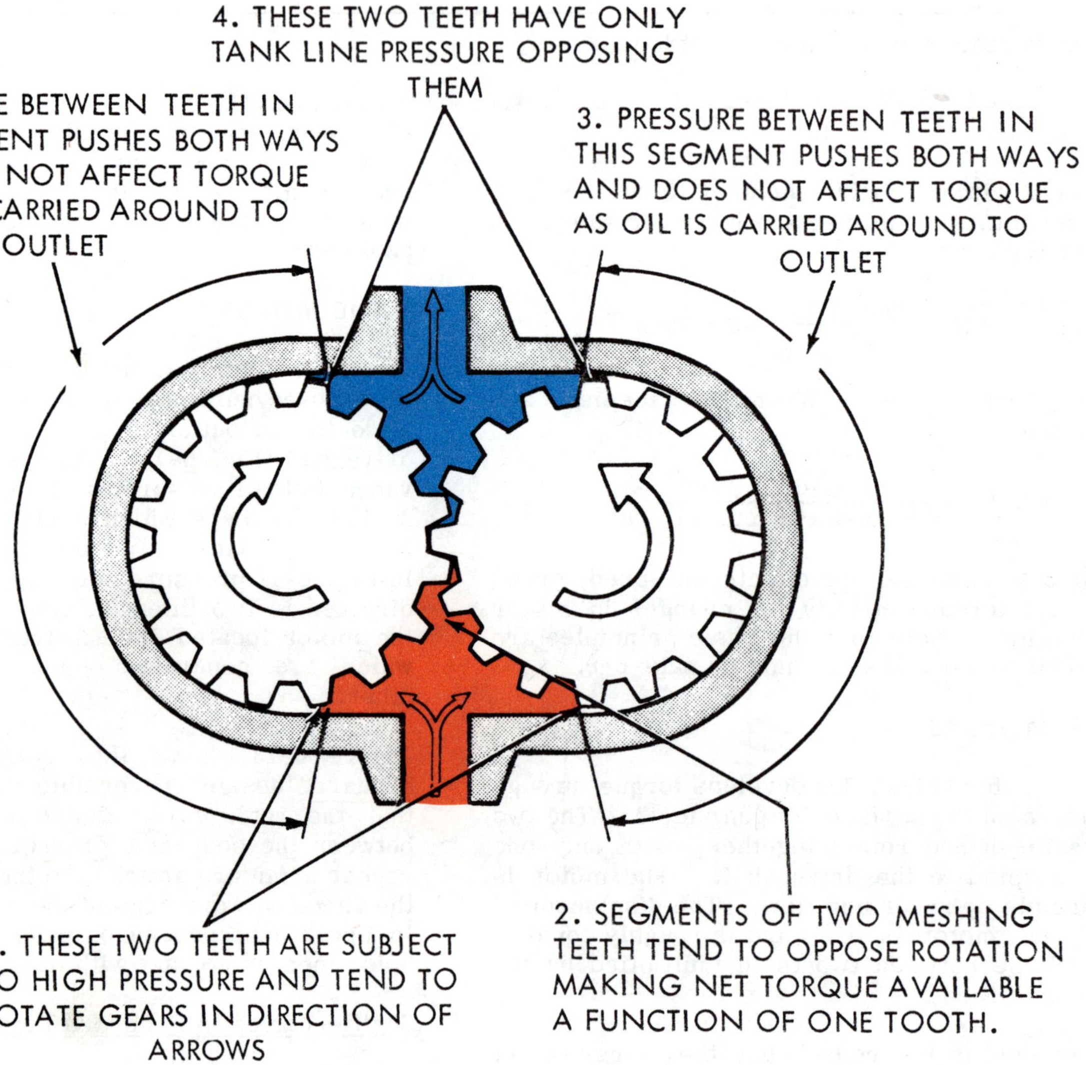

4. THESE TWO TEETH HAVE ONLY TANK LINE PRESSURE OPPOSING THEM

3. PRESSURE BETWEEN TEETH IN THIS SEGMENT PUSHES BOTH WAYS AND DOES NOT AFFECT TORQUE AS OIL IS CARRIED AROUND TO OUTLET

3. PRESSURE BETWEEN TEETH IN THIS SEGMENT PUSHES BOTH WAYS AND DOES NOT AFFECT TORQUE AS OIL IS CARRIED AROUND TO OUTLET

1. THESE TWO TEETH ARE SUBJECT TO HIGH PRESSURE AND TEND TO ROTATE GEARS IN DIRECTION OF ARROWS

2. SEGMENTS OF TWO MESHING TEETH TEND TO OPPOSE ROTATION MAKING NET TORQUE AVAILABLE A FUNCTION OF ONE TOOTH.

Fig. 6-11. Torque Developement in Gear Motor

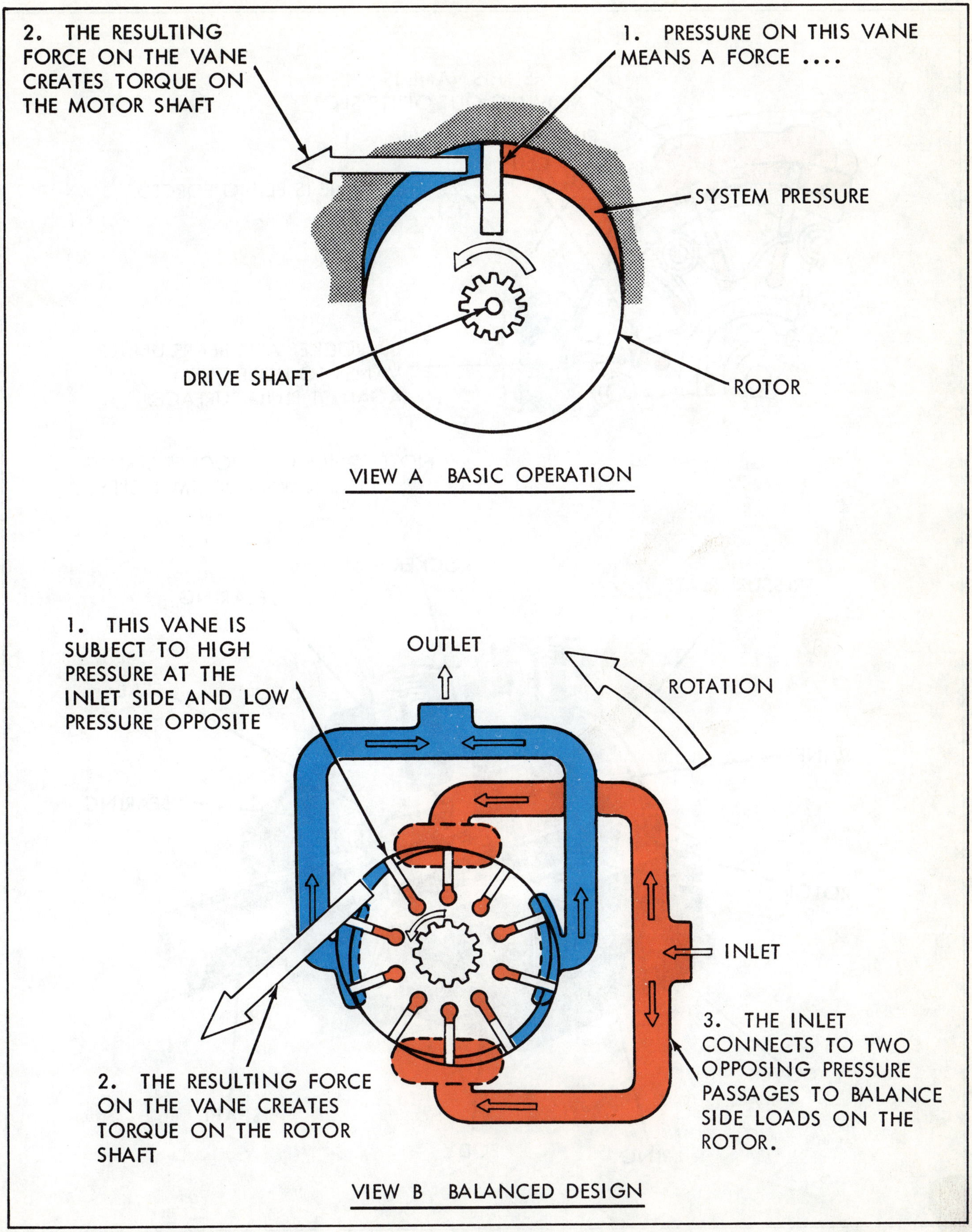

2. THE RESULTING FORCE ON THE VANE CREATES TORQUE ON THE MOTOR SHAFT

1. PRESSURE ON THIS VANE MEANS A FORCE ....

SYSTEM PRESSURE

DRIVE SHAFT

ROTOR

VIEW A BASIC OPERATION

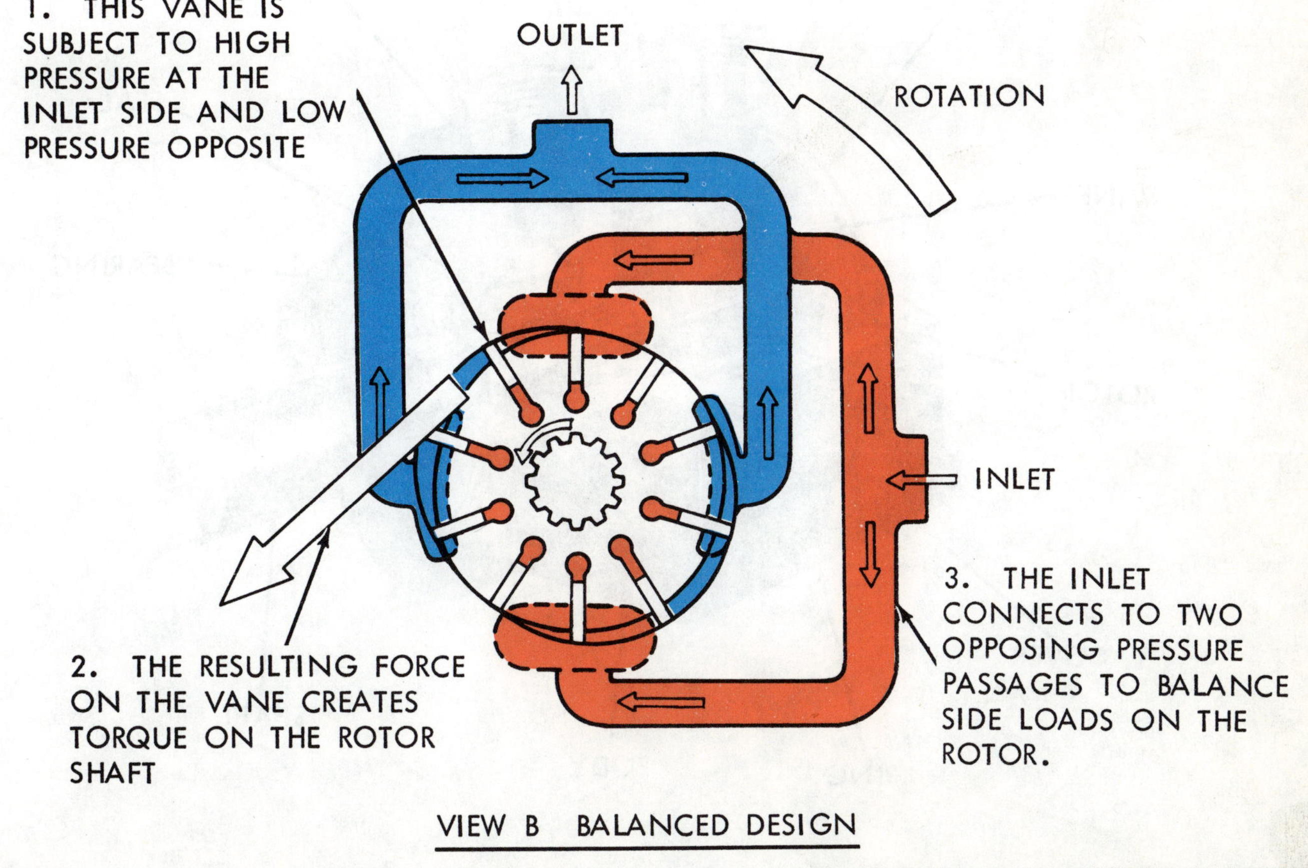

Fig. 6-12. Torque Development in Balanced Vane Motor

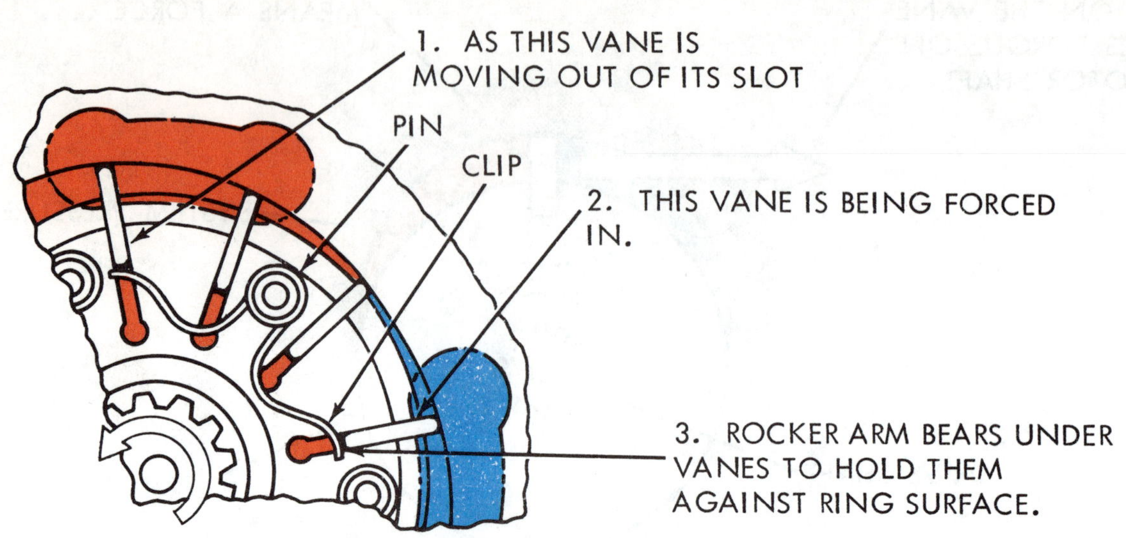

NOTE: ONLY ONE ROCKER SPRING IS SHOWN FOR SIMPLICITY.

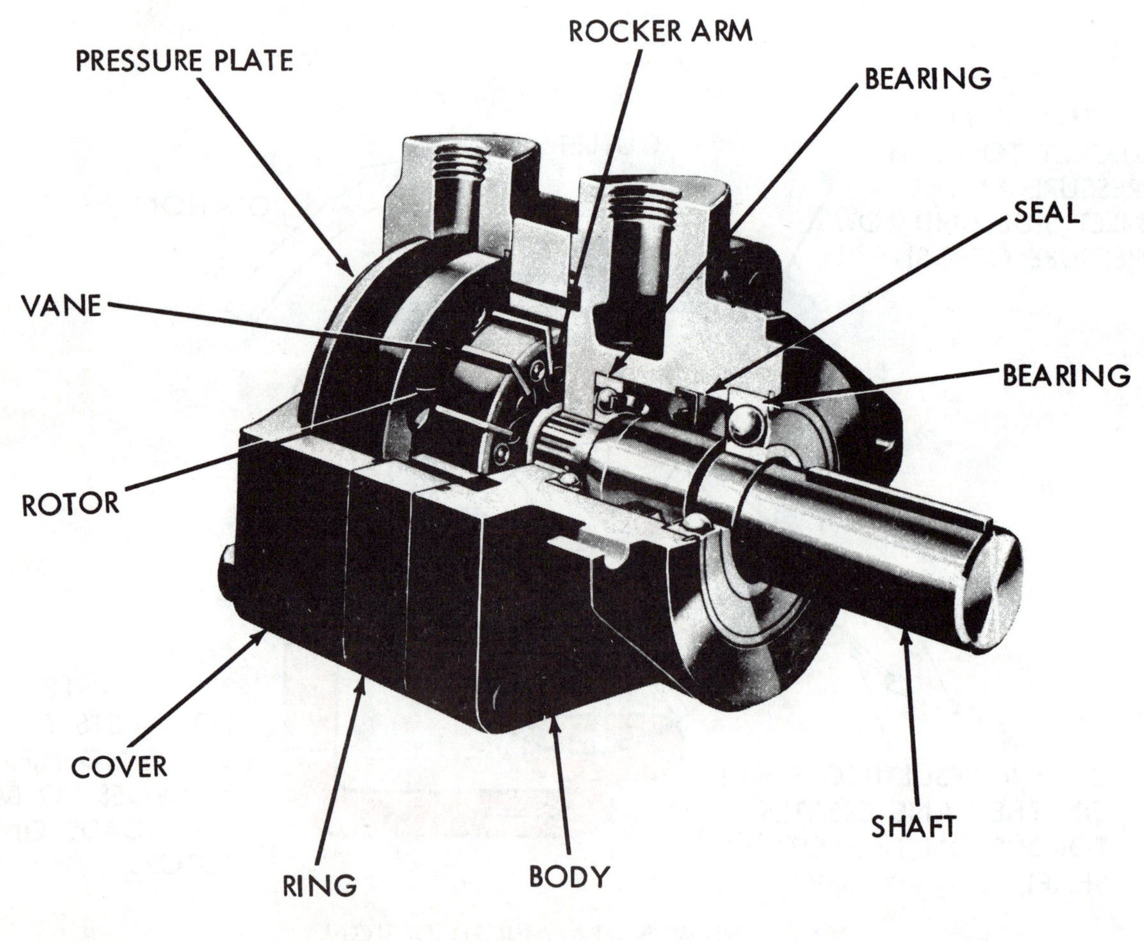

Fig. 6-13. Construction of "Square" Design Vane Motor

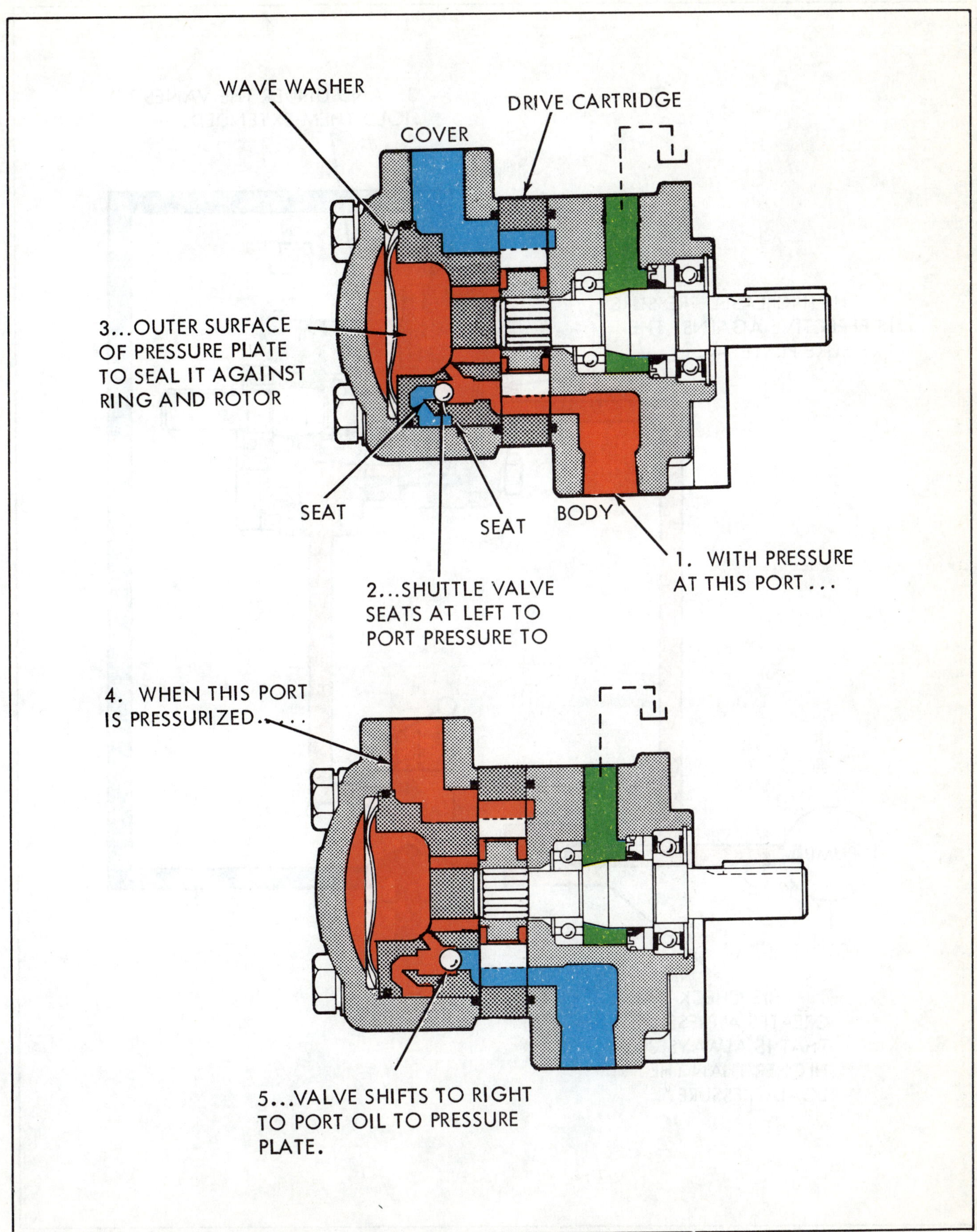

Fig. 6-14. Pressure Plate Seals Cartridge

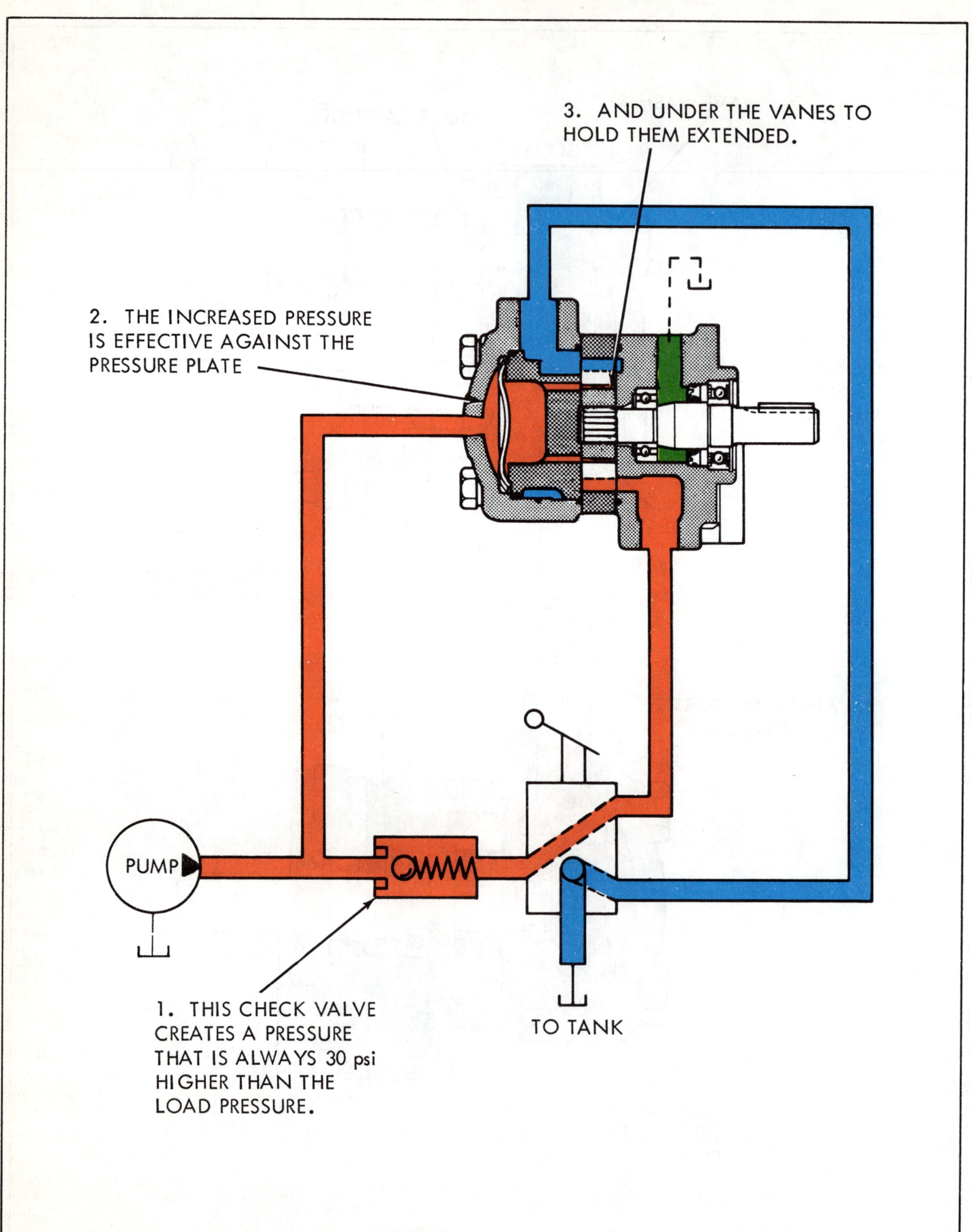

Fig. 6-15. "S2" Modification Eliminates Shuttle Valves and Rocker Arms

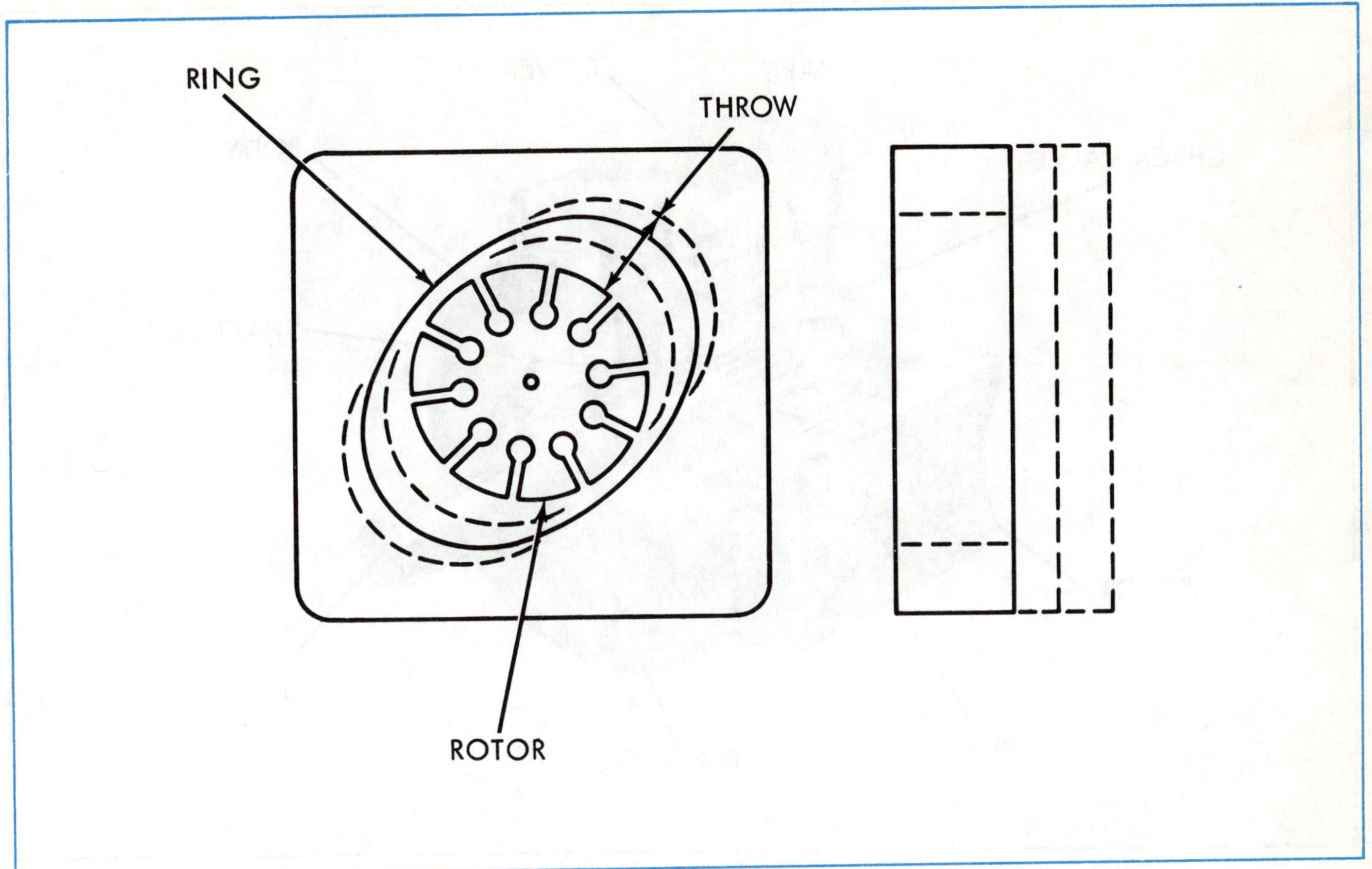

Fig. 6-16. Ring "Throw" Determines Displacement Within a Given Package Size

These motors have fixed displacements. The displacement of a given unit, however, can be changed by installing a cam ring with more or less "throw" (Fig. 6-16).

Another modification of this motor is a unidirectional or non-reversible design (Fig. 6-17). A check valve in its inlet port assures pressure to hold the vanes extended. Thus, this design does not require rocker arms, shuttle valves, or an external pressure source. Its application might be a fan drive or similar device which would rotate in only one direction.

HIGH-PERFORMANCE VANE MOTORS

The high performance vane motor (Fig. 6-18) is a later design of balanced vane motor. It develops torque in the same way as the "square" motor but has significant changes in construction.

In this design, the vanes are held out against the ring by coil springs. The entire assembly of ring, rotor, vanes and side plates is removable and replaceable as a unit (Fig. 6-19). In fact, preassembled and tested "cartridges" are available for field replacement.

These motors also are reversible by reversing flow to and from the ports. Both side plates function alternately as pressure plates (Fig. 6-20), depending on the direction of flow.

MHT HIGH TORQUE MOTOR

Another design of balanced vane motor is the MHT series high torque, low speed motor (Fig. 6-21). Available in several sizes, one size operates from 5 to 150 rpm and has an actual torque capacity of 4500 pound feet. A double version produces 9000 pound feet. It is adaptable to screw drives, mixer drives, heavy conveyors and turntables, dumping units, winches and others where their tremendous torque capabilities can be used to advantage.

INLINE PISTON MOTORS

Piston motors generate torque through pressure on the ends of reciprocating pistons operating in a cylinder block. In the inline design (Fig. 6-22), the motor driveshaft and cylinder block are centered on the same axis. Pressure at the ends of the pistons causes a reaction against a canted swash plate and drives the cylinder block and motor shaft in rotation. Torque is proportional to the area of the pistons and is a function of the angle at which the swash plate is positioned.

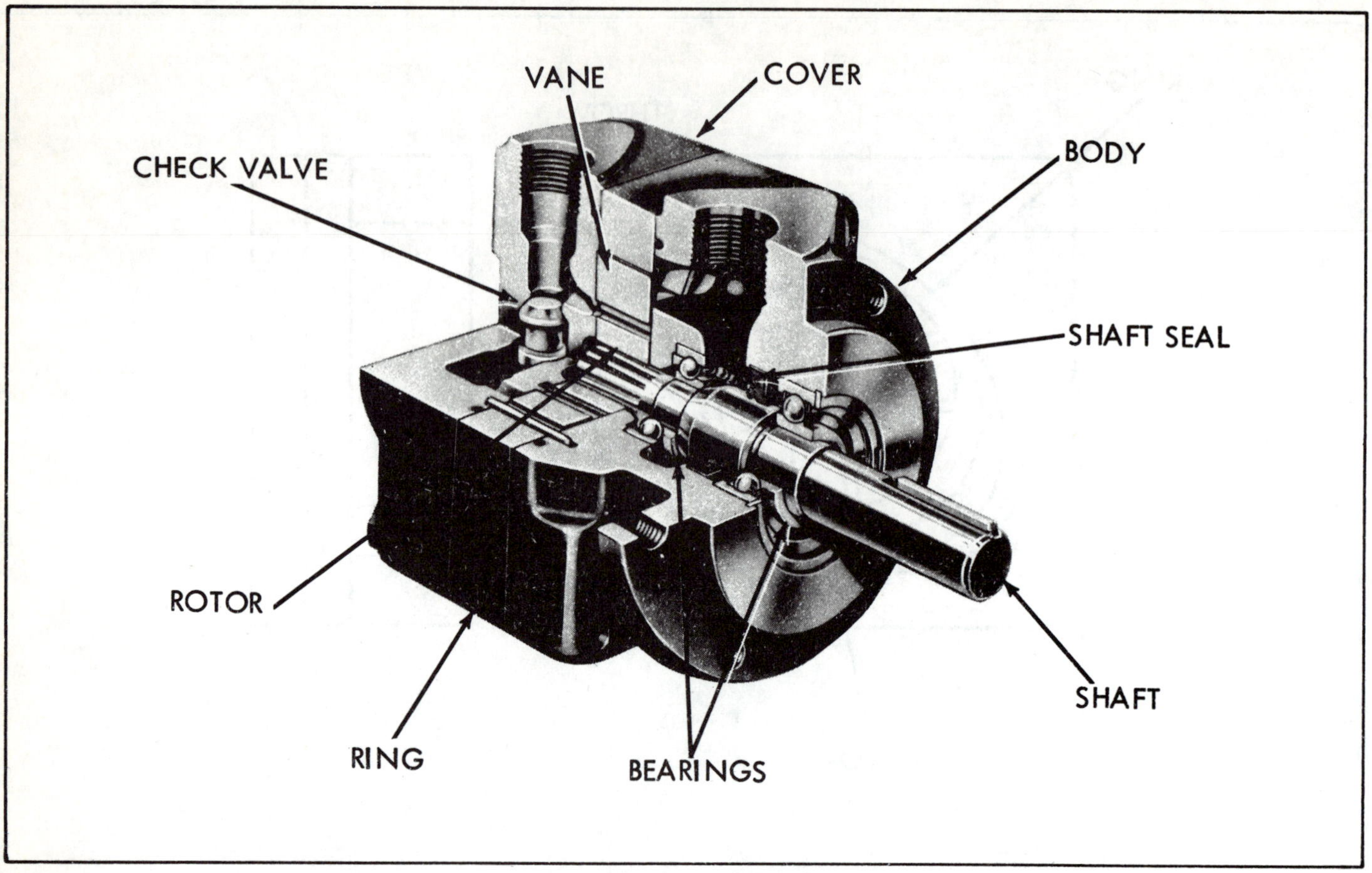

Fig. 6-17. Construction of Uni-Directional Vane Motor

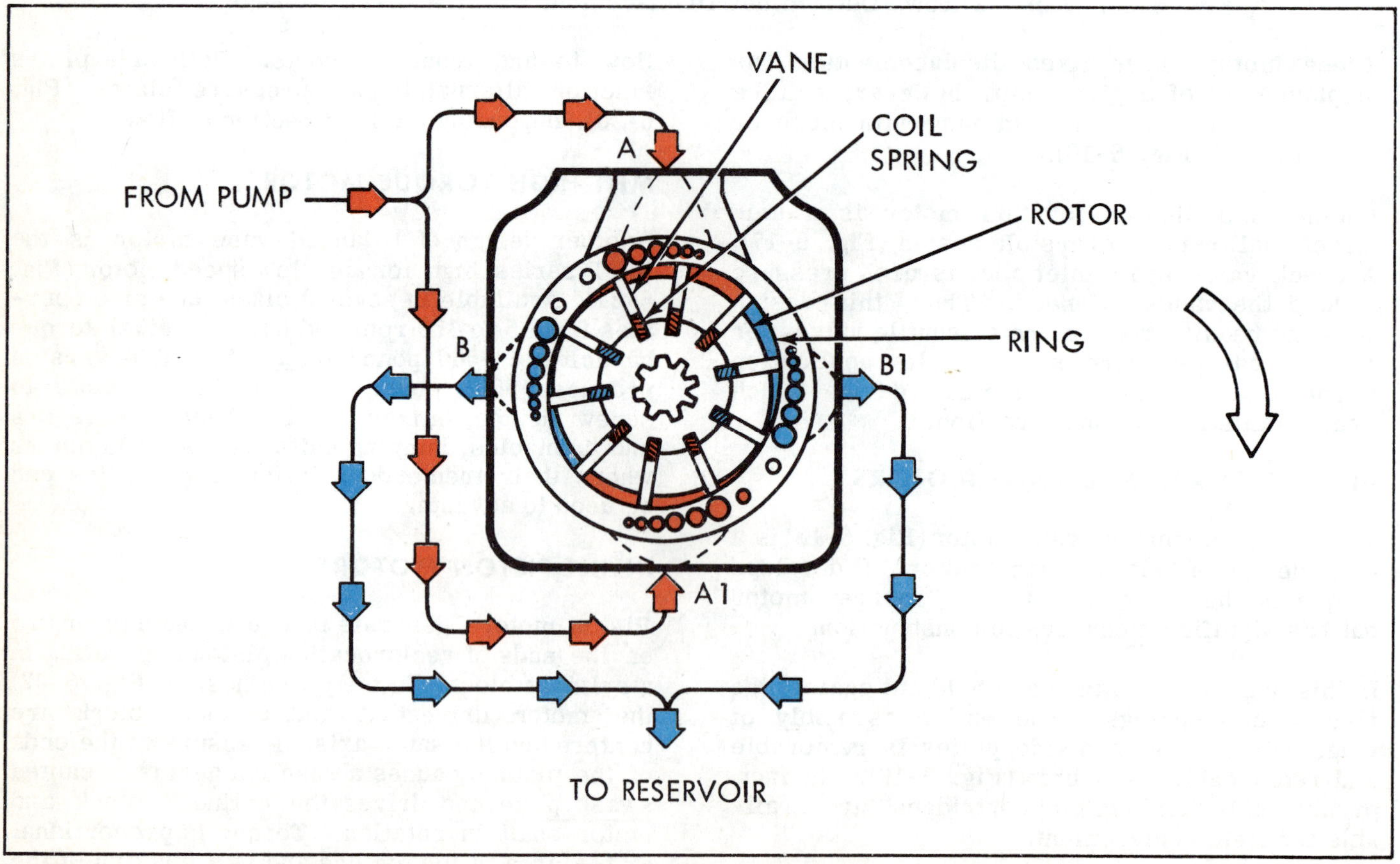

Fig. 6-18. Operation of High Performance Vane Motor

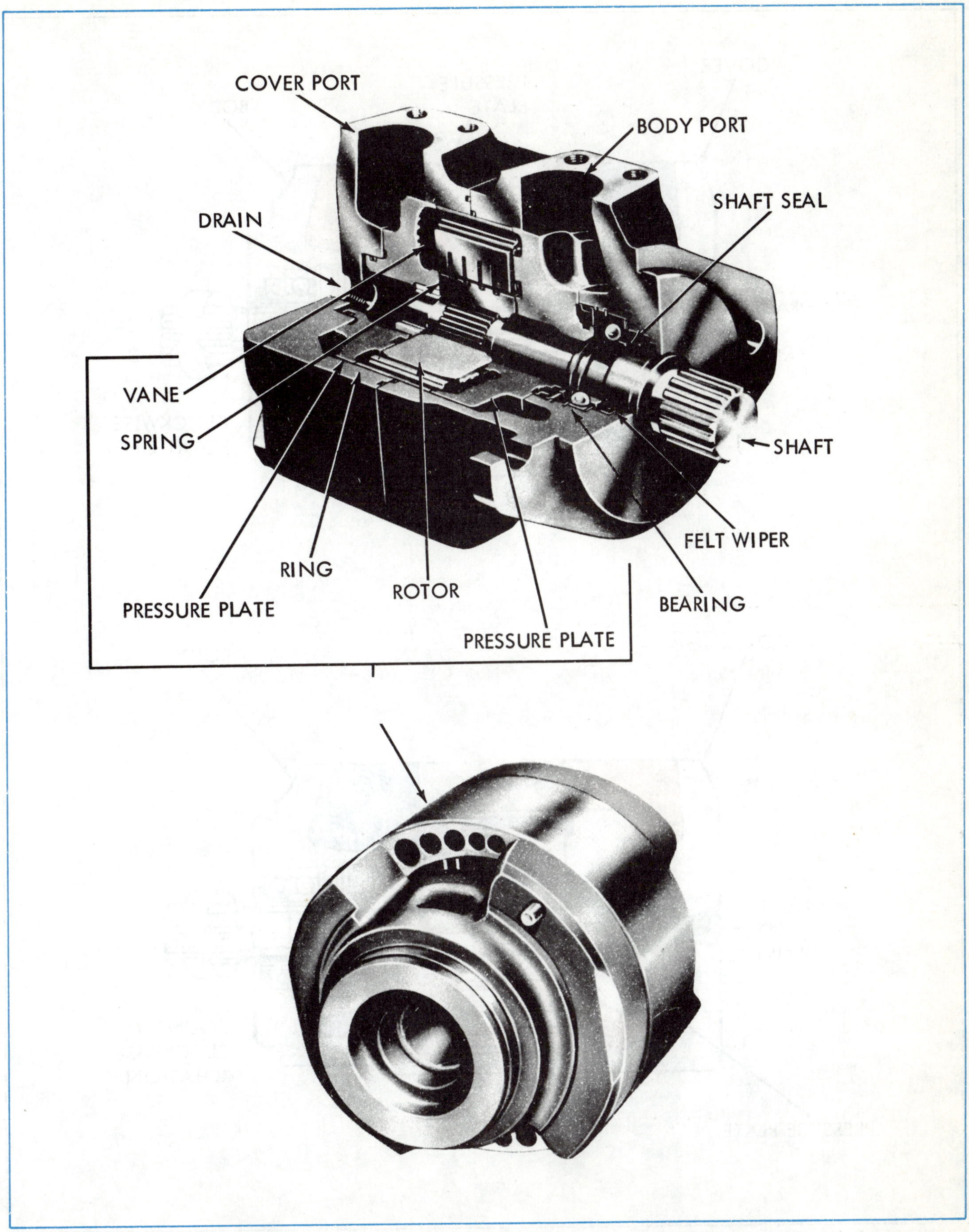

COVER PORT
BODY PORT
SHAFT SEAL
DRAIN
VANE
SPRING
SHAFT
FELT WIPER
RING
ROTOR
BEARING
PRESSURE PLATE
PRESSURE PLATE

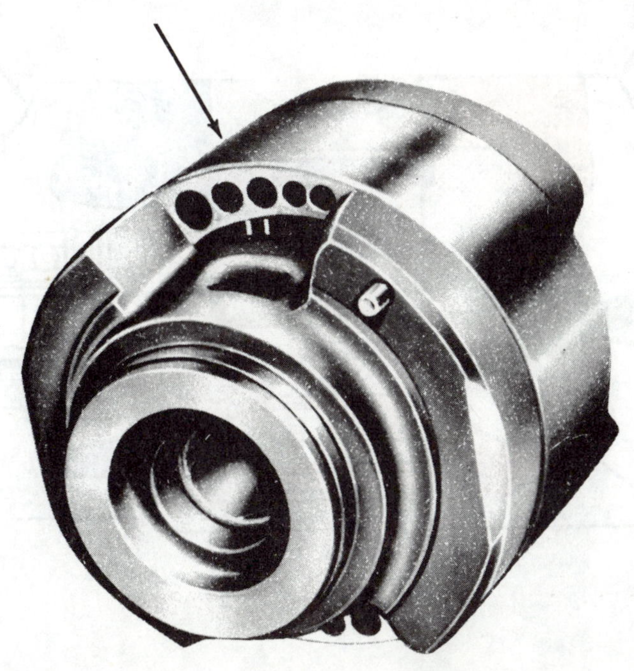

Fig. 6-19. Construction of High Performance Vane Motor

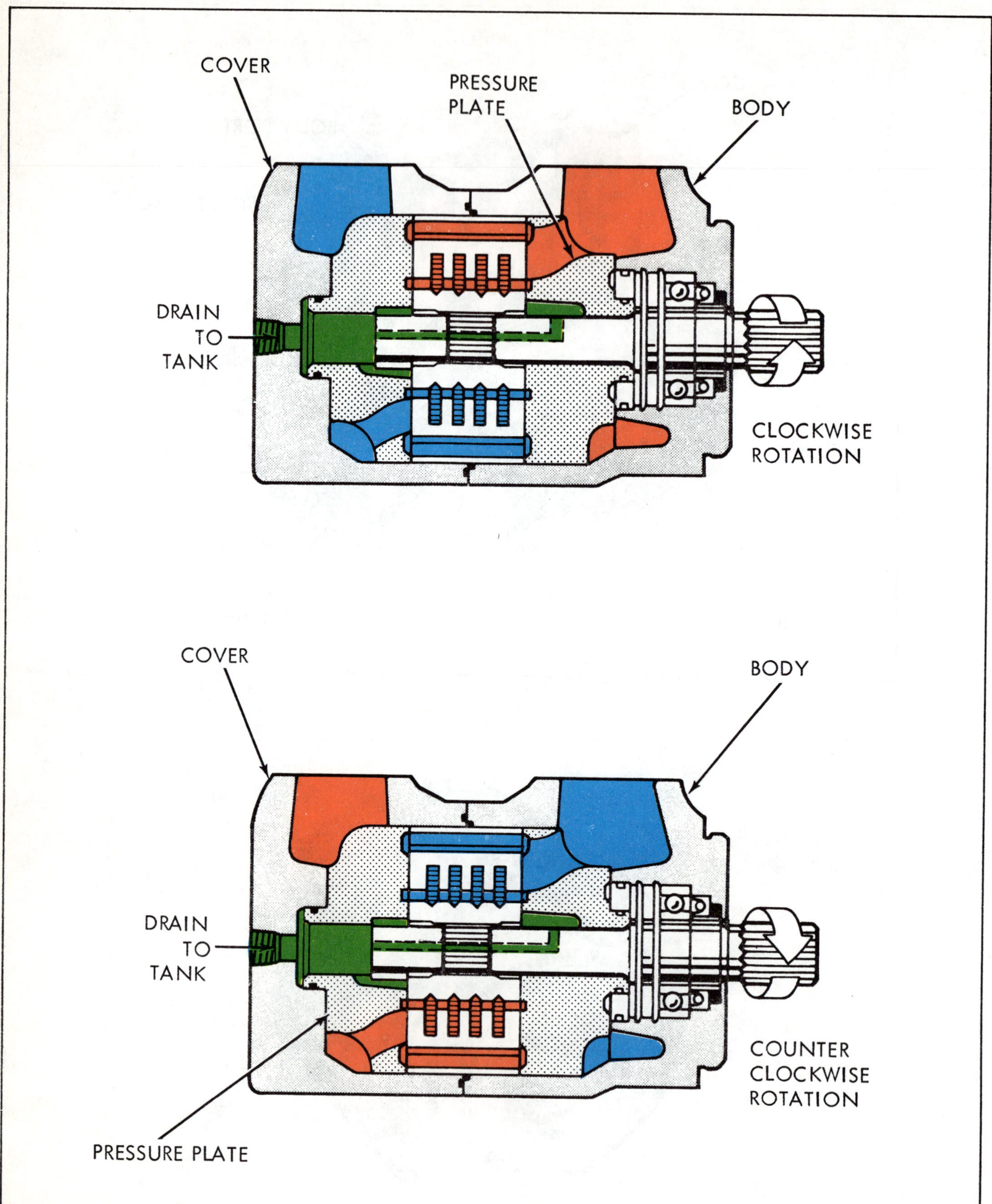

Fig. 6-20. Both Side Plates are Pressure Plates in High Performance Design

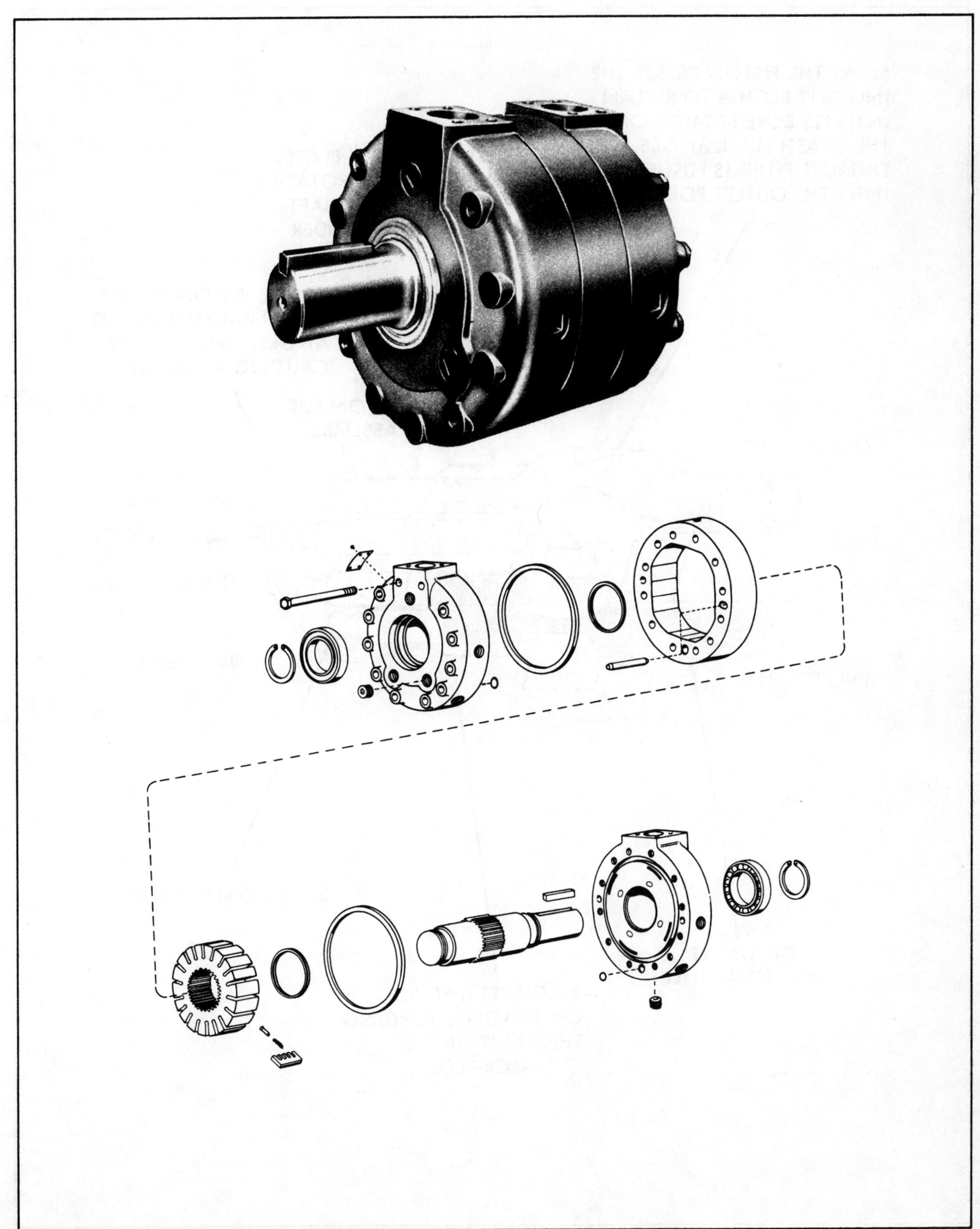

Fig. 6-21. High Torque Vane Motor

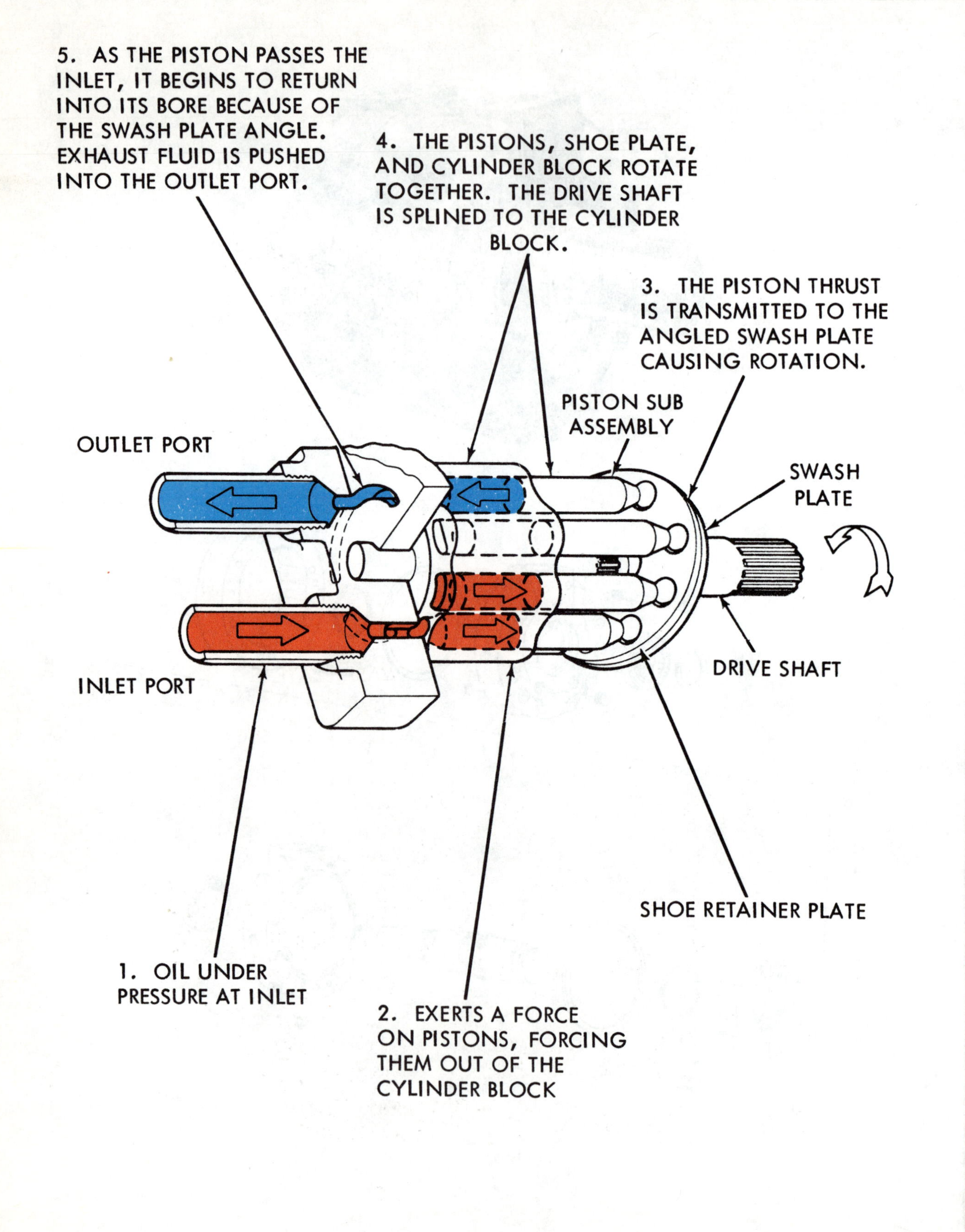

Fig. 6-22. Inline Piston Motor Operation

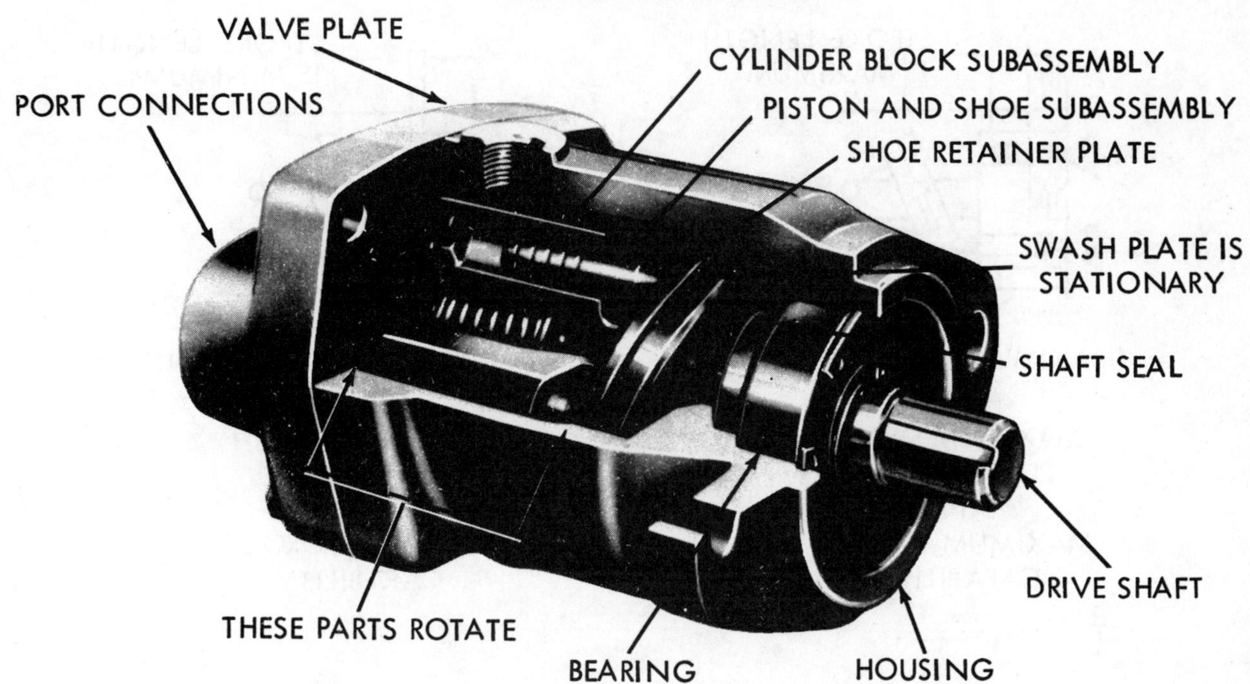

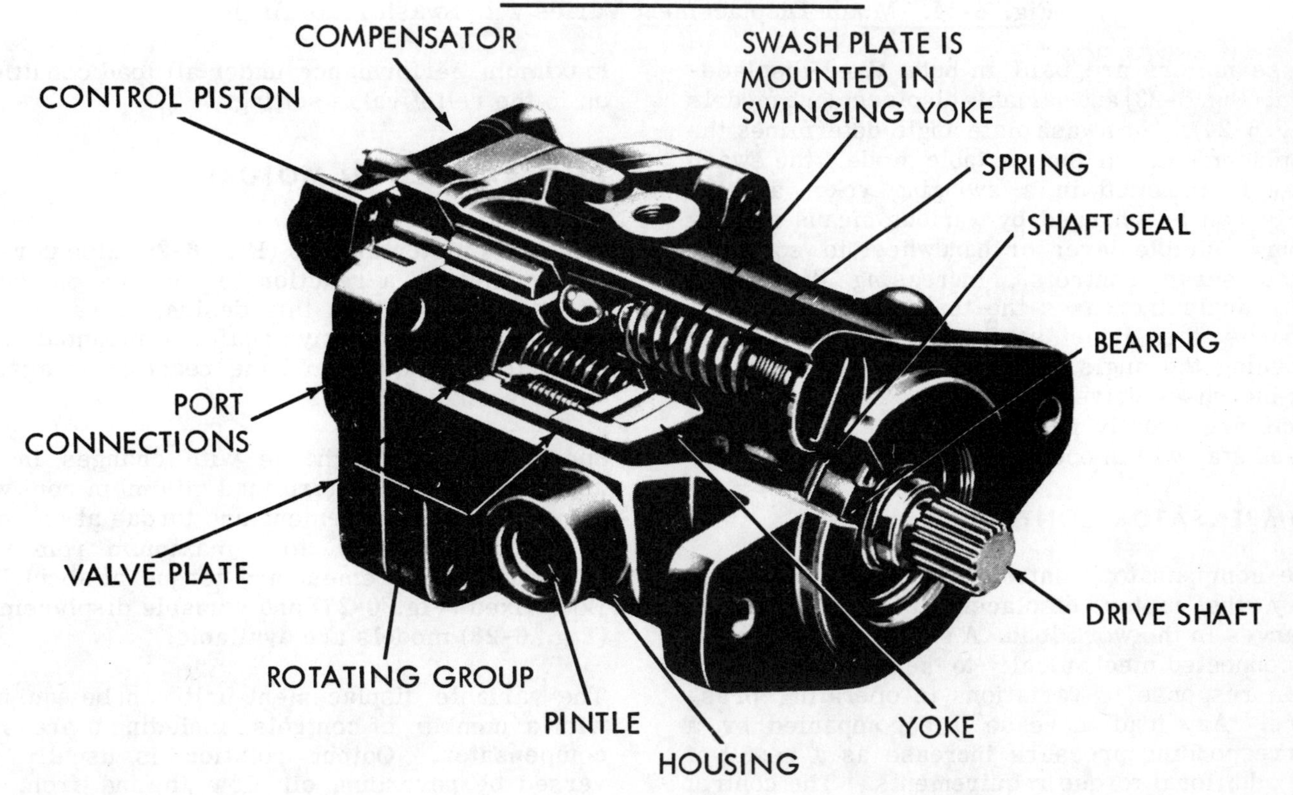

Fig. 6-23. Two Configurations of Inline Piston Motors

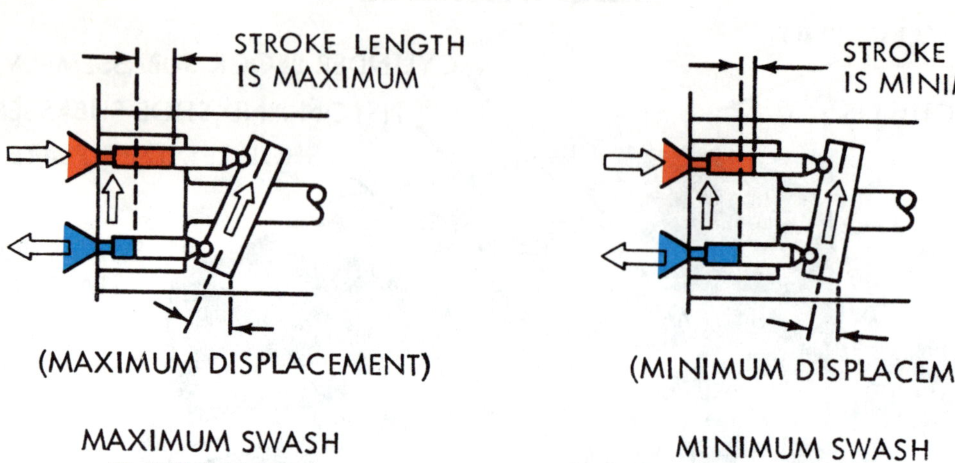

Fig. 6-24. Motor Displacement Varies with Swash Plate Angle

These motors are built in both fixed-displacement (Fig. 6-23) and variable displacement models (Fig. 6-24). The swash plate angle determines the displacement. In the variable model, the swash plate is mounted in a swinging yoke, and the angle can be changed by various means ranging from a simple lever or handwheel to sophisticated servo controls. Increasing the swash plate angle increases the torque capability but reduces the drive shaft speed. Conversely, reducing the angle reduces the torque capability but increases drive shaft speed. Minimum angle stops are usually provided so that torque and speed stay within operating limits.

COMPENSATOR CONTROL

The compensator control (Fig. 6-25) is used to vary the motor displacement in response to changes in the work load. A spring-loaded piston is connected mechanically to the yoke and moves it in response to variations in operating pressure. Any load increase is accompanied by a corresponding pressure increase as a result of the additional torque requirements. The control then automatically adjusts the yoke so that the torque increases under a heavy load and decreases when the load is light. Ideally, the compensator regulates the displacement for maximum performance under all load conditions up to the relief valve setting.

BENT-AXIS PISTON MOTORS

Bent-axis piston motors (Fig. 6-26) also develop torque through a reaction to pressure on reciprocating pistons. In this design, however, the cylinder block and drive shaft are mounted on an angle to each other and the reaction is against the driveshaft flange.

Speed and torque change with changes in the angle--from a predetermined minimum rpm with a maximum displacement and torque at an angle of approximately 30° to a maximum rpm with minimum displacement and torque at about $7\frac{1}{2}°$. Both fixed (Fig. 6-27) and variable displacement (Fig. 6-28) models are available.

The variable displacement unit can be equipped with a number of controls, including a pressure compensator. Output rotation is usually reversed by reversing oil flow to and from the motor. It is not practical to reverse a motor by swinging the yoke over center, since the torque would go to zero and its speed infinitely high (if it did not stall before reaching center).

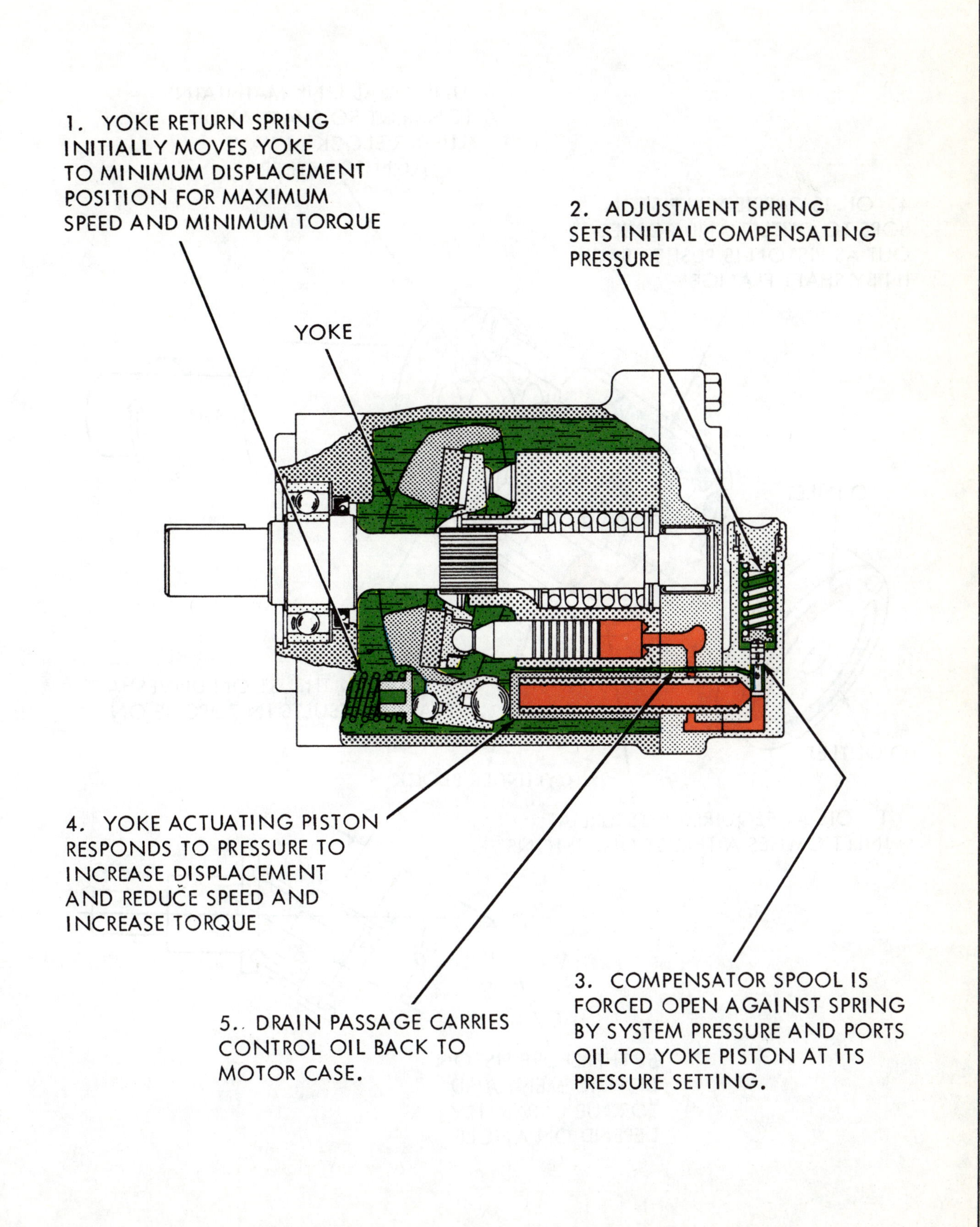

Fig. 6-25. Compensator Control Adjusts Speed to Load

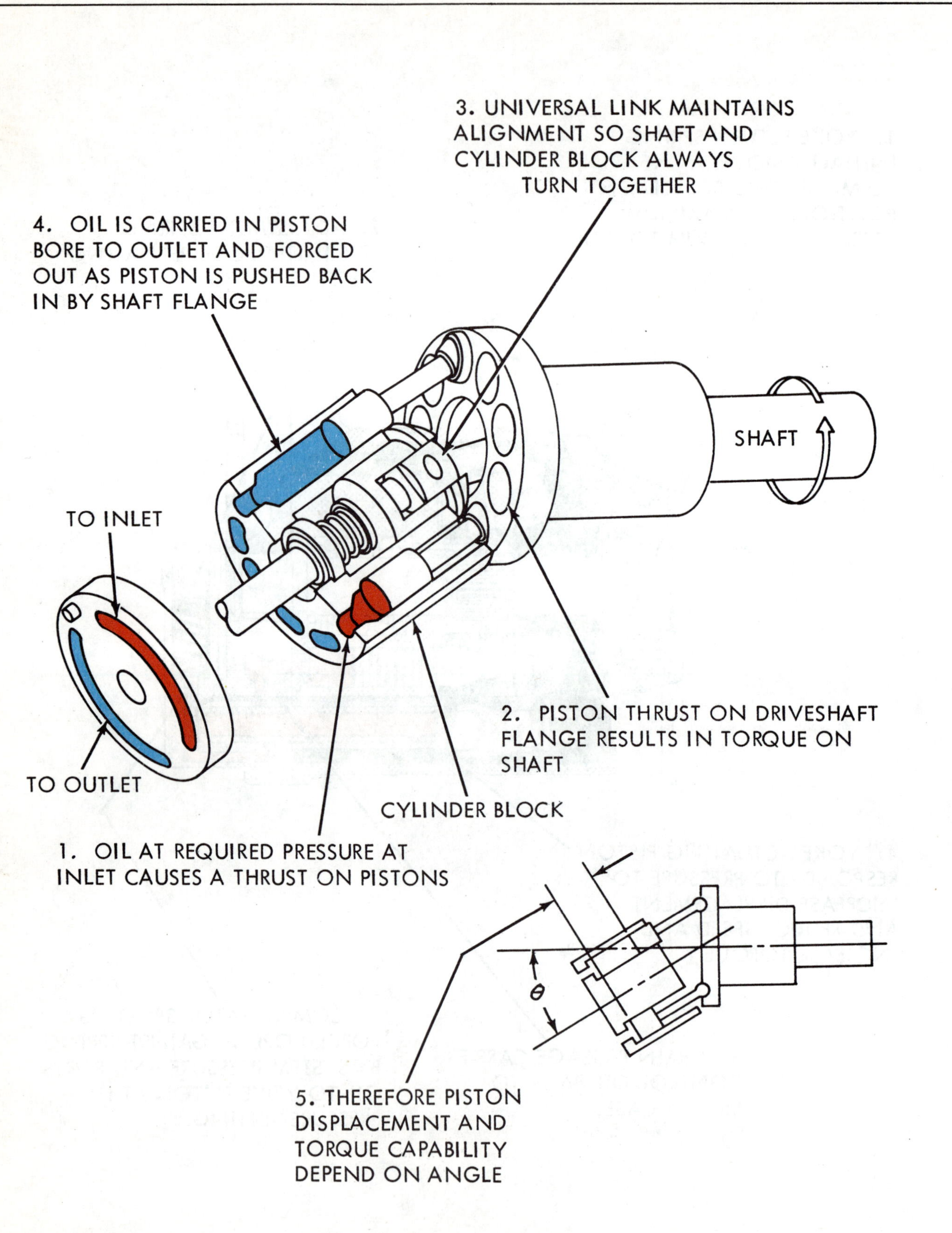

Fig. 6-26. Bent-Axis Piston Motor Operation

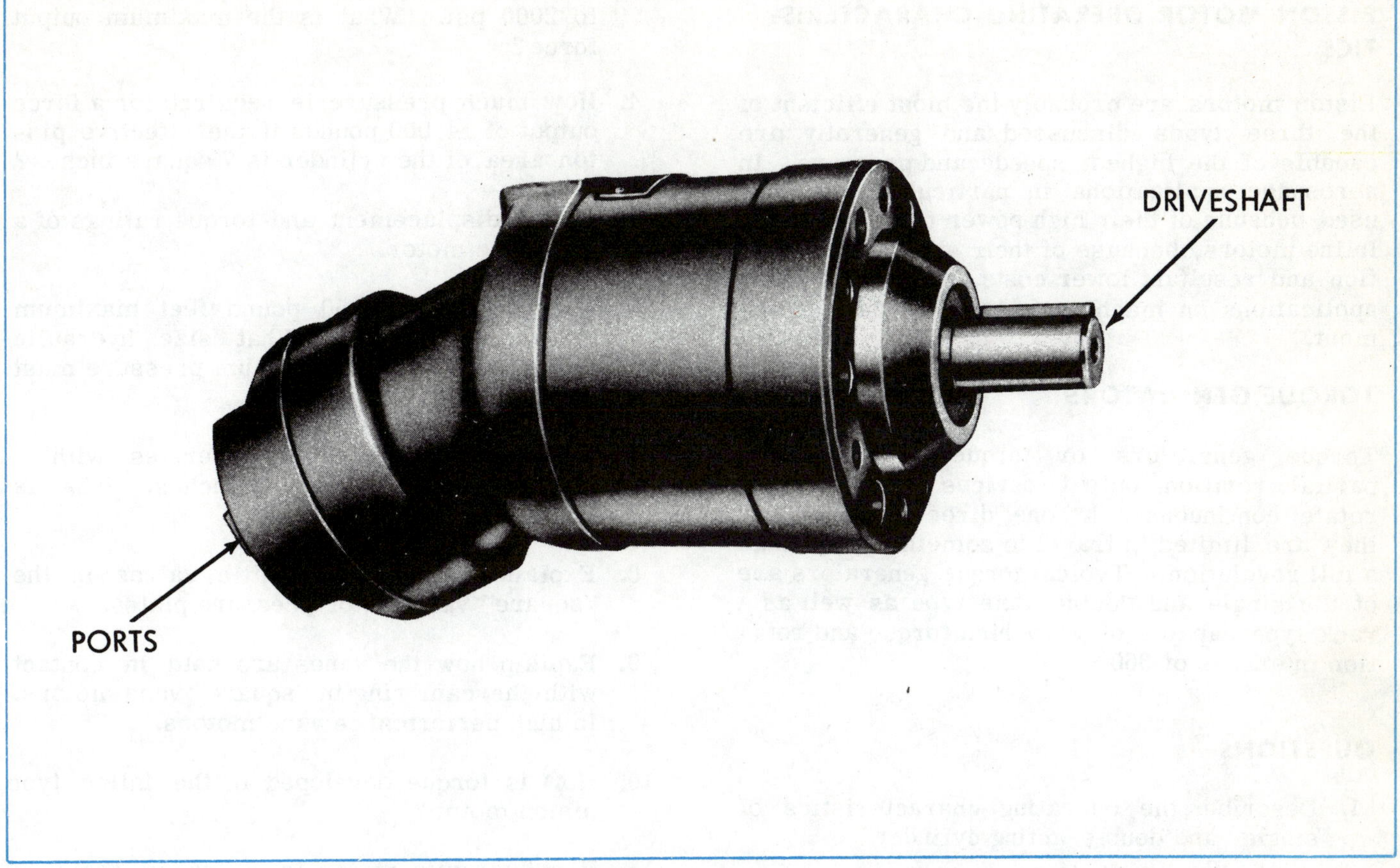

Fig 6-27. Typical Fixed Displacement Bent-Axis Motor

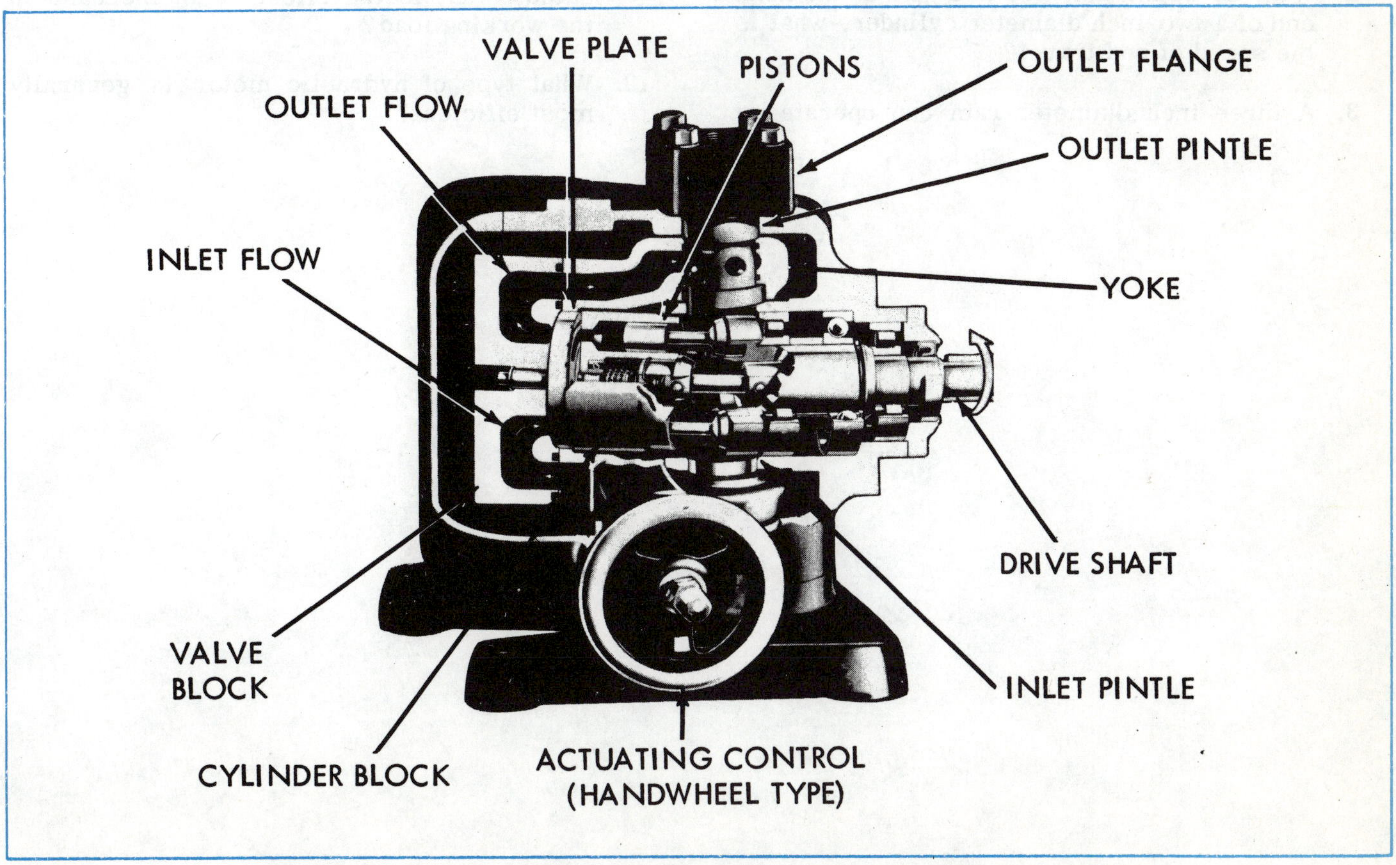

Fig. 6-28. Typical Variable Displacement Bent-Axis Piston Motor

PISTON MOTOR OPERATING CHARACTERISTICS

Piston motors are probably the most efficient of the three types discussed and generally are capable of the highest speeds and pressure. In aerospace applications in particular, they are used because of their high power to weight ratio. Inline motors, because of their simple construction and resultant lower costs, are finding many applications on machine tools and mobile equipment.

TORQUE GENERATORS

Torque generators or torque actuators are partial rotation output devices which cannot rotate continuously in one direction. Usually they are limited in travel to something less than a full revolution. Typical torque generators are of the single and double vane type as well as a rack type capable of very high torque and rotation in excess of 360°.

QUESTIONS

1. Describe the operating characteristics of single- and double-acting cylinder

2. With an actual delivery of 3 gpm to the head end of a two-inch diameter cylinder, what is the speed of rod travel?

3. A three-inch diameter ram can operate up to 2000 psi. What is the maximum output force?

4. How much pressure is required for a force output of 14,000 pounds if the effective piston area of the cylinder is 7 square inches?

5. Define displacement and torque ratings of a hydraulic motor.

6. A winch requires 50 pound feet maximum torque to operate. What size hydraulic motor is needed if maximum pressure must be limited to 1500 psi?

7. A 20 pound-inch motor operates with a torque load of 500 pound inches. What is the operating pressure?

8. Explain the use of shuttle valves in the "square" vane motor pressure plates.

9. Explain how the vanes are held in contact with the cam ring in "square" vane motors. In high performance vane motors.

10. How is torque developed in the inline type piston motor?

11. If a hydraulic motor is pressure compensated, what is the effect of an increase in the working load?

12. What type of hydraulic motor is generally most efficient?

CHAPTER 7

DIRECTIONAL CONTROLS

Directional valves, as the name implies, are used to control the direction of flow. Though sharing this common function, directional valves vary considerably in construction and operation. They are classified according to their principal characteristics, such as:

* Type of Internal Valving Element--poppet (piston or ball), rotary spool and sliding spool.

* Methods of Actuation--cams, plungers, manual lever, mechanical, electric solenoid, hydraulic pressure (pilot-operated) and others, including combinations of these.

* Number of Flow Paths--two-way, three-way, four-way, etc.

* Size--nominal size of pipe connections to valve or its mounting plate, or rated gpm flow.

* Connections--pipe thread, straight thread, flanged, and back-mounted (sometimes called gasket or subplate-mounted).

Finite Positioning

Most industrial directional valves are finite positioning. That is, they control where the oil goes by opening and closing flow paths in definite valve positions. You will note that the graphical symbol for a directional valve will contain a separate envelope (square) for each finite position, showing the flow paths in that position.

CHECK VALVES

A check valve can function as either a directional control or a pressure control. In its simplest form, however, a check valve is nothing more than a one-way directional valve (Fig. 7-1). It permits free flow in one direction and blocks flow in the other.

Note that the composite graphical symbol for a check valve indicates two valve positions--open and closed. This is a rather complicated diagram for such a simple valve, and is seldom used. Rather, the simple ball-and-seat symbol is used universally and therefore will be shown to designate a check valve throughout this manual.

INLINE CHECK VALVES

Inline check valves (Fig. 7-2) are so named because they are connected into the line and the oil flows straight through. The valve body is threaded for pipe or a tubing connector, and is machined inside to form a seat for the poppet or ball (Fig. 7-3). A light spring holds the poppet seated in the normal closed position permitting the valve to be mounted in any attitude.

In the free flow direction, the spring will be overcome and the valve will crack open at about 5 psi pressure drop. The springs are not adjustable, although a variety of sizes are available for special requirements such as creating pilot pressure or as a means of bypassing heat exchangers or oil filters in the event of high flow surges or clogging. In such instances they are not being used as check valves in their true sense but rather as sequence or relief valves.

Although operating pressures of 3000 psi are permissible, the inline check valves are not recommended for applications in which they could be subjected to high velocity return flow.

RIGHT ANGLE CHECK VALVES

A heavier-duty unit, the right angle valve, has a steel poppet and a hardened seat pressed into the iron body (Fig. 7-4 and 7-5). It gets its name from the angle between the flow passage to the poppet and the passage away from the poppet. These valves are built in threaded, flanged connected and back-connected versions. Sizes range from three gpm to 320 gpm with a wide range of cracking pressures.

RESTRICTION CHECK VALVE

A restriction check valve (Fig. 7-6) is a modification of a simple check valve. An orifice plug is placed in the poppet to permit a restricted flow in the normally closed position.

While their usage is somewhat limited, applications would include those which require a free

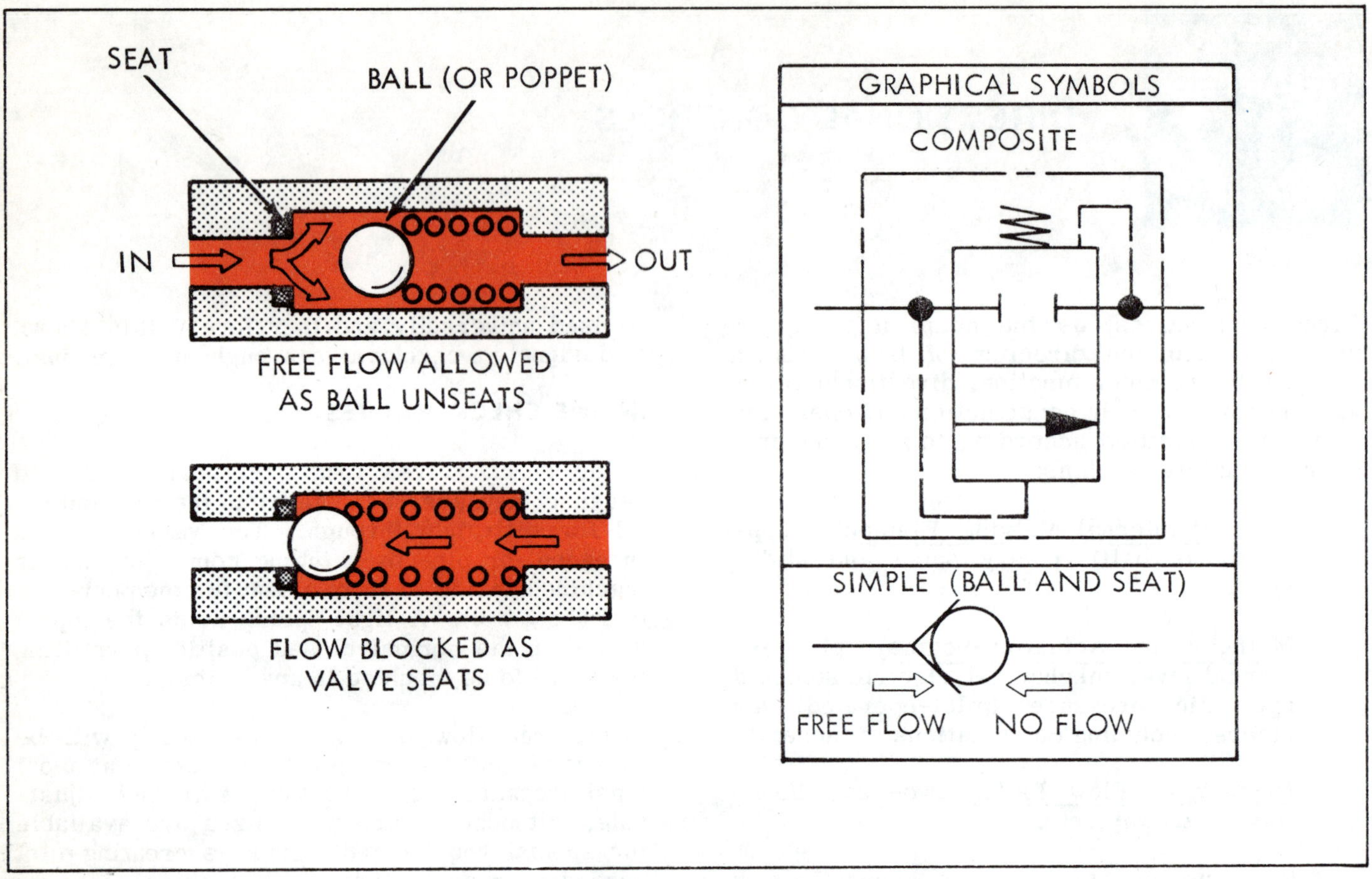

Fig. 7-1. A Check Valve is a One Way Valve

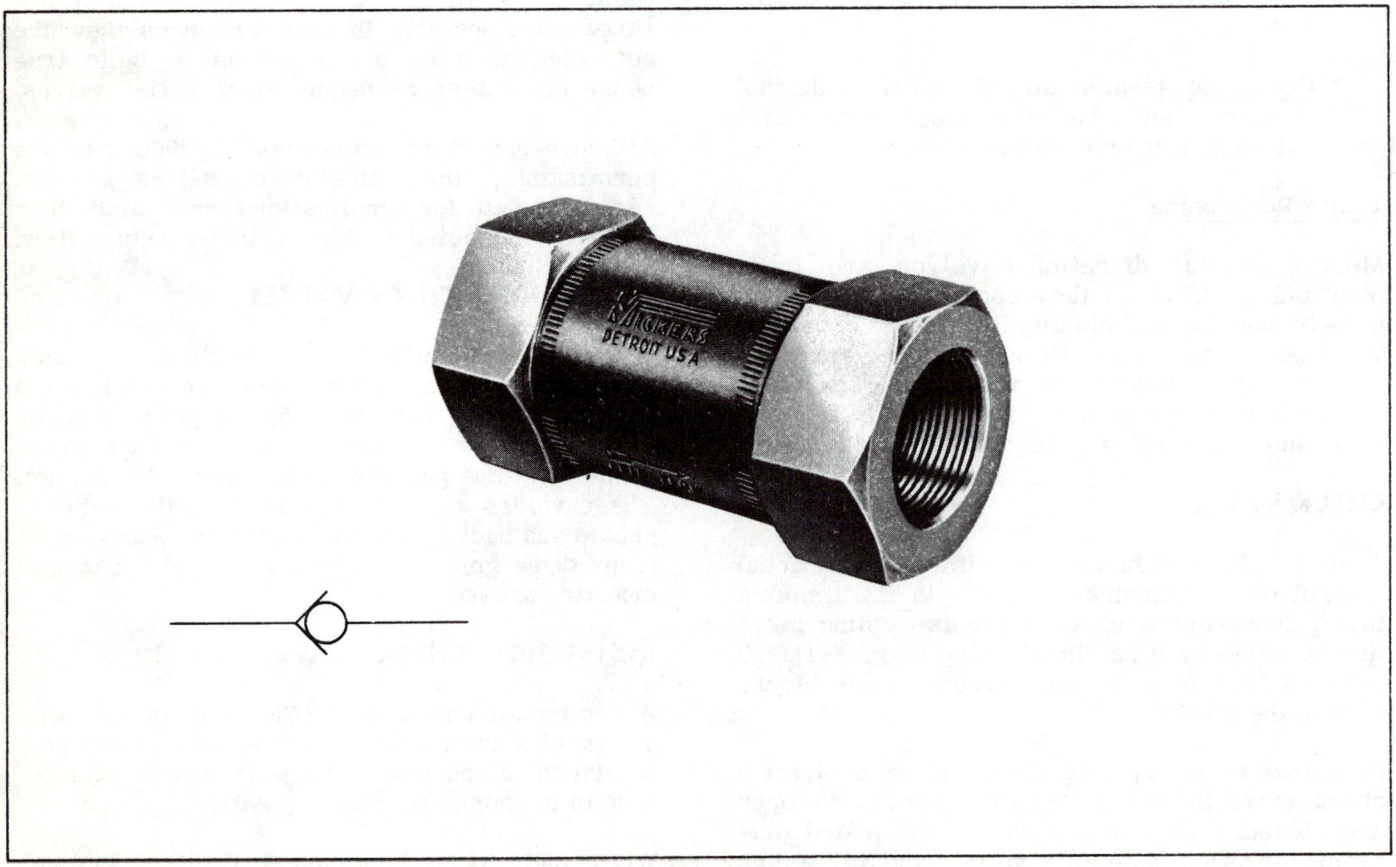

Fig. 7-2 Inline Check Valve

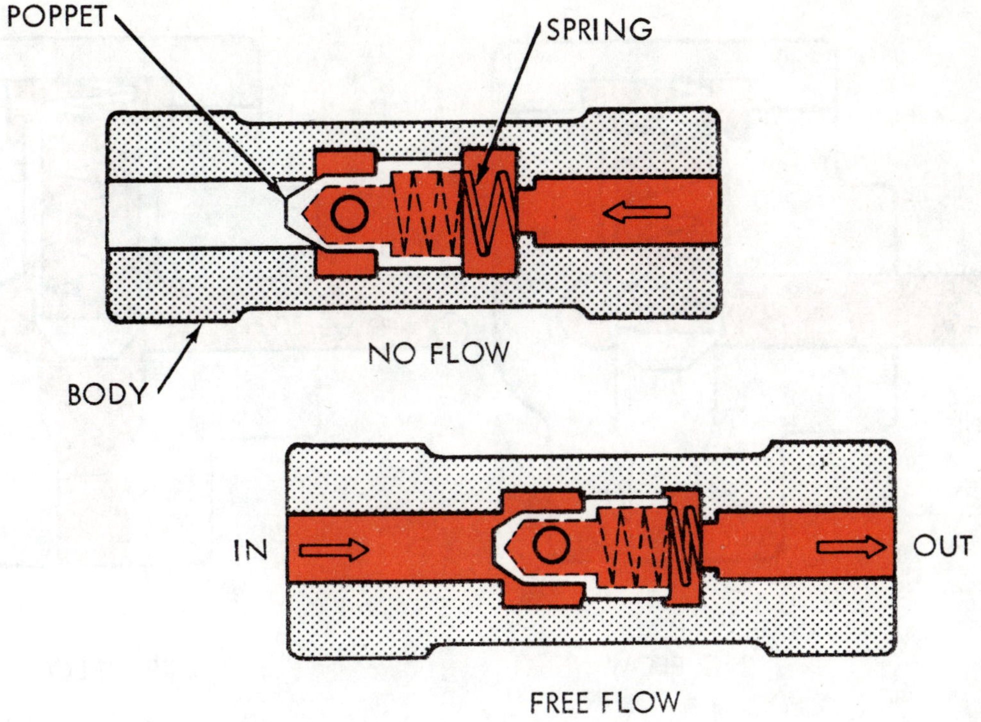

POPPET

SPRING

BODY

NO FLOW

IN

OUT

FREE FLOW

Fig. 7-3. Inline Check Valve Operation

Fig. 7-4. Typical "Right Angle" Check Valve

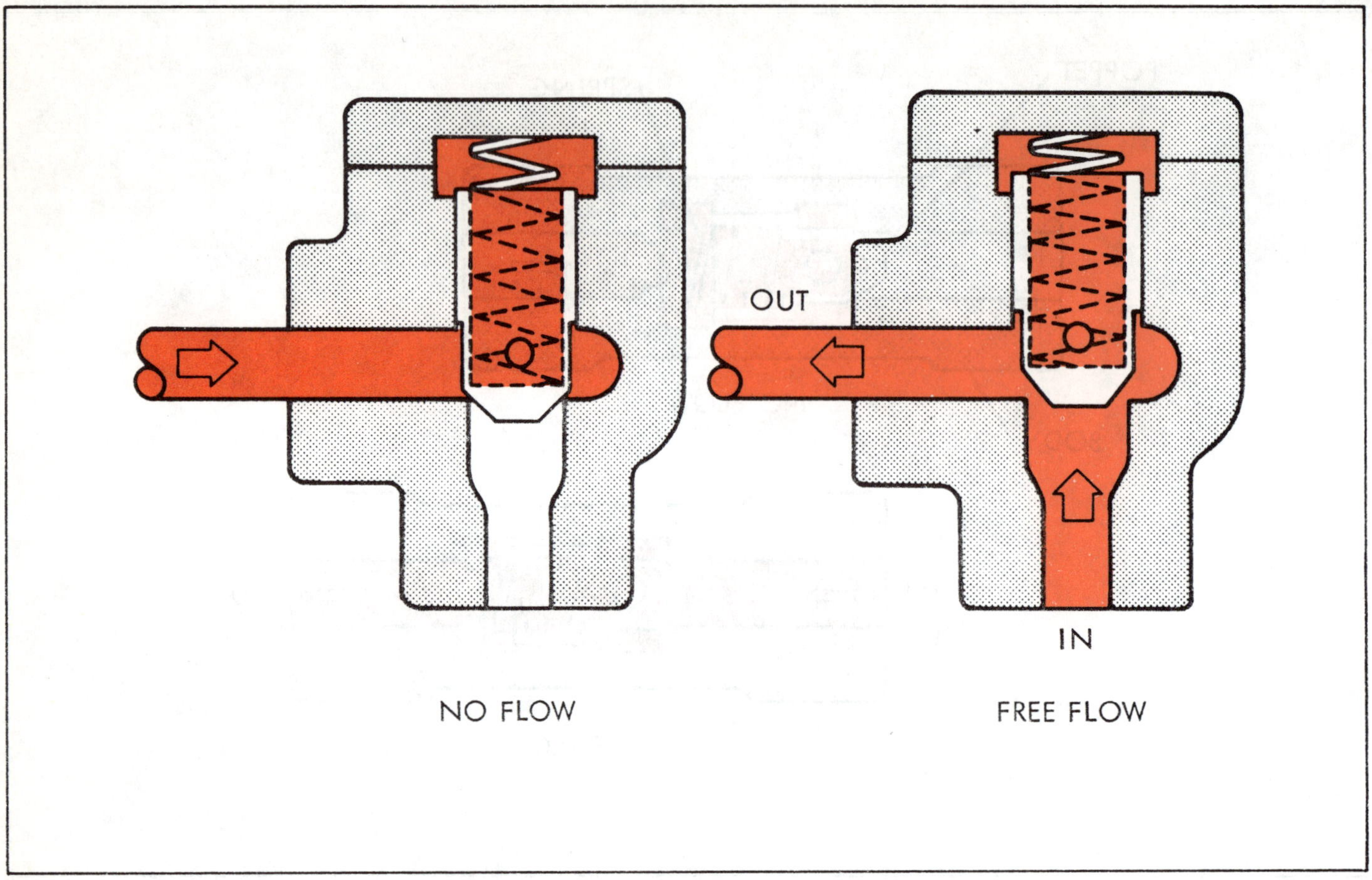

OUT

IN

NO FLOW

FREE FLOW

Fig. 7-5. Right Angle Check Valve

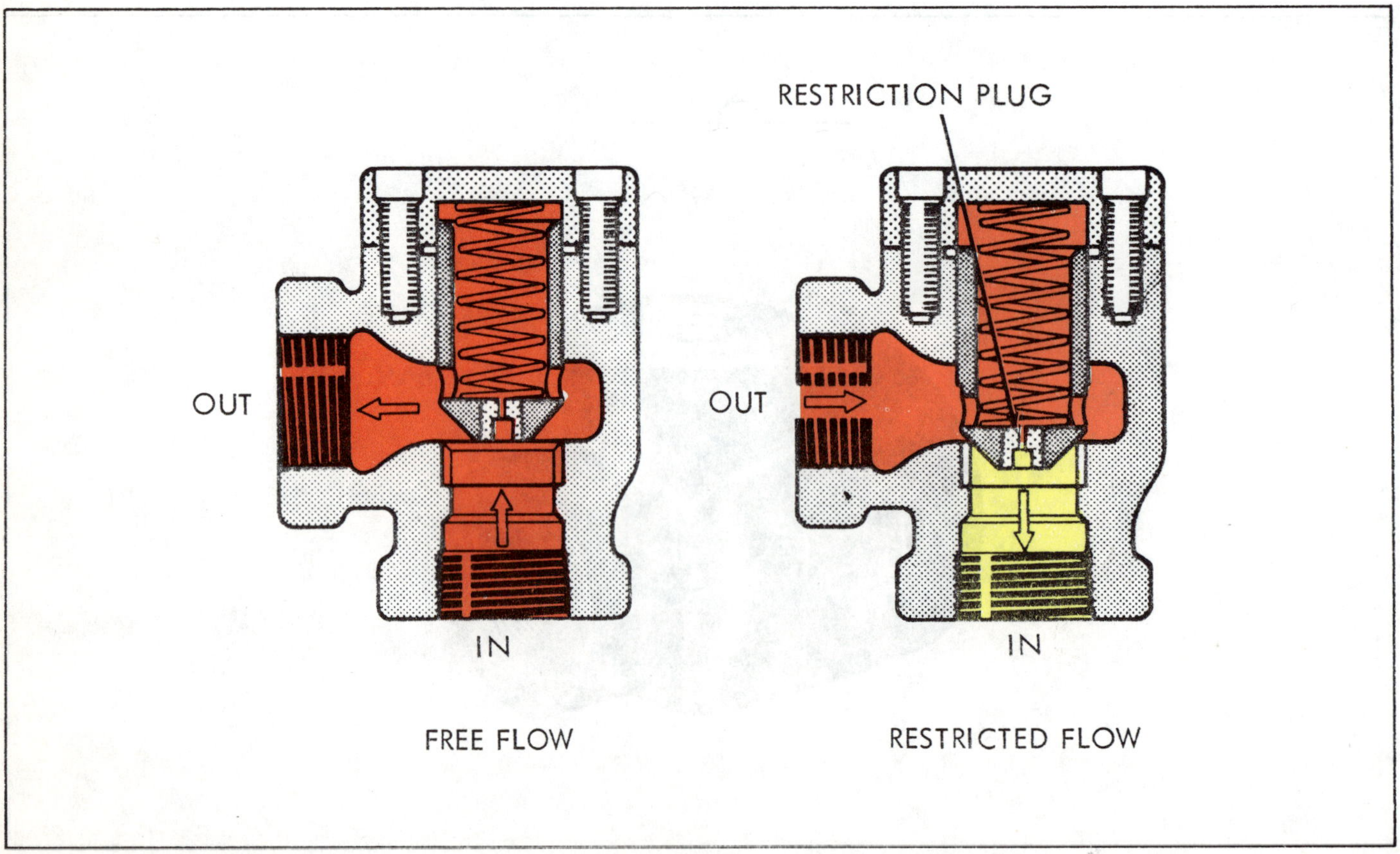

RESTRICTION PLUG

OUT

OUT

IN

IN

FREE FLOW

RESTRICTED FLOW

Fig. 7-6. Restriction Check Valve Allows Bleed in No Flow Direction

flow of fluid in one direction and a controlled amount in the other. One example would be in controlling the rate of decompression in a large press.

PILOT-OPERATED CHECK VALVES

Pilot-operated check valves are designed to permit free flow in one direction and to block return flow, until opened by a pressure (pilot) signal. They are used in hydraulic presses as prefill valves--to permit the main ram to fill by gravity during the "fast approach" part of the stroke. They also are used to support vertical pistons which otherwise might drift downward due to leakage past the directional valve spool.

Two designs of pilot-operated check valves are identified as "2C" or "4C" models.

4C Series

Figure 7-7 illustrates the construction of the "4C" type valve. The check valve poppet is lightly spring-loaded against its seat, which is integral with a sleeve that guides the pilot piston. A pilot pressure port in the valve end cover connects to a passage terminating at the head of the pilot piston.

Three operating conditions of the valve are shown in Figure 7-8. In views A and B there is no pilot pressure and the operation is the same as a conventional check valve. In A, pressure at the head of the poppet (valve inlet) has overcome spring force to open the valve and permit flow. In B, pressure is higher on the spring side of the poppet and reverse flow is blocked.

View C shows the condition when pilot pressure is applied to the head of the pilot piston. The stem pushes the poppet off its seat and allows reverse flow. The pilot pressure required to unseat the poppet this way must exceed 40 percent of the pressure in the "outlet" chamber.

2C Series

The "2C" type valve is illustrated in Figure 7-9. In this design, the check valve poppet resembles an automobile engine valve, and has the pilot piston attached to the threaded poppet stem by a nut. The light spring holds the poppet seated in a no-flow condition by pushing against the pilot piston. A separate drain port is provided to prevent oil from creating a pressure buildup on the underside of the piston.

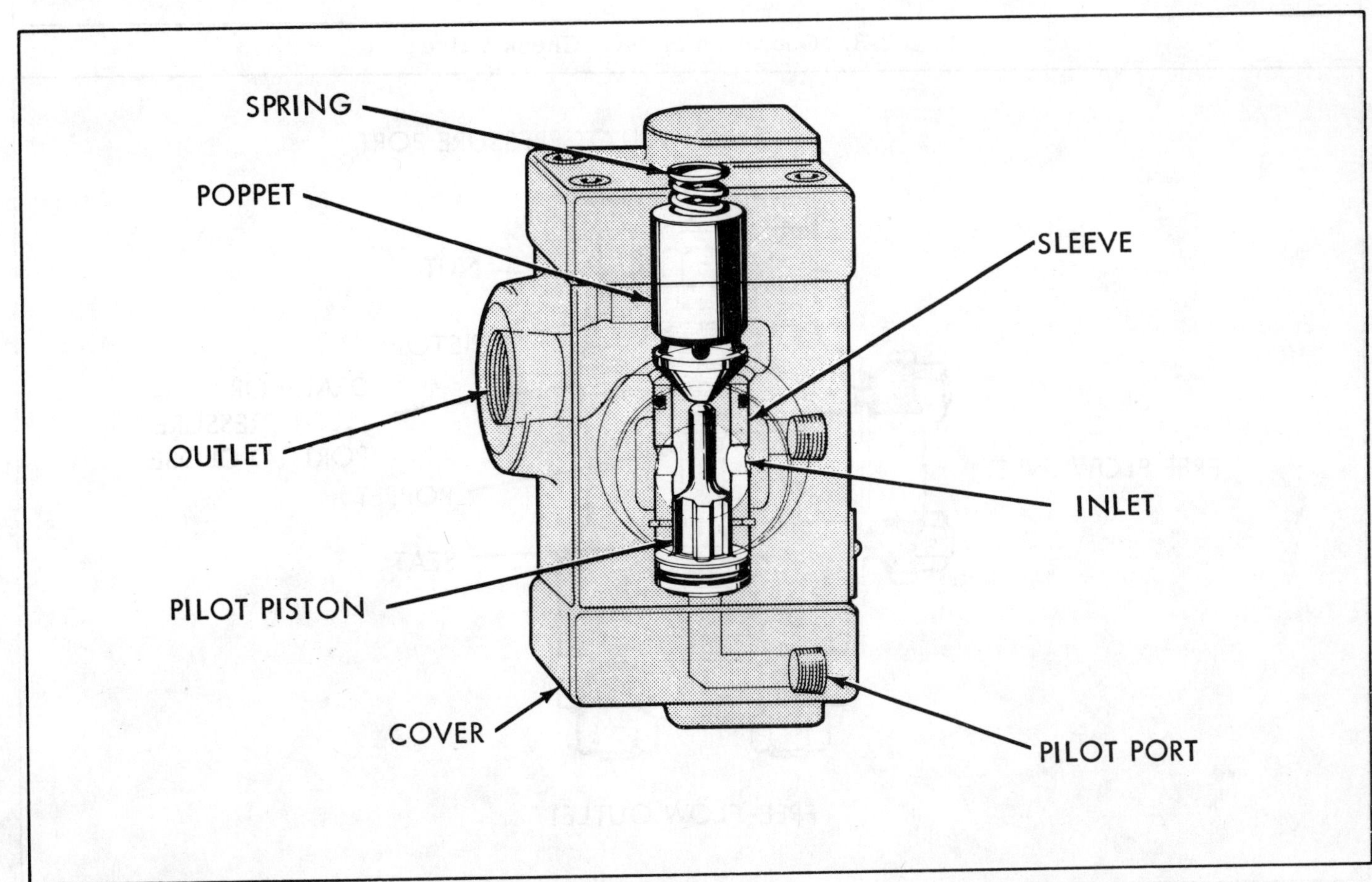

Fig. 7-7. Construction of "4C" Check Valve

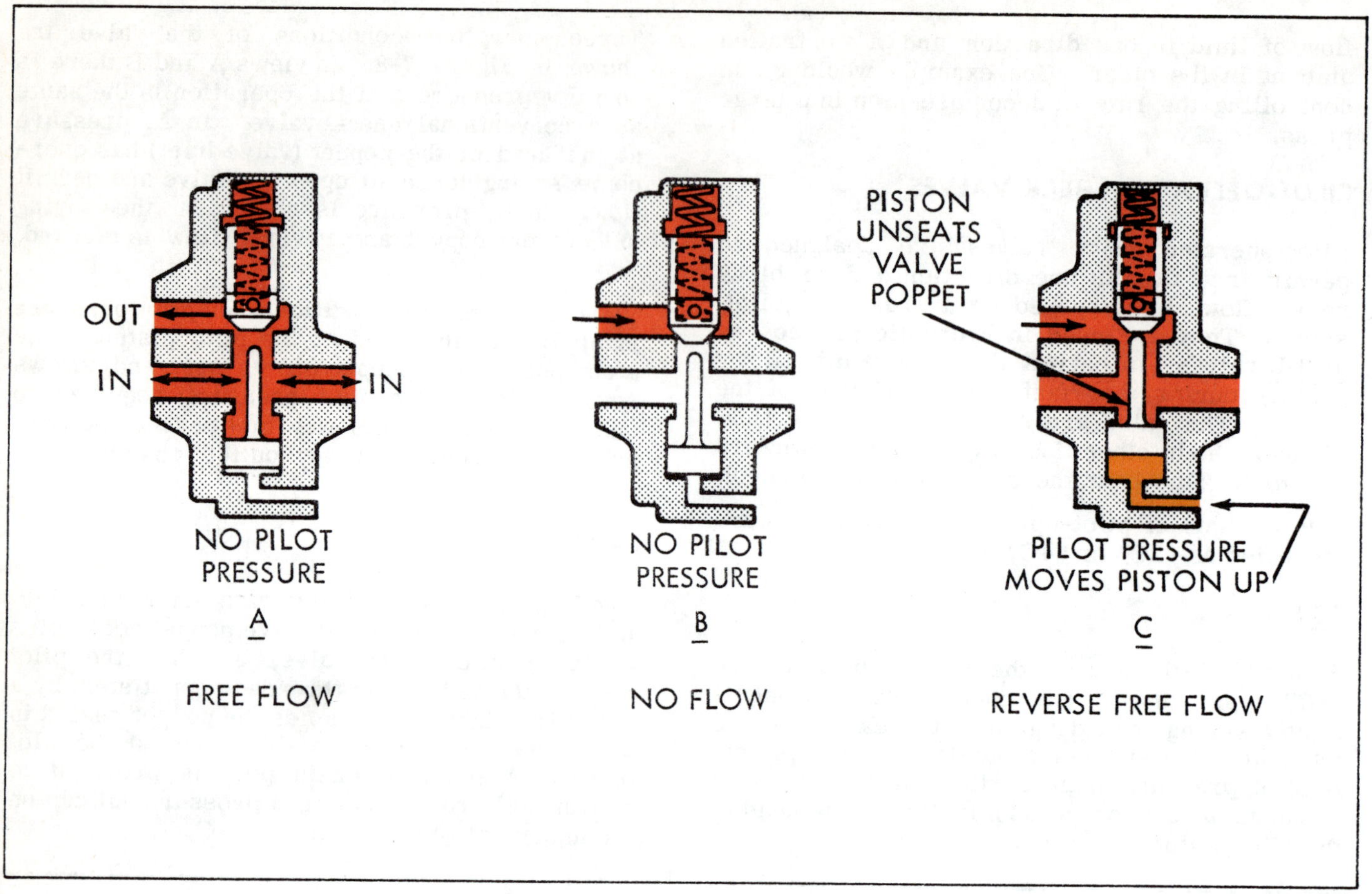

Fig. 7-8. Operation of "4C" Check Valve

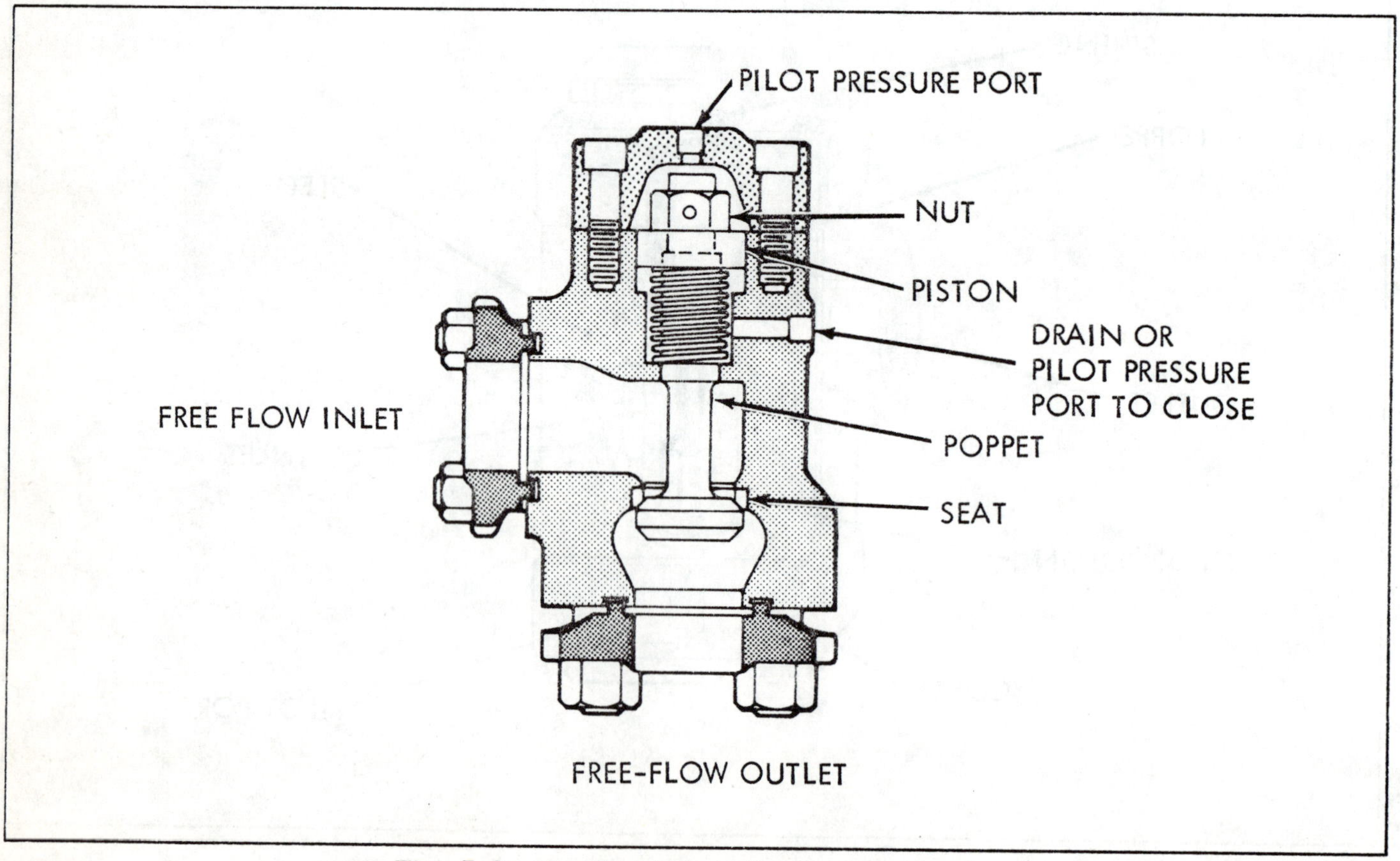

Fig. 7-9. Construction of "2C" Check Valve

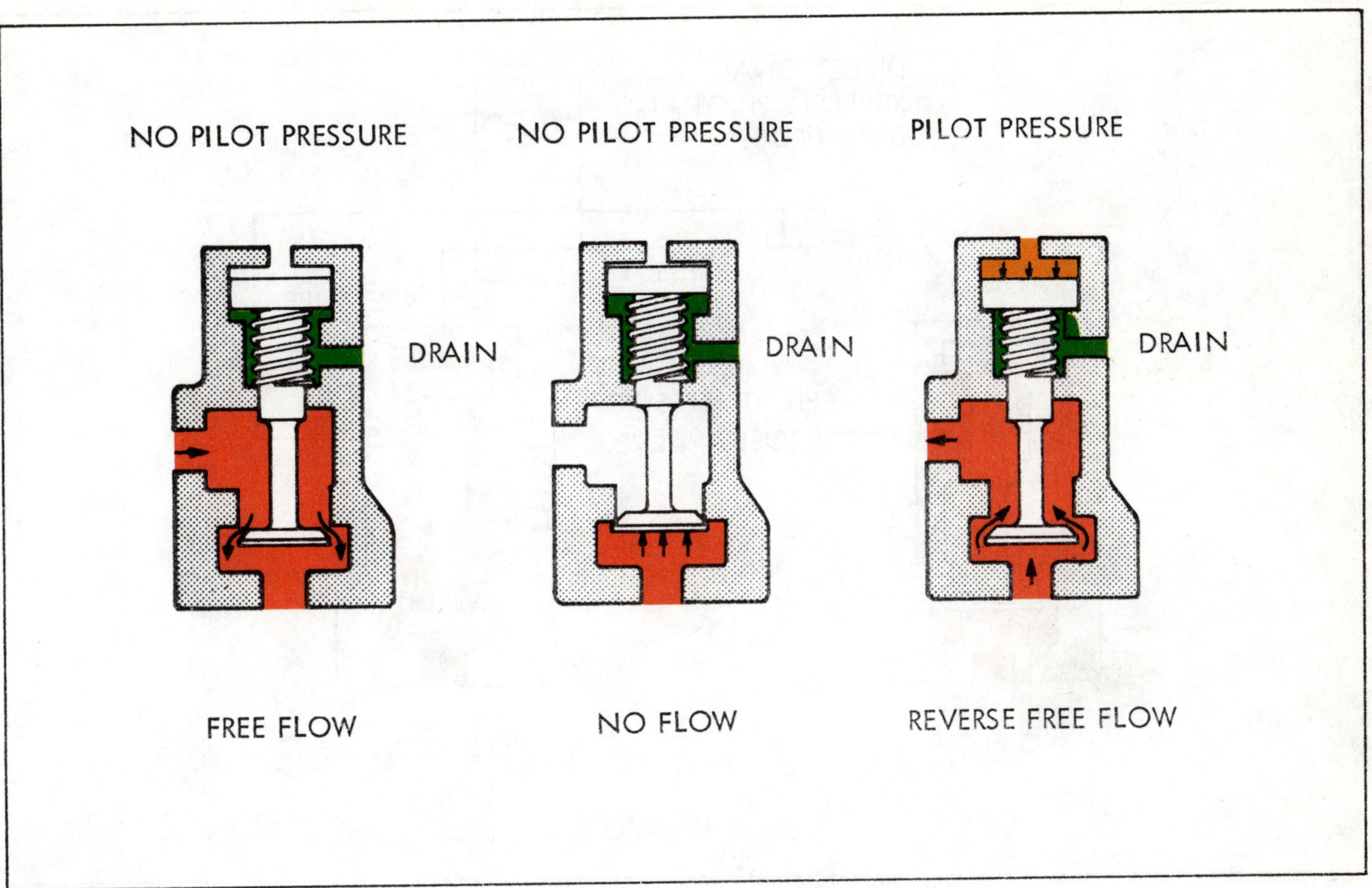

Fig. 7-10. "2C" Check Valve Operation

Figure 7-10, views A and B, show the operation as a conventional check valve with no pilot pressure imposed. Reverse flow (view C) can occur only when a pressure exceeding 80 percent of the pressure in the outlet chamber is effective against the pilot piston.

The valve is also built in a "no-spring" version (Fig. 7-11) for applications where it is desirable to hold the poppet either open or closed. In the r[illegible] spring design, the drain and pilot ports both [illegible] as pilot pressure ports and are reversed by use of a separate directional valve. Pilot pressure is used to hold the valve in the desired position.

Pilot-Operated Check Valve Applications

Figure 7-12 illustrates the basic operational difference between the "2C" and "4C" valves. In the "4C" type, pressure in the inlet chamber acts against the pilot piston to resist pilot operation. In the "2C" type, inlet pressure assists pilot actuation.

The "4C" valve thus is used in applications where the inlet port is connected to tank during reverse flow. Typical applications are blocking flow around a volume control valve during a feed cycle or preventing a cylinder from drifting due to leakage through a directional valve.

The "2C" valve is used effectively to intermittently block flow from an accumulator. It permits free flow to the accumulator, and can easily be pilot actuated to permit the accumulator to discharge even though pressure is present at both ports.

TWO-WAY AND FOUR-WAY VALVES

The basic function of two-way and four-way valves is to direct inlet flow to either of two outlet ports. As shown in Figure 7-13, flow to the "P" (pump) port of the valve can be directed to either outlet port (labeled A and B for convenience). In the four-way valve the alternate port is open to the tank port permitting return flow to the reservoir. In two-way valves the alternate port is blocked and the tank port serves only to drain leakage from within the valve.

Most of these valves are the sliding spool type, although there are rotary valves which are used principally for pilot control. They are built in two-position or three-position versions. The three-position valve has a center or neutral position. Methods of actuation include manual

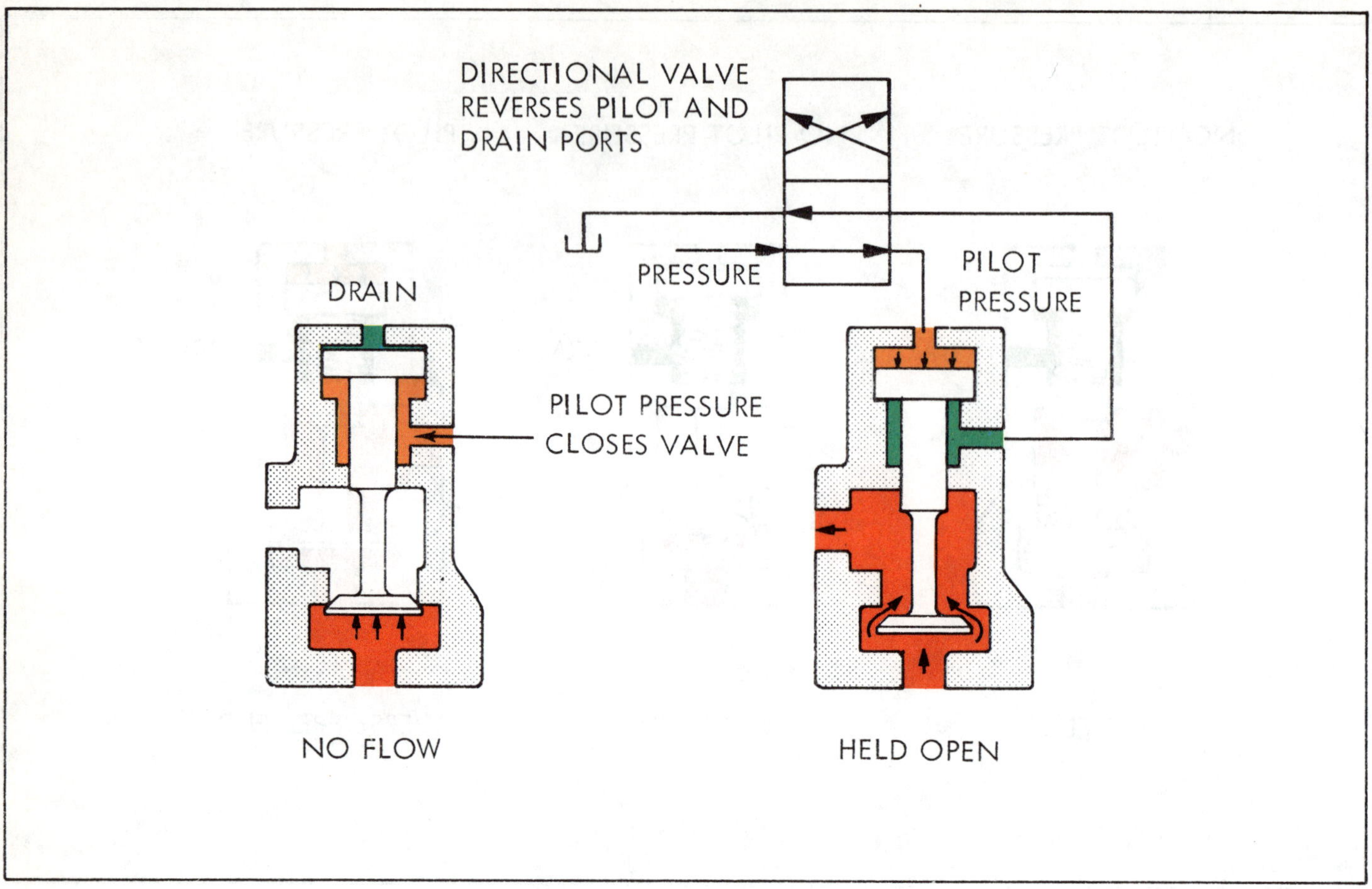

DIRECTIONAL VALVE
REVERSES PILOT AND
DRAIN PORTS
PRESSURE
PILOT
PRESSURE
DRAIN
PILOT PRESSURE
CLOSES VALVE
NO FLOW
HELD OPEN

Fig. 7-11. No-Spring "2C" Check Valve

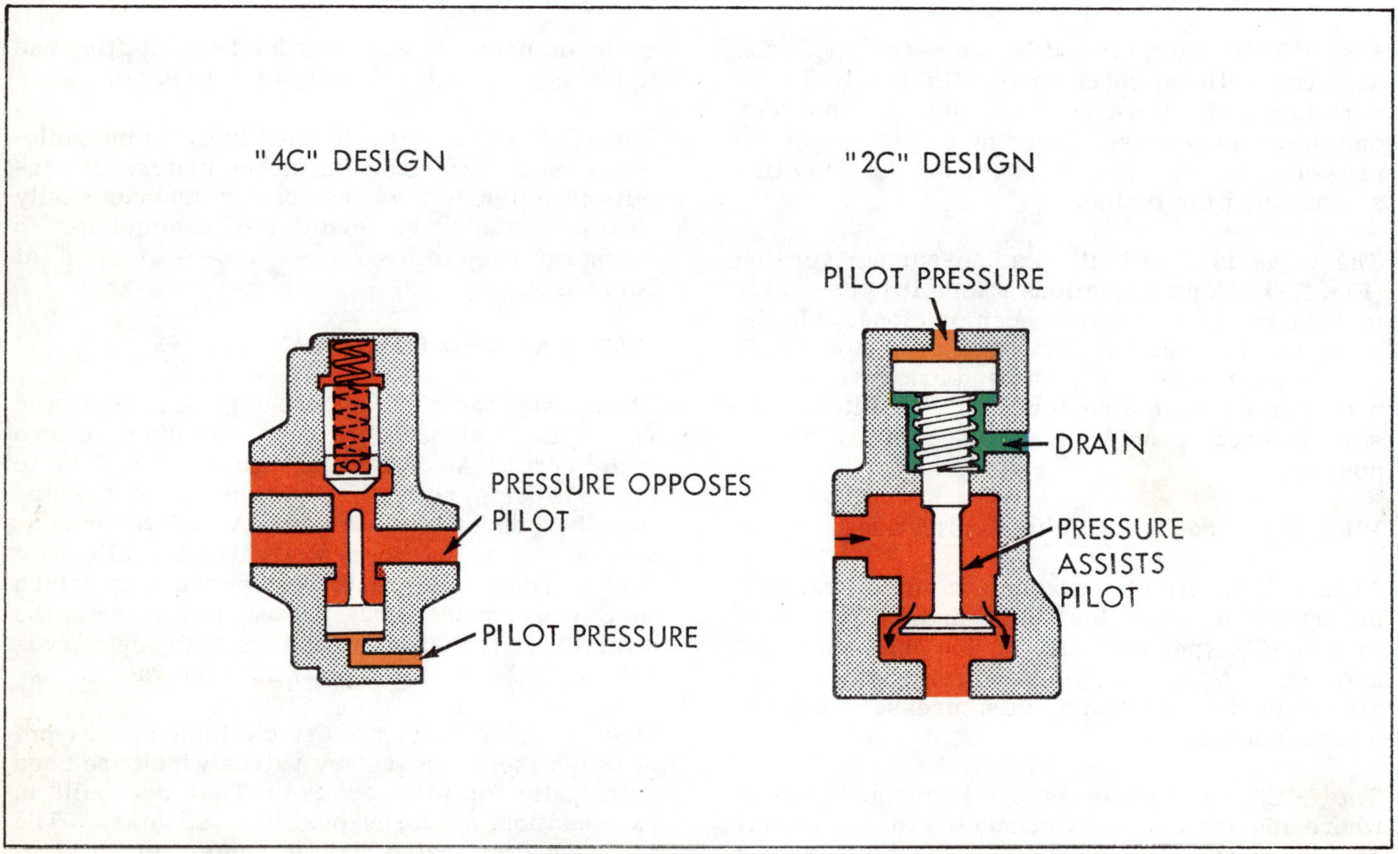

"4C" DESIGN
PRESSURE OPPOSES
PILOT
PILOT PRESSURE
"2C" DESIGN
PILOT PRESSURE
DRAIN
PRESSURE
ASSISTS
PILOT

Fig. 7-12. Effects of Pressure on Pilot Control

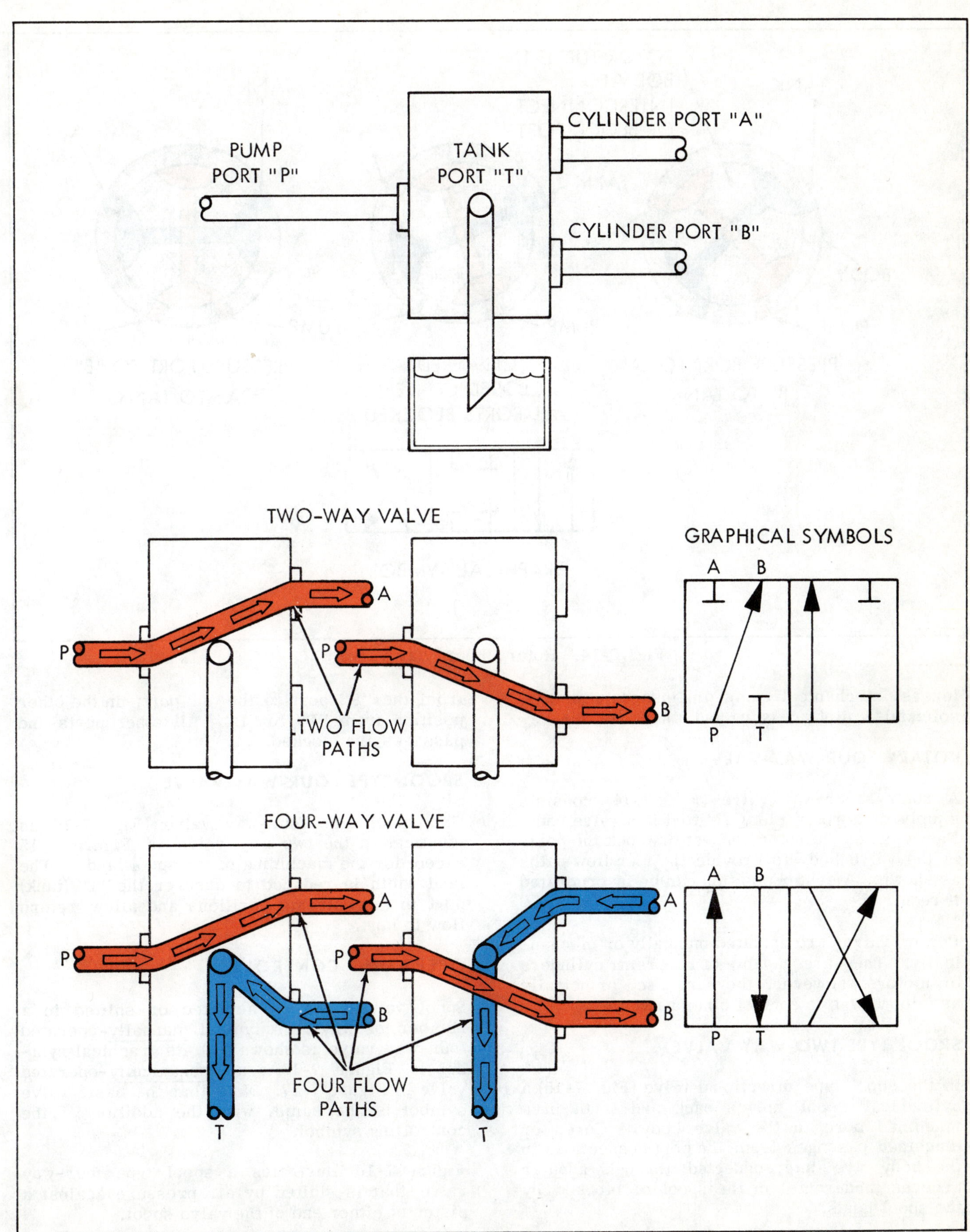

Fig. 7-13. Flow Paths in Two Way and Four-Way Valves

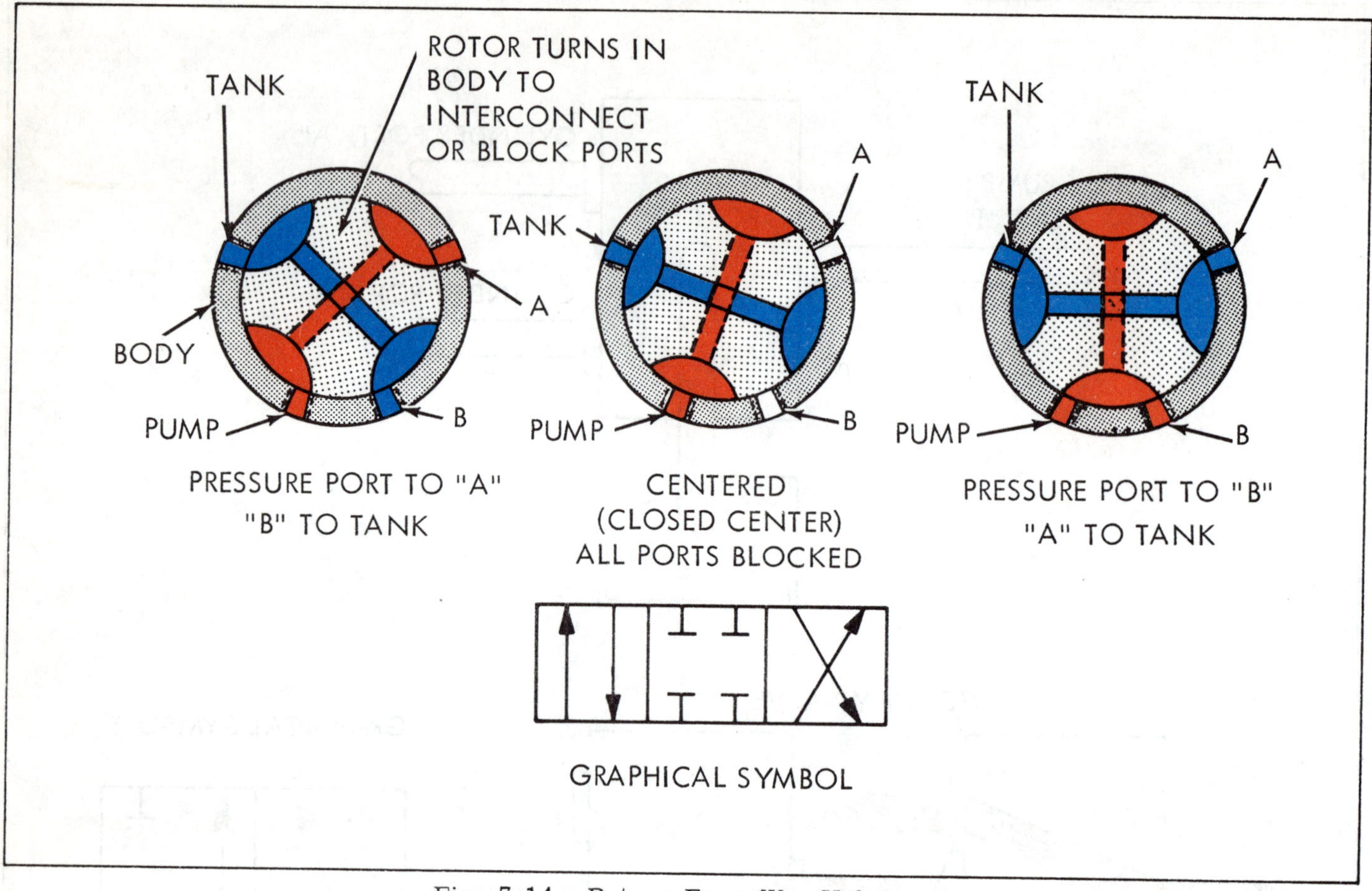

Fig. 7-14. Rotary Four-Way Valve

levers, mechanical cams or linkages, springs, solenoids, pilot pressure and others.

ROTARY FOUR-WAY VALVE

A rotary four-way valve (Fig. 7-14) consists simply of a rotor closely fitted in a valve body. Passages in the rotor connect or block the ports in the valve body to provide the four flow paths as shown. A center position can be incorporated if required.

Rotary valves are actuated manually or mechanically. They are capable of reversing cylinders or motors; however, they are used principally as pilot valves to control other valves.

SPOOL TYPE TWO-WAY VALVE

In the spool type directional valve (Fig. 7-15) a cylindrical spool moves back and forth in a machined bore in the valve body. Cored or machined passages from the port connections in the body are interconnected through annular grooves (undercuts) in the spool or blocked by the spool lands.

The two-way version permits selection of two flow paths. In one position, flow is permitted from the "P" port to the "A" port; in the other position from "P" to "B". All other ports and passages are blocked.

SPOOL TYPE FOUR-WAY VALVE

The spool type four-way valve (Fig. 7-16) is identical to the two-way valve in Figure 7-15 except for the machining of the spool lands. The land width is reduced to uncover the "T" (tank) port in the extreme positions and allow return flow to tank.

OPERATING CONTROLS

Spool valves can be actuated or shifted in a number of ways. A typical manually-operated four-way valve is shown with its graphical symbol in Figure 7-16A; a mechanically-operated valve in Figure 7-17. Note that the basic valve symbol is the same, with the addition of the controlling symbol.

Figure 7-18 illustrates a spool type four-way valve that is shifted by air pressure against a piston at either end of the valve spool.

A very common method of actuating a small spool valve is with a solenoid (Fig. 7-19). Elec-

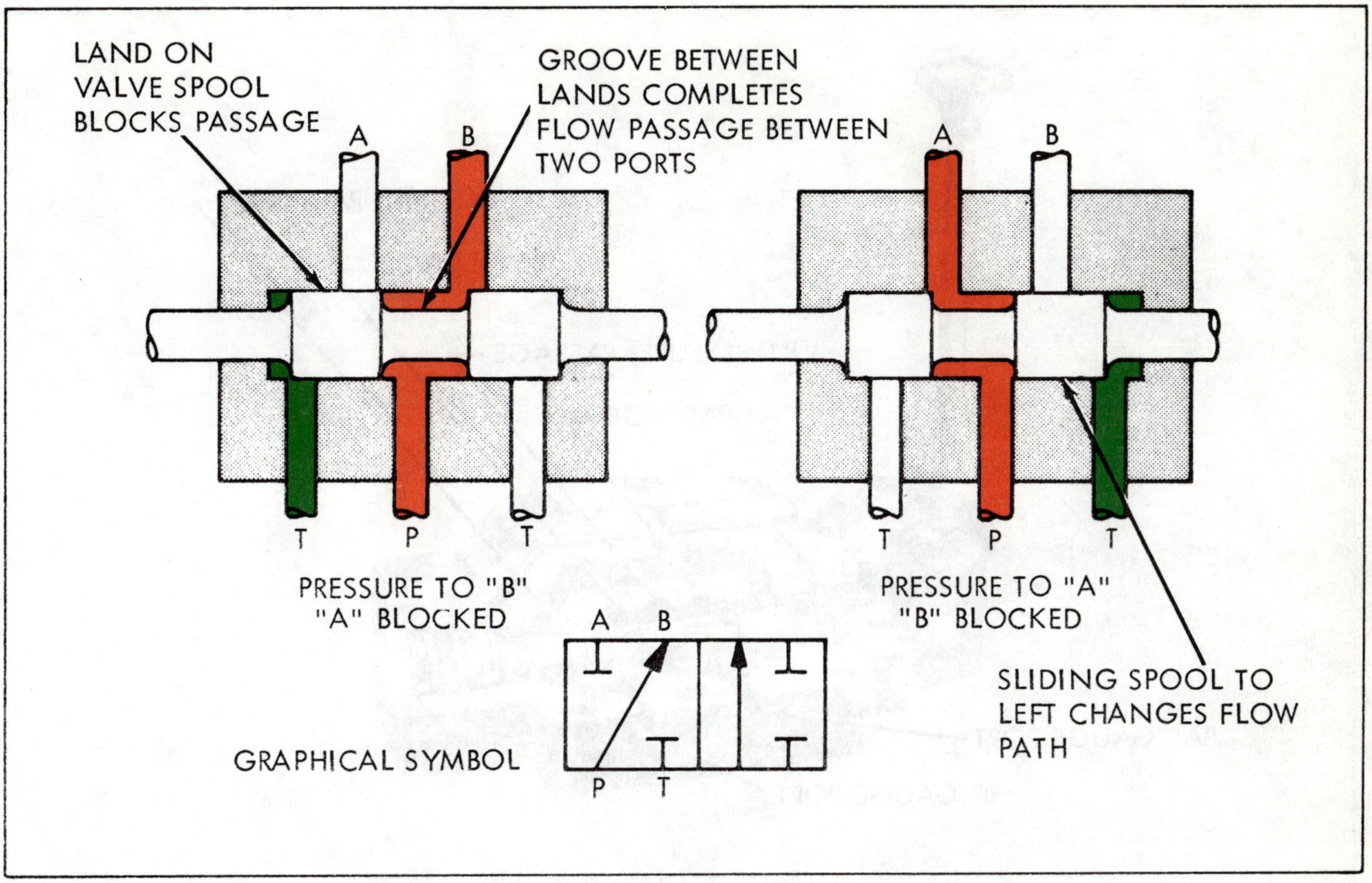

LAND ON
VALVE SPOOL
BLOCKS PASSAGE

GROOVE BETWEEN
LANDS COMPLETES
FLOW PASSAGE BETWEEN
TWO PORTS

A B A B

T P T T P T

PRESSURE TO "B"
"A" BLOCKED

PRESSURE TO "A"
"B" BLOCKED

GRAPHICAL SYMBOL

A B
P T

SLIDING SPOOL TO
LEFT CHANGES FLOW
PATH

Fig. 7-15. Two-Way Spool Valve Slides in Machined Bore

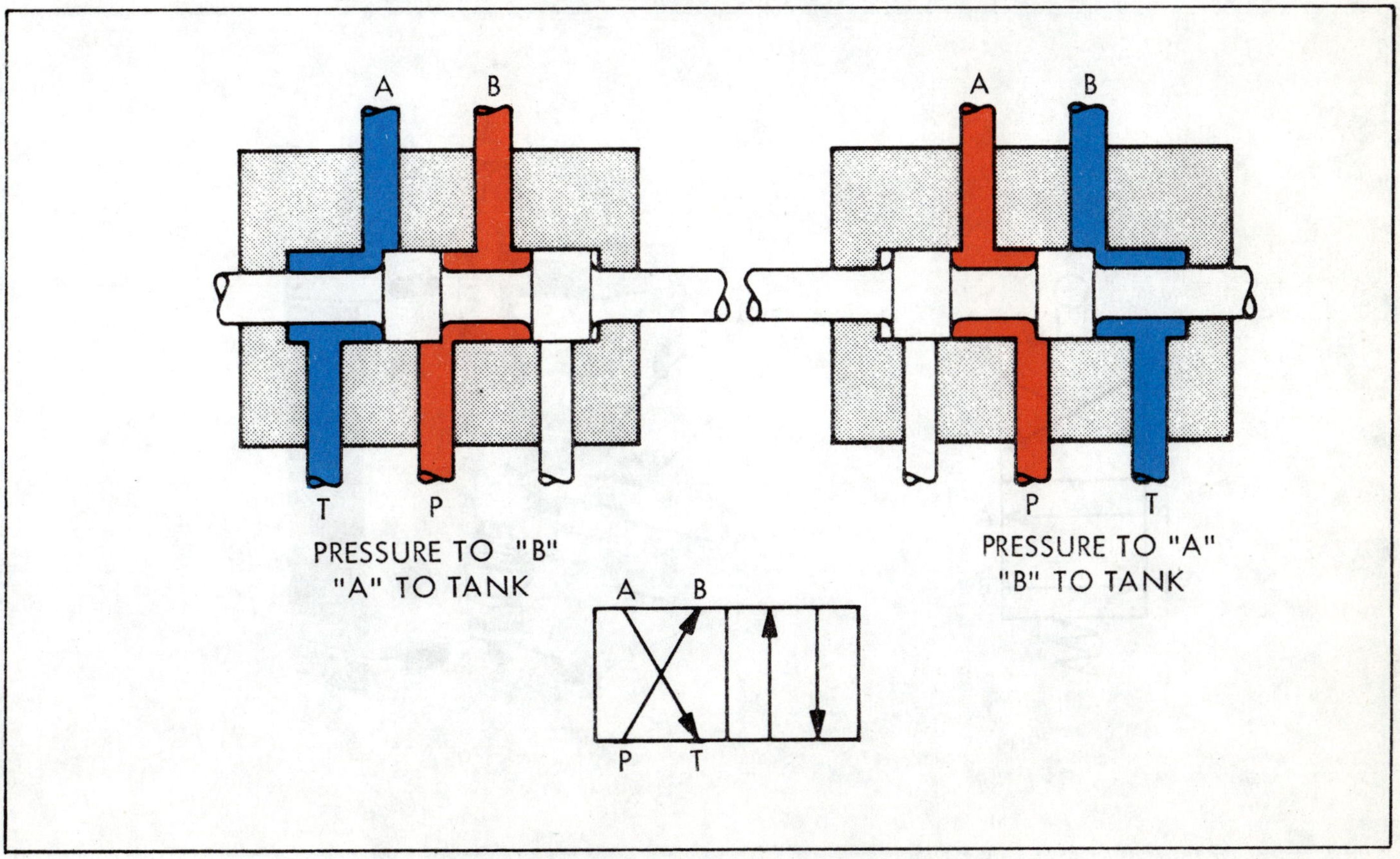

A B A B

T P P T

PRESSURE TO "B"
"A" TO TANK

PRESSURE TO "A"
"B" TO TANK

A B
P T

Fig. 7-16. Spool Type Four-Way Valve

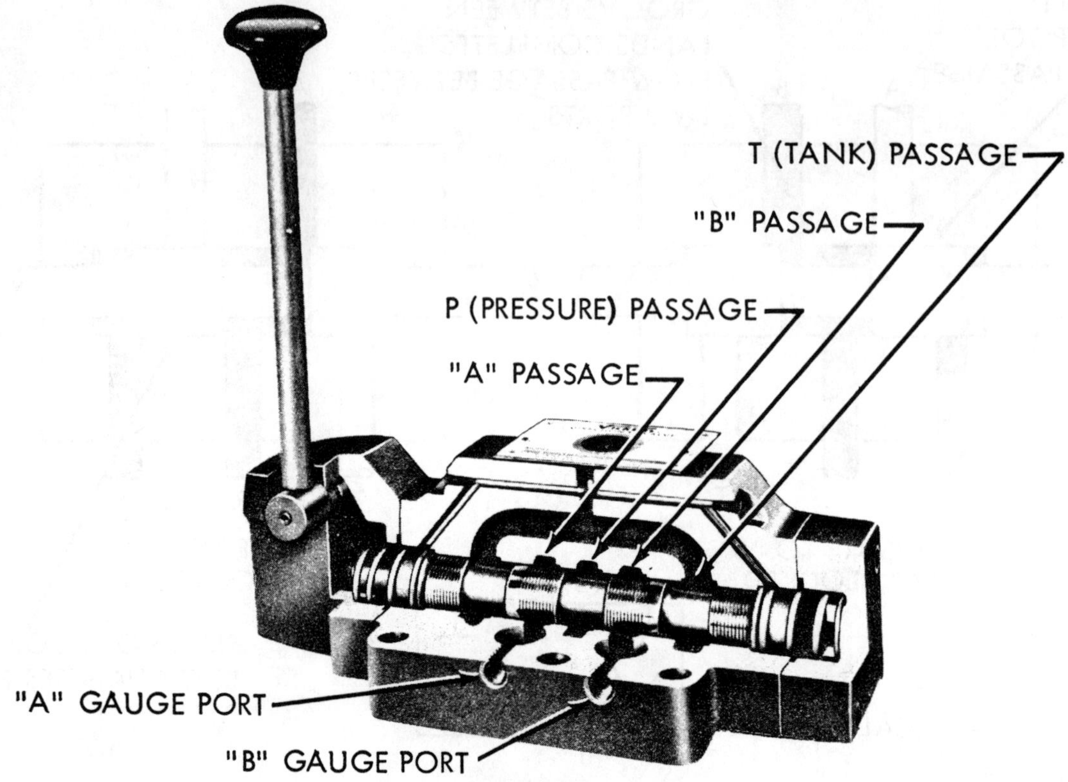

T (TANK) PASSAGE
"B" PASSAGE
P (PRESSURE) PASSAGE
"A" PASSAGE
"A" GAUGE PORT
"B" GAUGE PORT

Fig. 7-16A. Manually Operated Four-Way Valve

P
A
T
B
"A"
"P"
"B"
"T"

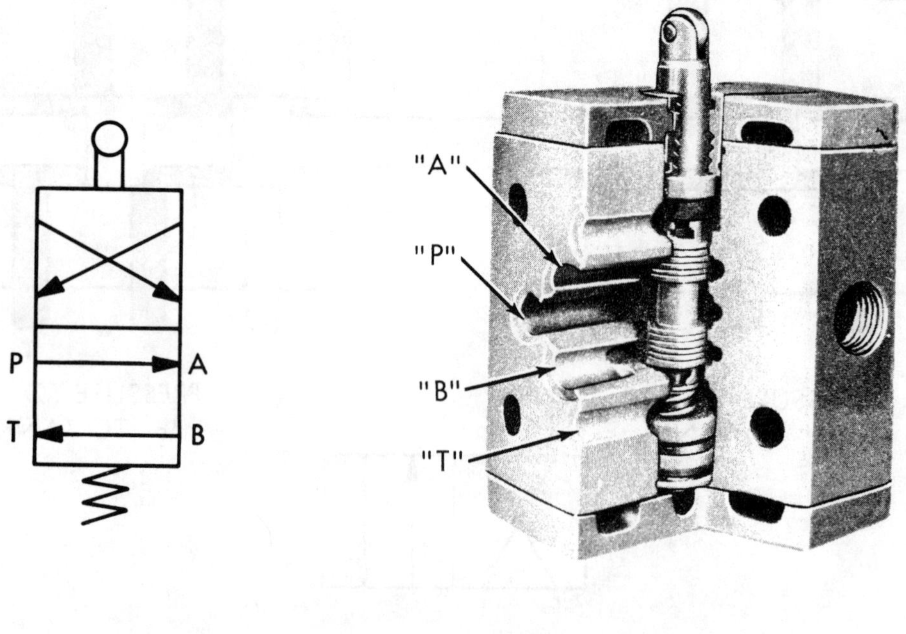

Fig. 7-17. Mechanically Operated Four-Way Valve

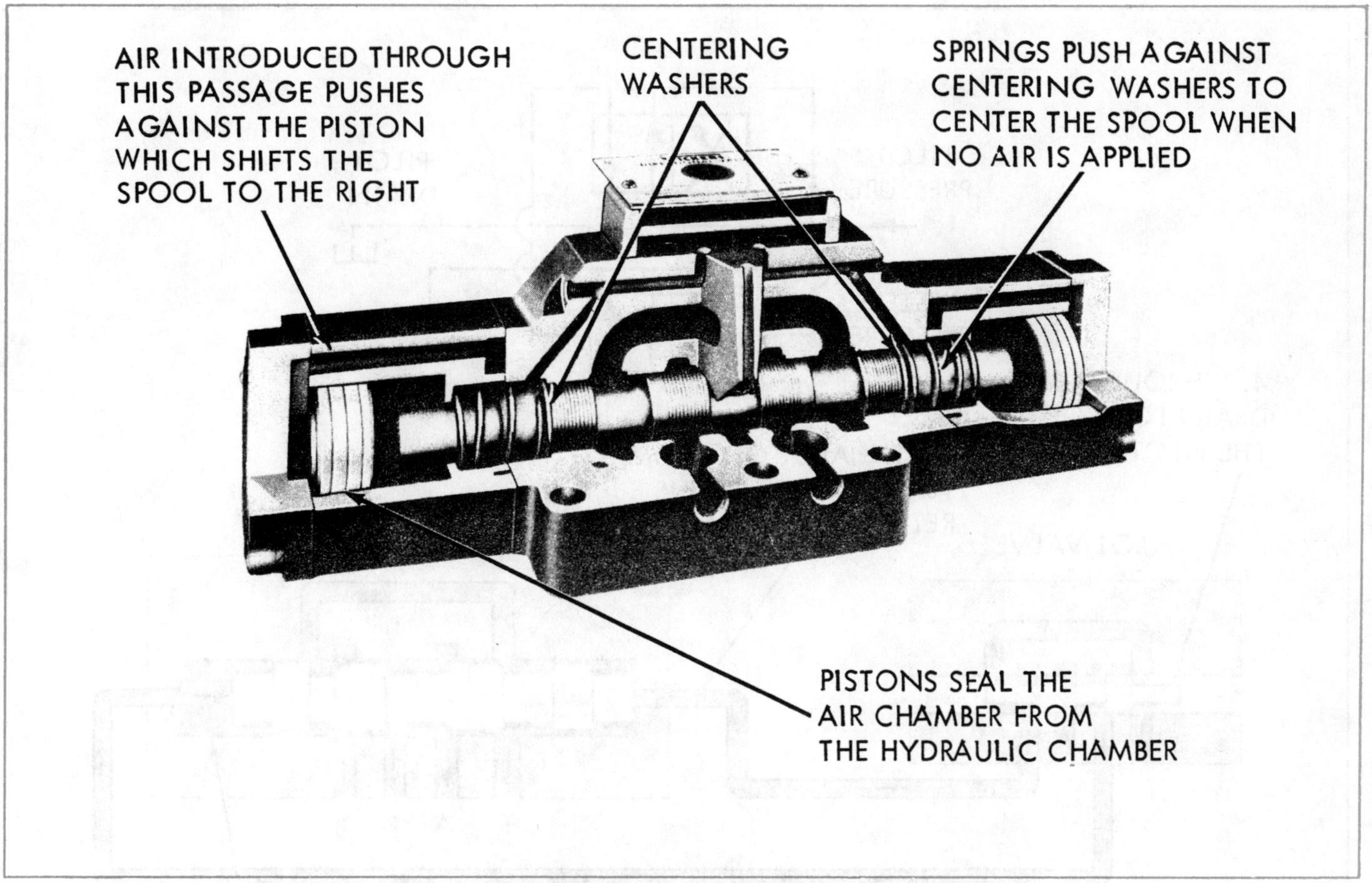

AIR INTRODUCED THROUGH THIS PASSAGE PUSHES AGAINST THE PISTON WHICH SHIFTS THE SPOOL TO THE RIGHT

CENTERING WASHERS

SPRINGS PUSH AGAINST CENTERING WASHERS TO CENTER THE SPOOL WHEN NO AIR IS APPLIED

PISTONS SEAL THE AIR CHAMBER FROM THE HYDRAULIC CHAMBER

Fig. 7-18. Pneumatically-Shifted Four-Way Valve

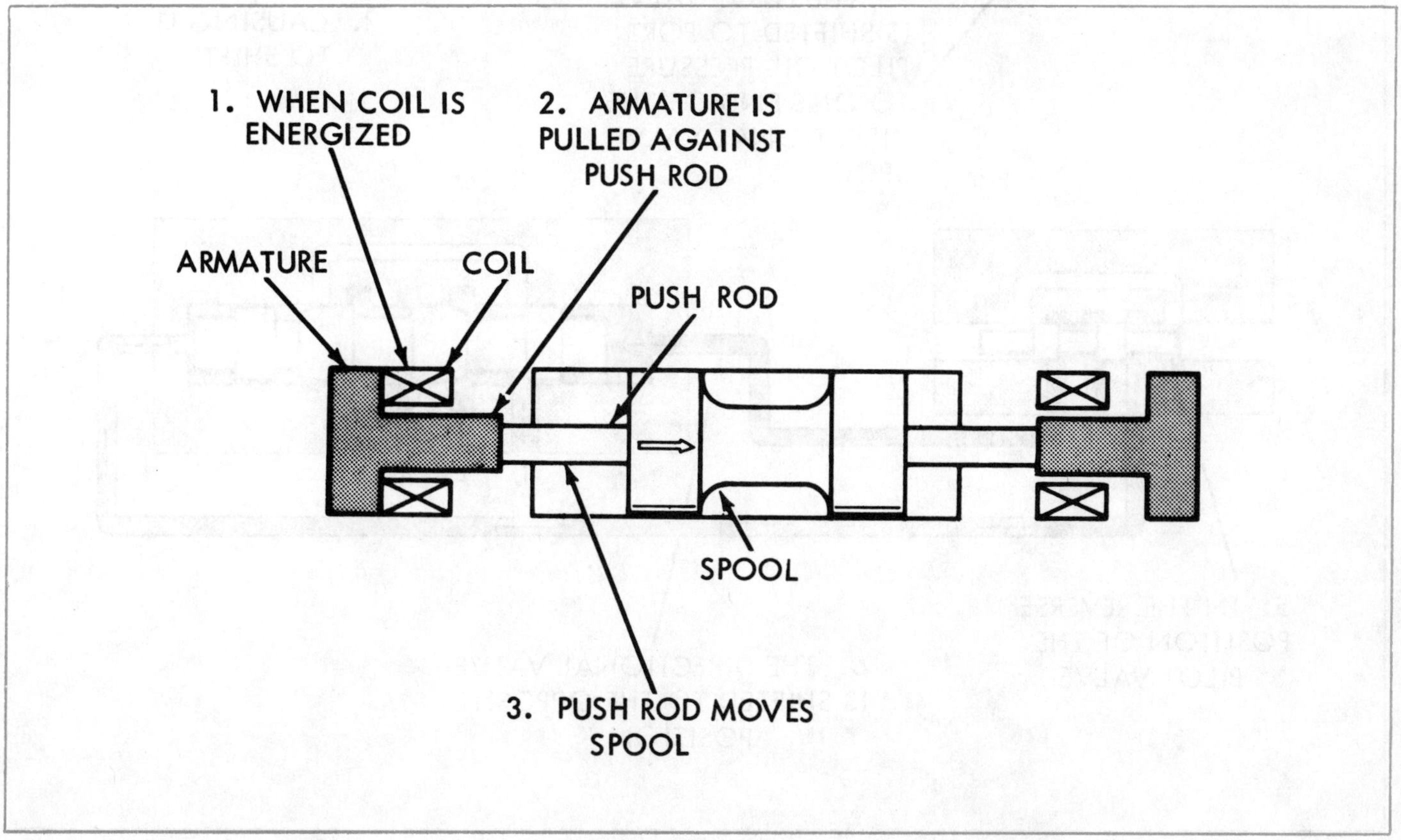

1. WHEN COIL IS ENERGIZED

2. ARMATURE IS PULLED AGAINST PUSH ROD

ARMATURE

COIL

PUSH ROD

SPOOL

3. PUSH ROD MOVES SPOOL

Fig. 7-19. Push-Type Solenoids Shift Many Small Valve Spools

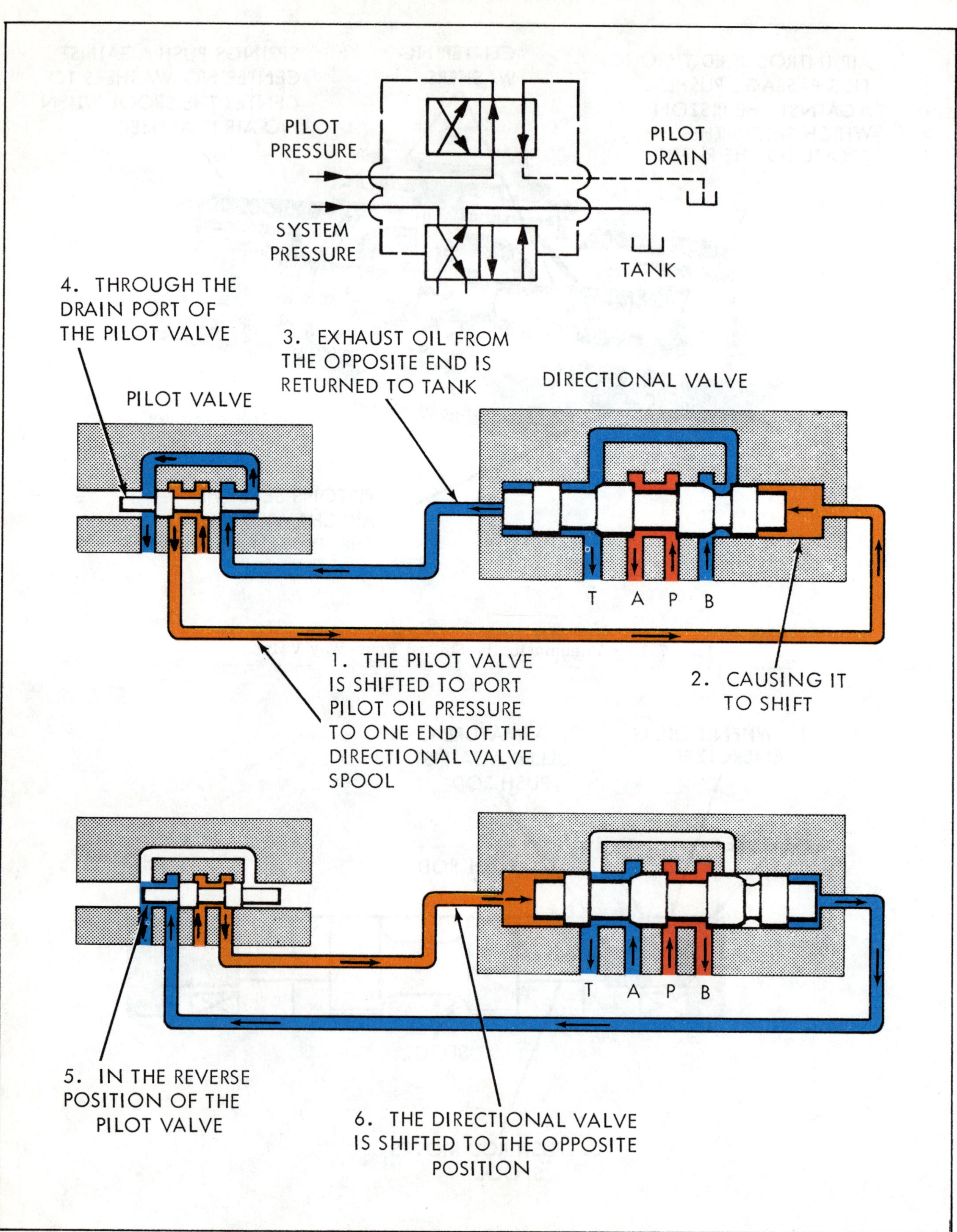

Fig. 7-20. Pilot Pressure is Used To Shift Large Directional Valves

tric energy applied to the solenoid coil creates a magnetic field that draws the armature into the coil. The armature motion is transmitted through a push rod which in turn moves the spool.

In larger valves, the force required to shift the spool is more than it is practical to obtain from a solenoid. Most large directional valves are actuated by pilot pressure against either spool end (Fig. 7-20). The pilot oil is furnished from a smaller four-way valve--referred to as a pilot valve, which while usually actuated by a solenoid may use any of the methods shown in Figures 7-16 through 7-18.

"SPRING CENTERED," "SPRING OFFSET" AND "NO SPRING"

The terms "spring-centered" and "spring-offset" refer to the use of springs returning valve spools to their normal position.

A spring centered valve is returned to the center position by spring force whenever the actuating effort is released. The air-operated valve illustrated in Figure 7-18 is spring-centered.

A spring-offset valve (Fig. 7-21) is a two position valve returned to one extreme position by a spring whenever the actuating effort is released. It is shifted to the opposite position by one of the above methods.

A "no-spring" valve must be actuated entirely by an external control and may "float" between its two positions when the control is released unless it is retained by detents or friction pads. For this reason it is good practice to maintain control on the valve throughout the cycle.

SPOOL CENTER CONDITIONS

Most three-position valves are available with a variety of interchangeable spools. All of the four-way spools provide identical flow patterns in the shifted positions, with different centered conditions as illustrated in Figure 7-22. The open-center type interconnects all ports and the pump delivery can flow to tank at low pressure. The closed center has all ports blocked, so that pump delivery may be used for other operations within the circuit. Otherwise it is forced over the relief valve. Other center conditions permit blocking of selected ports with others open. The tandem type has both cylinder ports blocked in neutral, but the pressure port is open to tank,

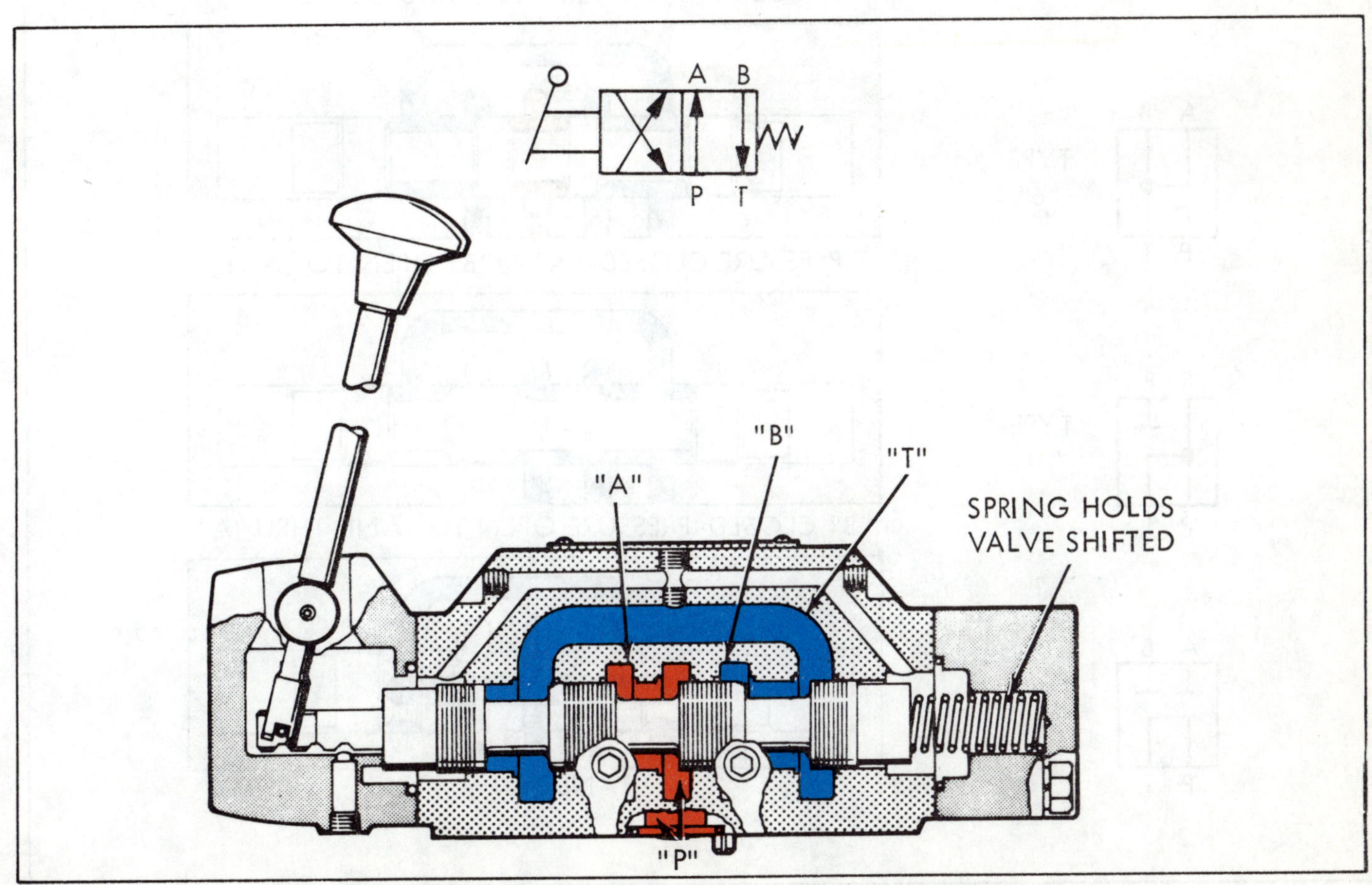

Fig. 7-21. Spring Offset Valve Has Two Positions

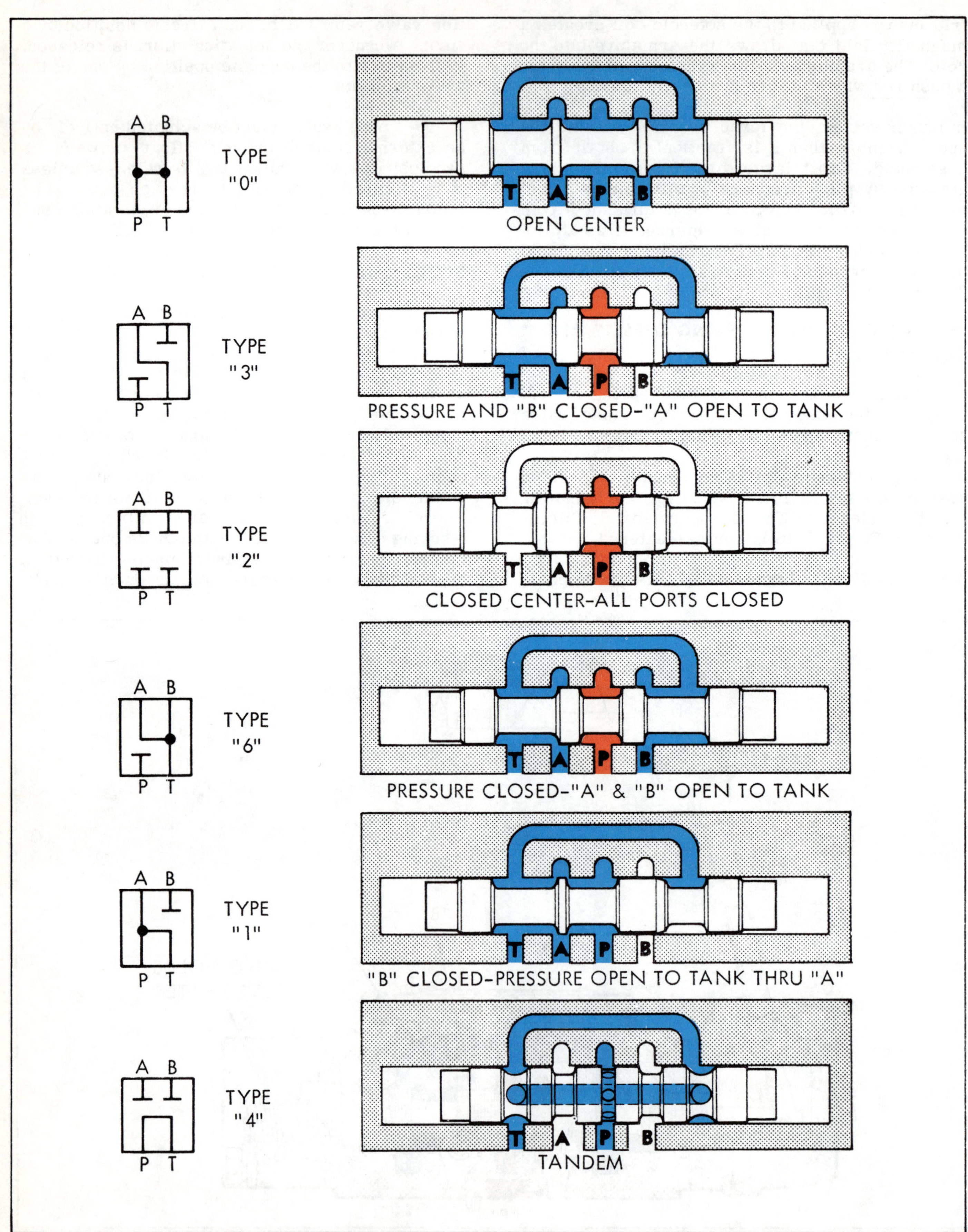

Fig. 7-22. Various Center Conditions for Four-Way Valves

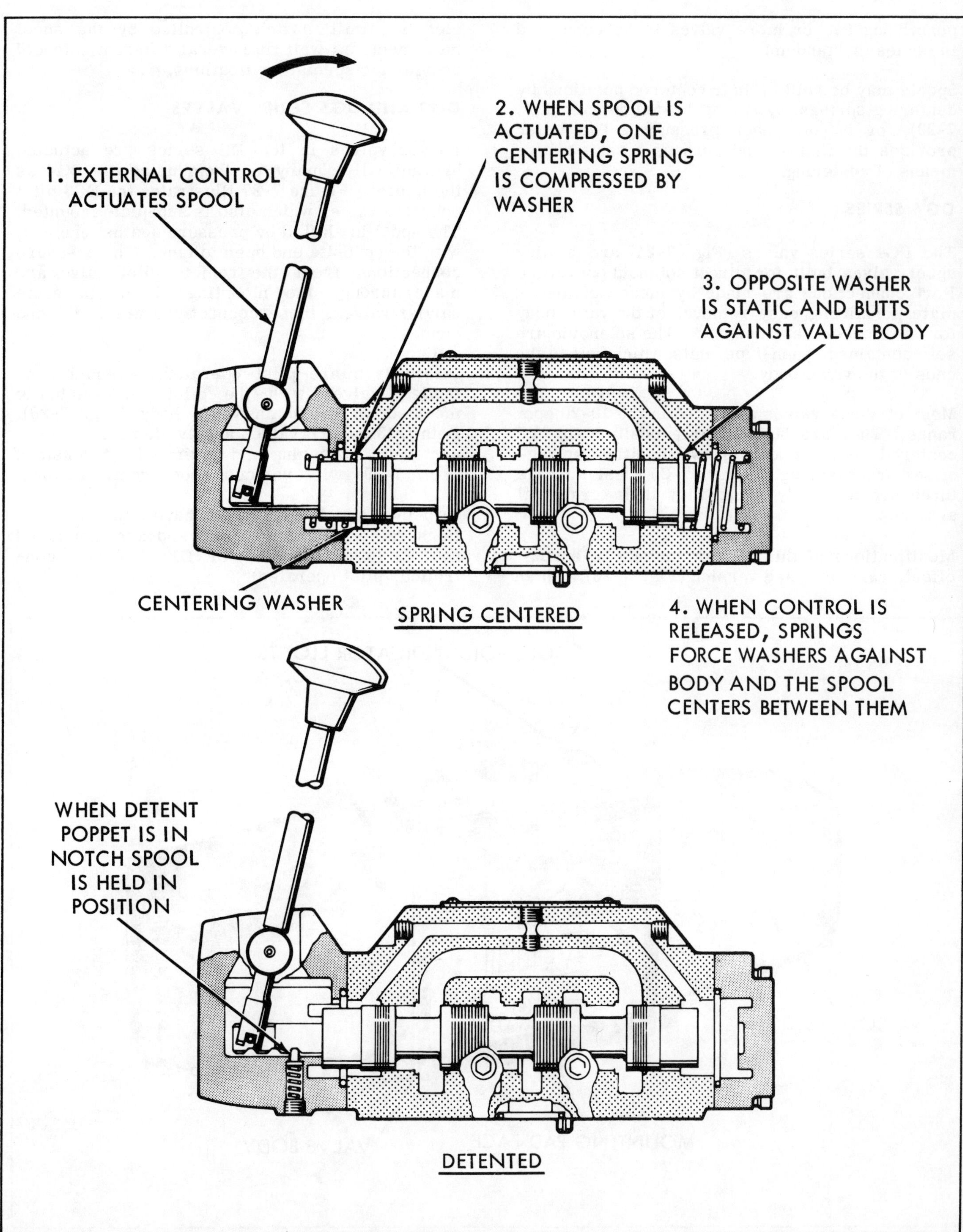

Fig. 7-23. Spool Centering Devices

permitting two or more valves to be connected in series or "tandem."

Spools may be held in their centered positions by centering springs, by spring-loaded detents (Fig. 7-23), or by oil under pressure. The latter provides the fastest and perhaps most positive means of centering.

DG4 SERIES

The DG4 series valves (Fig. 7-24) are sliding spool valves built for direct solenoid operation. Port connections are made by means of a sub-plate permitting easy removal of the valve body for service or replacement. The solenoids are self-contained, push-type units which bolt to the ends of the valve body.

Most of these valves are rated in the 10-20 gpm range. They are built in three-position spring-centered versions and in two position spring-offset or no-spring. Figure 7-25 illustrates the three schematically along with their graphical symbols.

Modifications of this design include a spring-offset, cam-operated version (Fig. 7-26) with an electric limit switch controlled by the spool movement, as well as several different solenoid designs for special applications.

DG3 AND DG5 SERIES VALVES

Larger valves in the DG series are actuated hydraulically, many of them using the DG4 as their pilot. Figure 7-27 illustrates the DG3 pilot operated valve, which also is sub-plate mounted. The spool is shifted by pressure against one end, with the opposite end open to tank. The pressure connections from the remote pilot valve are made through the mounting plate. In some earlier valves, these connections are in the end caps.

DG5 valves are pilot operated, solenoid controlled valves with the pilot valve actually mounted on the main valve body (Fig. 7-28). Both DG3 and DG5 valves are available in spring-centered, no-spring and spring-offset versions (Fig. 7-29) with various spool configurations.

Two-inch and larger valves have flange connections rather than sub-plates and are designated as DF3 (pilot operated) or DF5 (solenoid-controlled, pilot operated).

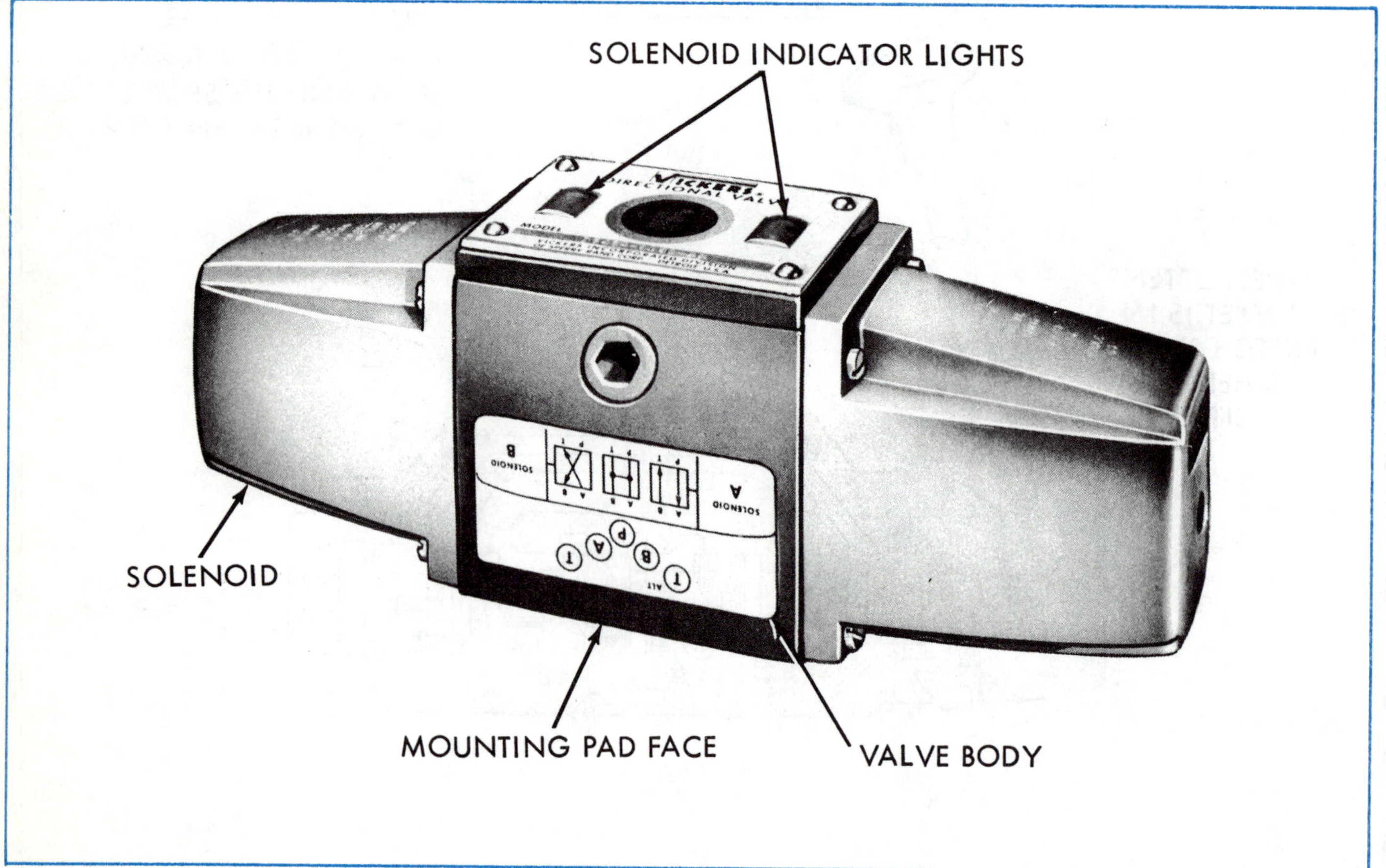

Fig. 7-24. DG4 Series Valves are Solenoid Operated

SPRING CENTERED

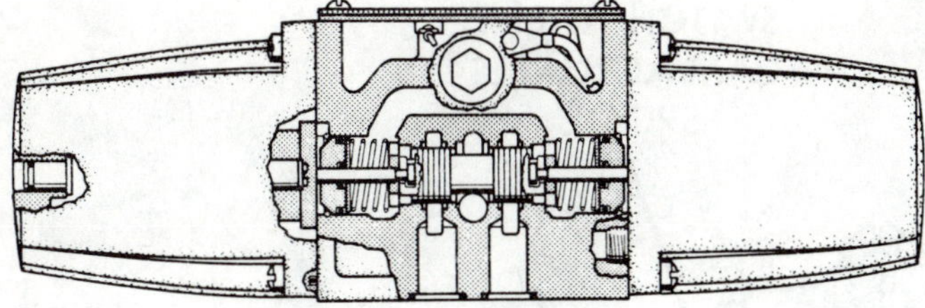

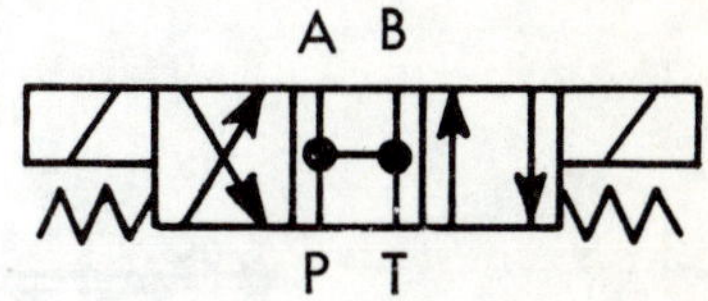

SPRING OFFSET

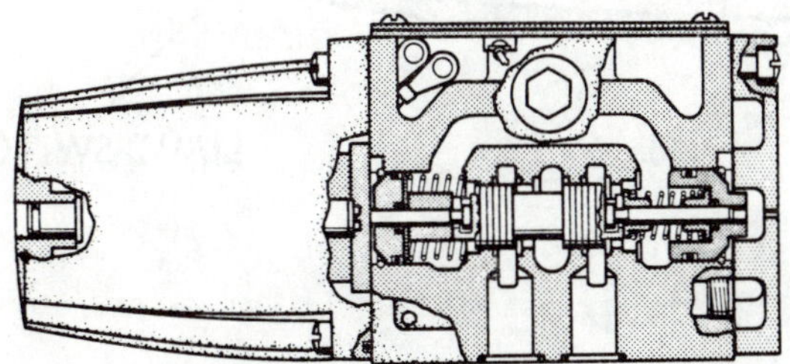

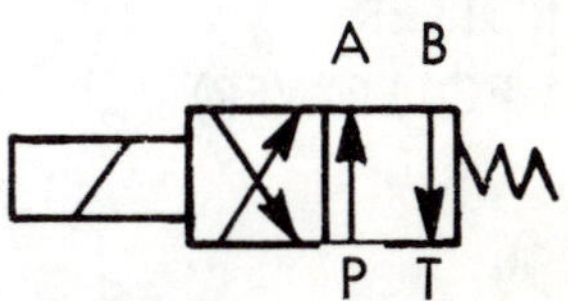

DETENTED

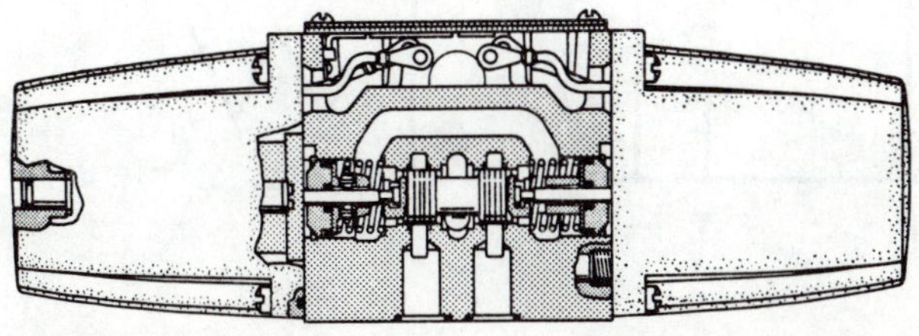

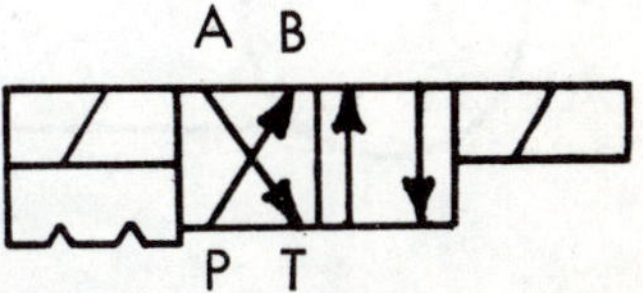

Fig. 7-25. Three Versions of DG Type Four-Way Valve

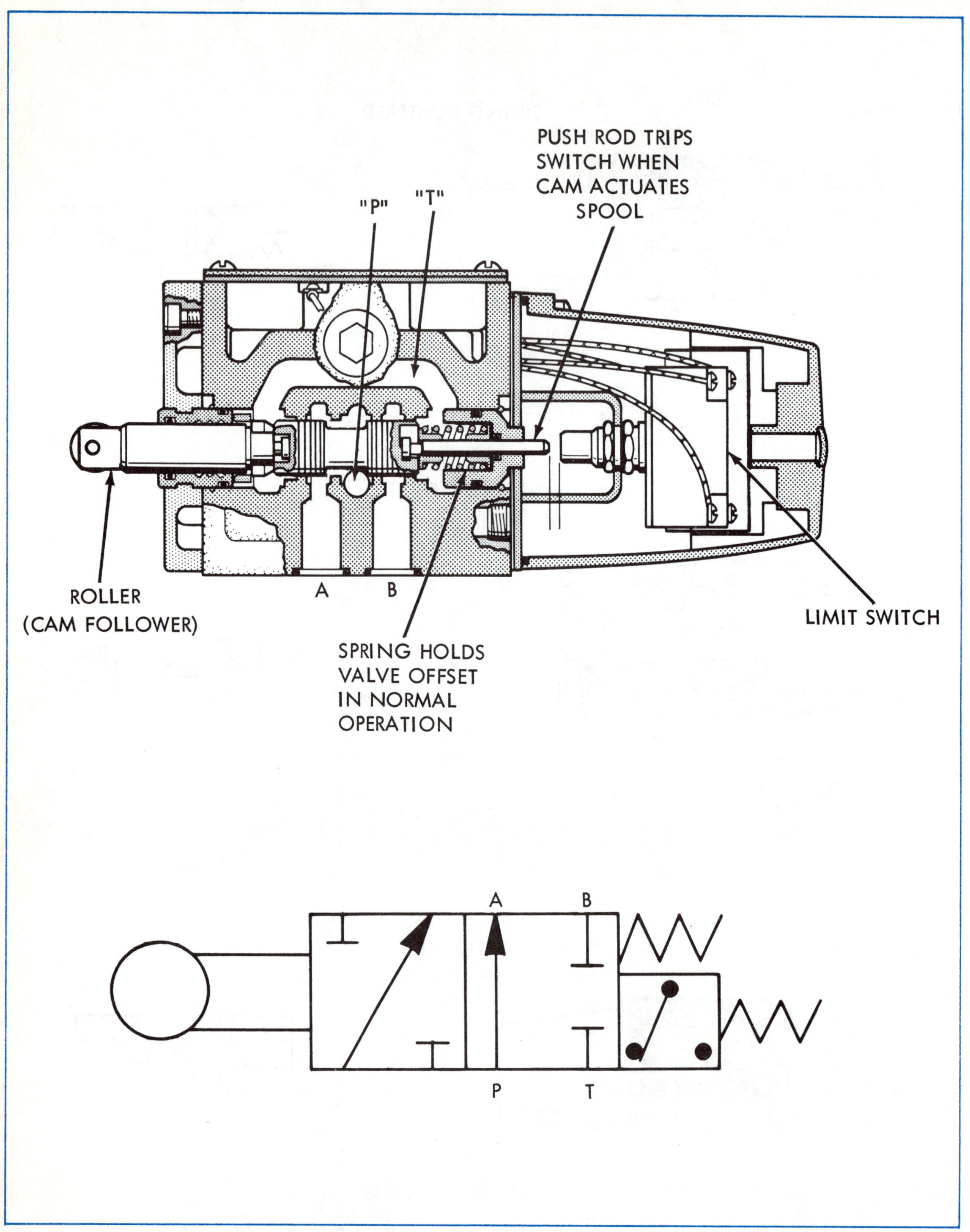

Fig. 7-26. Cam-Operated "DG" Valve With Electric Switch

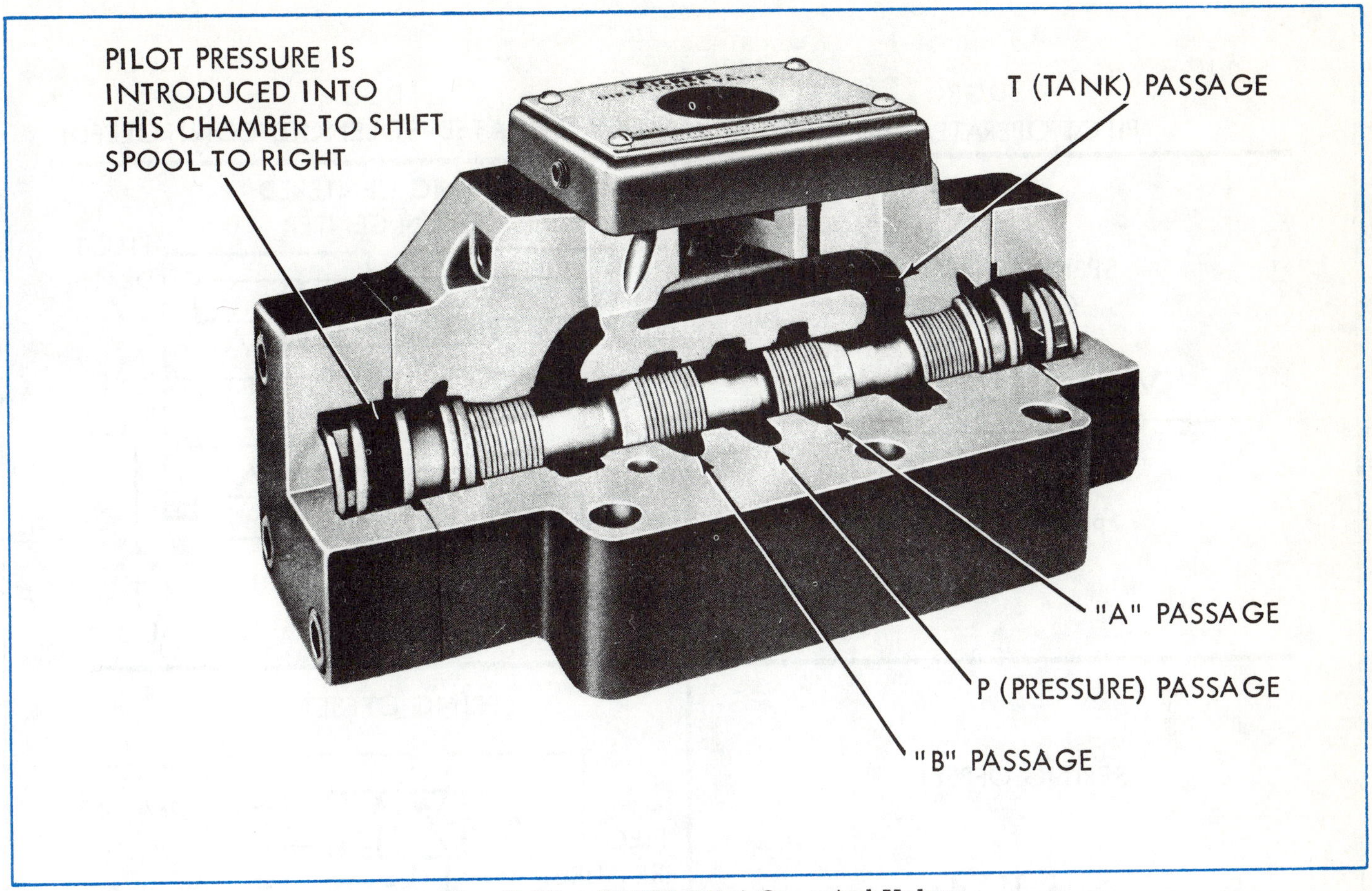

PILOT PRESSURE IS INTRODUCED INTO THIS CHAMBER TO SHIFT SPOOL TO RIGHT

T (TANK) PASSAGE

"A" PASSAGE

P (PRESSURE) PASSAGE

"B" PASSAGE

Fig. 7-27. "DG3" Pilot Operated Valve

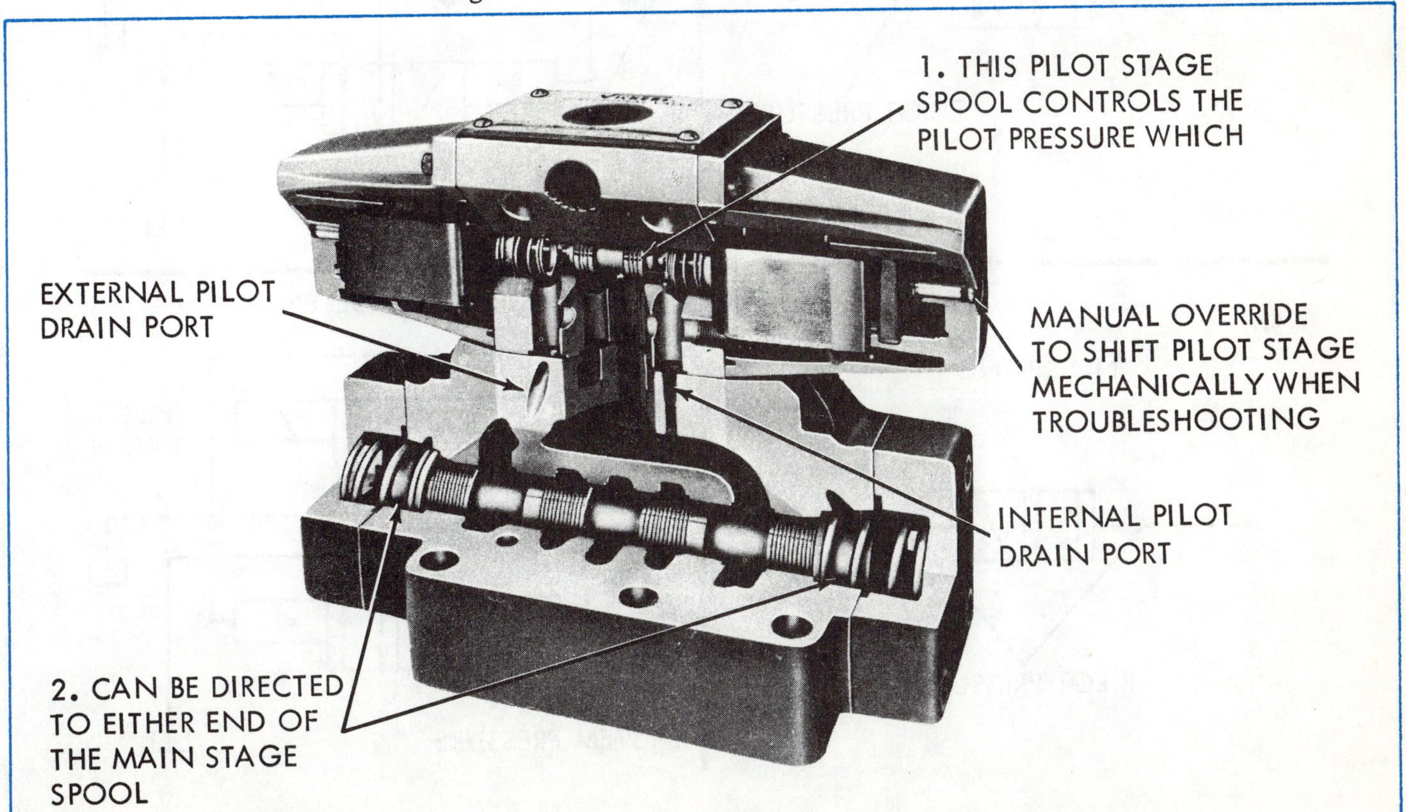

1. THIS PILOT STAGE SPOOL CONTROLS THE PILOT PRESSURE WHICH

EXTERNAL PILOT DRAIN PORT

MANUAL OVERRIDE TO SHIFT PILOT STAGE MECHANICALLY WHEN TROUBLESHOOTING

INTERNAL PILOT DRAIN PORT

2. CAN BE DIRECTED TO EITHER END OF THE MAIN STAGE SPOOL

Fig. 7-28. Typical "DG5" Type Solenoid Controlled, Pilot Operated Valve

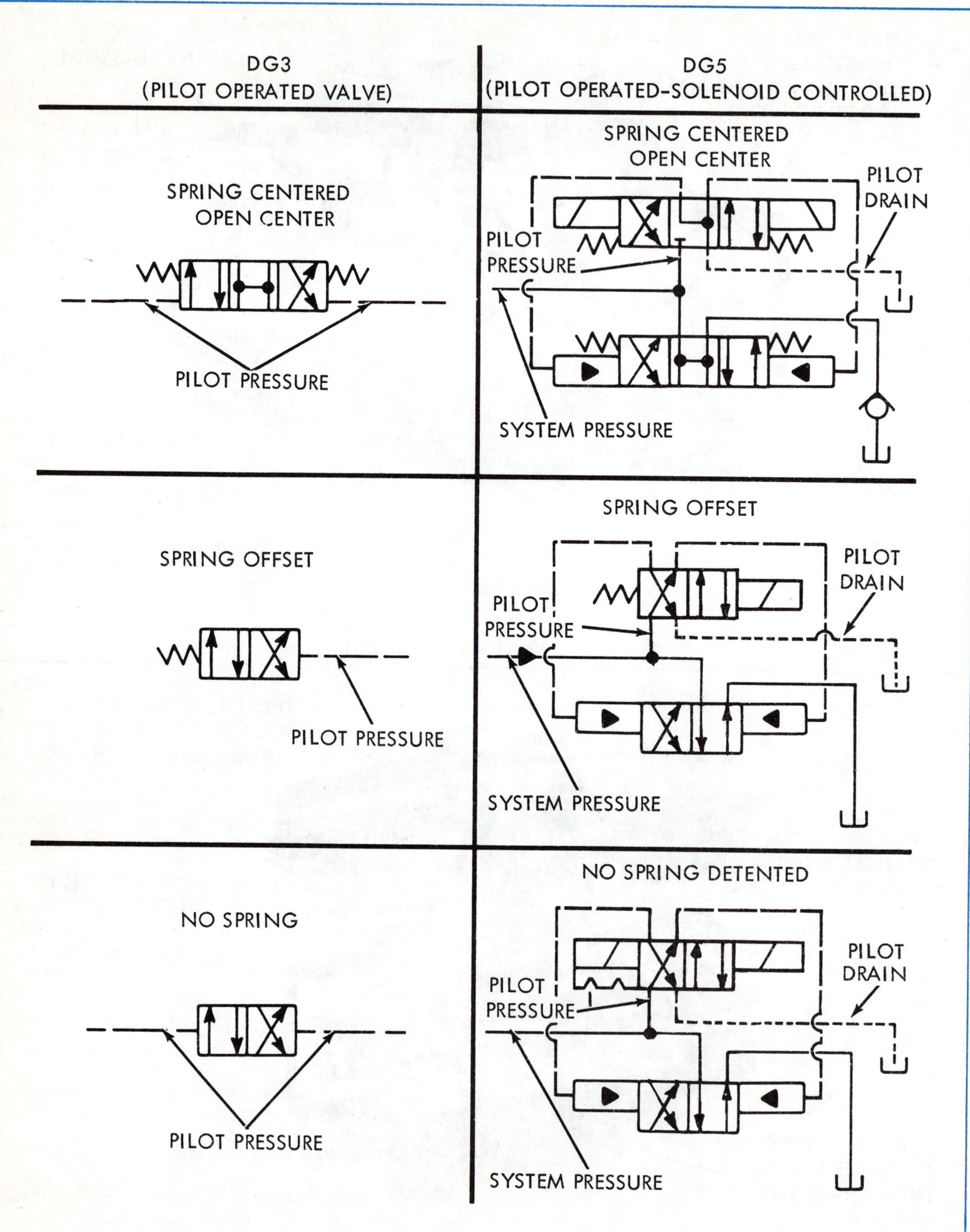

Fig. 7-29. Variations of DG3 and DG5 Valves

PILOT PRESSURE SOURCES

Normally, pilot pressure is supplied in the DG5 valves through an internal passage from the "P" port of the main valve (Fig. 7-29).

Where the pump port is open to tank in the center position, it may be necessary to install a check valve in the tank (return) line to create pilot pressure (Fig. 7-30).

There also are conditions which make it desirable or necessary to use an external source of pilot pressure. In this case, the internal pilot connection is plugged, and pilot oil is supplied through a separate port in the sub-plate. A connection ahead of a check valve installed upstream from the directional valve as shown in Figure 7-30 is one method of accomplishing this. However, some models of the DG5 valve are available with a check valve installed in the pressure port of the body for this purpose. The pilot oil is then available internally. See Figure 7-31.

PILOT CHOKE

A pilot choke (Fig. 7-32) may be incorporated to slow the spool travel for smoother reversals or to provide a brief time delay or dwell period before the actuator is reversed.

The pilot choke functions in effect as a meter-out restriction valve. It allows free pilot flow to the end of the main spool, but restricts flow out of opposite end; thus reversing flow gradually and cushioning the spool's contact when it shifts. The controlling orifices are adjustable. Free flow in is accomplished by check valves.

A pilot choke assembly is available for mounting directly on a DG3 or DF3 valve, or between the pilot and main valves in a DG5 or DF5 valve (Fig. 7-33). In other valves the choke may be built into the end caps.

PILOT PISTONS

Pilot pistons (Fig. 7-34) are sometimes used when large valve spools are shifted hydraulically. Since it is only necessary to fill the volume displaced by the small piston, less pilot oil is required and faster shift times can be attained.

A differential piston is simply the incorporation of a single piston on one end to provide differential areas. Constant pressure applied to the smaller area of the differential piston may be used to bias the spool to one side in place of a spring.

DECELERATION VALVES

Hydraulic cylinders often have cushions built in to slow down the cylinder pistons at the extreme ends of their travel. When it is necessary to decelerate a cylinder at some intermediate position or to slow down or stop a rotary actuator (motor), an external valve is required.

Most deceleration valves are cam operated valves with tapered spools. They are used to gradually decrease flow to or from an actuator for smooth stopping or deceleration. A "normally open" valve cuts off flow when its plunger is depressed by a cam. It may be used to slow a drill head cylinder down at the transition from rapid traverse to feed or to stop heavy index tables and large presses smoothly.

Some applications require a valve to permit flow when it is actuated and to cut off flow when the plunger is released. In this case a "normally closed" valve is used. This type valve often is used to provide an interlocking arrangement whereby flow can be directed to another branch of the circuit when the actuator or load reaches a certain position. Both the "normally open" and "normally closed" type valves are available with integral check valves to permit reverse free flow.

TAPERED PLUNGER DESIGN

An early design of deceleration valve (Fig. 7-35) uses a tapered plunger to reduce flow as it is actuated by the cam. Before the plunger is depressed (view A), free flow is permitted from the inlet to the outlet. Depressing the plunger gradually cuts the flow off (view B). Reverse free flow (view C) is permitted by the integral check valve.

The control range of this valve depends on the volume of flow and on the cam rise. At nearly maximum volume; that is, with an initial pressure drop through the valve, there is control throughout the plunger stroke. At low flow rates, only part of the travel is available for control--from the point where a pressure drop is created.

This drawback has been overcome in the adjustable orifice design valve which permits tailoring of the valve to any given flow.

ADJUSTABLE ORIFICE DESIGN

The adjustable-orifice design valve, model series DT15S2, is illustrated in Figure 7-36. In this valve, a closely-fitted plunger and sleeve with rectangular ports or windows are used to control flow.

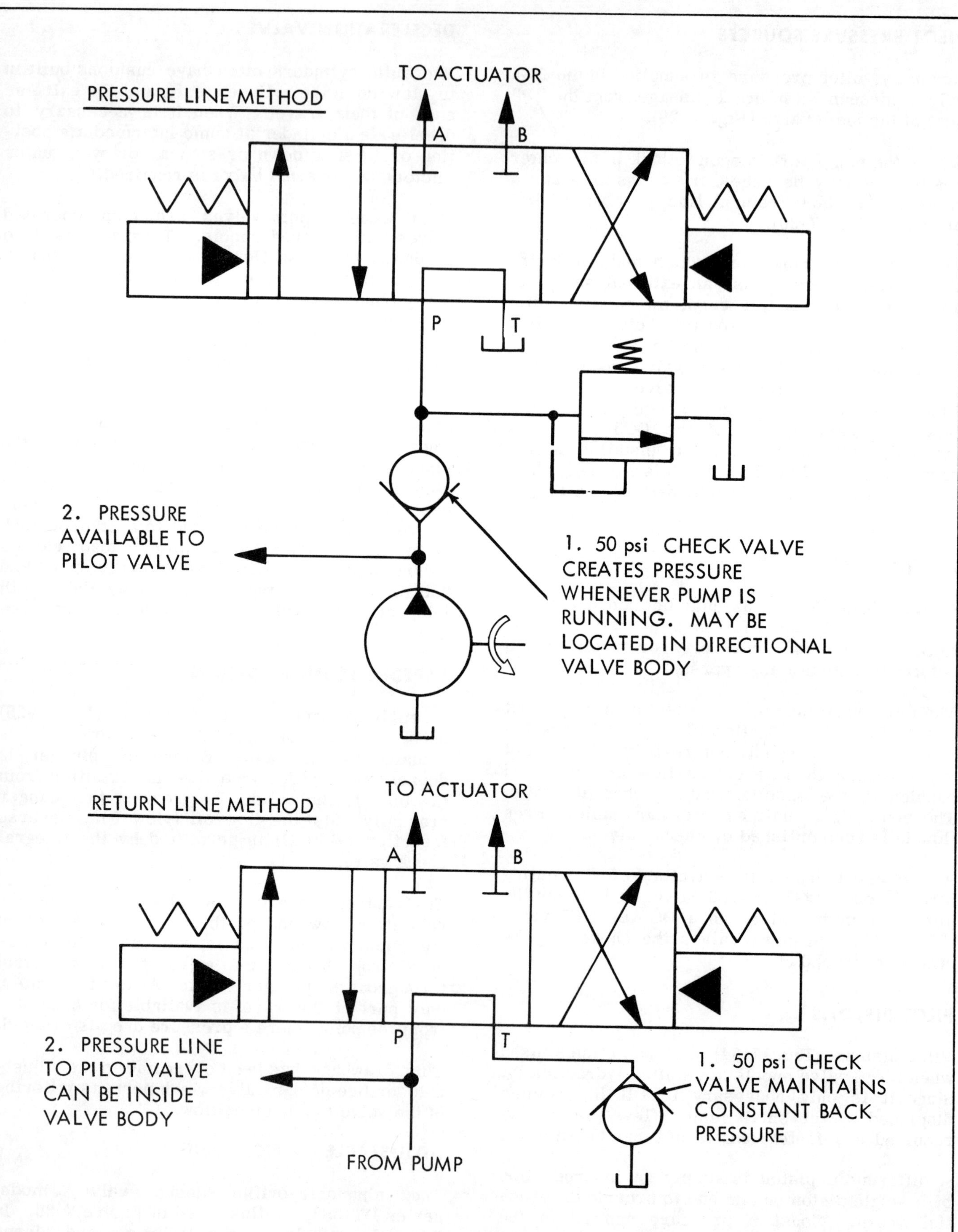

Fig. 7-30. Pilot Pressure Check Valve

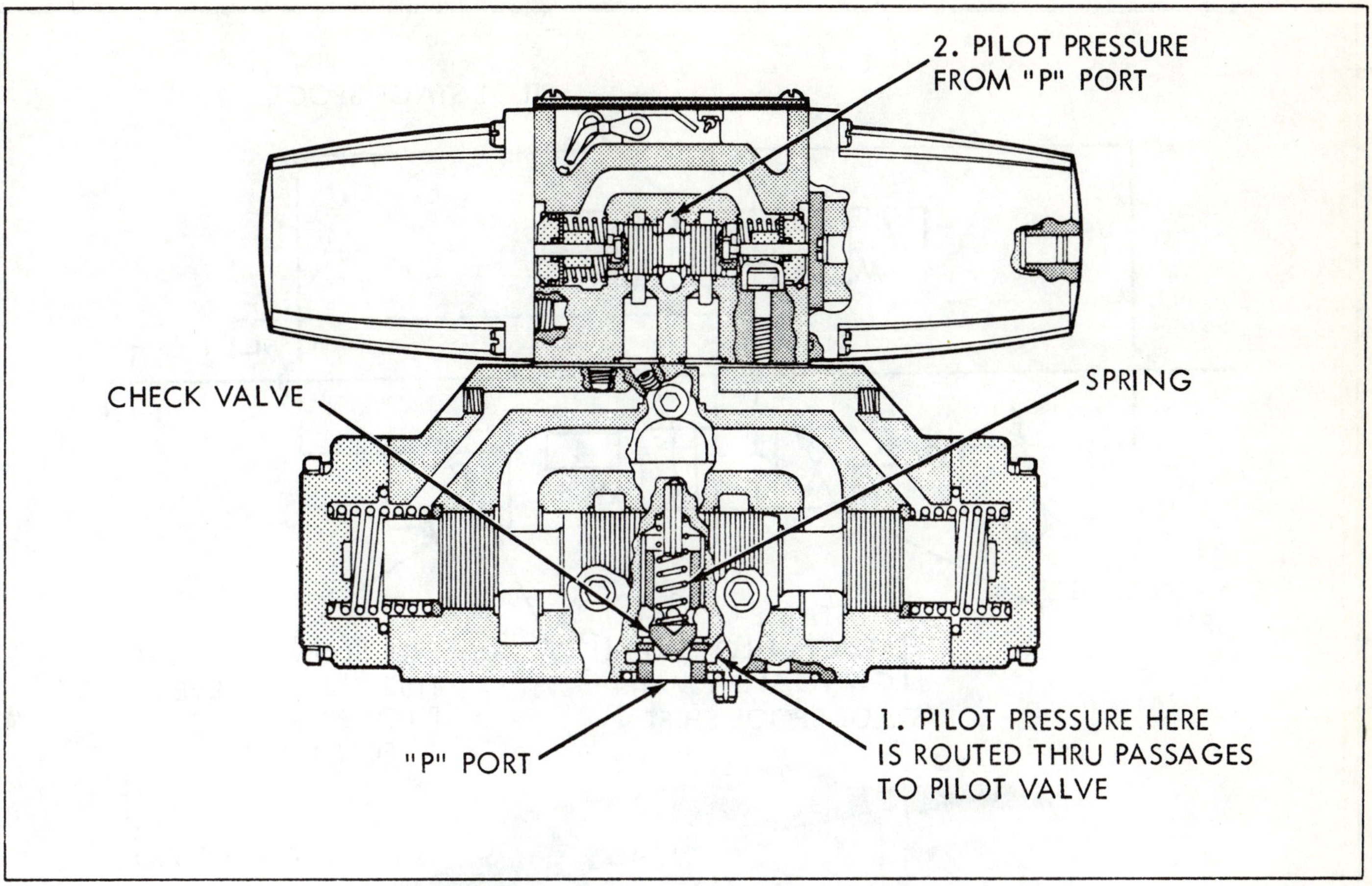

Fig. 7-31 -- Integral Pilot Pressure Check Valve

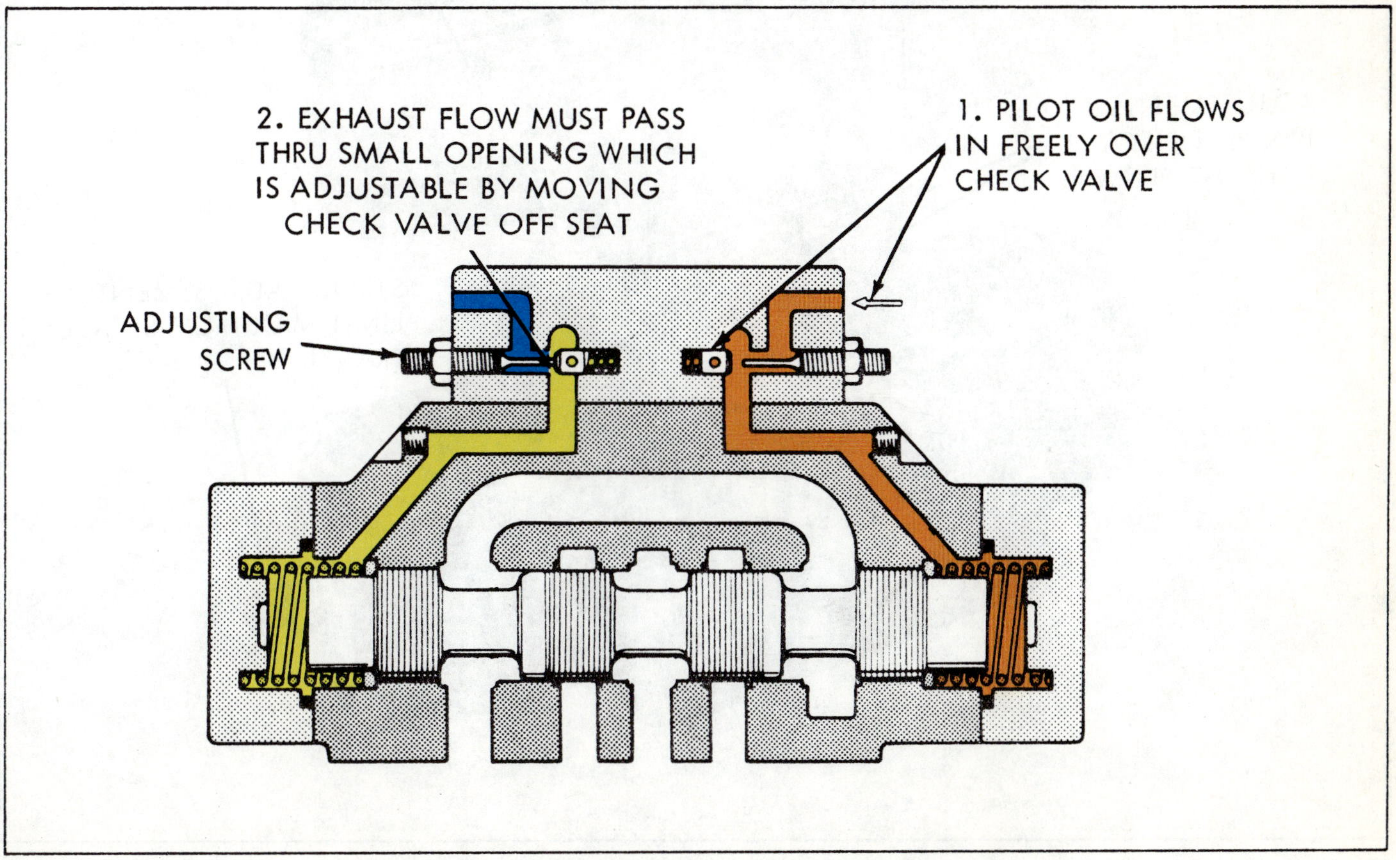

Fig. 7-32. Pilot Choke Controls Spool Movement

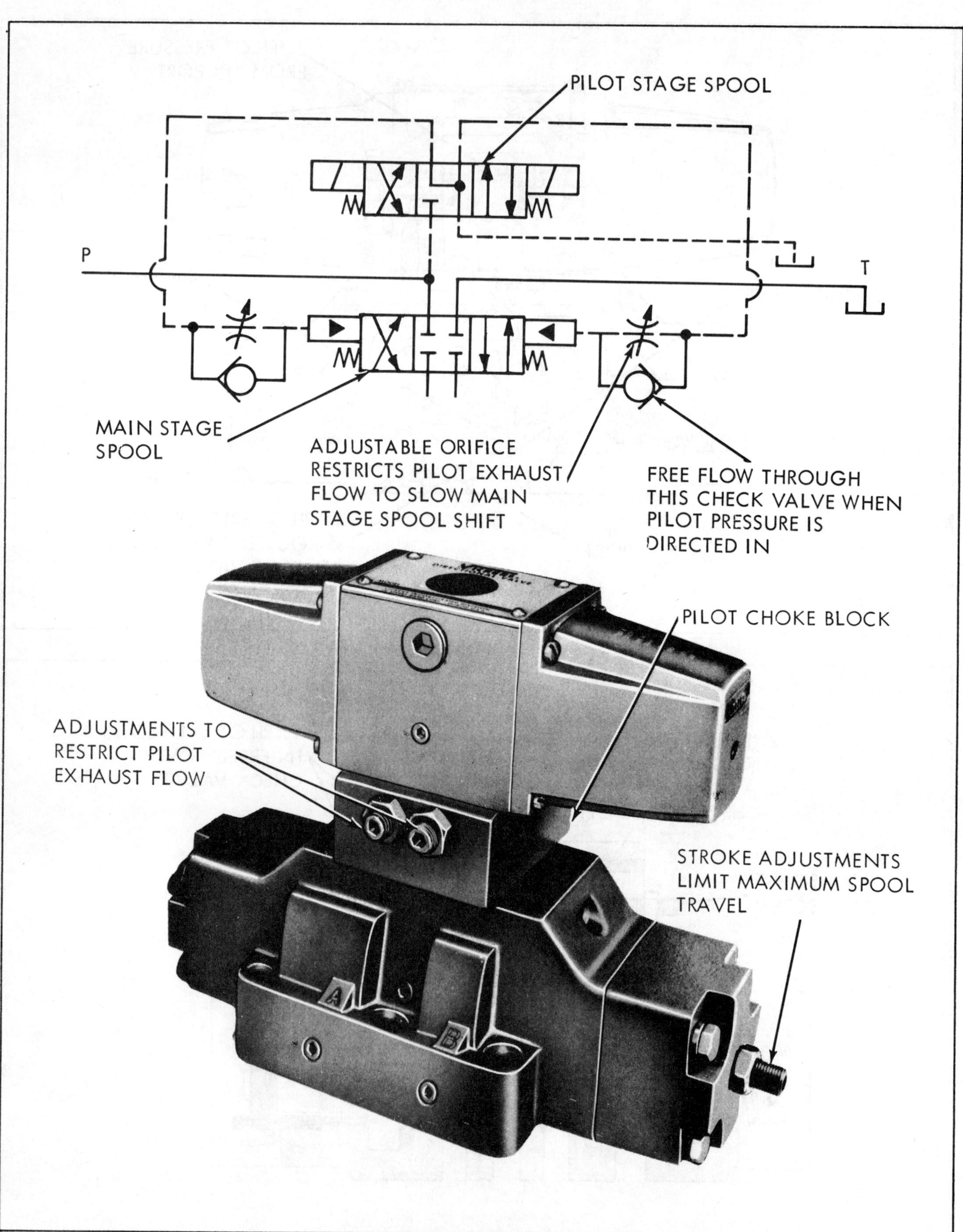

Fig. 7-33. Pilot Choke Mounts on DG3 or DG5 Valve

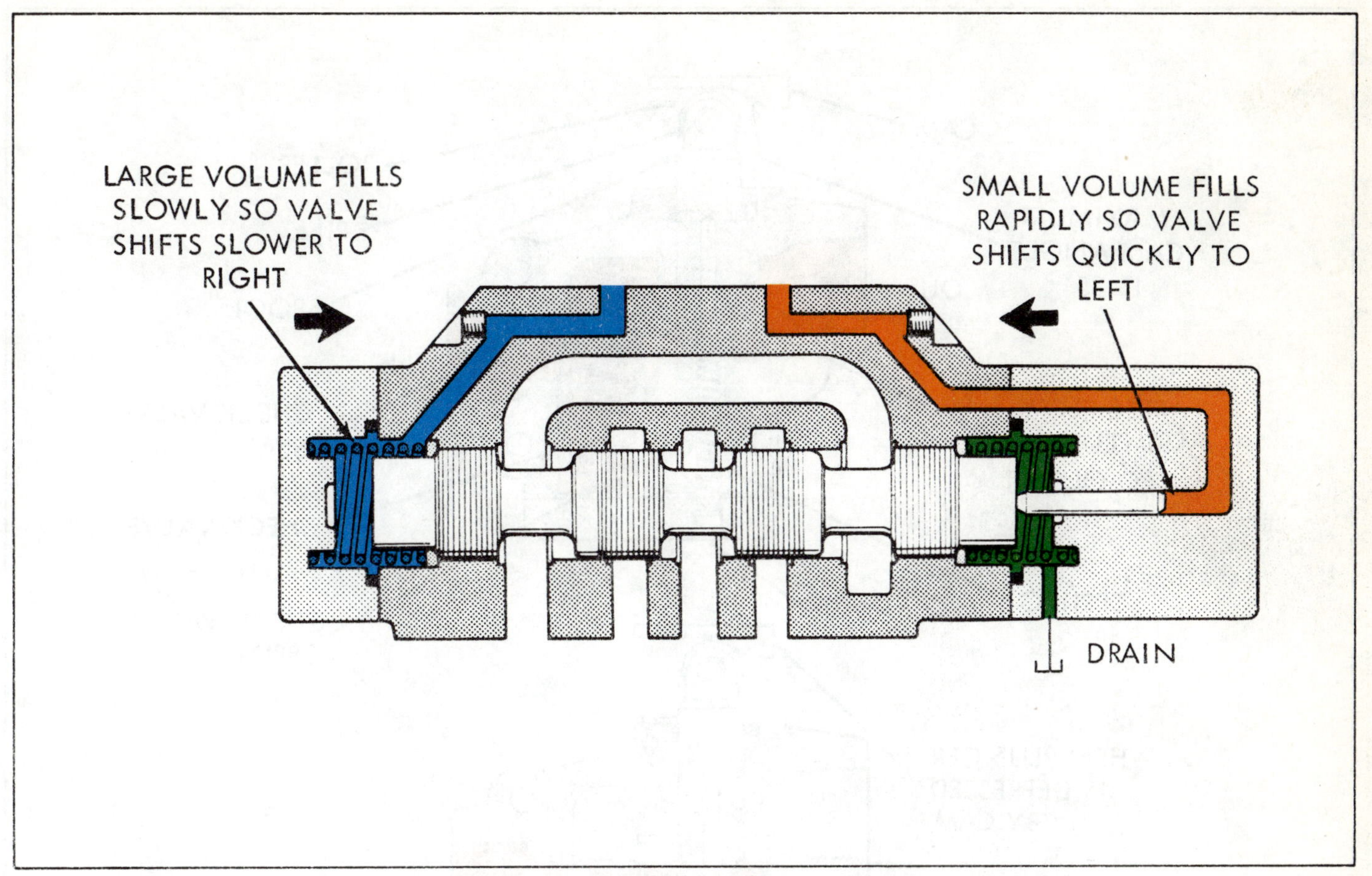

Fig. 7-34. Pilot Piston Speeds Valve Shifting

The plunger moves inside the sleeve, and the ports in each coincide in the open position. Oil entering the inlet flows through the small upper ports in the sleeve and plunger, down through the center of the plunger and out the large ports to the outlet (view A). When the plunger is depressed, the "window" area is gradually cut off to stop flow (view B). Reverse free flow is allowed by the integral check valve.

Initial Pressure Drop Adjustment

For precise control throughout the plunger stroke, the width of the "window" openings is controlled by adjusting screws which turn the sleeve. For low flow rates, the opening is narrow; for higher flow rates, it is wide. The adjustment is made by attaching a pressure gauge at the side of the valve and turning the screws to obtain the desired initial pressure drop.

This valve also includes an adjustable orifice which allows some flow with the plunger fully depressed. This permits the load to creep to its final position in indexing and similar applications. It consists of a small plunger with a chamfered end and "vee" notch which can be set to by-pass the spool-sleeve closure.

The window orifice valve is built in both pipe-threaded and back-mounted versions. Both valves require a drain to permit leakage oil to escape from beneath the plunger.

TYPICAL APPLICATIONS

Figure 7-37 illustrates a typical application of a deceleration valve. Here it is used to slow a drill head cylinder from rapid advance speed to feed speed at a preset point. View A shows rapid advance, with exhaust flow from the cylinder passing through the deceleration valve unrestricted.

In view B, we see the valve plunger being depressed by a cam. Exhaust flow is cut off at the deceleration valve and must pass through the volume control valve, which sets the feed speed. In view C, the directional valve has been reversed to return the cylinder. Oil from the directional valve passes through the deceleration valve freely over the check valve, whether or not the plunger is depressed.

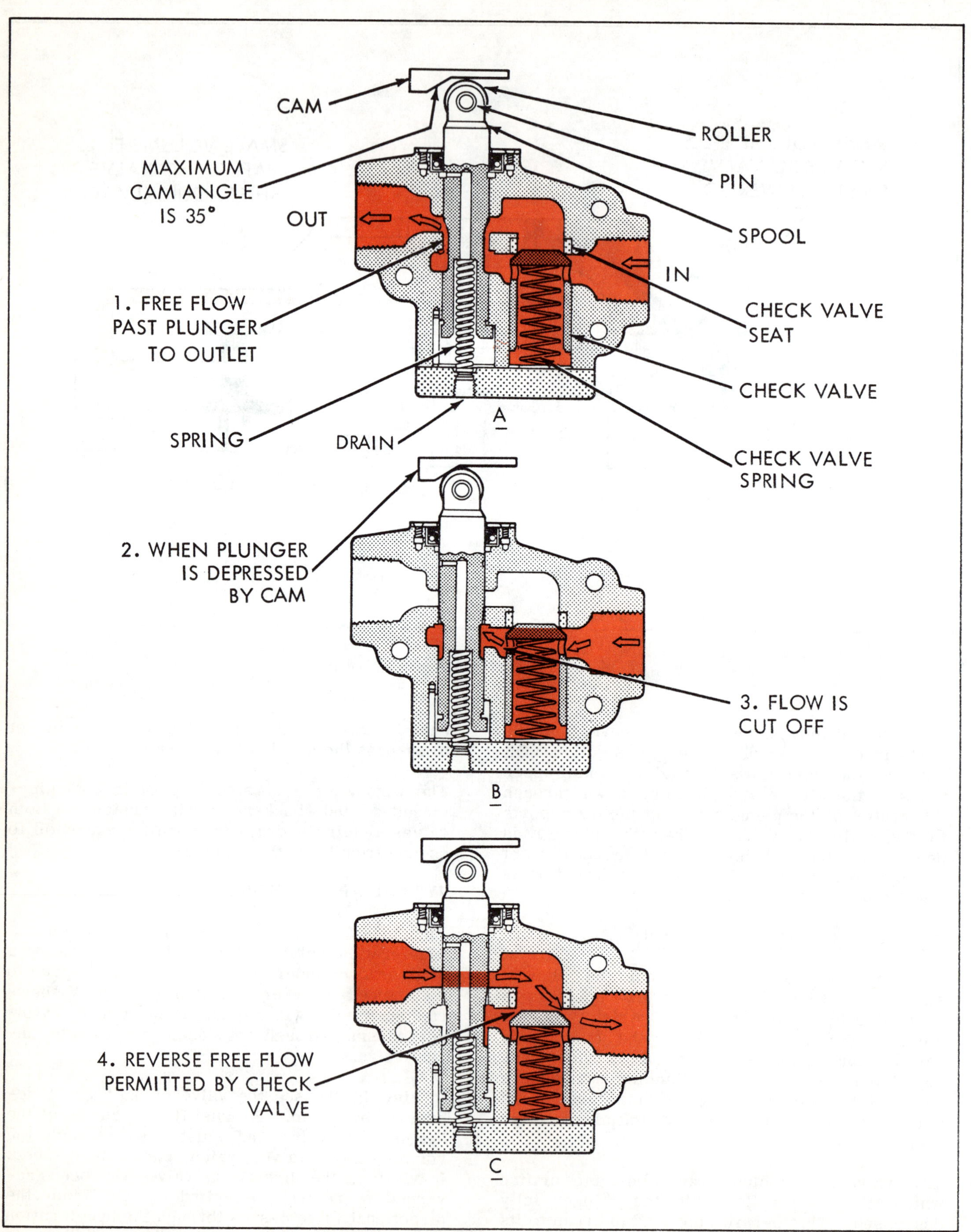

Fig. 7-35. C-700 Series Deceleration Valve Has Tapered Plunger

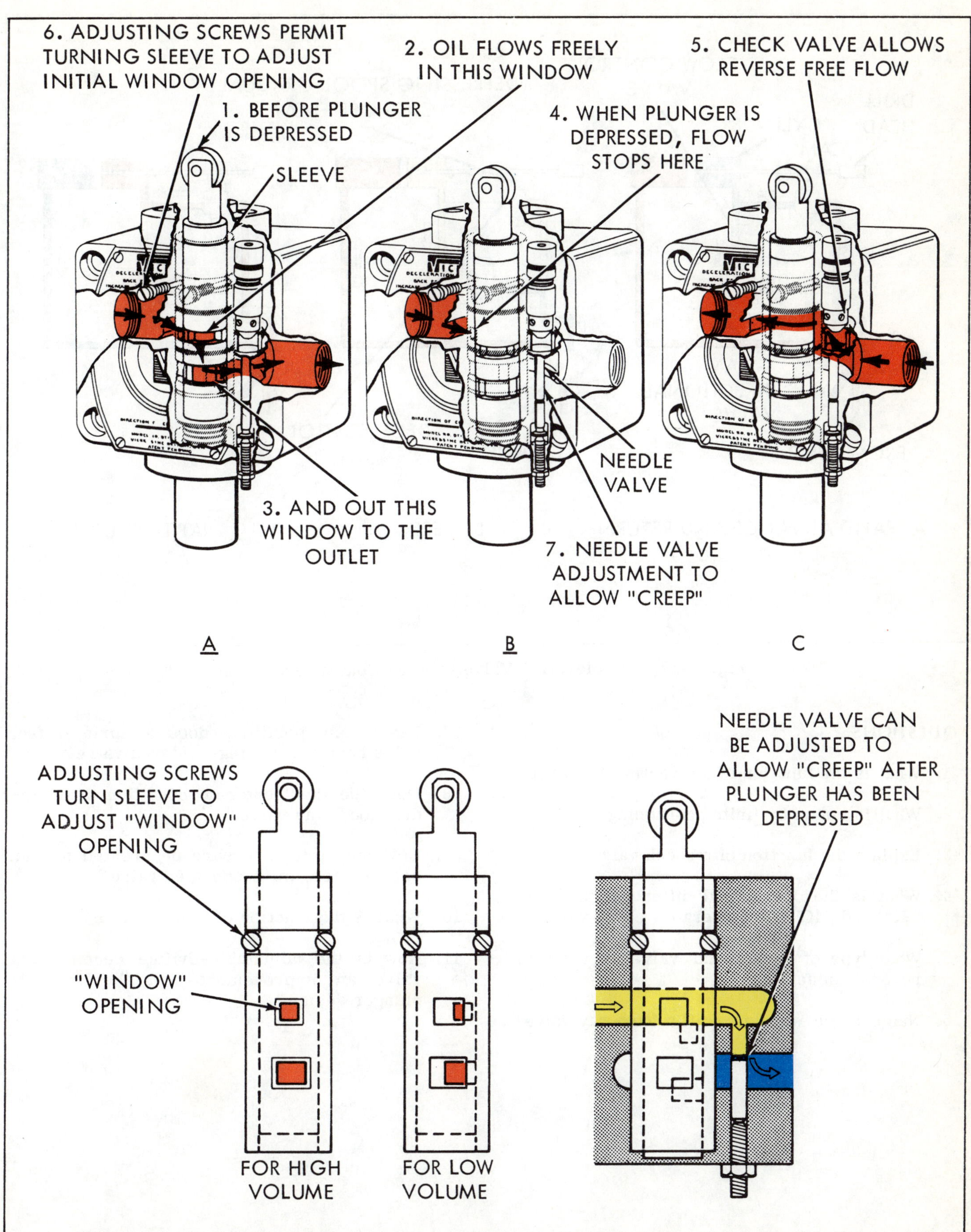

Fig. 7-36. This Deceleration Valve Has a "Window" Orifice In Place of Plunger Taper

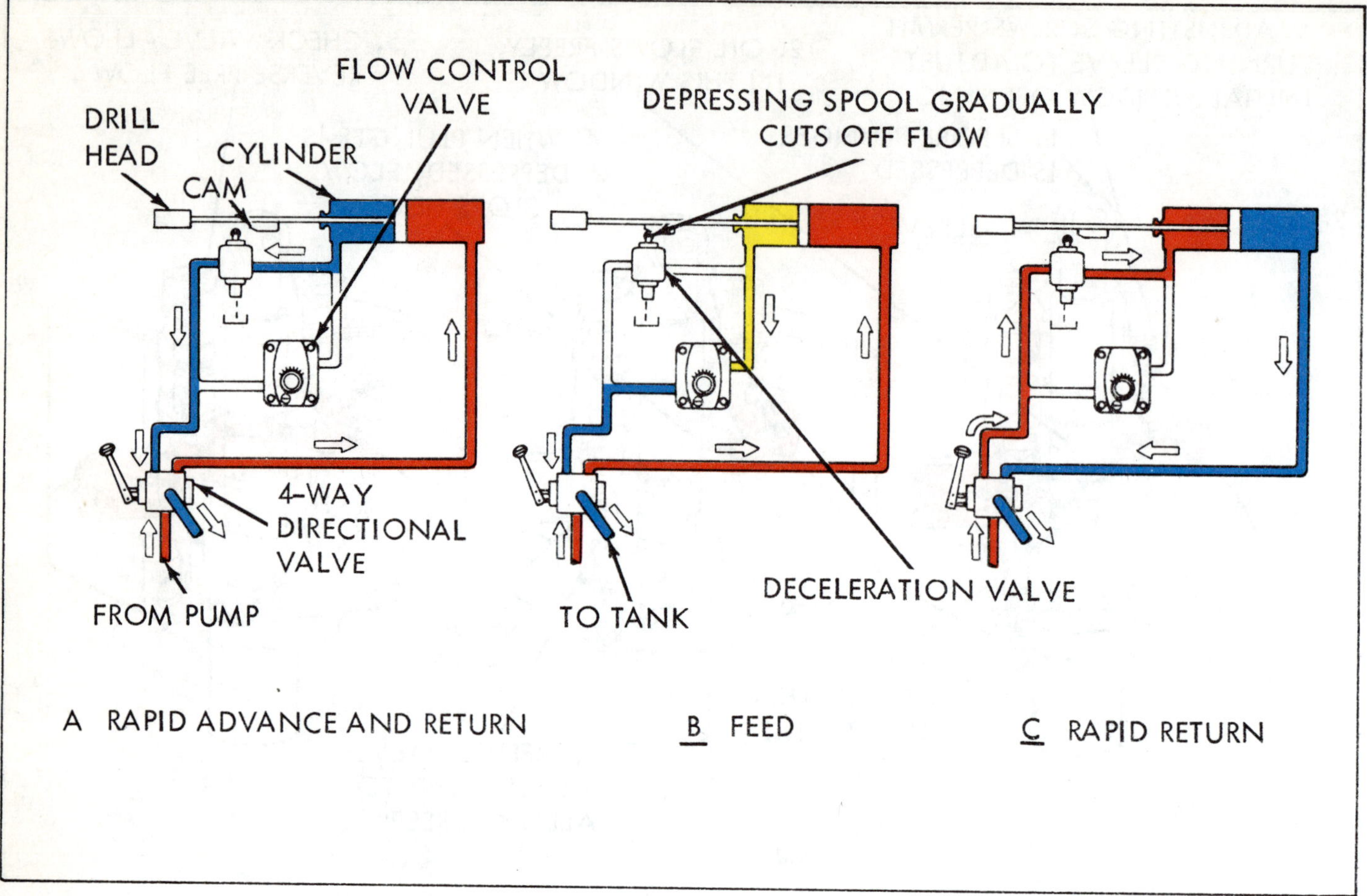

Fig. 7-37. Deceleration Valve In Feed Control System

QUESTIONS

1. What is the function of a directional control?
2. What is meant by finite positioning?
3. Explain the function of a check valve.
4. What is the operational difference between "2C" and "4C" pilot-operated check valves?
5. What type of directional valve is used to reverse an actuator?
6. Name three ways to shift a four-way valve.
7. How many positions does a spring-offset valve have? A spring-centered valve?
8. Describe the "centered" condition of a tandem four-way valve.
9. How may pilot pressure be created for an open-center, pilot-operated valve?
10. What is the function of a pilot choke?
11. How is the adjustable-orifice deceleration valve an improvement over the tapered-plunger design?

SERVO VALVES

A servo valve is a directional valve which may be infinitely positioned to provide the additional feature of controlling the amount as well as the direction of fluid flow. When coupled with the proper feedback sensing devices very accurate control of the position, velocity or acceleration of an actuator may be obtained.

The mechanical servo valve or follow valve has been in use for several decades. The electro-hydraulic servo valve is a more recent arrival on the industrial scene.

MECHANICAL SERVO

The mechanical servo is essentially a force amplifier used for positioning control. It is illustrated schematically in Figure 8-1.

The control handle or other mechanical linkage is connected to the valve spool. The valve body is connected to and moves with the load. When the spool is actuated, it ports fluid to a cylinder or piston to move the load in the same direction as the spool is actuated. The valve body thus "follows" the spool. Flow continues until the body is centered or neutral with the spool. The effect is that the load always moves a distance proportional to spool movement. Any tendency to move farther would reverse oil flow to move it back into position.

The mechanical servo unit is often referred to as a booster; the hydraulic "boost" being capable of considerably greater force than the mechanical input, with precise control of the distance moved.

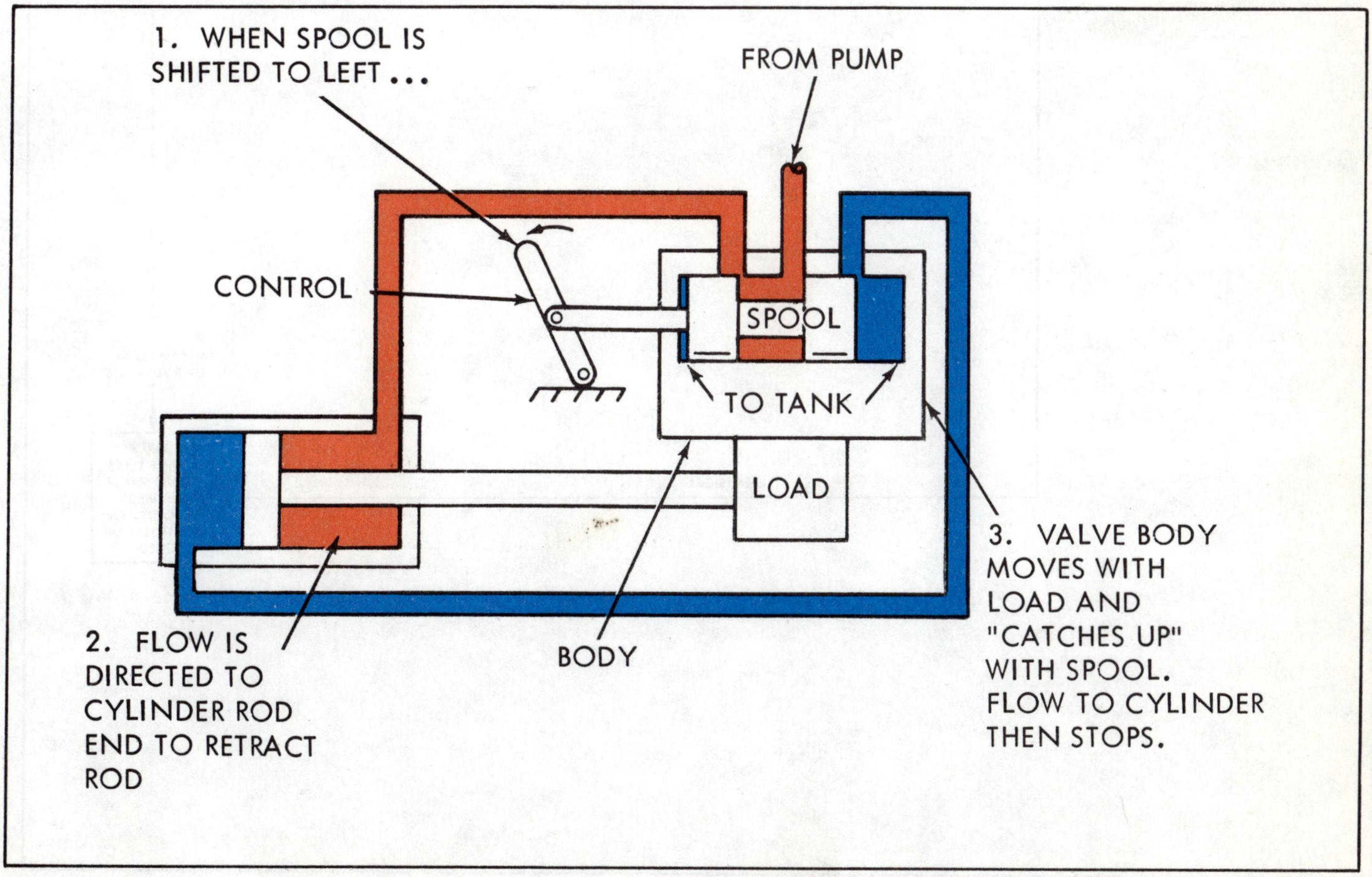

Figure 8-1. Mechanical Servo Uses "Follow" Valve

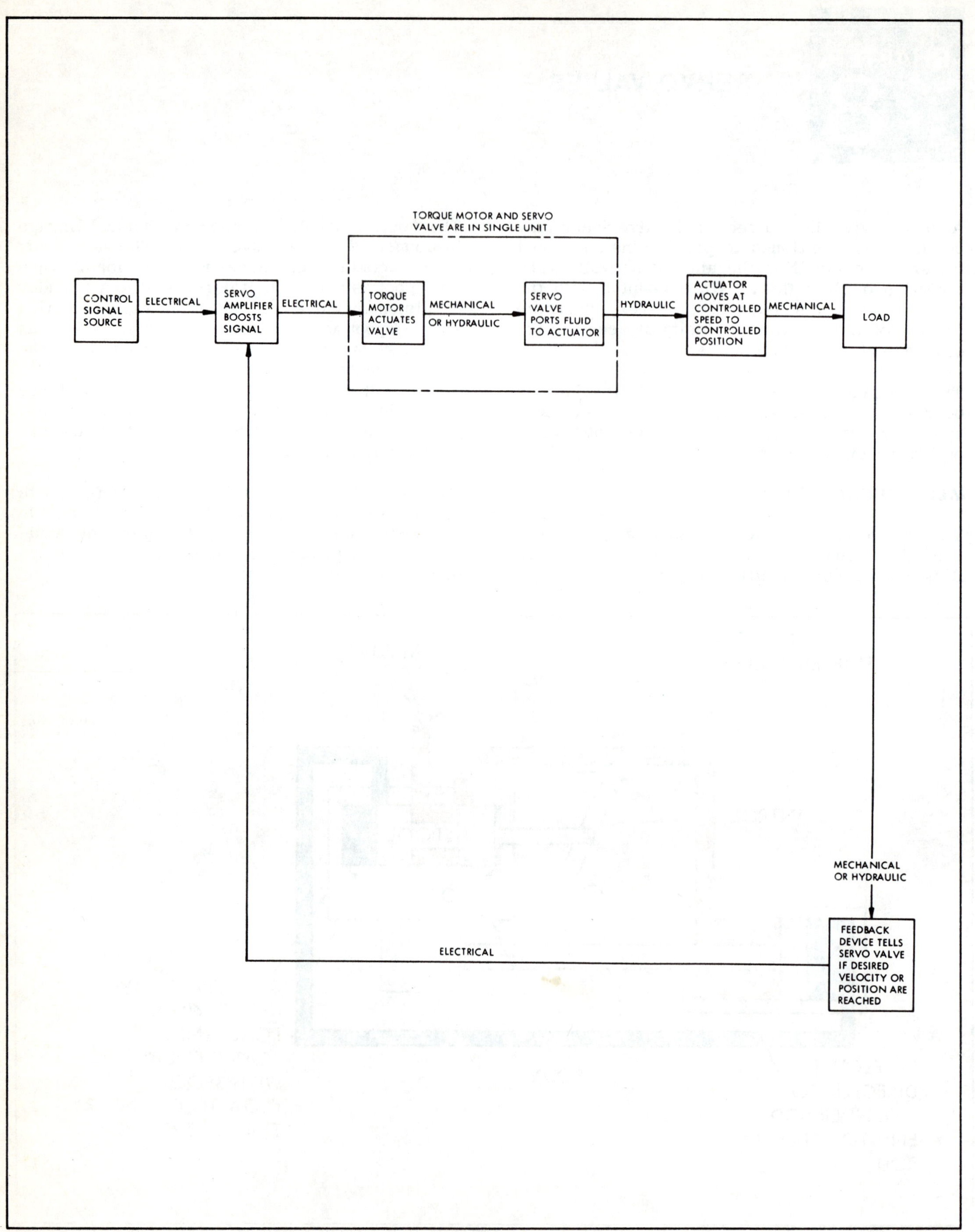

Figure 8-2. Block Diagram of Servo Valve System

Perhaps the most significant application of the mechanical servo is in power steering. Some of the first hydraulic steering units were developed by Harry Vickers, founder of Vickers, Inc. Power steering today is almost universal on full-size passenger cars and widely used on trucks, busses and other large vehicles. There are now many design variations of power steering systems but all operate on this same principle.

ELECTRO-HYDRAULIC SERVO VALVES

Electro-hydraulic servo valves operate essentially from an electrical signal to a torque motor or similar device which directly or indirectly positions a valve spool. The signal to the torque motor (Fig. 8-2) may come from a simple potentiometer, a magnetic or punched tape or other source. This signal, fed to the servo valve through a servo amplifier, "commands" the load to move to a specific position or assume a specific velocity. The amplifier also receives an electrical signal fed back by a tachometer generator, potentiometer or other transducer connected to the load. This feedback is compared with the original "command" input, and any resultant deviation is relayed to the torque motor as an error signal causing a correction to be made. The various types of electro-hydraulic servos can provide very precise control of positioning or velocity. Most often, the servo valve controls a cylinder or motor; but when volume requirements are large, it may be used to operate the displacement control of a variable delivery pump.

SINGLE STAGE SPOOL SERVO VALVE

Figure 8-3 shows the construction and operation of the single stage, spool type servo valve. The sliding spool is actuated directly by the torque motor, and opens the valve ports in proportion to the electric signal. Flow capacity of such valves is usually small due to the low forces and limited travel of the torque motor armature.

This type valve is back-mounted with O-ring seals. It can be bolted to a mounting plate or to a manifold attached to a hydraulic motor. "Manifolding" the valve this way reduces the amount of oil under compression; a critical factor in servo circuits.

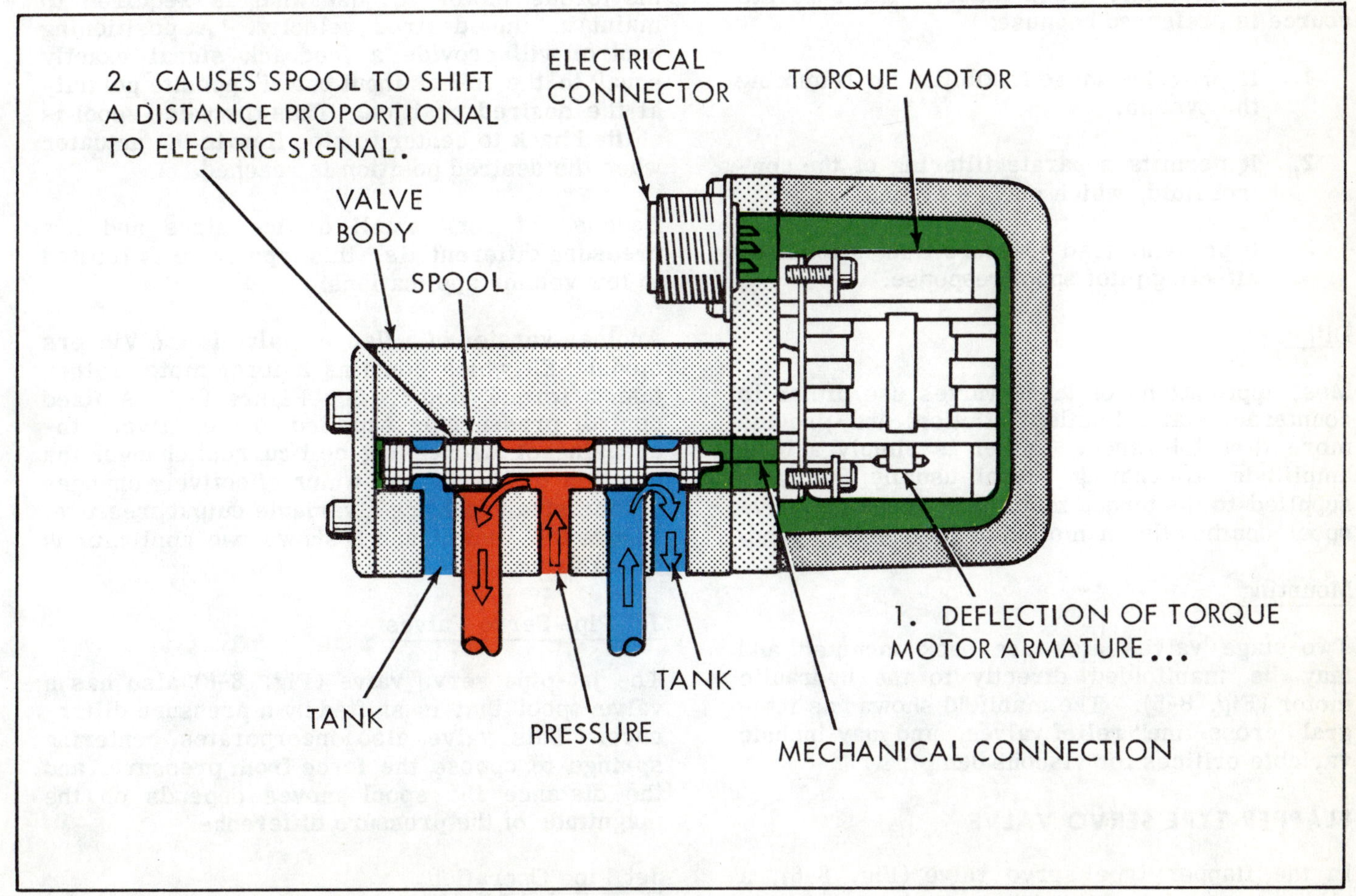

Figure 8-3. Single Stage Spool Type Servo Valve is Shifted Directly

TWO-STAGE SPOOL TYPE SERVO VALVES

Two-stage spool servo valves (Fig. 8-4) are used where larger flow rates are desired. In this design, the torque motor actuates a pilot valve inside a ported sleeve. The pilot valve, when shifted, directs fluid to shift the main valve spool. The main valve spool ports fluid to the actuator.

Mechanical Feedback

The mechanical feedback linkage in this valve lets the pilot valve act as a "follow" valve. Movement of the main spool is transmitted back to the pilot valve sleeve to effectively "center" the pilot valve when the main spool has moved the desired increment. The feedback linkage fulcrum is variable so that the ratio of main spool movement to that of the pilot spool can be as much as 5-1/2 to one.

Control Pressure

Control pressure for this valve is usually taken from a separate source. It can be taken from the supply pressure by incorporating a pressure reducing valve and accumulator. The separate source is preferred because:

1. It provides more flexibility for trimming the system.
2. It permits separate filtering of the control fluid, which may be critical.
3. It prevents load pressure fluctuation from affecting pilot spool response.

Dither

Most applications of these valves use dither to counteract static friction (stiction) and provide more dirt tolerance. Dither is simply a low-amplitude alternating signal usually 60 cycle supplied to the torque motor that keeps the valve spool continually in motion to reduce dead band.

Mounting

Two-stage valves also are back mounted and may be manifolded directly to the hydraulic motor (Fig. 8-5). The manifold shown has integral "cross-line" relief valves, and may include variable orifices for viscous damping.

FLAPPER TYPE SERVO VALVE

In the flapper type servo valve (Fig. 8-6), a sliding spool is actuated by a pressure difference on the two ends. Normally, a control pressure is equal at both ends of the spool. A controlled amount of oil continuously flows through orificed passages to nozzles that terminate at the flapper, then to exhaust.

When a signal to the torque motor moves the armature, the flapper moves toward one or the other of the nozzles. The balance of flow is changed through the orifices and nozzles, causing pressure to increase on one end of the spool and decrease on the other. The spool then moves until the pressure difference is balanced by the tension of the spool springs. Internal feedback is provided by a mechanical linkage from the spool to the flapper.

The distance the spool moves, and therefore the amount of oil it meters, depends on how far the flapper is deflected. This in turn depends on the size of the electrical signal to the torque motor. A high input signal results in a high volume of flow, a low signal in a low volume.

In a velocity control, the valve will be initially actuated by a large signal during acceleration. As load speed increases, an opposing signal from the load will reduce the effective signal to the torque motor to just what is required to maintain the desired velocity. A positioning control will provide a feedback signal exactly equal to the initial input but of opposite polarity at the desired position. Thus the valve spool is shifted back to center to stop flow to the actuator when the desired position is reached.

Because of very small orifice sizes and low pressure differentials, this type valve is limited to low volume applications.

Another version of a flapper valve is the Vickers model SE3 which contains a force motor rather than a torque motor. See Figure 8-7. A fixed control pressure is supplied to the valve. Increasing or decreasing coil current changes the position of the flapper which effectively changes orifice size providing a variable output pressure. Figure 8-7 View A and B shows two applications of the valve.

Jet Pipe Servo Valves

The jet-pipe servo valve (Fig. 8-8) also has a valve spool that is shifted by a pressure difference. This valve also incorporates centering springs to oppose the force from pressure, and the distance the spool moves depends on the magnitude of the pressure difference.

Jet Pipe Operation

The pilot section of the valve consists of the jet

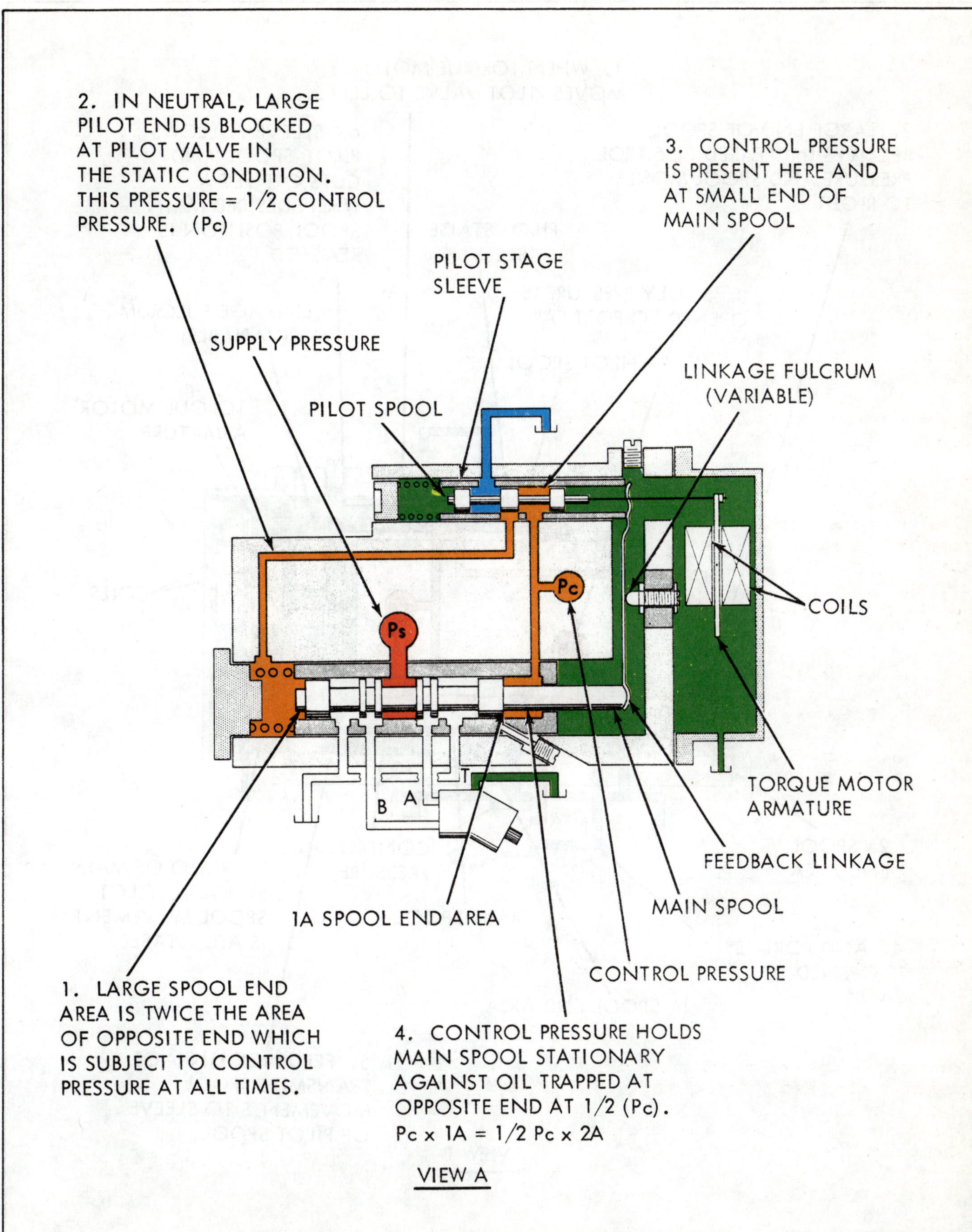

Figure 8-4. Two Stage Servo Valve is Pilot Operated

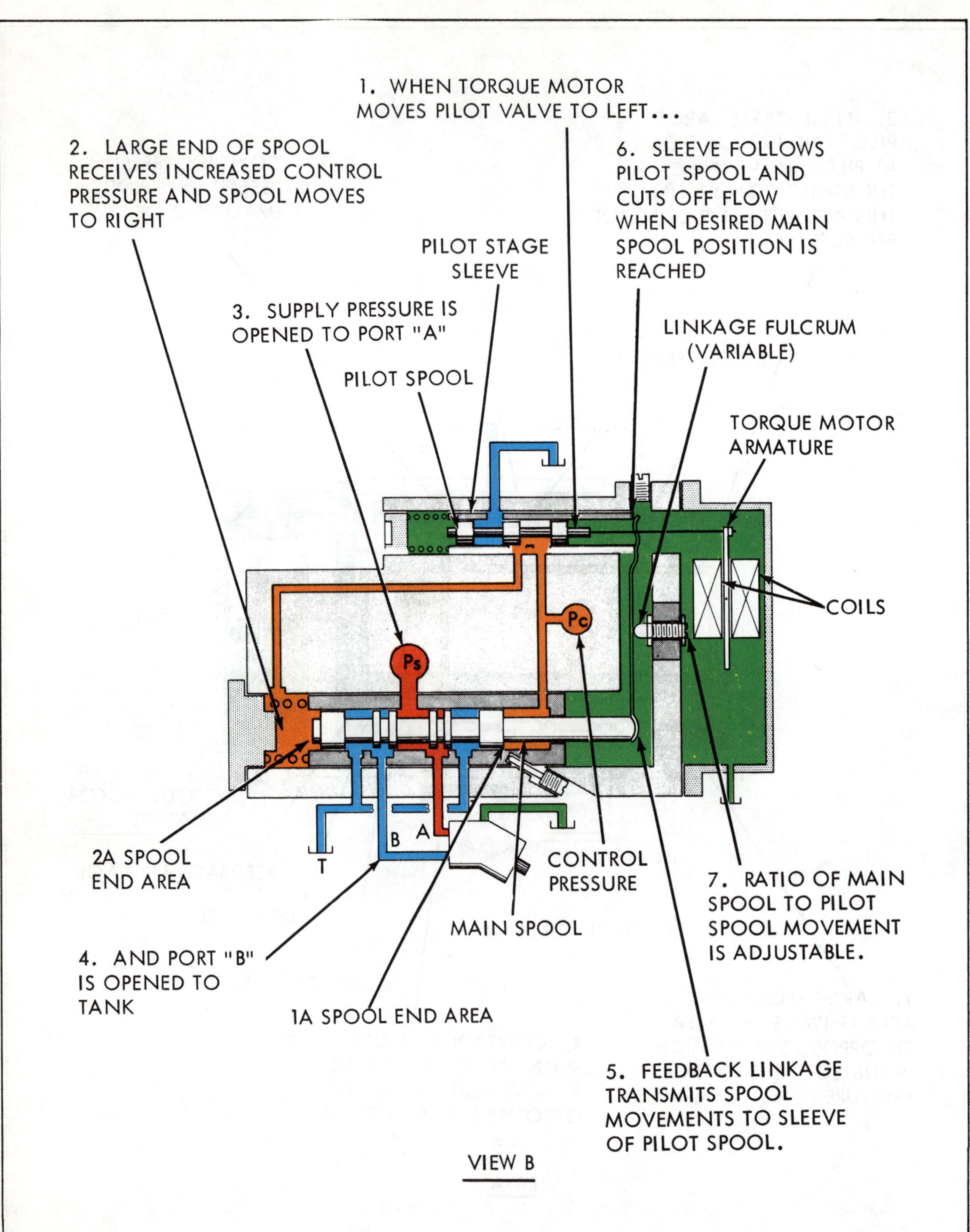

Figure 8-4. Two Stage Servo Valve is Pilot Operated (Cont'd)

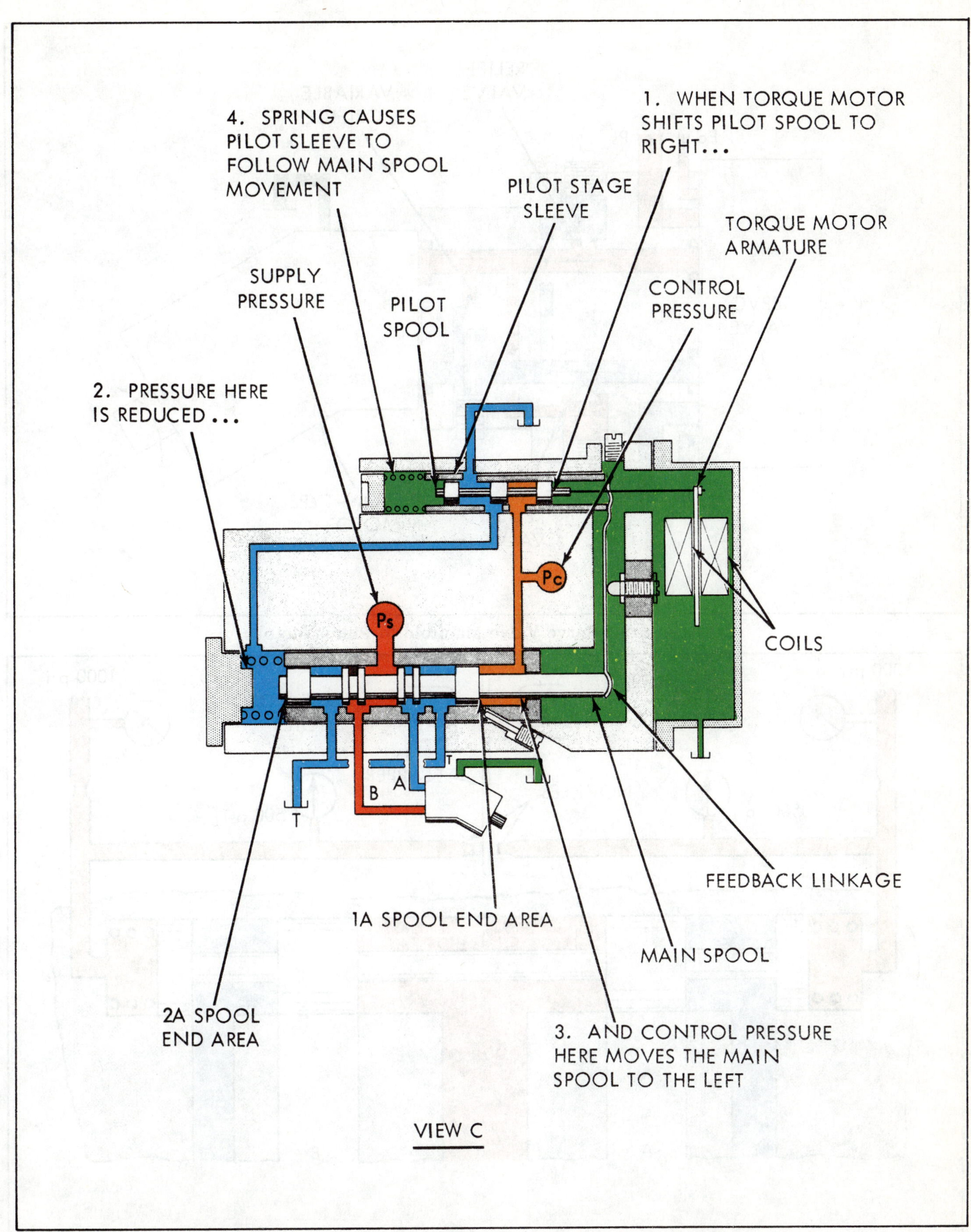

Figure 8-4. Two Stage Servo Valve is Pilot Operated (Cont'd)

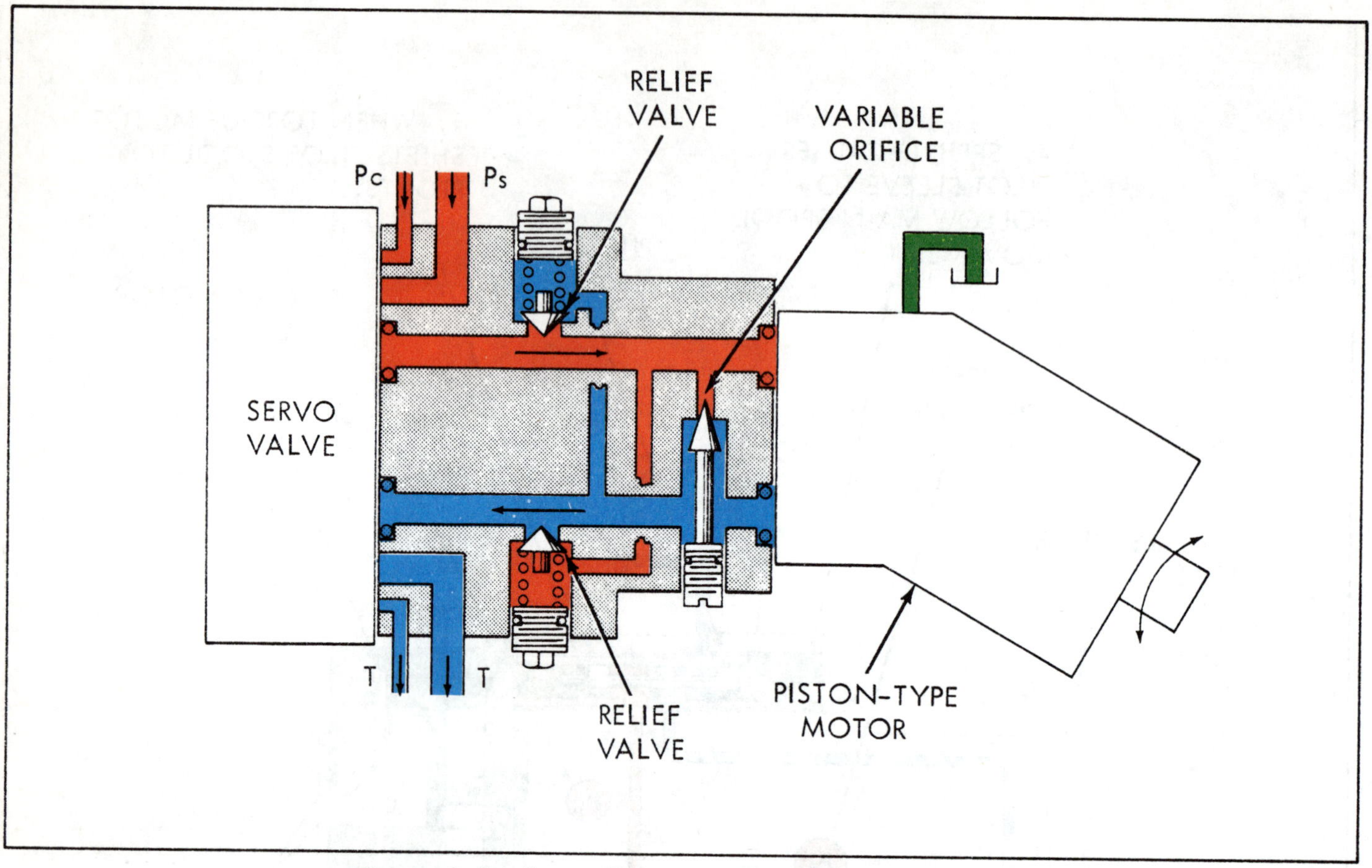

Figure 8-5. Servo Valve Manifold to Piston Motor

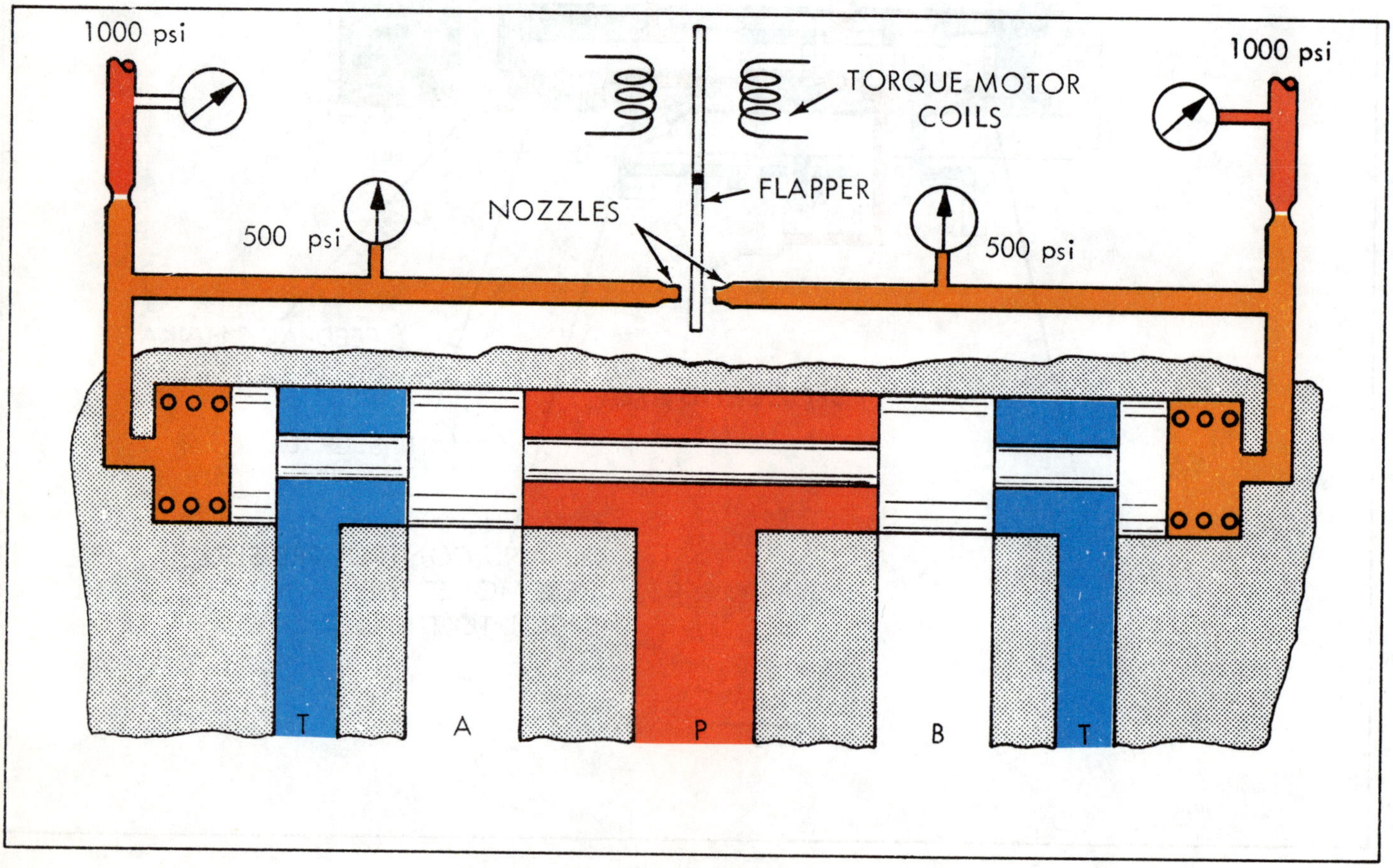

Figure 8-6. Flapper Type Servo Valve

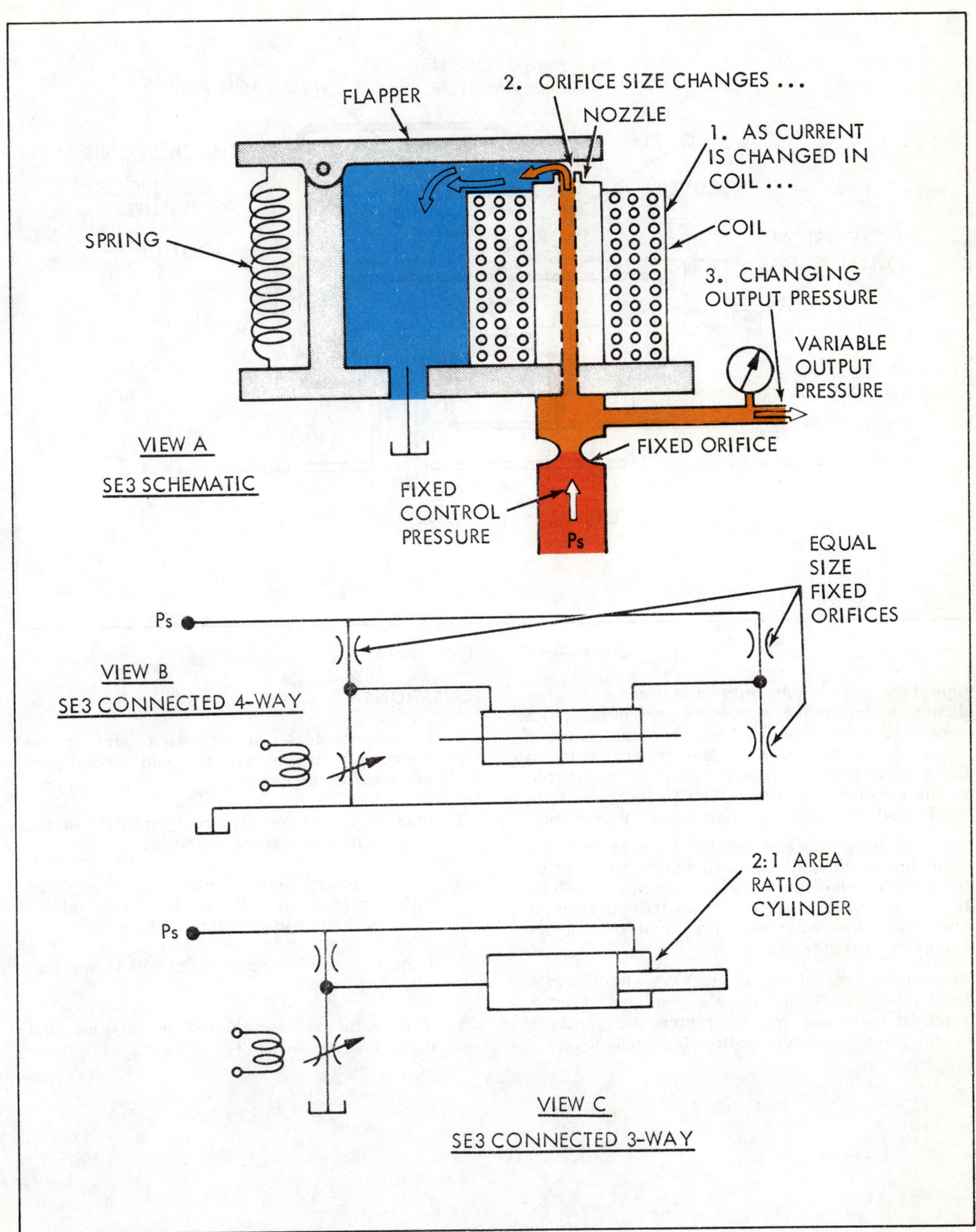

Figure 8-7. Functional Schematic of SE3 Servo Valve

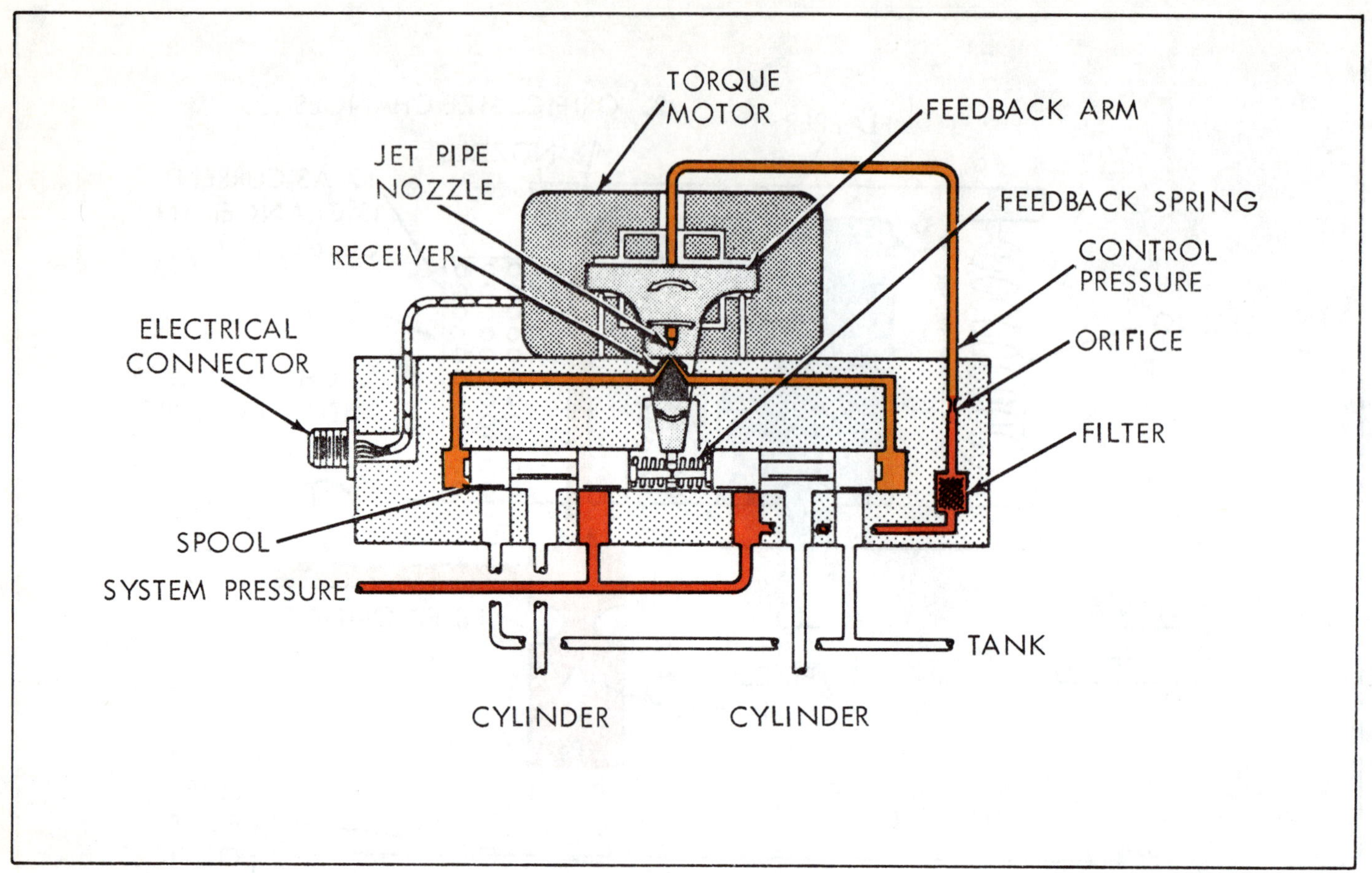

Figure 8-8. Jet Pipe Servo

pipe (Fig. 8-8), a tube with an orificed end which directs a continuing stream of control oil into a receiver. The receiver has two outlet ports connected to the valve spool ends. Pressure in these ports is equal when the jet pipe is centered in the receiver opening. With pressure at both ends equal, the spool's springs hold it centered.

The torque motor can deflect the pipe in either direction an amount proportional to the positive or negative electric signal it receives. Deflection of the pipe causes a pressure differential that shifts the valve spool against one of its centering springs.

Positioning and velocity control are effected with input and feedback signals that maintain a selected valve opening or return the spool to center when a desired position is reached.

QUESTIONS

1. In a mechanical servo, what part of the servo valve moves with the load? What part moves with the control?

2. In a single-stage electro-hydraulic servo, how is the valve spool actuated?

3. What primary feature makes a servo valve different from an ordinary directional valve? What is the purpose of this feature?

4. Explain dither--what it is, how it is applied, why it is needed.

5. How is the spool actuated in jet-pipe and flapper servo valves?

CHAPTER 9

PRESSURE CONTROLS

Pressure control valves perform functions such as limiting maximum system pressure or regulating reduced pressure in certain portions of a circuit, and other functions wherein their actuation is a result of a change in operating pressure. Their operation is based on a balance between pressure and spring force. Most are infinite positioning; that is, the valves can assume various positions between fully closed and fully open, depending on flow rate and pressure differential.

Pressure controls are usually named for their primary function, such as relief valve, sequence valve, brake valve, etc. They are classified by type of connections, size and pressure operating range. The valves covered in this chapter are typical of the pressure controls in most industrial systems.

RELIEF VALVES

The relief valve is found in virtually every hydraulic system. It is a normally-closed valve connected between the pressure line (pump outlet) and the reservoir. Its purpose is to limit pressure in the system to a preset maximum by diverting some or all of the pump's output to tank when the pressure setting is reached.

SIMPLE RELIEF VALVE

A simple or direct acting relief valve (Fig. 9-1) may consist of nothing but a ball or poppet held seated in the valve body by a heavy spring. When pressure at the inlet is insufficient to overcome the force of the spring, the valve remains closed. When the preset pressure is reached, the ball or poppet is forced off its seat and allows flow through the outlet to tank for as long as pressure is maintained.

In most of these valves, an adjusting screw is provided to vary the spring force. Thus the valve can be set to open at any pressure within its specified range.

Pressure Override

The pressure at which the valve first begins to divert flow is called the cracking pressure. As flow through the valve increases, the poppet is forced farther off its seat causing increased compression of the spring. Thus, when the valve is bypassing its full rated flow, the pressure can be considerably higher than the cracking pressure.

Pressure at the inlet when the valve is passing its maximum volume is called full-flow pressure. The difference between full-flow pressure and cracking pressure is sometimes called pressure override.

In some cases, pressure override may not be objectionable. In others, it can result in considerable wasted power due to the fluid lost through the valve before its maximum setting is reached. It can permit maximum system pressure to exceed the ratings of other components. Where it is desirable to minimize override, a compound relief valve should be used.

COMPOUND RELIEF VALVE

A compound relief valve (Fig. 9-2) operates in two stages. The pilot stage in the upper valve body contains the pressure limiting valve, a poppet held against a seat by an adjustable spring. The port connections are made to the lower body, and diversion of the full-flow volume is accomplished by the balanced piston in the lower body.

Balanced Piston

The balanced piston is so named because in normal operation (Fig. 9-3, View A), it is in hydraulic balance. Pressure at the inlet port acting under the piston is also sensed on its top by means of an orifice drilled through the large land. At any pressure less than the valve setting, the piston is held on its seat by a light spring.

When pressure reaches the setting of the adjustable spring, the poppet is forced off its seat limiting pressure in the upper chamber.

The restricted flow through the orifice into the upper chamber results in an increase in pressure in the lower chamber. This unbalances the hydraulic forces and tends to raise the piston off

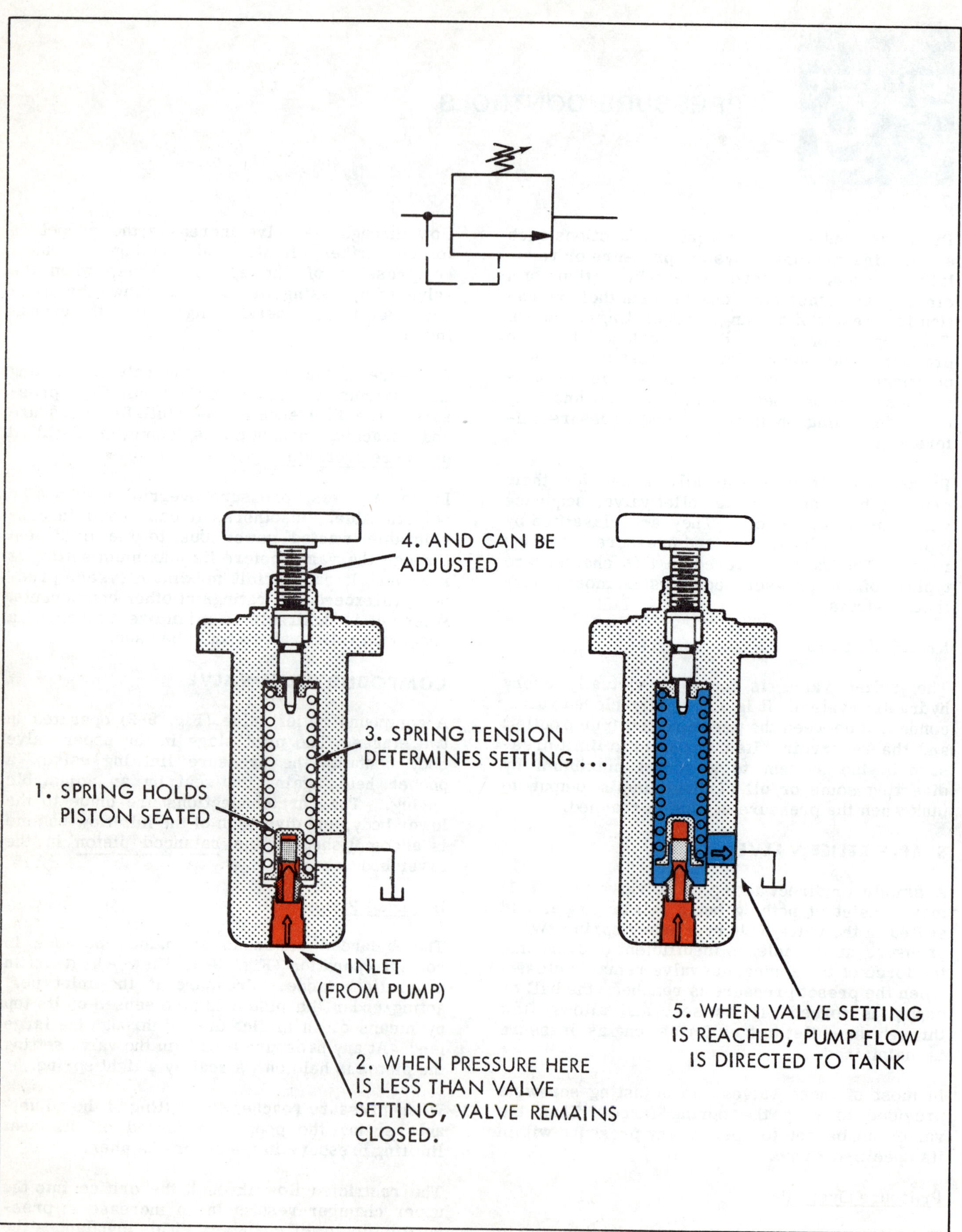

Fig. 9-1. Simple Relief Valve

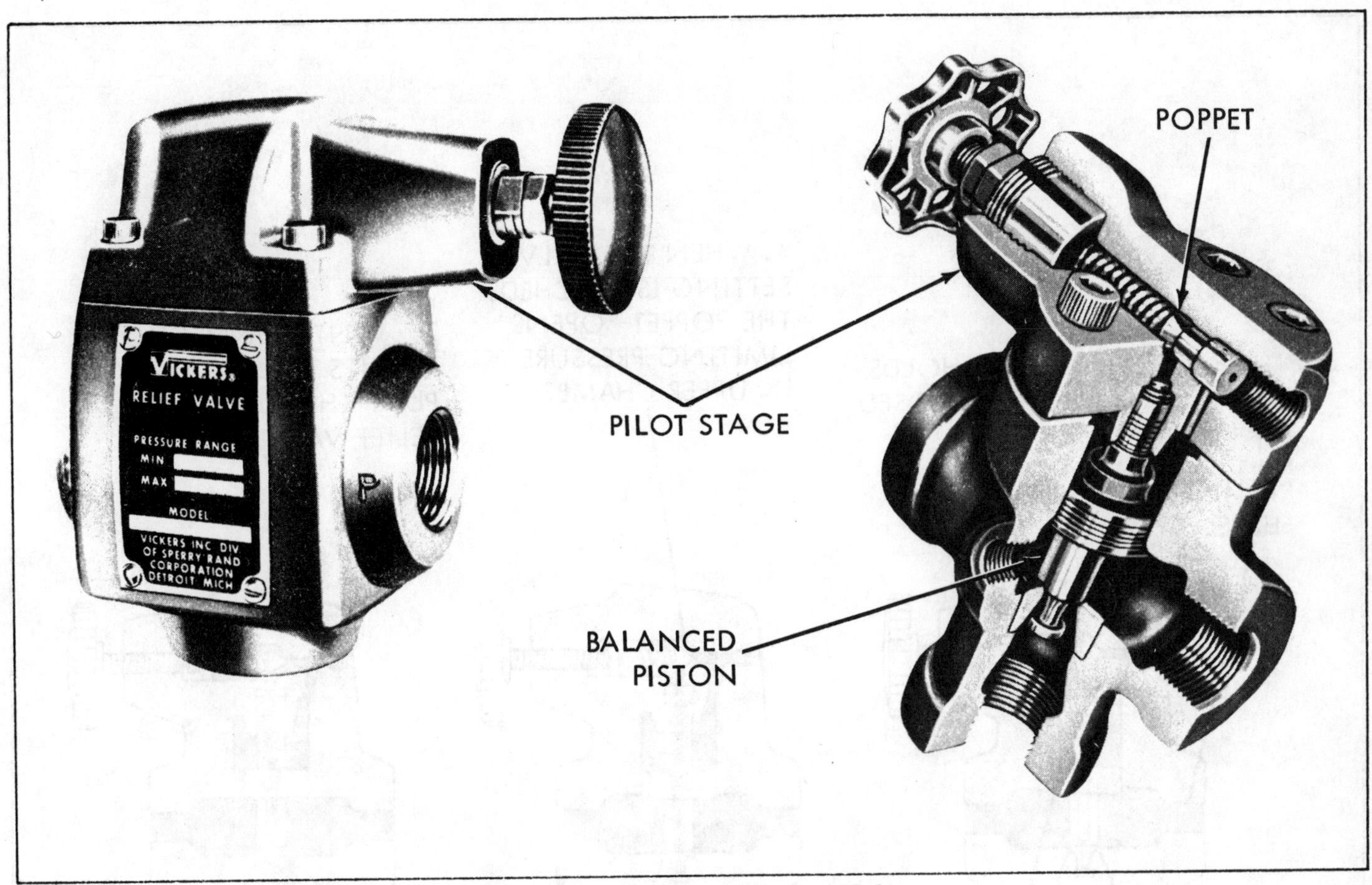

Fig. 9-2. Compound Relief Valve

its seat. When the difference in pressure between the upper and lower chambers is sufficient to overcome the force of the light spring (approximately 20 psi), the large piston unseats permitting flow directly to tank. Increased flow through the valve causes the piston to lift further off its seat but since this compresses only the light spring very little override is encountered.

Vent Connection

Compound relief valves may be remotely controlled by means of an outlet port from the chamber above the piston. When this chamber is "vented" to tank, the only force holding the piston on its seat is that of the light spring; and the valve will open fully at approximately 20 psi. See figure 9-4.

Occasionally, this standard spring is replaced by a heavier one permitting "vent" pressures of 50-70 psi when required for pilot pressure. A second benefit of the high vent spring is that it causes faster and more positive seating of the piston.

It also is possible to connect a simple relief valve to the venting port to control pressure from a remote location (Fig. 9-5). To exercise control, the remote valve must be set for a lower pressure than the integral pilot stage. An application of remote pressure control is illustrated in Chapter 13.

"R" TYPE VALVES

The "R" type valve (Fig. 9-6) is a direct-acting sliding spool type pressure control valve. The spool operates within a valve body and is held in the closed position by an adjustable spring. Operating pressure sensed through a passage in the bottom cover opposes the spring load. The spool area is such that with the heaviest spring normally used, the valve would open at approximately 125 psi. To extend their pressure range, most models include a small piston or plunger in the bottom cover to reduce the pressure reaction area to 1/8 (1/16 in the 2000 psi range) of the area of the spool end. When operating pressure exceeds the valve setting, the spool is raised and oil can flow from the primary to the secondary port.

A drain passage is provided in the top cover to drain the spring chamber. This drain also removes leakage oil from the space between the spool and piston by means of a passage drilled lengthwise through the spool.

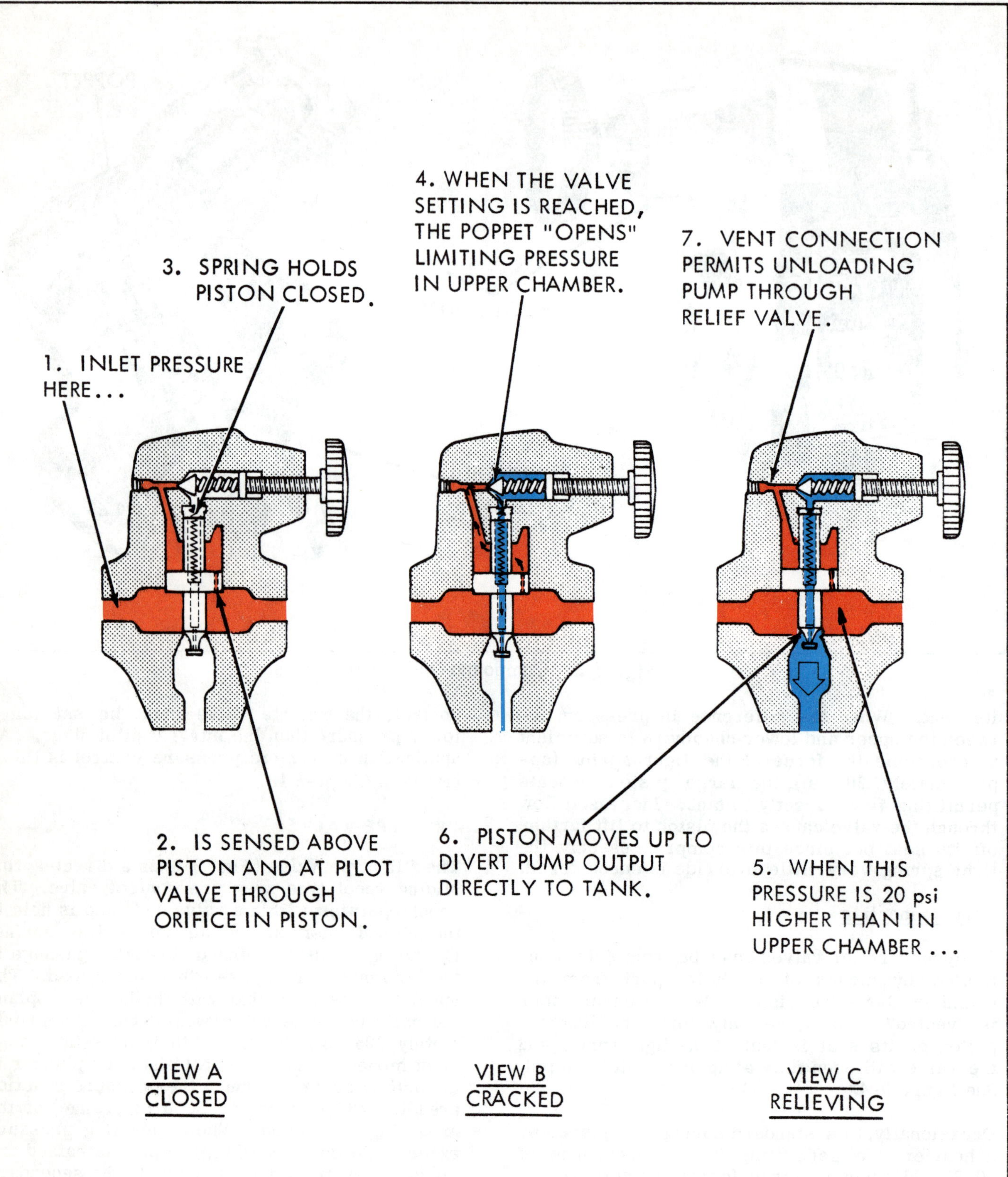

Fig. 9-3. Operation of Balanced Piston Relief Valve

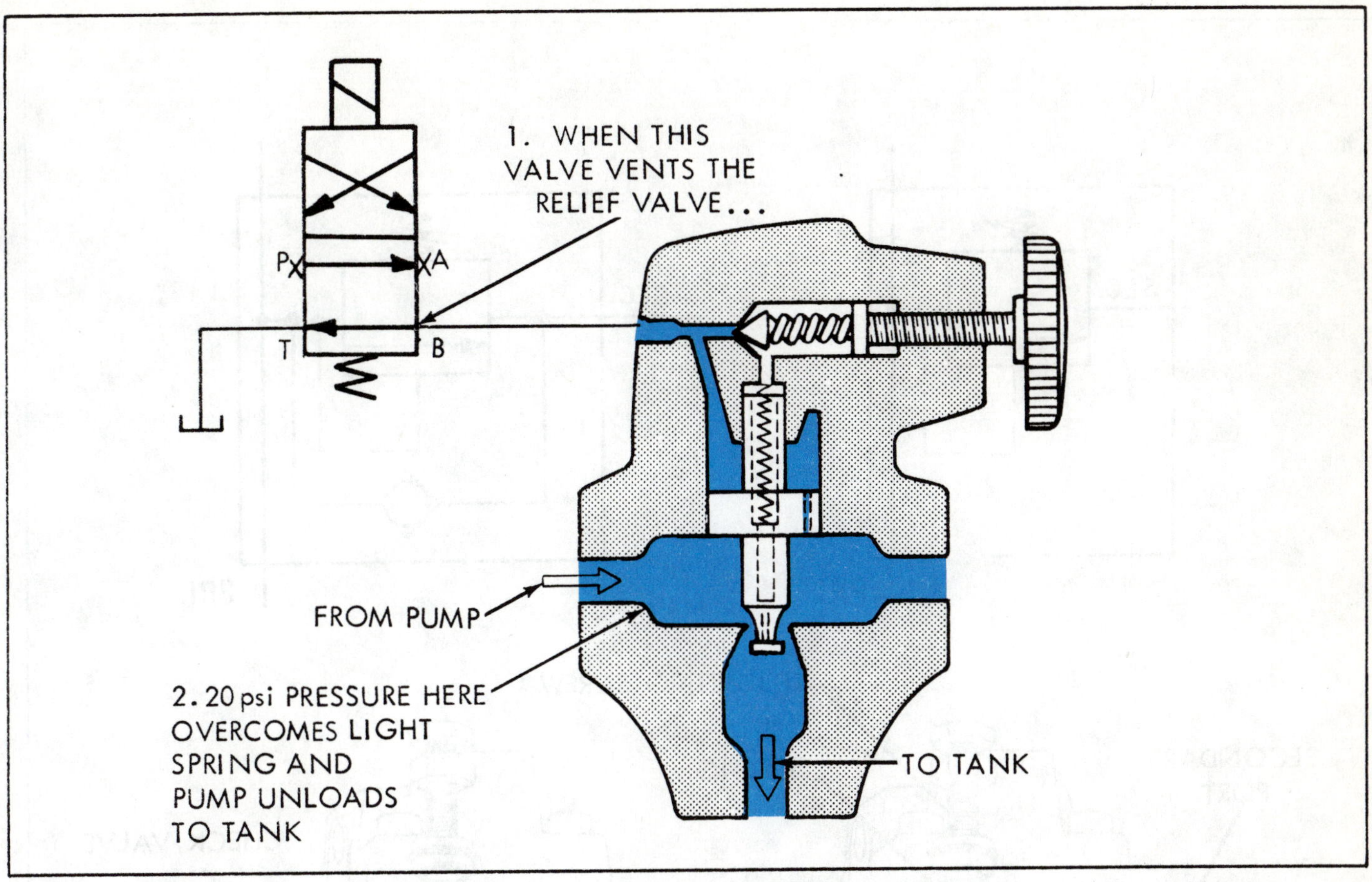

Fig. 9-4. Venting the Relief Valve

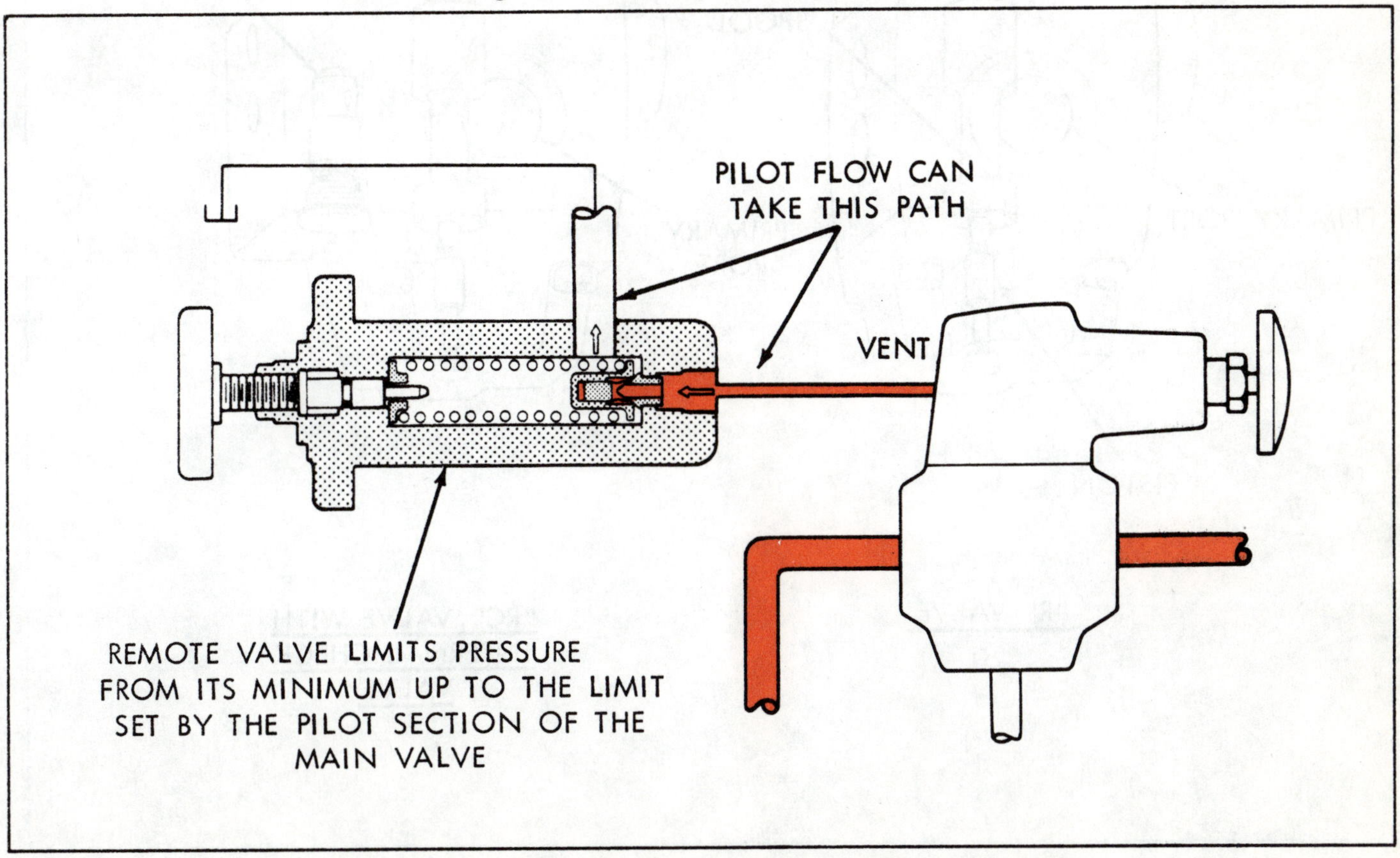

Fig. 9-5. Simple Relief Valve Connected to Venting Port

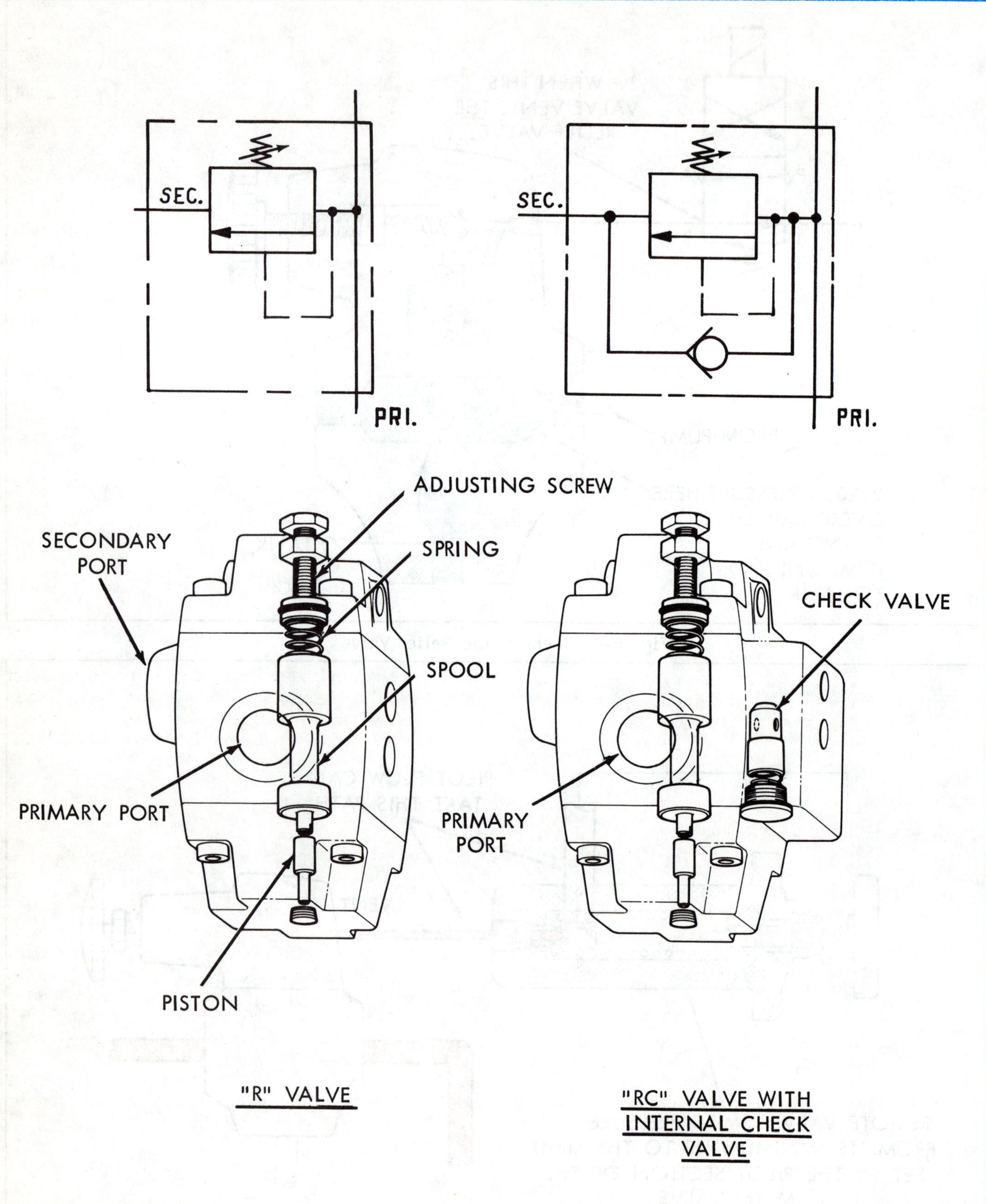

Fig. 9-6. "R" Type Valve

Depending on the assembly of the top and bottom covers, this valve can be used as a relief valve, sequence valve or unloading valve. It is also built with an integral check valve ("RC" type) to permit reverse flow when used as a sequence, counterbalance or brake valve.

"R" TYPE RELIEF VALVES

Figure 9-7 illustrates the "R" valve assembled for relief valve operation. The pressure line is connected to the primary port and the secondary port is connected to tank. This application permits the valve to be internally drained and the upper cover is assembled with the drain passage aligned with the secondary port. The lower cover is assembled so that operating pressure is sampled internally from the primary port making it necessary to maintain maximum system pressure to keep the valve open.

In view A, the system pressure against the piston is too low to overcome the spring and the valve remains closed. In view B, pressure has shifted the spool to allow flow to the secondary port and to tank at the pressure determined by the spring setting.

With the small piston, this valve is capable of operation at higher pressures. However, because of its relatively high override characteristics, it is not recommended for use as a relief valve above 500 psi.

"R" TYPE UNLOADING VALVE

To use the same valve as an unloading valve (Fig. 9-8), the lower cover is assembled to block the internal operating pressure passages. An external pressure source is used to move the spool and divert pump delivery to the secondary port. The drain connection remains internal, since the secondary port is still connected to the tank.

Note the operating difference between the unloading and relief valves (Fig. 9-7, view B). The relief valve operates in balance, being held open at one of an infinite number of positions by the flow of oil through it. Maximum pressure maintained at the primary port is determined by the spring adjustment. With the unloading valve, however, the primary port pressure is independent of the spring force because the remote pressure source operates the spool. As long as the control pressure is at least 150 psi above the spring setting, free flow is permitted from the primary to the secondary port.

"R" TYPE SEQUENCE VALVE

A sequence valve is used to cause actions to take place in a system in a definite order, and to maintain a pre-determined minimum pressure in the primary line while the secondary operation occurs. Figure 9-9 shows the "R" valve assembled for sequencing. Fluid flows freely through the primary passage to operate the first phase until the pressure setting of the valve is reached. As the spool lifts (view B), flow is diverted to the secondary port to operate a second phase. A typical application is clamping from the primary port and feeding a drill head from the secondary after the work piece is firmly clamped.

To maintain pressure in the primary system, the valve is internally operated. However, the drain connection must be external, since the secondary port is under pressure when the valve "sequences." If this pressure were allowed in the drain passage, it would add to the spring force and raise the pressure required to open the valve.

"RC" TYPE SEQUENCE VALVE

The "R" type sequence valve is suitable for systems where it can be installed upstream from the directional valve. If it is installed downstream (in a cylinder line), some provision must be made for return free flow when the cylinder is reversed. A bypass check valve can be used, or the "R" valve can be replaced with the "RC" valve (Fig. 9-10), which has an integral check valve for return flow. The operation otherwise is identical.

REMOTELY OPERATED SEQUENCE VALVE

In some systems, it is desirable to provide an interlock so that sequencing does not occur until the primary actuator reaches a definite position. In these applications, the bottom cover on the sequence valve is assembled for remote operation. A cam-operated directional valve blocks the control pressure from the piston in the bottom cover until the clamp cylinder reaches the prescribed position. Only then is the sequence valve permitted to shift and direct flow to the second operation.

"RC" TYPE COUNTERBALANCE VALVE

A counterbalance valve is used to maintain control over a vertical cylinder so that it will not fall freely because of gravity. The primary port

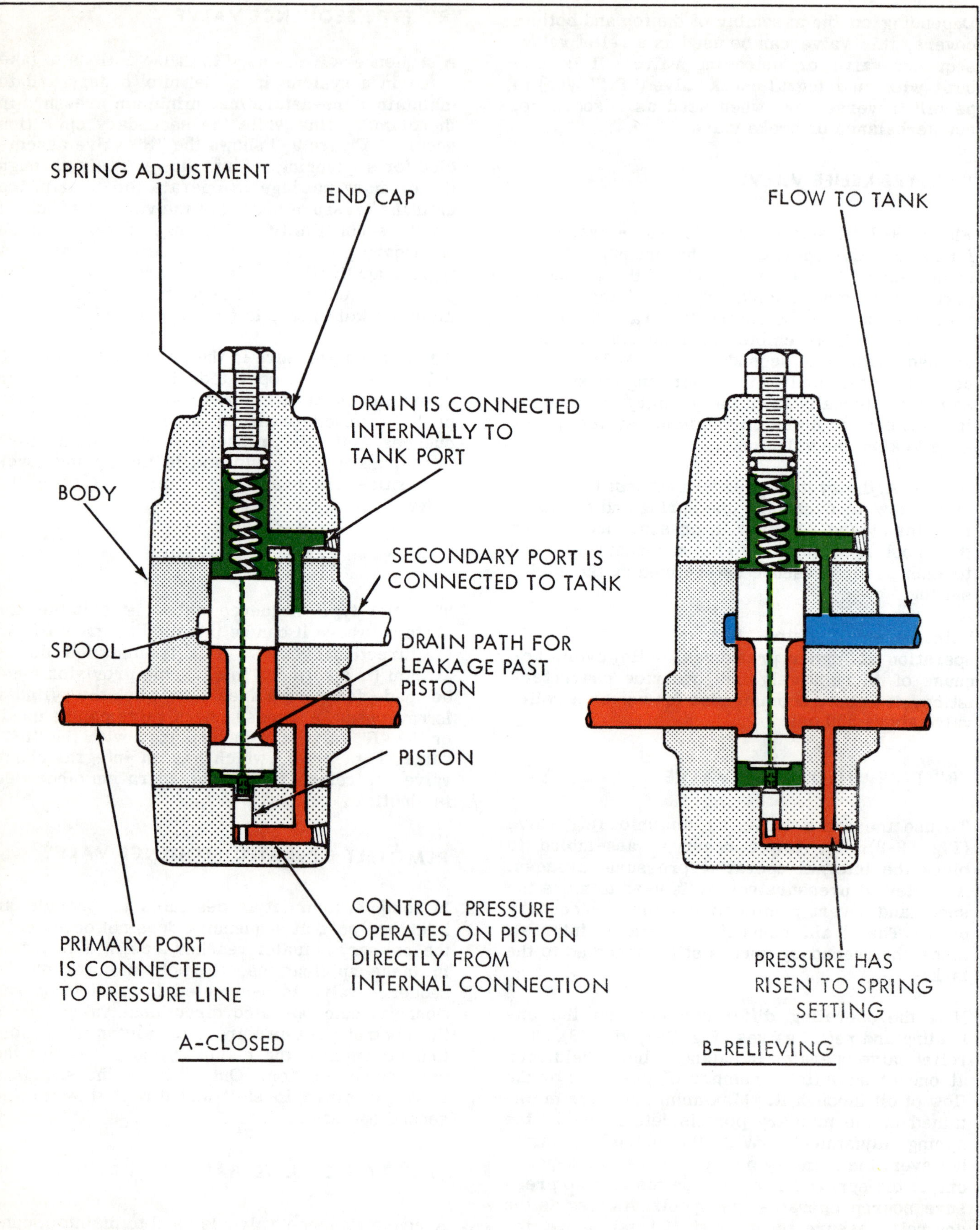

Fig. 9-7. "R" Type Relief Valve

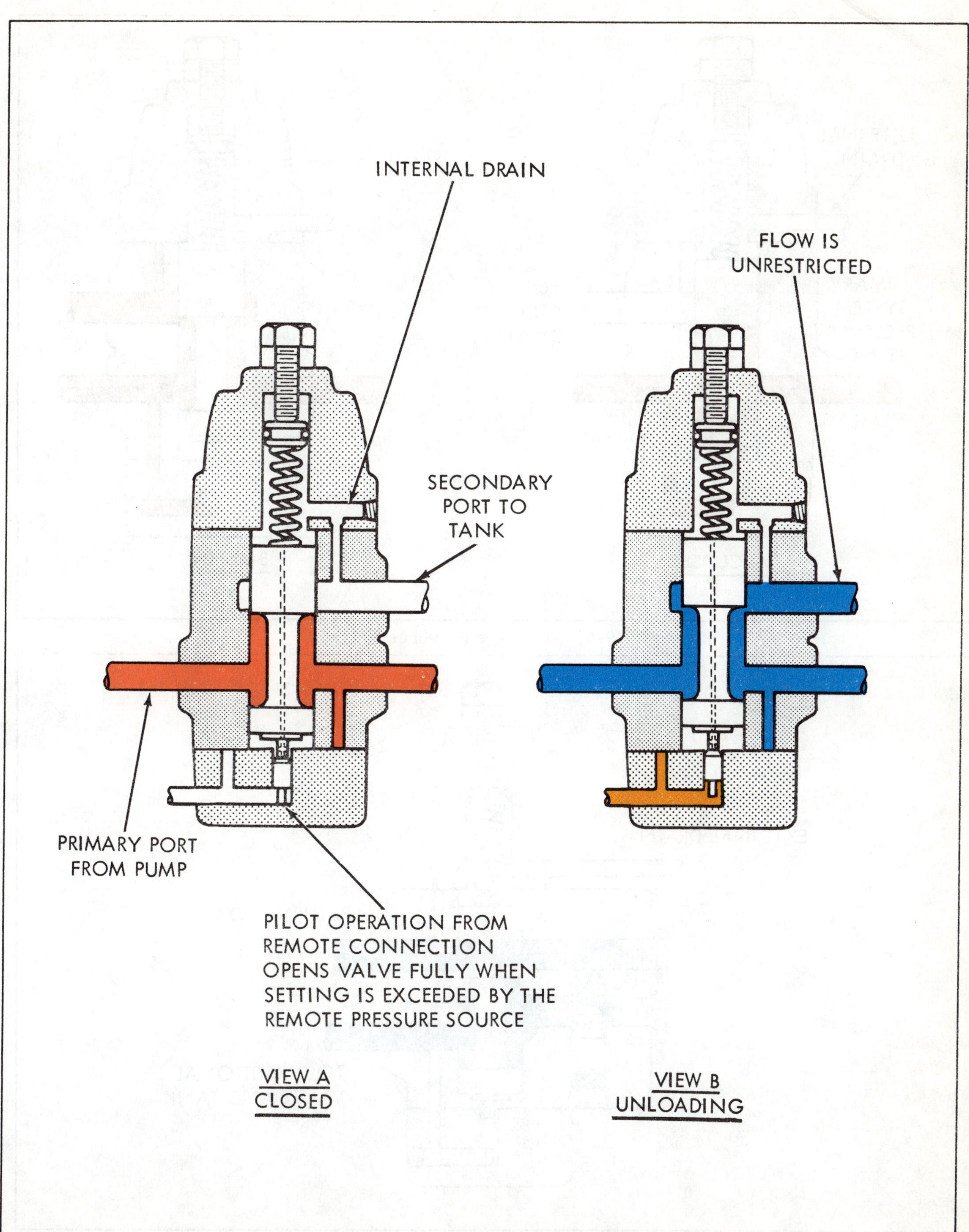

Fig. 9-8. "R" Type Unloading Valve

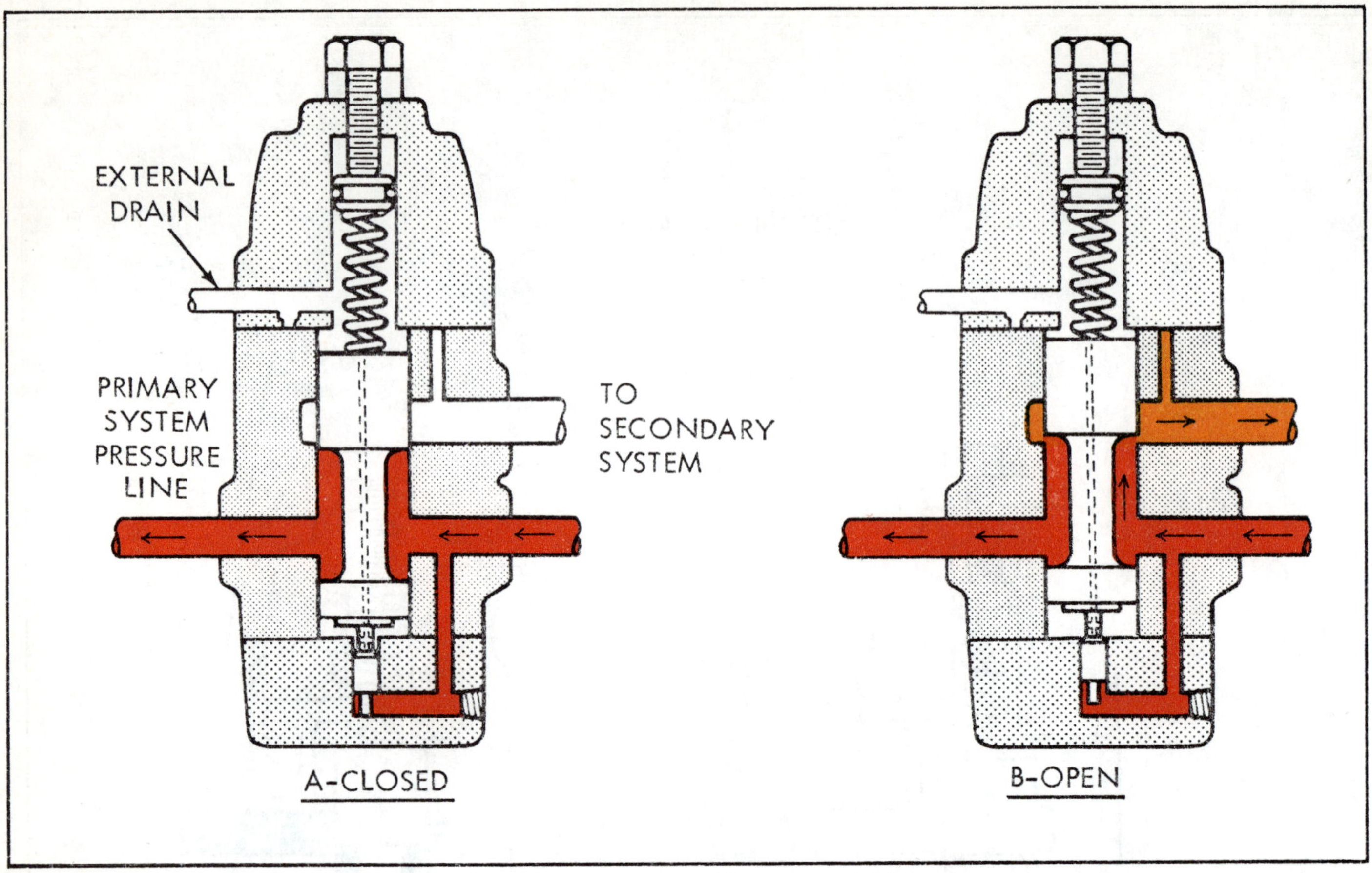

Fig. 9-9. "R" Type Sequence Valve

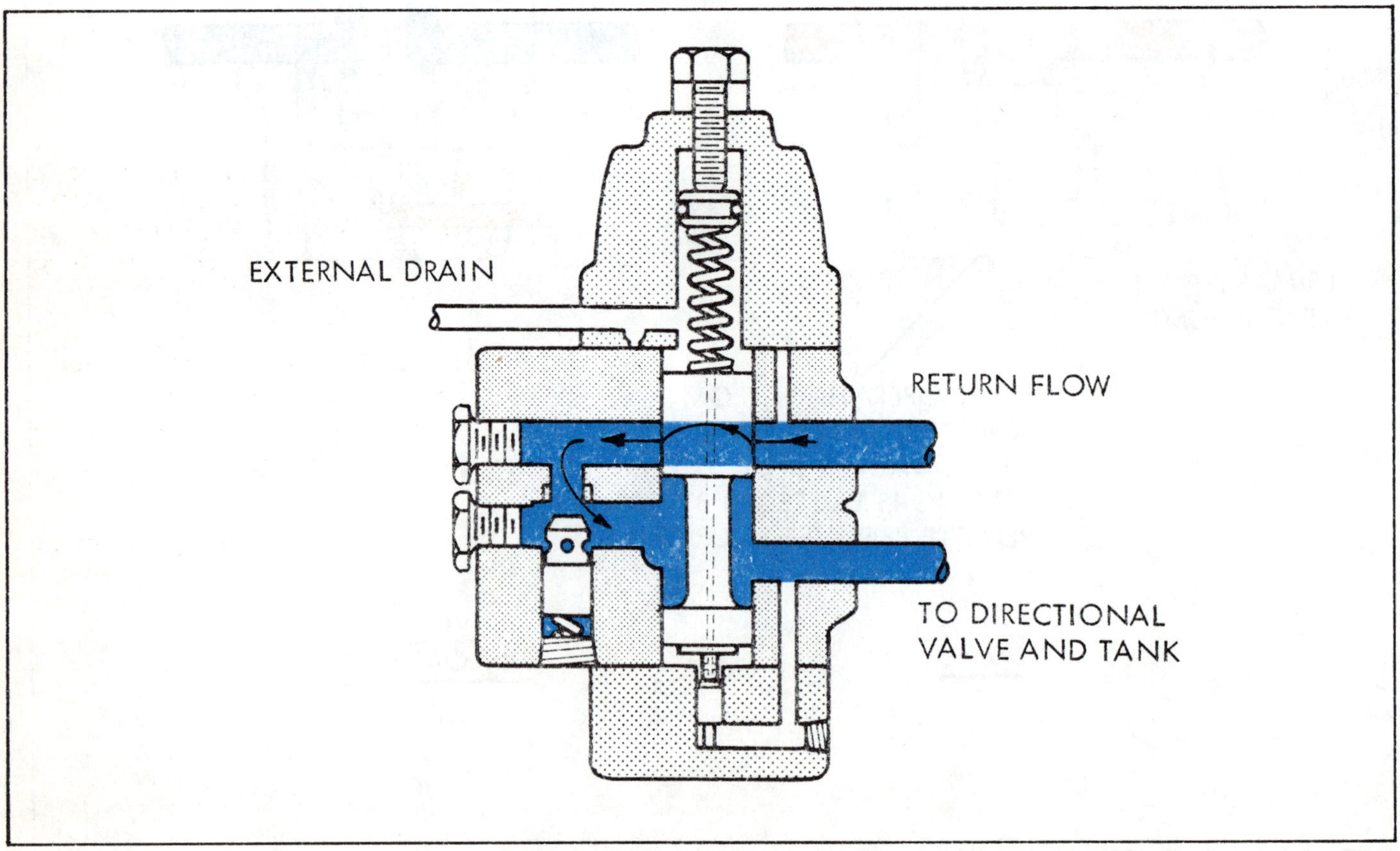

Fig. 9-10. "RC" Type Sequence Valve Permits Reverse Free Flow

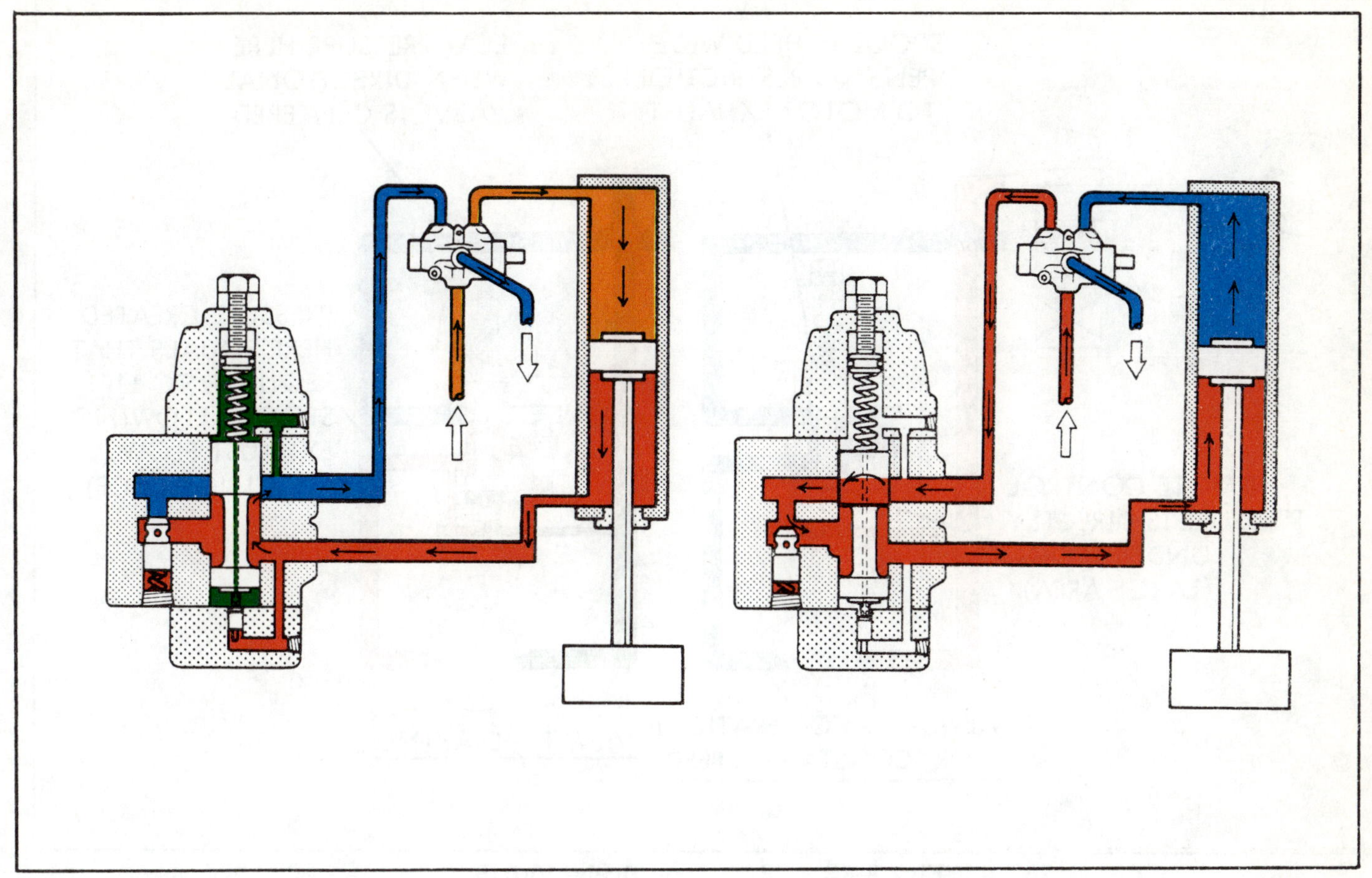

Fig. 9-11. "RC" Type Counterbalance Valve

of the "RC" valve is connected to the lower cylinder port and the secondary port to the directional valve (Fig. 9-11). The pressure setting is slightly higher than is required to hold the load from falling.

When the pump delivery is directed to the top of the cylinder, the cylinder piston is forced down causing pressure at the primary port to increase and raise the spool, opening a flow path for discharge through the secondary port to the directional valve and subsequently to tank. In cases where it is desired to remove back pressure at the cylinder and increase the force potential at the bottom of the stroke, this valve too can be operated remotely.

When the cylinder is being raised (view B), the integral check valve opens to permit free flow for returning the cylinder.

The counterbalance valve can be internally drained. In the lowering position (view A), when the valve must open, its secondary port is connected to tank. In the reverse condition, it does not matter that load pressure is effective in the drain passage, because the check valve bypasses the spool.

"RC" TYPE BRAKE VALVE

A brake valve is used in the exhaust line of a hydraulic motor to (1) prevent overspeeding when an overrunning load is applied to the motor shaft and (2) prevent excessive pressure build up when decelerating or stopping a load.

When the "RC" valve is used as a brake valve, it has a solid spool (no drain hole through center); and there is a remote operating pressure connection in the bottom cover directly under the spool (Fig. 9-12). This connection is teed into the supply line to the motor. The internal control connection also is used under the small piston and senses pressure from the primary port of the "RC" valve which is connected to the motor exhaust port.

Accelerating the Load

When the load is being accelerated, pressure is maximum at the motor inlet and under the large area of the brake valve spool holding it in the full open position permitting free flow from the exhaust port of the motor.

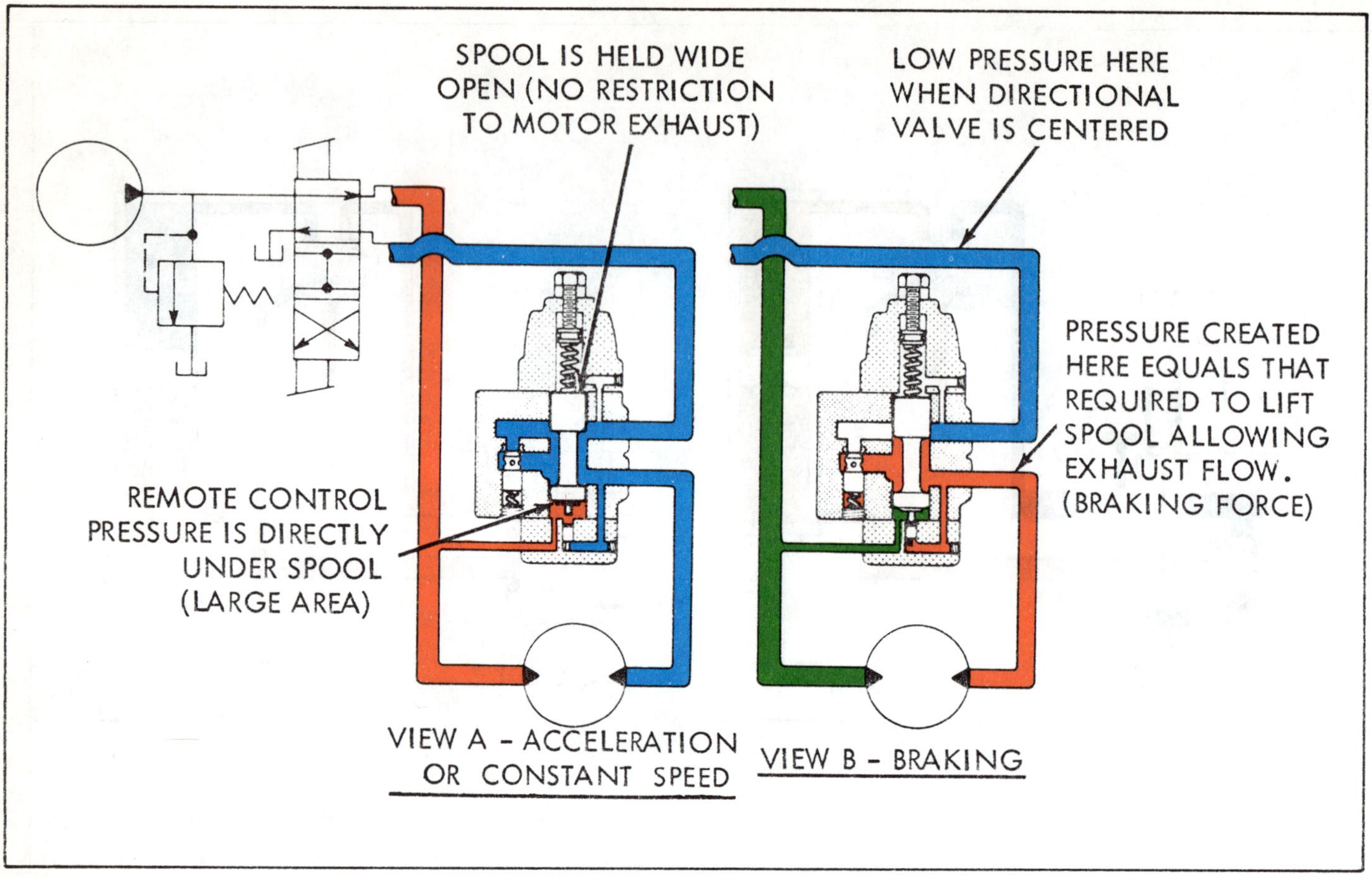

Fig. 9-12. "RC" Type Brake Valve

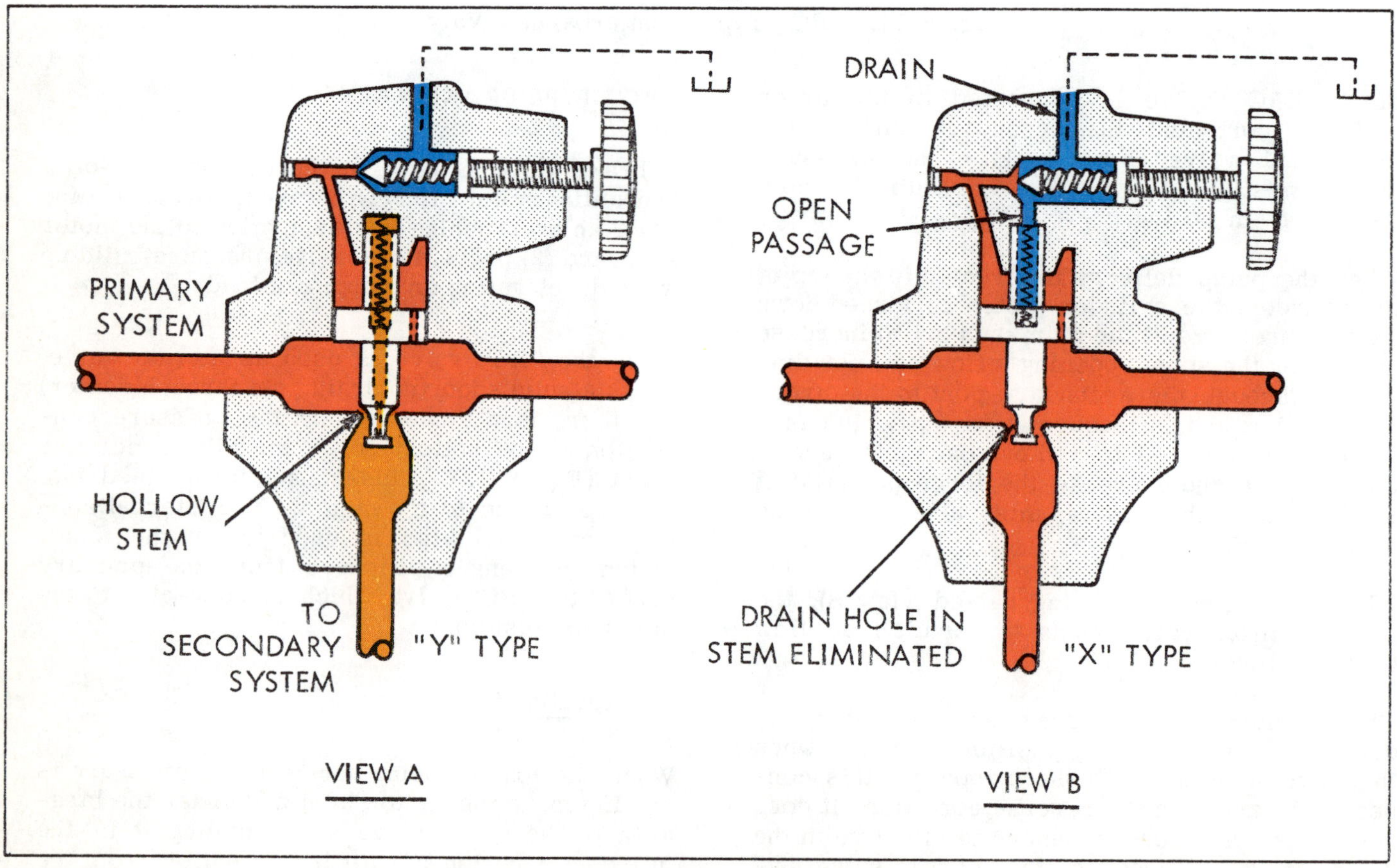

Fig. 9-13. Two Stage Sequence Valve

At Operating Speed

When the motor gets up to speed, load pressure still holds the brake valve open unless the load tries to run away. If this happens, the pressure falls off at the motor inlet and in the remote control pressure passage (view B). The spring force tends to close the valve thus increasing the back pressure. This in turn raises the drive line pressure to the motor and under the small piston holding the valve at the proper metering position to maintain constant motor speed.

Braking

When the directional valve is shifted to neutral, inertia causes the motor to continue rotating. Until the motor stops turning, it will operate as a pump, drawing fluid from the reservoir through the directional valve and circulating it back through the brake valve.

At this time,pressure at the motor outlet tending to bring it to a stop will be whatever is required under the small piston to overcome the brake valve setting.

VALVE MOUNTINGS AND RATINGS

"R" and "RC" valves are built in 3/8" to 2" sizes with pipe threaded, gasket mounted or flange connections. Rated flow capacity ranges from 12 gpm for the 3/8" size to 125 gpm for the 2" models. Pressure adjustments are limited to 2000 psi although 3000 psi operating pressure is permissible.

COMPOUND SEQUENCE VALVES

Sequence valves also are built in a two-stage version which is similar to the balanced piston relief valve (Fig. 9-13).

Construction differs from the relief valve in that the drain passage from the pilot stage is external rather than through the stem of the balanced piston. In operation, the primary system passage is connected below the piston land, with the secondary system connected to the bottom port. Sequencing occurs when the primary system pressure is about 20 psi higher than the pilot valve cracking pressure.

"X" AND "Y" TYPE SEQUENCE VALVES

Two modifications of the compound sequence valve provide different throttling characteristics to the secondary system. In the "Y" type (view A, Fig. 9-13), the stem of the balanced piston is hollow. Secondary system pressure below the piston is sensed at the top of the stem, and balances the pressure under the stem. Secondary system pressure then has no effect on the piston movement. The piston thus remains infinite positioning and maintains the preset pressure in the primary system. As the piston opens at the preset pressure, flow is routed to the secondary system. Reverse flow is not possible. If required, a check valve is used to permit flow from secondary to primary.

In the "X" type (view B), the piston center is solid and the spring chamber at the top of the piston is open to drain. When the valve opens at its adjusted setting, pressure below the stem upsets the pressure balance and forces the piston to a wide open position. Pressure then equalizes in both systems at secondary system operating pressure. The valve will open whenever the secondary system pressure exceeds the equivalent force of the piston spring. This in effect provides free flow from secondary to primary since the spring equivalent is 20 psi.

PRESSURE REDUCING VALVES

Pressure reducing valves are normally-open pressure controls used to maintain reduced pressures in certain portions of the system. They are actuated by pressure sensed in the branch circuit and tend to close as it reaches the valve setting, thus preventing further buildup. Both direct acting and pilot operated versions are in use.

DIRECT-ACTING PRESSURE REDUCING VALVE

A typical direct-acting valve is shown in Figure 9-14. It uses a spring loaded spool to control the downstream pressure.

If the main supply pressure is below the valve setting, fluid will flow freely from the inlet to the outlet. An internal connection from the outlet passage transmits the outlet pressure to the spool end opposite the spring.

When the outlet pressure rises to the valve setting (Fig. 9-14, view B), the spool moves to partly block the outlet port. Only enough flow is passed to the outlet to maintain the preset pressure. If the valve closes completely, leakage past the spool could cause pressure to build up in the branch circuit. Instead a continuous bleed to tank is permitted to keep it slightly open and prevent downstream pressure from rising above the valve setting. A separate drain passage is provided to return this leakage flow to tank.

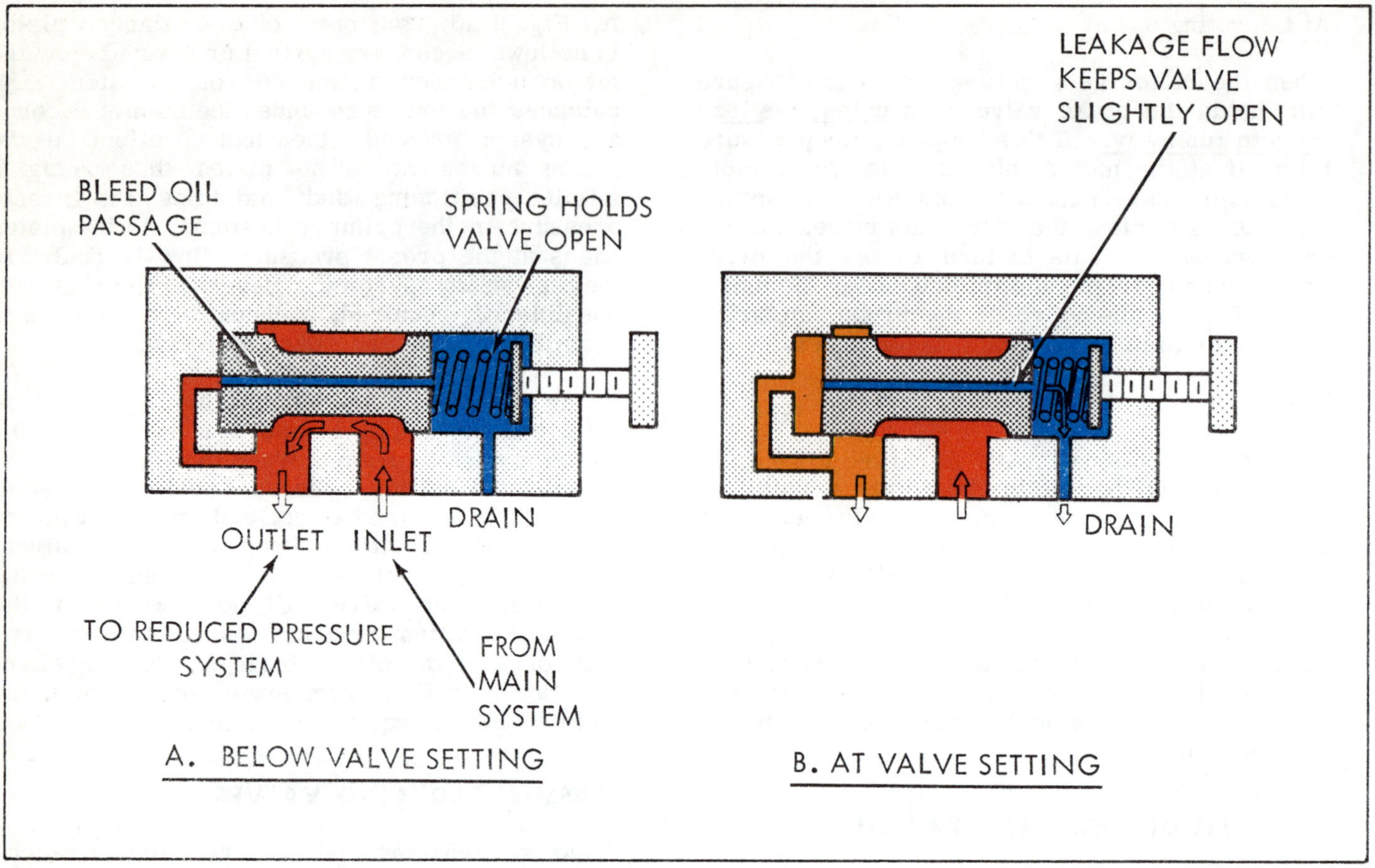

Fig. 9-14. Direct Acting Pressure Reducing Valve

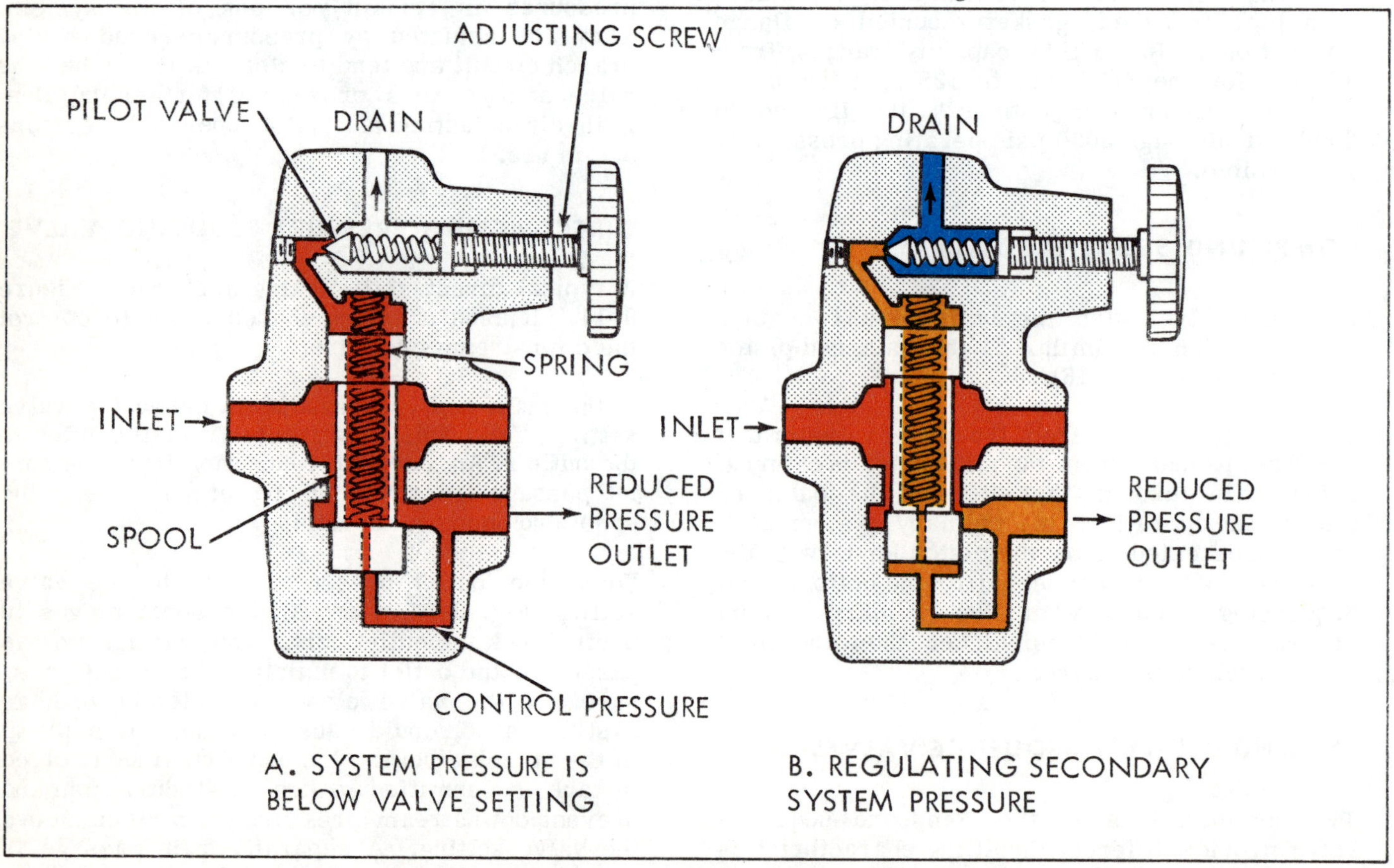

Fig. 9-15. Pilot-Operated Pressure Reducing Valve

PILOT-OPERATED PRESSURE REDUCING VALVES

The pilot-operated pressure reducing valve (Fig. 9-15) has a wider range of adjustment and generally provides more accurate control. The operating pressure is set by an adjustable spring in the pilot stage in the upper body. The valve spool in the lower body functions in essentially the same manner as the direct acting valve discussed previously.

Figure 9-15, view A, shows the condition when supply pressure is less than the valve setting. The spool is hydraulically balanced through an orifice in its center, and the light spring holds it in the wide-open position.

In view B, pressure has reached the valve setting and the pilot valve is diverting flow to the drain passage limiting pressure above the spool. Flow through the orifice in the spool creates a pressure difference that moves the spool up against the spring force. The spool partially closes the outlet port to create a pressure drop from the supply to the branch system.

Again, the outlet port is never entirely closed. When no flow is called for in the branch system, there is still a continuous flow of some 60-90 cubic inches per minute--through the spool orifice and the pilot valve to drain.

REVERSE FREE FLOW

The valve illustrated in Figure 9-15 will handle reverse flow only if the system pressure is less than the valve setting. If reverse flow pressure is higher, a bypass check valve is required. This is an integral part of the valve shown in Figure 9-16.

UNLOADING RELIEF VALVE

An unloading relief valve (Fig. 9-17) is used in accumulator charging circuits to (1) limit maximum pressure and (2) unload the pump when the desired accumulator pressure is reached.

In construction, it contains a compound, balanced piston relief valve, a check valve to prevent reverse flow from the accumulator and a pressure operated plunger which vents the relief valve at the selected pressure.

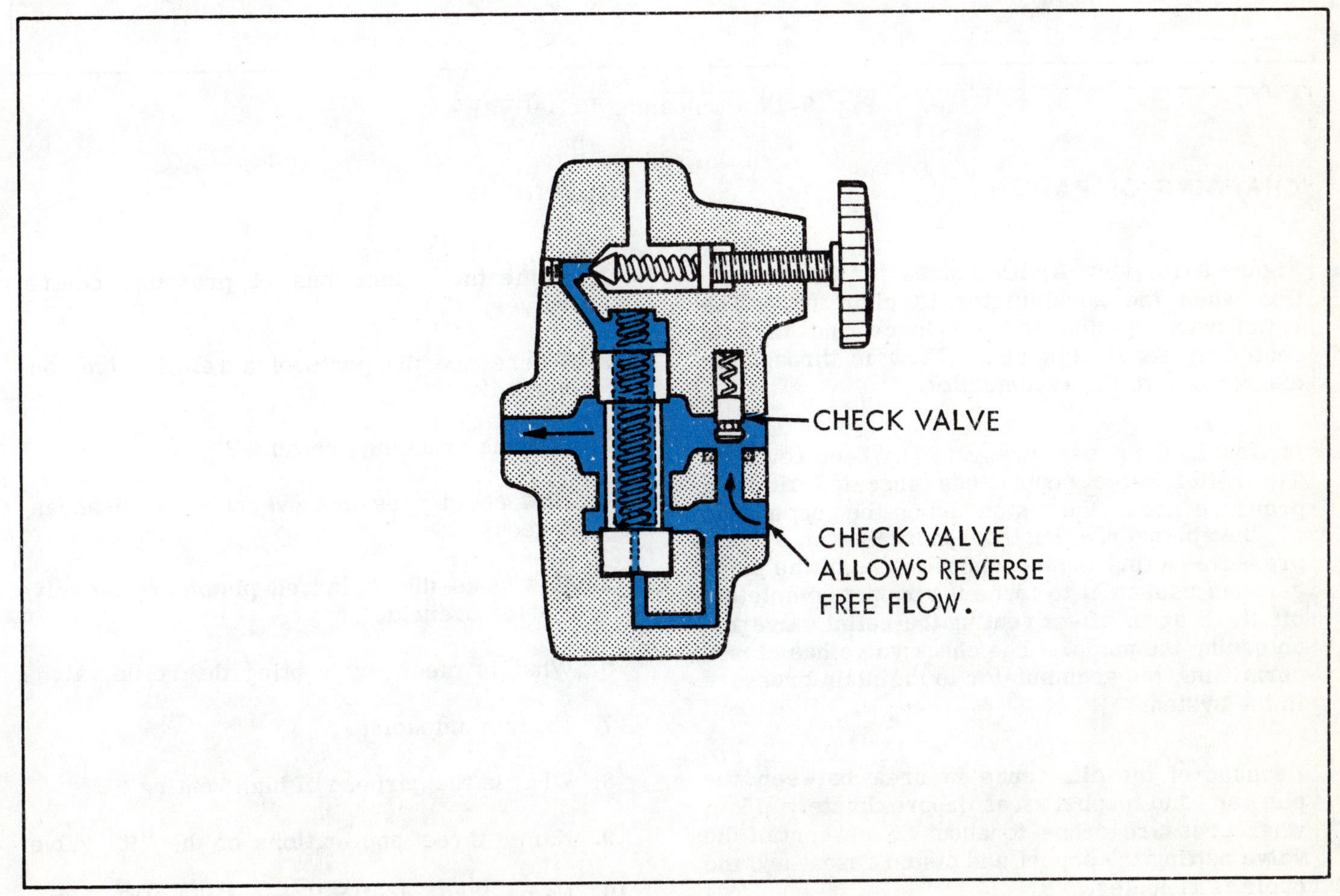

Fig. 9-16. Pressure Reducing Valve with Internal Check Valve

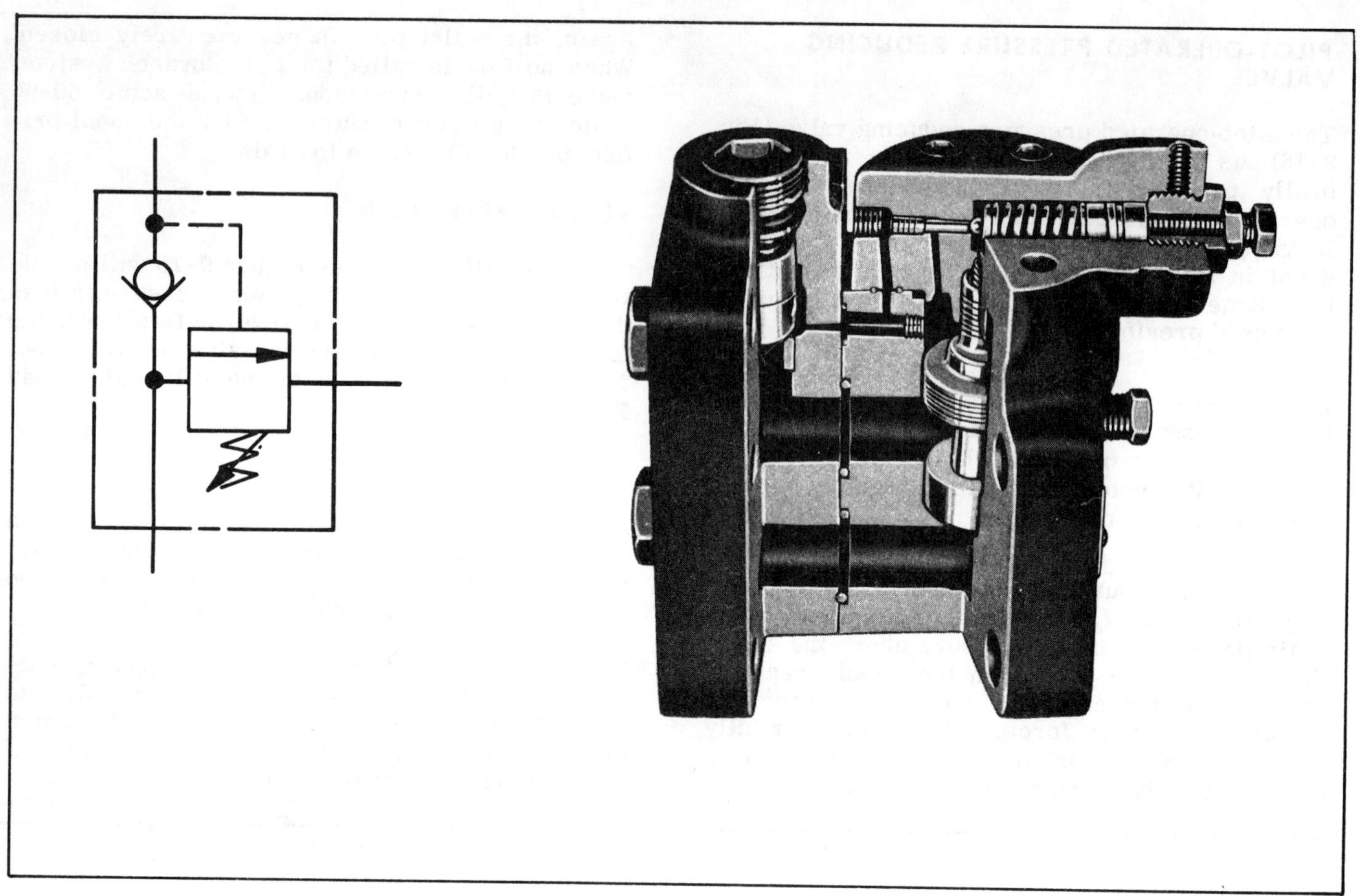

Fig. 9-17. Unloading Relief Valve

CHARGING OPERATION

Figure 9-18, view A, illustrates the flow condition when the accumulator is charging. The relief valve piston is in balance and is held seated by its light spring. Flow is through the check valve to the accumulator.

In view B, the preset pressure has been reached. The relief valve poppet has unseated limiting pressure above the piston and on the poppet side of the plunger. Further increase in system pressure acting on the opposite end of the plunger has caused it to force the poppet completely off its seat in effect venting the relief valve and unloading the pump. The check valve has closed permitting the accumulator to maintain pressure in the system.

Because of the difference in area between the plunger and poppet seat (approximately 15%), when pressure drops to about 85 percent of the valve setting, the poppet and piston reseat and the cycle is repeated.

QUESTIONS

1. Name three functions of pressure control valves.
2. Where are the ports of a relief valve connected?
3. What is cracking pressure?
4. How could pressure override be disadvantageous?
5. How does the "balanced piston" relief valve reduce override?
6. What is meant by venting the relief valve?
7. Explain unloading.
8. What is the purpose of high venting?
9. Name three applications of the "R" valve.
10. Name three applications of the "RC" valve.

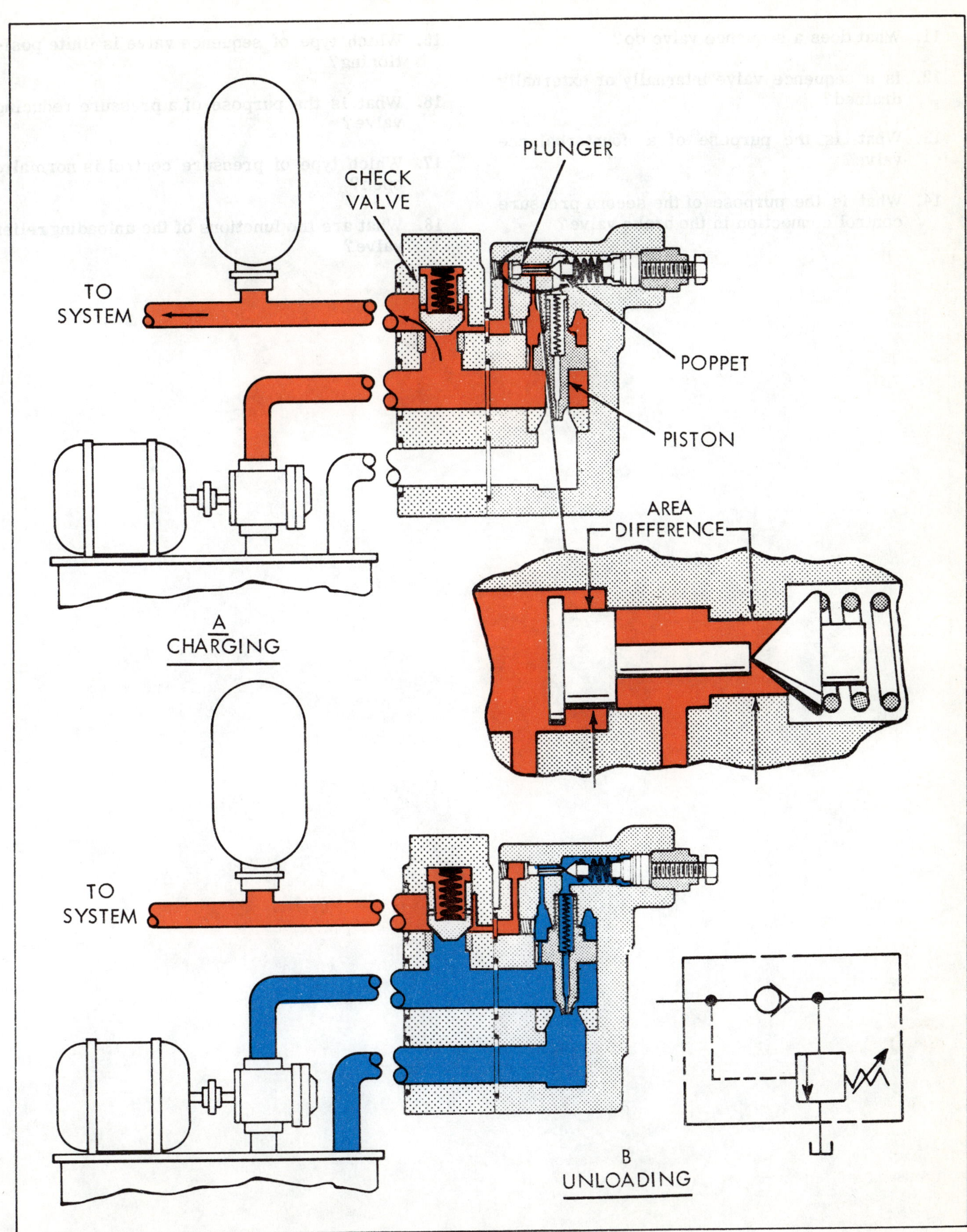

Fig. 9-18. Operation of Unloading Relief Valve

11. What does a sequence valve do?

12. Is a sequence valve internally or externally drained?

13. What is the purpose of a counterbalance valve?

14. What is the purpose of the second pressure control connection in the brake valve?

15. Which type of sequence valve is finite positioning?

16. What is the purpose of a pressure reducing valve?

17. Which type of pressure control is normally open?

18. What are the functions of the unloading relief valve?

CHAPTER 10

VOLUME CONTROLS

Volume or flow control valves are used to regulate speed. As was developed in earlier chapters, the speed of an actuator depends on how much oil is pumped into it per unit of time. It is possible to regulate flow with a variable displacement pump, but in many circuits it is more practical to use a fixed displacement pump and regulate flow with a volume control valve.

FLOW CONTROL METHODS

There are three basic methods of applying volume control valves to control actuator speeds. They are meter-in, meter-out and bleed-off.

Meter-In Circuit

In meter-in operation, the flow control valve is placed between the pump and actuator (Fig. 10-1). In this way, it controls the amount of fluid going into the actuator. Pump delivery in excess of the metered amount is diverted to tank over the relief valve.

With the flow control valve installed in the cylinder line as shown, flow is controlled in one direction. A check valve must be included in the flow control or placed in parallel with it for return flow. If it is desired to control speed in both directions, the flow control can be installed in the pump outlet line prior to the directional valve.

The meter-in method is highly accurate. It is used in applications where the load continually resists movement of the actuator, such as rais-

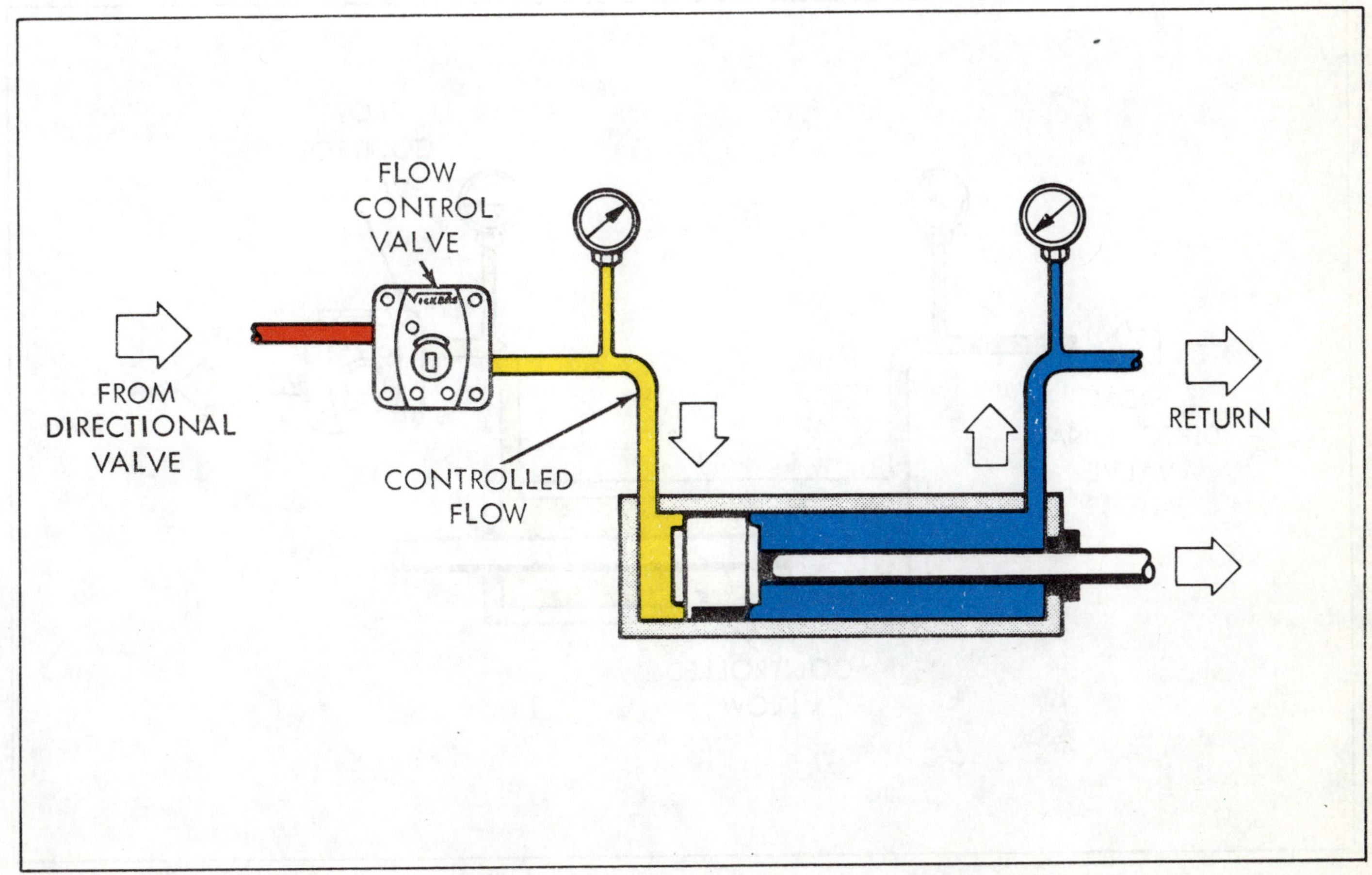

Figure 10-1. Meter In Flow Control

ing a vertical cylinder under load or pushing a load at a controlled speed.

Meter-Out Circuit

Meter-out control (Fig. 10-2) is used where the load might tend to "run away." The flow control is located where it will restrict exhaust flow from the actuator.

To regulate speed in both directions, the valve is installed in the tank line from the directional valve. More often control is needed in only one direction and it is placed in the line between the actuator and directional valve. Here too a by-pass check valve would be required for a rapid return stroke.

Bleed-Off Circuit

In a bleed-off arrangement (Fig. 10-3), the flow control is teed off the supply line from the pump and determines the actuator speed by metering a portion of the pump delivery to tank. The advantage is that the pump operates at the pressure required by the work, since excess fluid returns to tank through the flow control instead of through the relief valve.

Its disadvantage is some loss of accuracy because the measured flow is to tank rather than into the cylinder, making the latter subject to variations in the pump delivery due to changing work loads.

Bleed-off circuits should not be used in applications where there is a possibility of the load running away.

TYPES OF FLOW CONTROLS

Flow control valves fall into two basic categories: pressure compensated and non-pressure compensated. The latter being used where load pressures remain relatively constant and feed rates are not too critical. They may be as simple as a fixed orifice or an adjustable needle valve, although more sophisticated units may even include a check valve (Fig. 10-4) for free flow in the reverse direction. Use of non-pressure compensated valves is somewhat limited, since flow through an orifice is essentially proportional to the square root of the pressure drop (Δ P) across it. This means that any appreciable change in the work load would affect the feed rate.

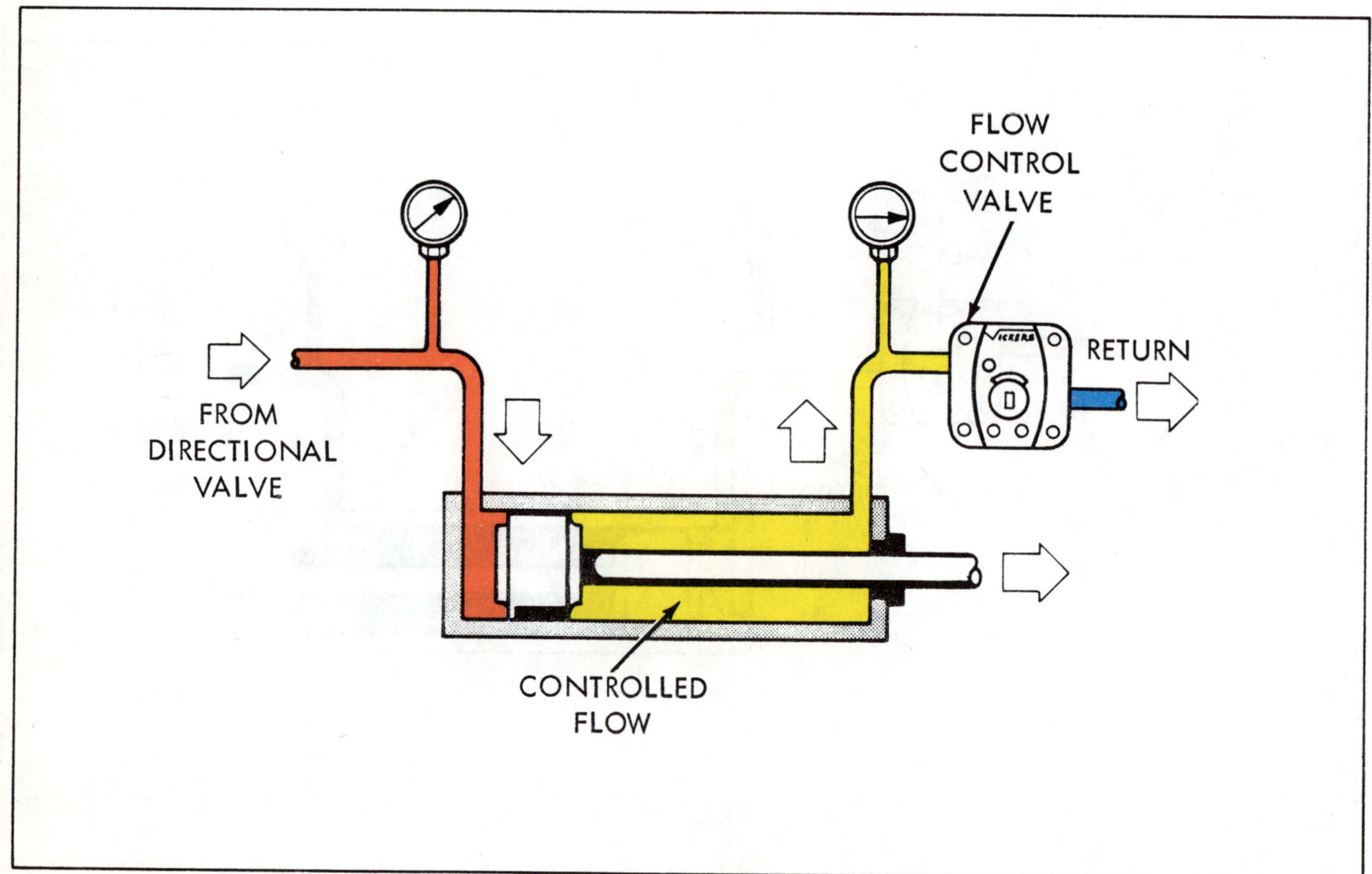

Figure 10-2. Meter Out Flow Control

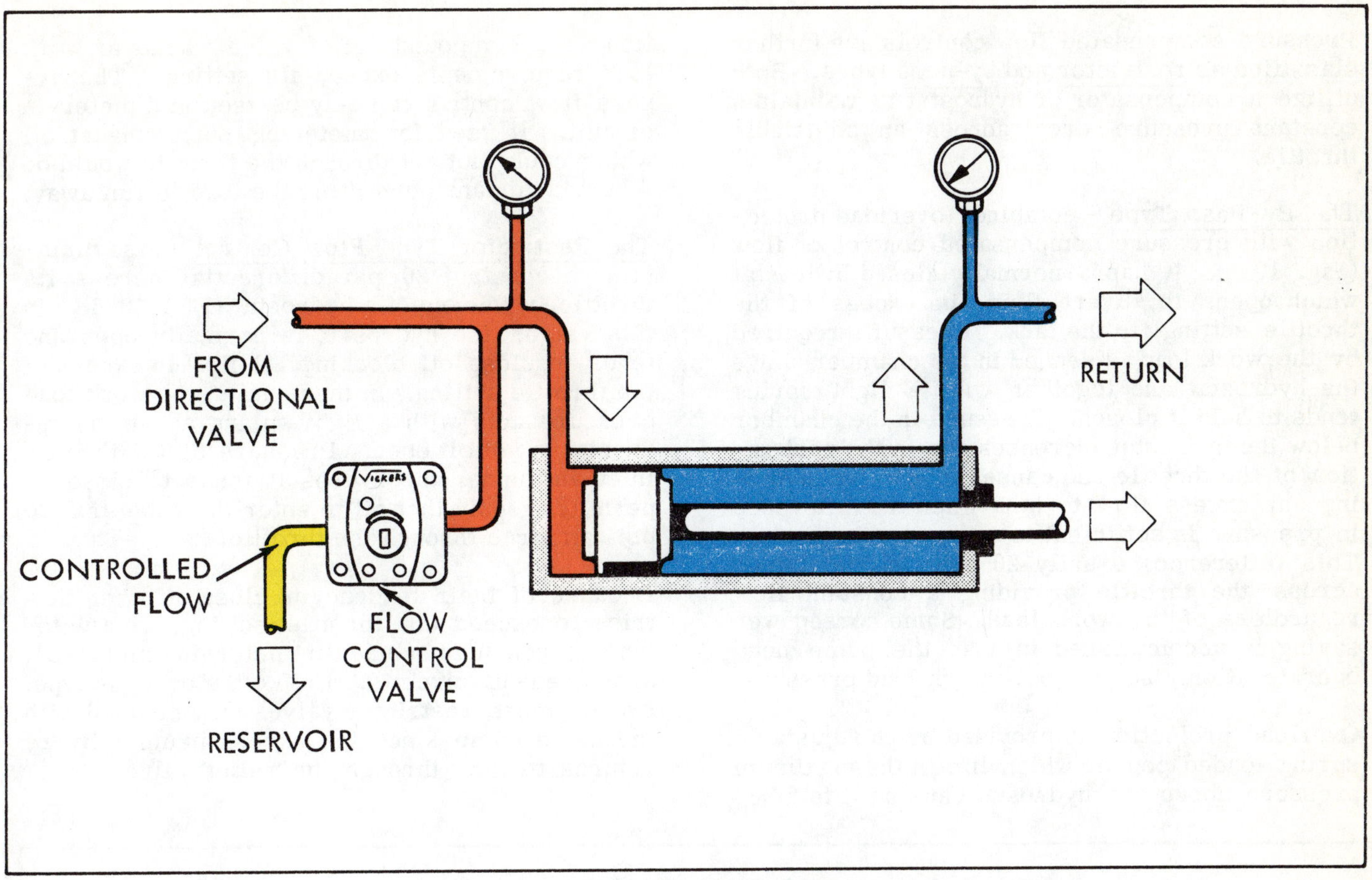

Figure 10-3. Bleed-Off Flow Control

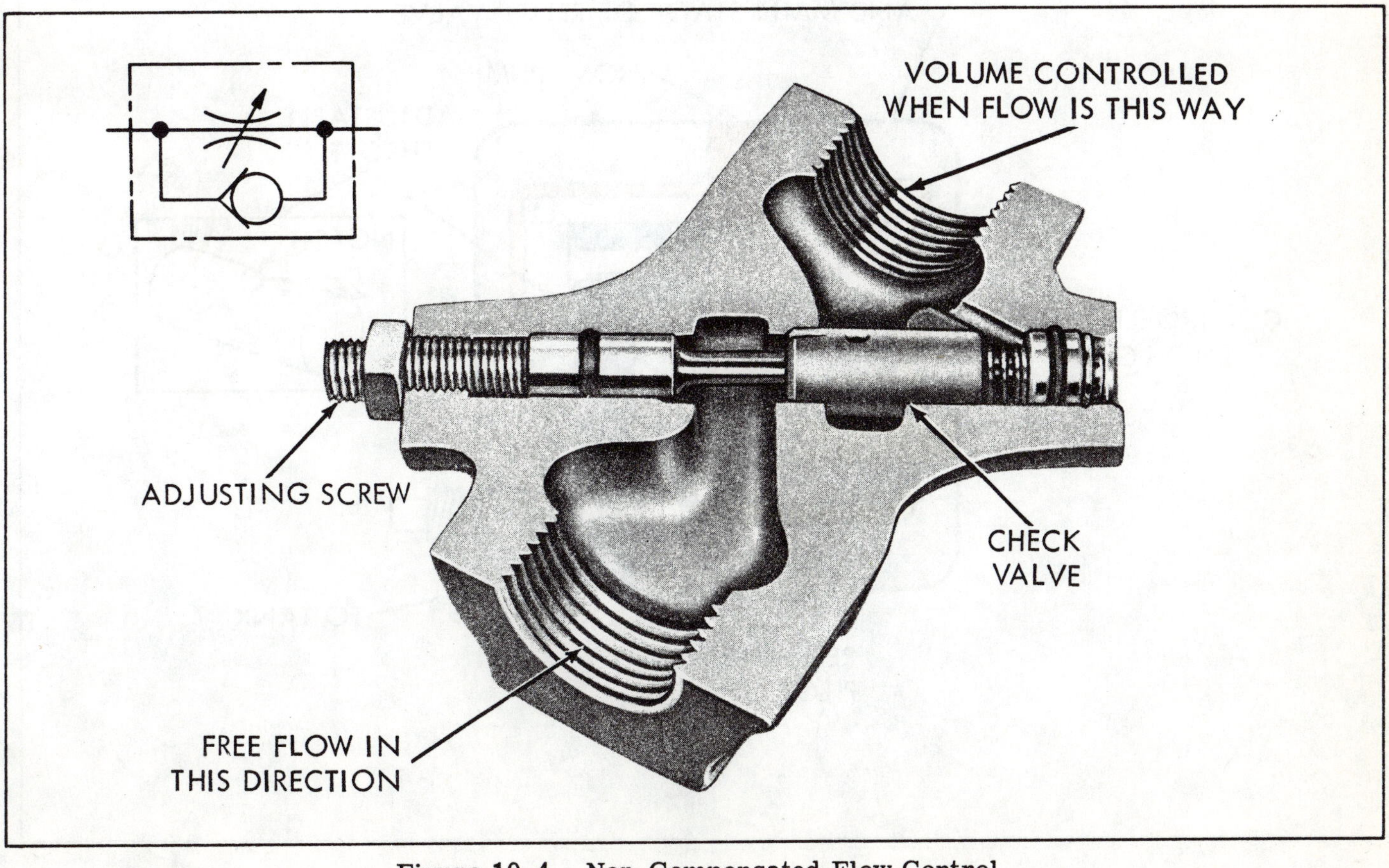

Figure 10-4. Non-Compensated Flow Control

Pressure compensated flow controls are further classified as restrictor and by-pass types. Both utilize a compensator or hydrostat to maintain a constant pressure drop across an adjustable throttle.

The By-Pass Type - combines overload protection with pressure compensated control of flow (Fig. 10-5). It has a normally closed hydrostat which opens to divert fluid, in excess of the throttle setting, to the tank. Pressure required by the work load is sensed in the chamber above the hydrostat and together with a light spring tends to hold it closed. Pressure in the chamber below the hydrostat increases due to the restriction of the throttle and causes it to raise diverting any excess flow to tank when the difference in pressure is sufficient to overcome the spring. This difference, usually 20 psi, is maintained across the throttle providing a constant flow regardless of the work load. Some horsepower saving is accomplished in that the pump need operate at only 20 psi above work load pressure.

Overload protection is provided by an adjustable spring loaded poppet which limits the maximum pressure above the hydrostat,causing it to function as a compound relief valve whenever work load requirements exceed its setting. The by-pass flow control can only be used in a meter-in circuit. If used for metering out, exhaust oil which could not get through the throttle would be diverted to tank permitting the load to run away.

The Restrictor Type Flow Control - also maintains a constant 20 psi differential across its throttle by means of a hydrostat (Fig. 10-6). In this valve, the hydrostat is normally open and tends to close off blocking all flow in excess of the throttle setting. In these units, the work load pressure acts with a light spring above the hydrostat to hold it open. Pressure at the throttle inlet and under the hydrostat tends to close it, permitting only that oil to enter the valve that 20 psi can force through the throttle.

Because of their tendency to close off when flow tries to exceed the throttle setting, restrictor type valves may be used in meter-in, meter-out and bleed-off circuits. Unlike the by-pass type, two or more restrictor valves may be used with the same pump since the excess pump delivery returns to tank through the relief valve.

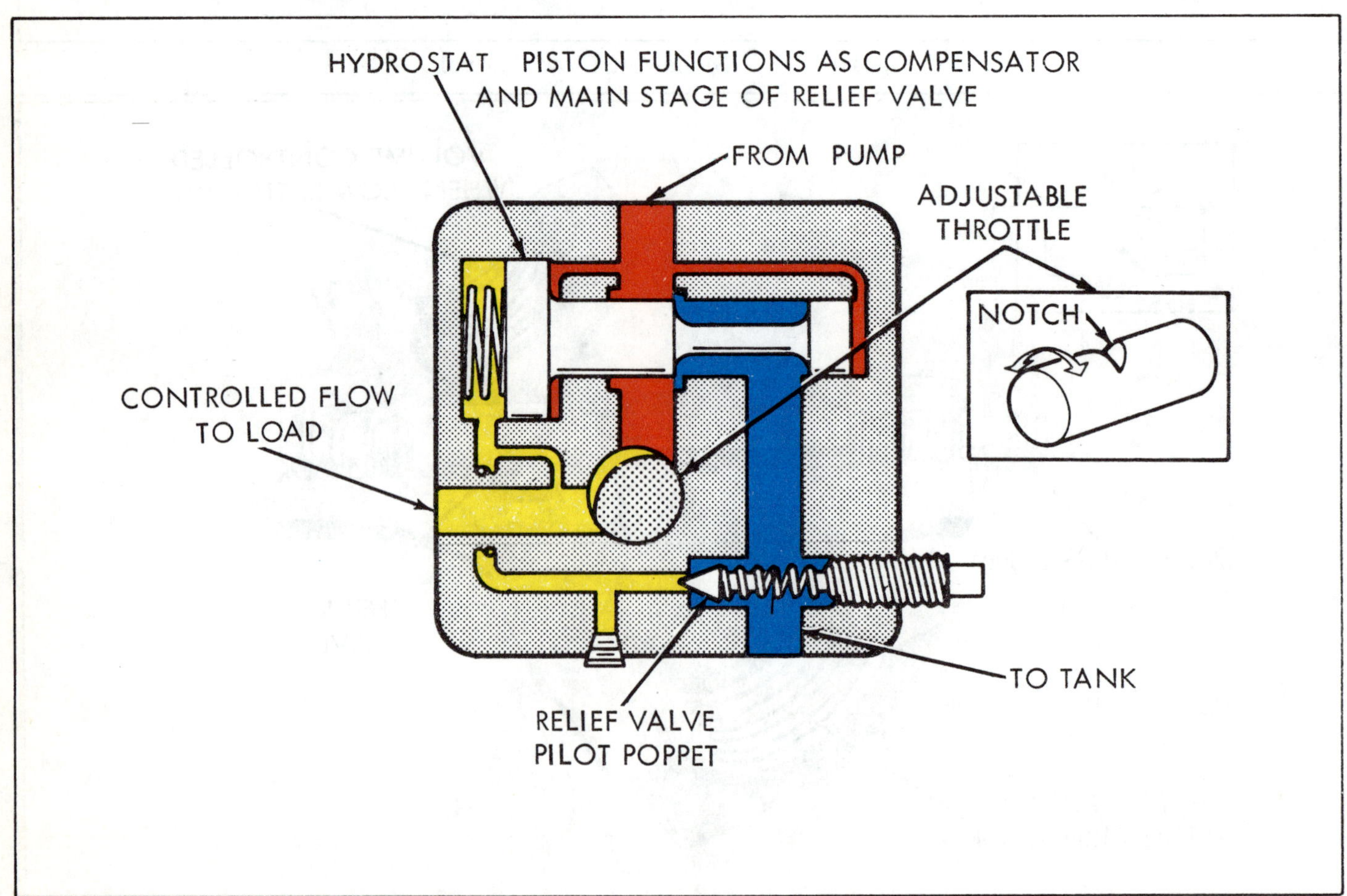

Figure 10-5. Flow Control and Relief Valve Meters In

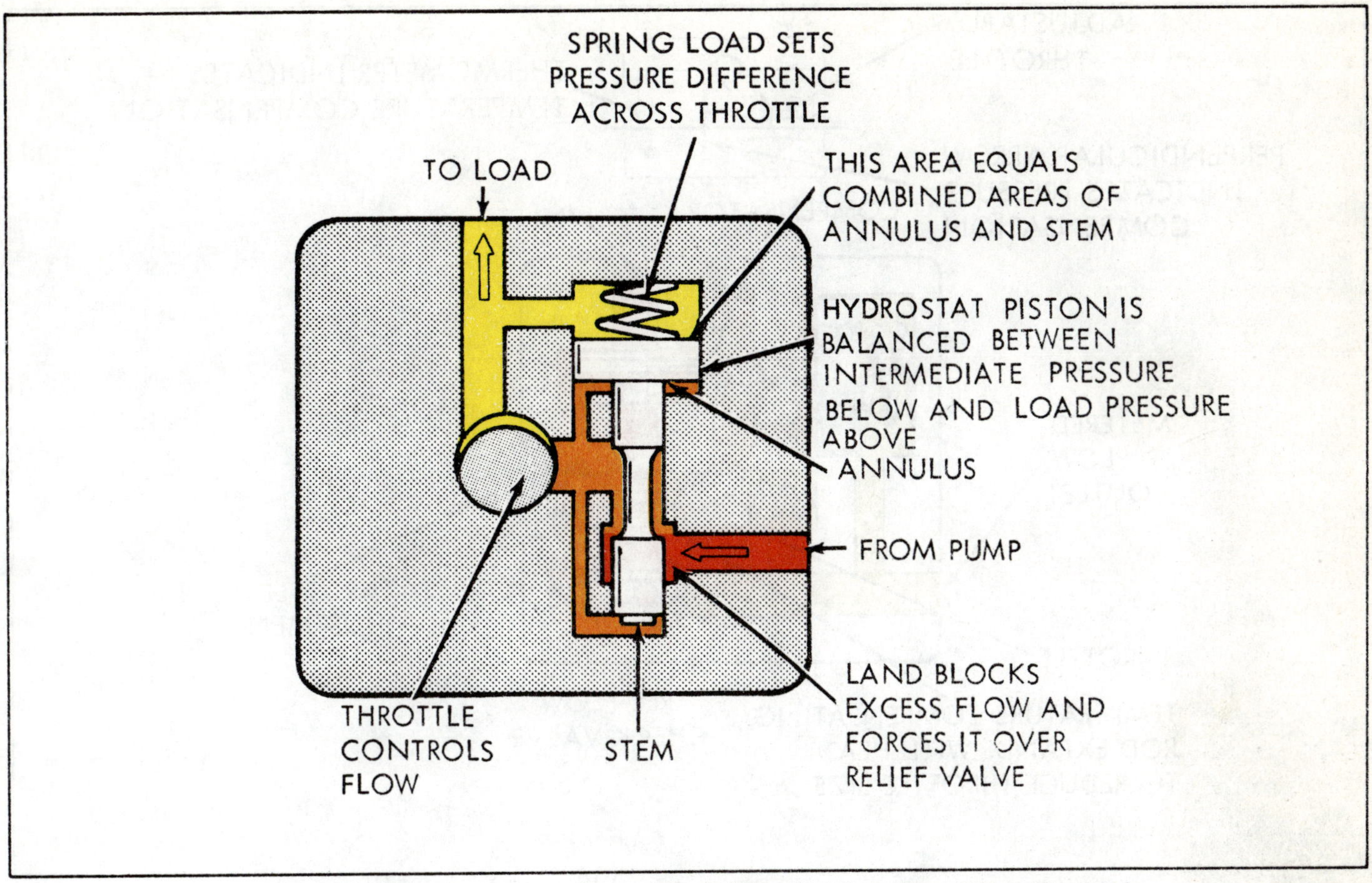

Figure 10-6. Pressure Compensated Restrictor Type Flow Control

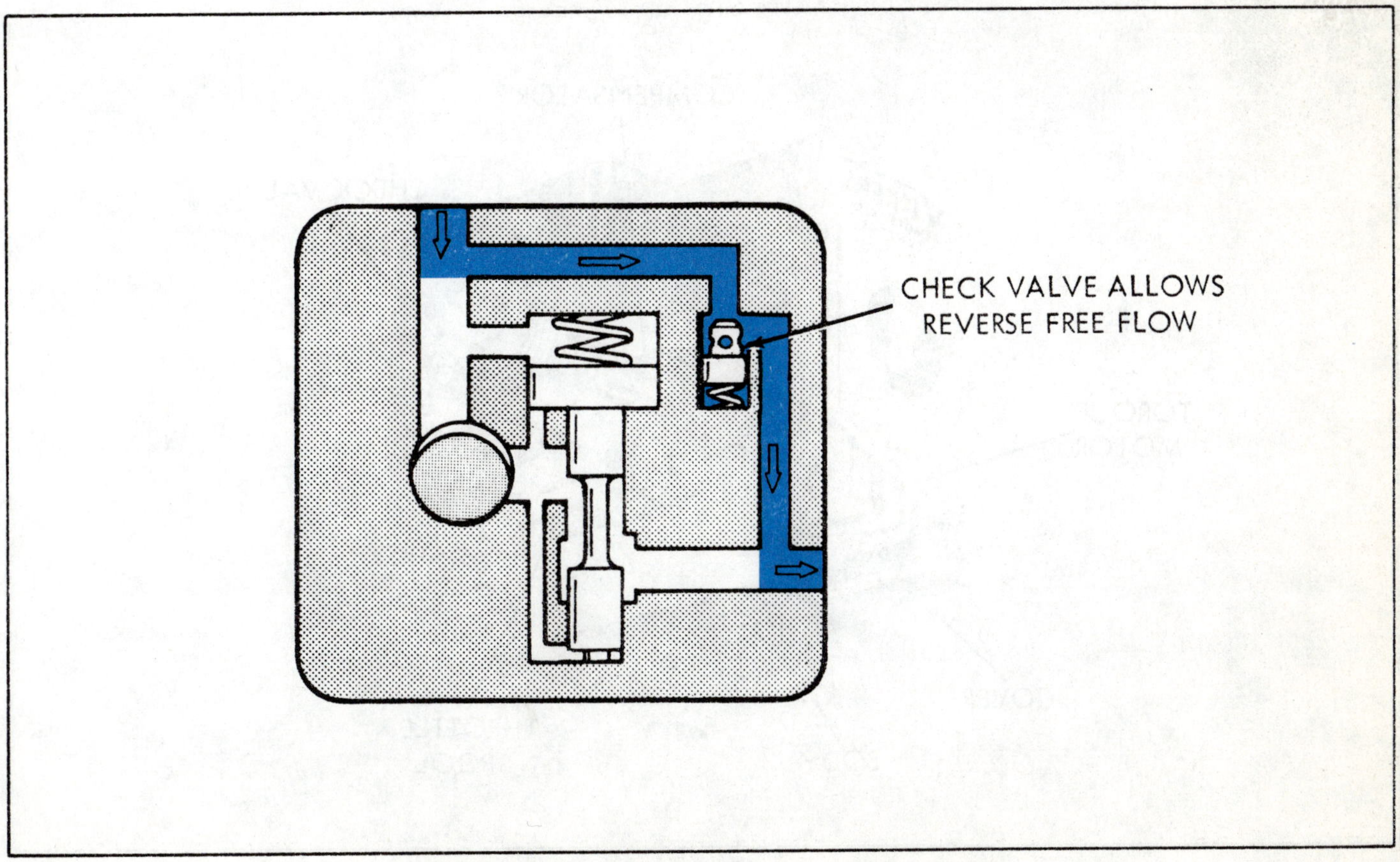

Figure 10-7. Flow Control and Check Valve

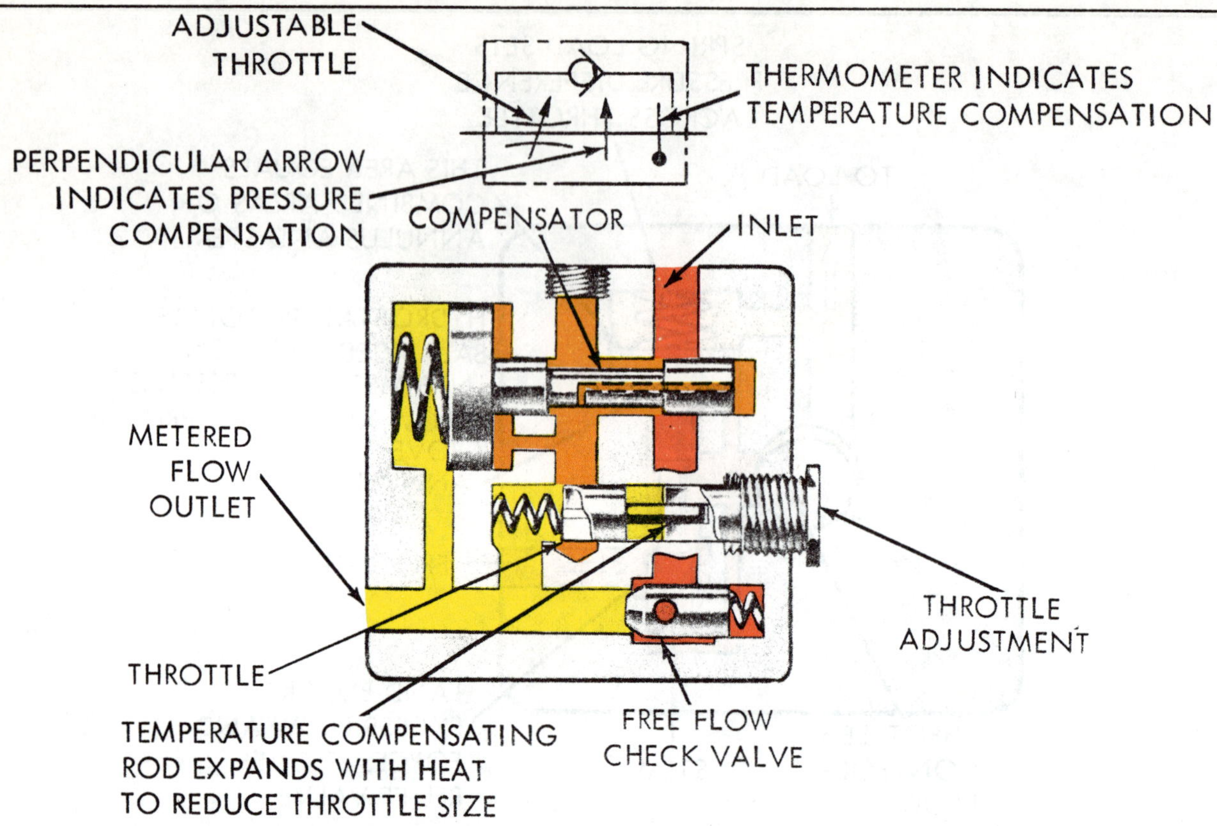

Figure 10-8. Operation of Pressure and Temperature Compensated Flow Control

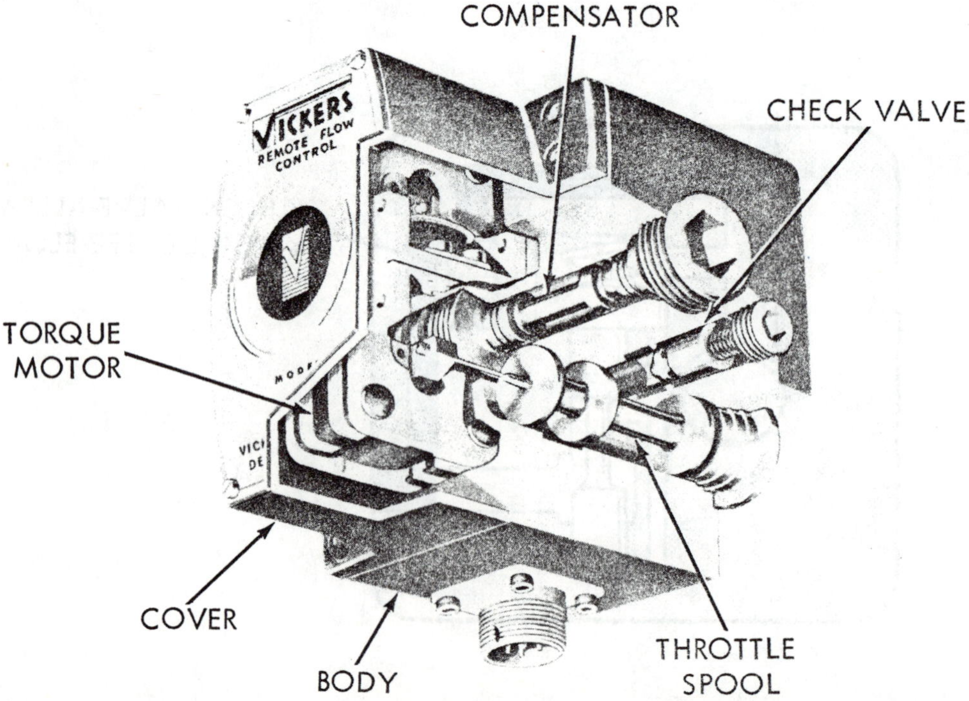

Figure 10-9. Remote Flow Control Valve

When placed in cylinder lines an integral check valve is optional to provide free flow for a rapid return stroke (Fig. 10-7). One would not be required for valves placed in the main supply line, the tank line of a directional valve or when they are used in bleed-off circuits.

Temperature Compensated Flow Control Valve

Flow through a pressure compensated flow control valve is subject to change with variations in oil temperature. Later design Vickers valves incorporate a temperature compensating feature. Although oil flows more freely when it is hot, constant flow can be maintained by decreasing the size of the throttle opening as the temperature rises.

This is accomplished through a compensating rod which lengthens with heat and contracts when cold (Fig. 10-8). The throttle is a simple plunger that is moved in and out of the control port. The compensating rod is installed between the throttle and its adjuster.

This design also is available with a reverse free-flow check valve.

Remote Flow Control Valves

Remote flow control valves (Fig. 10-9) permit adjustment of the throttle size by an electrical signal. The throttle spool is linked to the armature of a torque motor and moves in response to signals to the torque motor. Operation is otherwise the same as a pressure compensated flow control valve.

QUESTIONS

1. Name two ways of regulating flow to an actuator.
2. What are the three methods of applying flow control valves?
3. Under what conditions would you use each?
4. How can the same valve control flow in both directions of actuator movement?
5. What is the difference between a by-pass and restrictor type flow control?
6. What is pressure compensation?
7. How is temperature compensation indicated in a valve symbol?
8. When might temperature compensation be needed?
9. What is the advantage of the flow control and relief valve over a conventional flow control?
10. How is the throttle positioned in a remote flow control valve?

CHAPTER 11

HYDRAULIC PUMPS

The pump is probably the most important and least understood component in the hydraulic system. Its function is to convert mechanical energy to hydraulic energy by pushing the hydraulic fluid into the system. Pumps are made in many sizes and shapes--mechanical and manual--with many different pumping mechanisms and for many different purposes. All pumps, however, fall into one of two basic categories, hydrodynamic or hydrostatic.

HYDRODYNAMIC

Hydrodynamic, or non-positive displacement pumps such as centrifugal or turbine designs are used primarily in the transfer of fluids where the only resistance encountered is that created by the weight of the fluid itself and friction.

Most non-positive displacement pumps (Fig. 11-1) operate by centrifugal force whereby fluids entering the center of the pump housing are thrown to the outside by means of a rapidly driven impeller. There is no positive seal between the inlet and outlet ports and pressure capabilities are a function of drive speed.

While they provide smooth continuous flow their output is reduced as resistance is increased. It is, in fact, possible to completely block off the outlet while the pump is running. For this and other reasons non-positive displacement pumps are seldom used in hydraulic systems as we know them today.

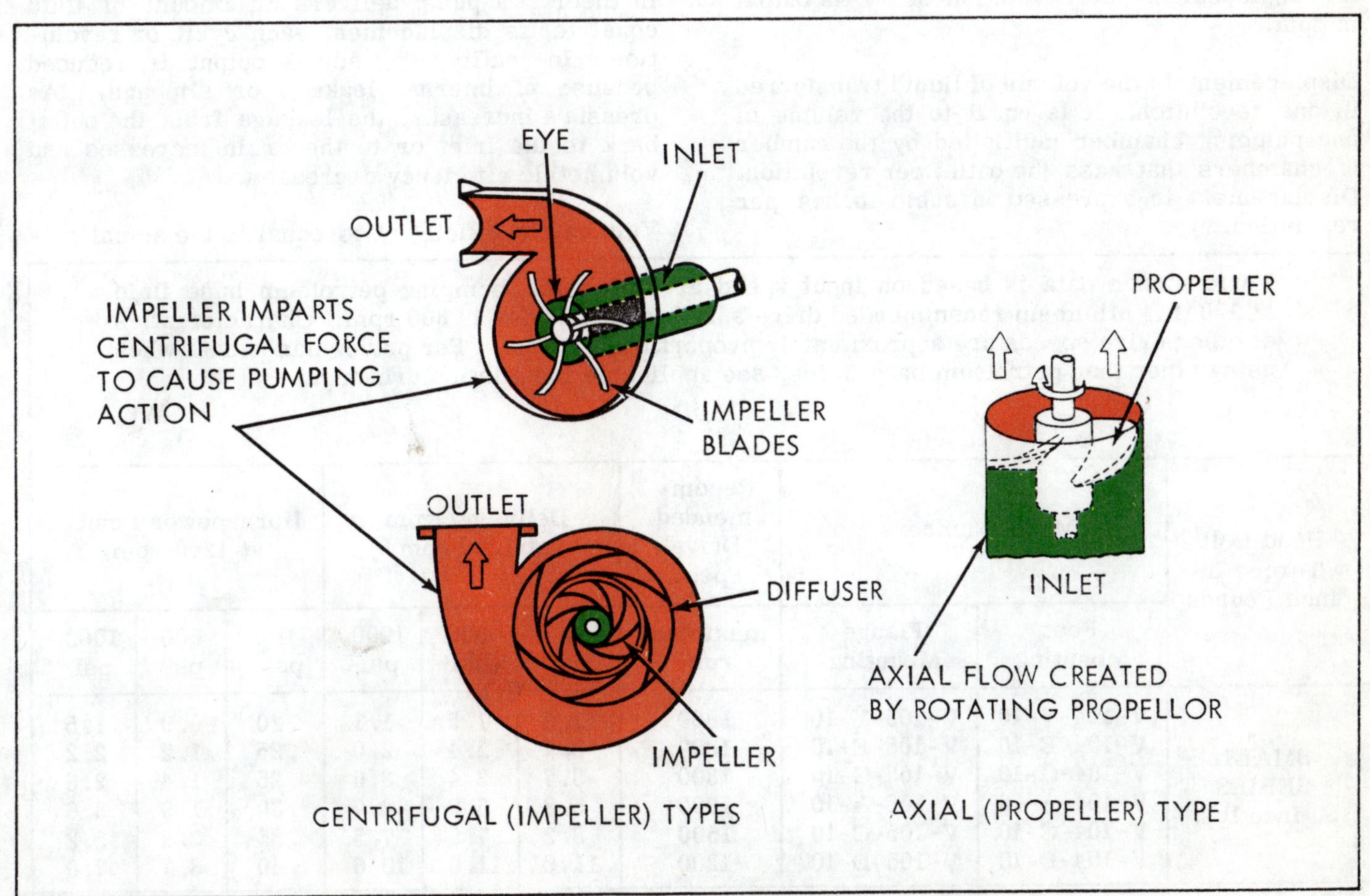

Fig. 11-1. Non-Positive Displacement Pumps

HYDROSTATIC

Hydrostatic or positive displacement pumps as their name implies provide a given amount of fluid for every stroke, revolution, or cycle. Their output except for leakage losses is independent of outlet pressure making them well suited for use in the transmission of power.

Pump Ratings

Pumps are generally rated by their maximum operating pressure capability and their output in gpm at a given drive speed.

Pressure Ratings

The pressure rating of a pump is determined by the manufacturer based upon reasonable service life expectancy under specified operating conditions. It is important to note that there is no standard industry wide safety factor in this rating. Operating at higher pressure may result in reduced pump life or more serious damage.

Displacement

The flow capacity of a pump can be expressed as its displacement per revolution or by its output in gpm.

Displacement is the volume of liquid transferred in one revolution. It is equal to the volume of one pumping chamber multiplied by the number of chambers that pass the outlet per revolution. Displacement is expressed in cubic inches per revolution.

Most pumps have a fixed displacement which cannot be changed except by replacing certain components. It is possible in some, however, to vary the size of the pumping chamber and thereby the displacement by means of external controls.

Certain unbalanced vane pumps and many piston units can be varied from maximum to zero delivery, with some being capable of reversing their flow as the control crosses a center or neutral position.

Delivery in gpm

A pump may be nominally rated as a 10 gpm unit. Actually it may pump more than that under no-load conditions and less than that at its rated operating pressure. Its delivery too will be proportional to drive shaft speed. Most manufacturers provide a table or graph (Fig. 11-2) showing pump deliveries and horsepower requirements under specific test conditions as to drive speeds and pressures.

Volumetric Efficiency

In theory, a pump delivers an amount of fluid equal to its displacement each cycle or revolution. In reality, the actual output is reduced because of internal leakage or slippage. As pressure increases, the leakage from the outlet back to the inlet or to the drain increases and volumetric efficiency decreases.

Volumetric efficiency is equal to the actual out-

Performance data is based on input speed at 1200 rpm, pumping petroleum base fluid at 120°F. Minimum recommended drive speed for all series is 600 rpm. Characteristics at other drive speeds are approximately proportional to rpm. For performance data when using other than petroleum base fluids, see applicable installation drawing.

Head Bolt Torque in Inch Pounds	Model Numbers		Recommended Drive Speed	Delivery, gpm at 1200 rpm			Horsepower Input at 1200 rpm		
	Foot Mounting	Flange Mounting	maximum rpm	0 psi	500 psi	1000 psi	0 psi	500 psi	1000 psi
SMALL SERIES 50 Inch lbs.	V-104-Y-10	V-105-Y-10	1800	1.8	1.5	1.1	.20	0.9	1.5
	V-104-E-10	V-105-E-10	1800	2.7	2.4	2.0	.25	1.2	2.2
	V-104-G-10	V-105-G-10	1800	3.7	3.4	3.0	.25	1.4	2.6
	V-104-A-10	V-105-A-10	1800	5.3	5.0	4.7	.30	1.9	3.6
	V-104-C-10	V-105-C-10	1500	8.2	7.9	7.5	.35	2.8	5.2
	V-104-D-10	V-105-D-10	1200	11.5	11.0	10.6	.40	3.7	7.0

Figure 11-2. Typical Specification Table

put divided by the theoretical output. It is expressed as a percentage.

$$\text{Efficiency} = \frac{\text{Actual output}}{\text{Theoretical output}}$$

For example, if a pump theoretically should deliver 10 gpm but delivers only 9 gpm at 1000 psi its volumetric efficiency at that pressure is 90%.

$$\text{Efficiency} = 9/10 = .9 \text{ or } 90\%$$

GEAR PUMPS

A gear pump (Fig. 11-3) develops flow by carrying fluid between the teeth of two meshed gears. One gear is driven by the drive shaft and turns the other. The pumping chambers formed between the gear teeth are enclosed by the pump housing and side plates (often called wear or pressure plates).

A partial vacuum is created at the inlet as the gear teeth unmesh. Fluid flows in to fill the space and is carried around the outside of the gears. As the teeth mesh again at the outlet the fluid is forced out.

High pressure at the pump outlet imposes an unbalanced load on the gears and the bearings supporting them.

Figure 11-4 illustrates a typical internal gear pump; in this design, the pumping chambers also are formed between gear teeth. A crescent seal is machined into the valve body between the inlet and outlet where clearance between the teeth is maximum.

Also in the general family of gear pumps is the lobe or rotor pump (Fig. 11-5). This pump operates on the same principle as the external gear pump, but has a higher displacement.

The gerotor pump (Fig. 11-6) operates much like the internal gear pump. The inner rotor is driven and carries the outer rotor around in mesh. Pumping chambers are formed between the rotor lobes. The crescent seal is not used. Rather, the tips of the inner rotor contact the outer rotor to seal the chambers from each other.

Gear Pump Characteristics

Most gear type pumps are fixed displacement.

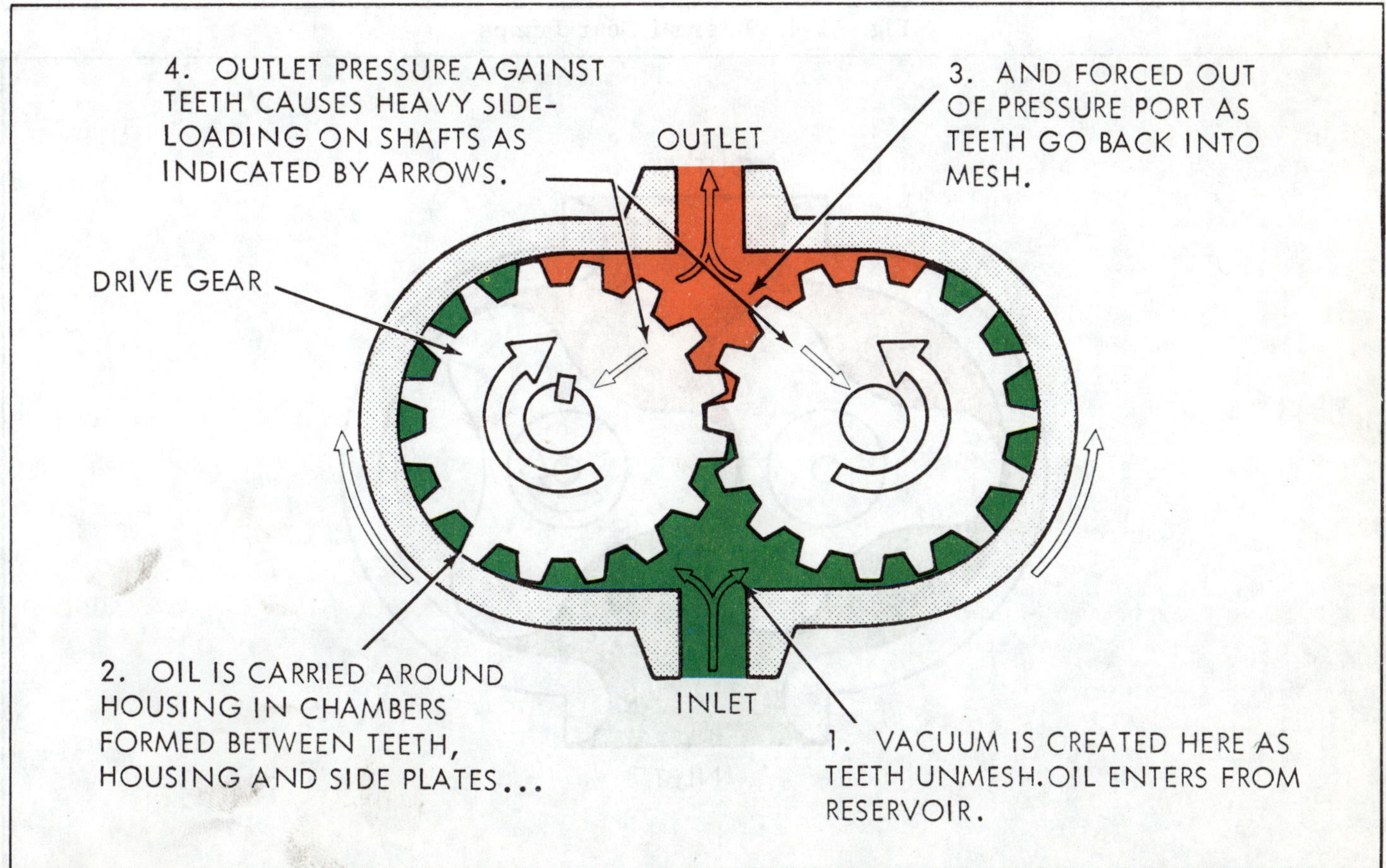

Fig. 11-3. External Gear Pumps

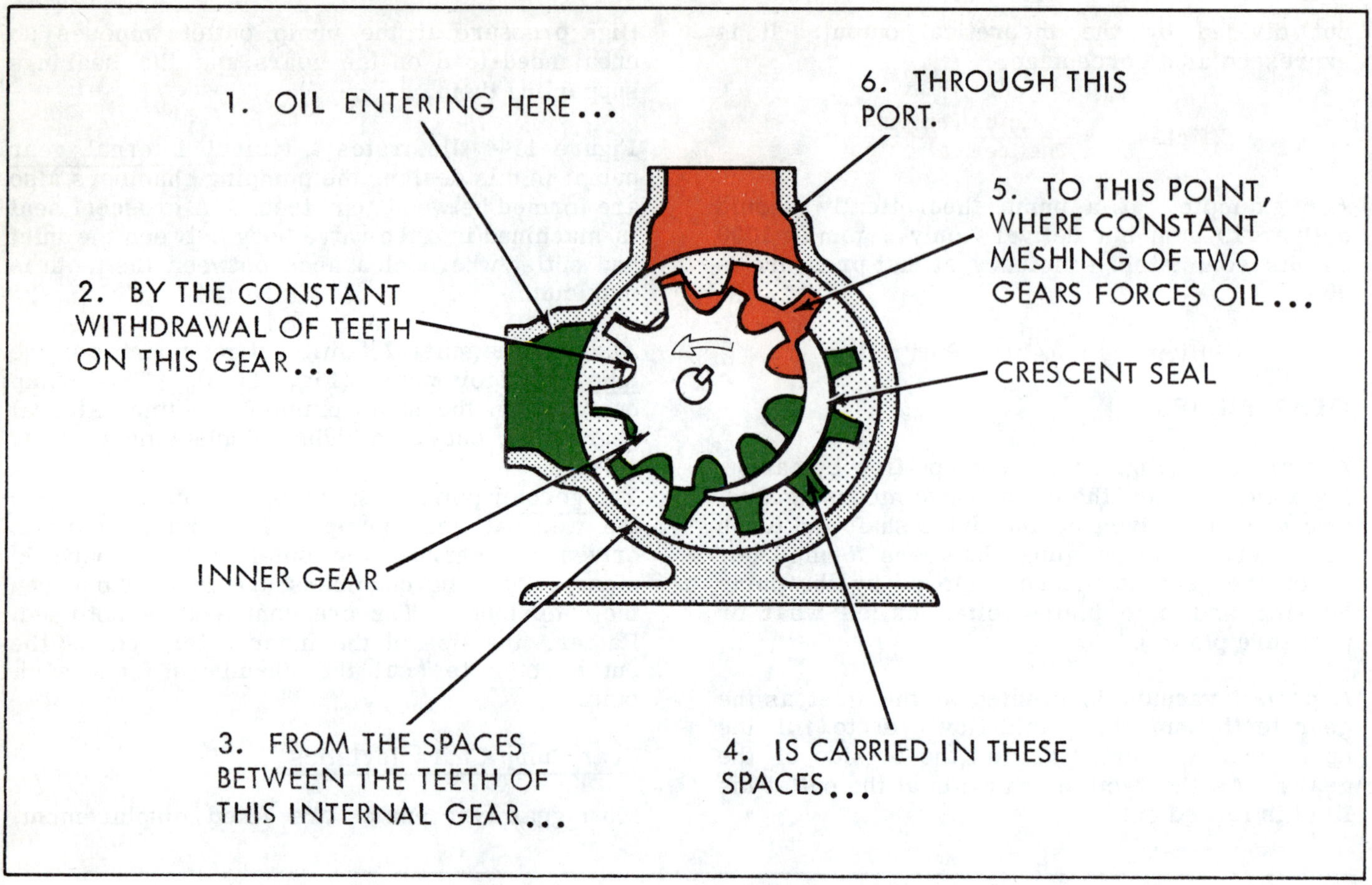

Fig. 11-4. Internal Gear Pumps

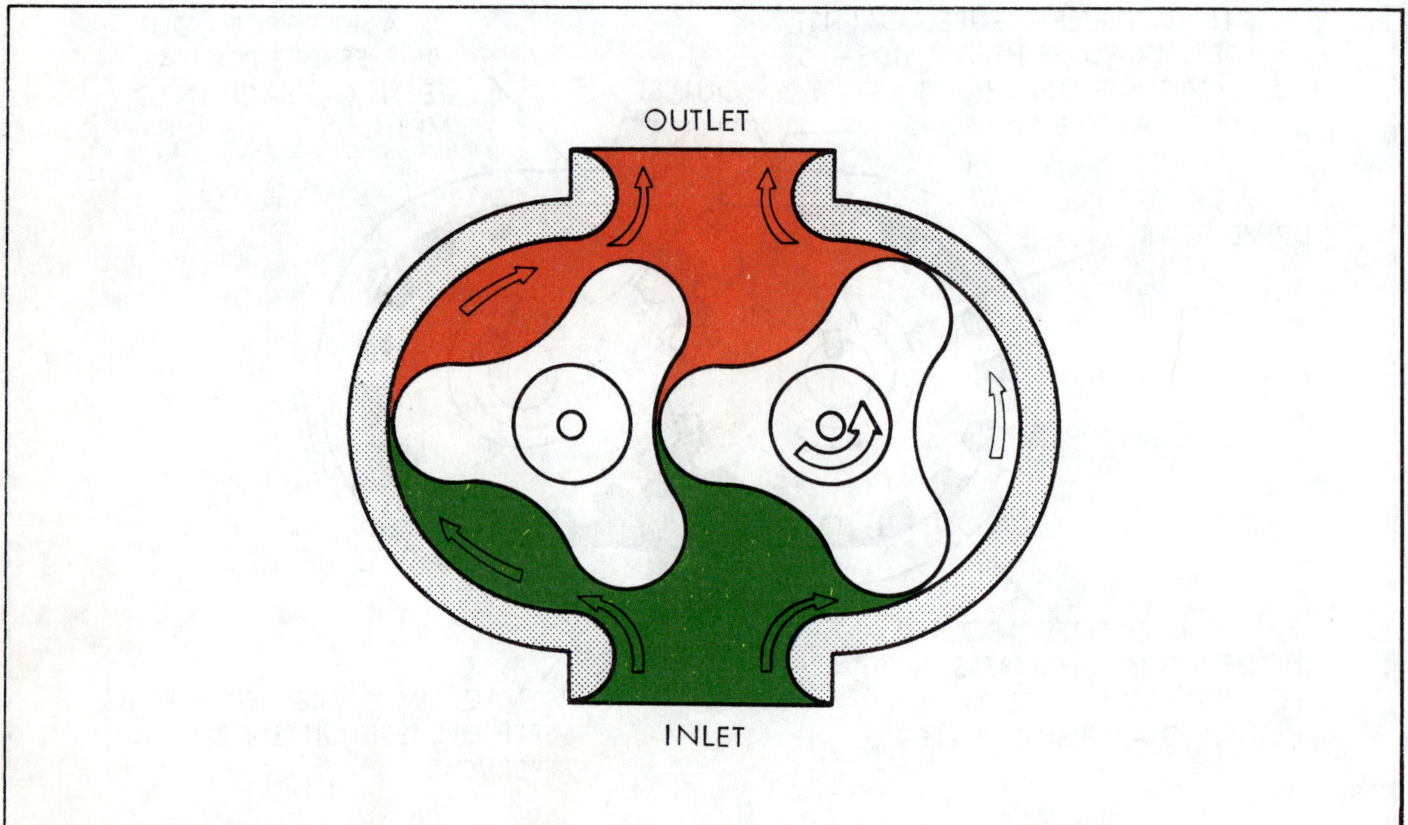

Fig. 11-5. Lobe Pump Operates on External Gear Pump Principle

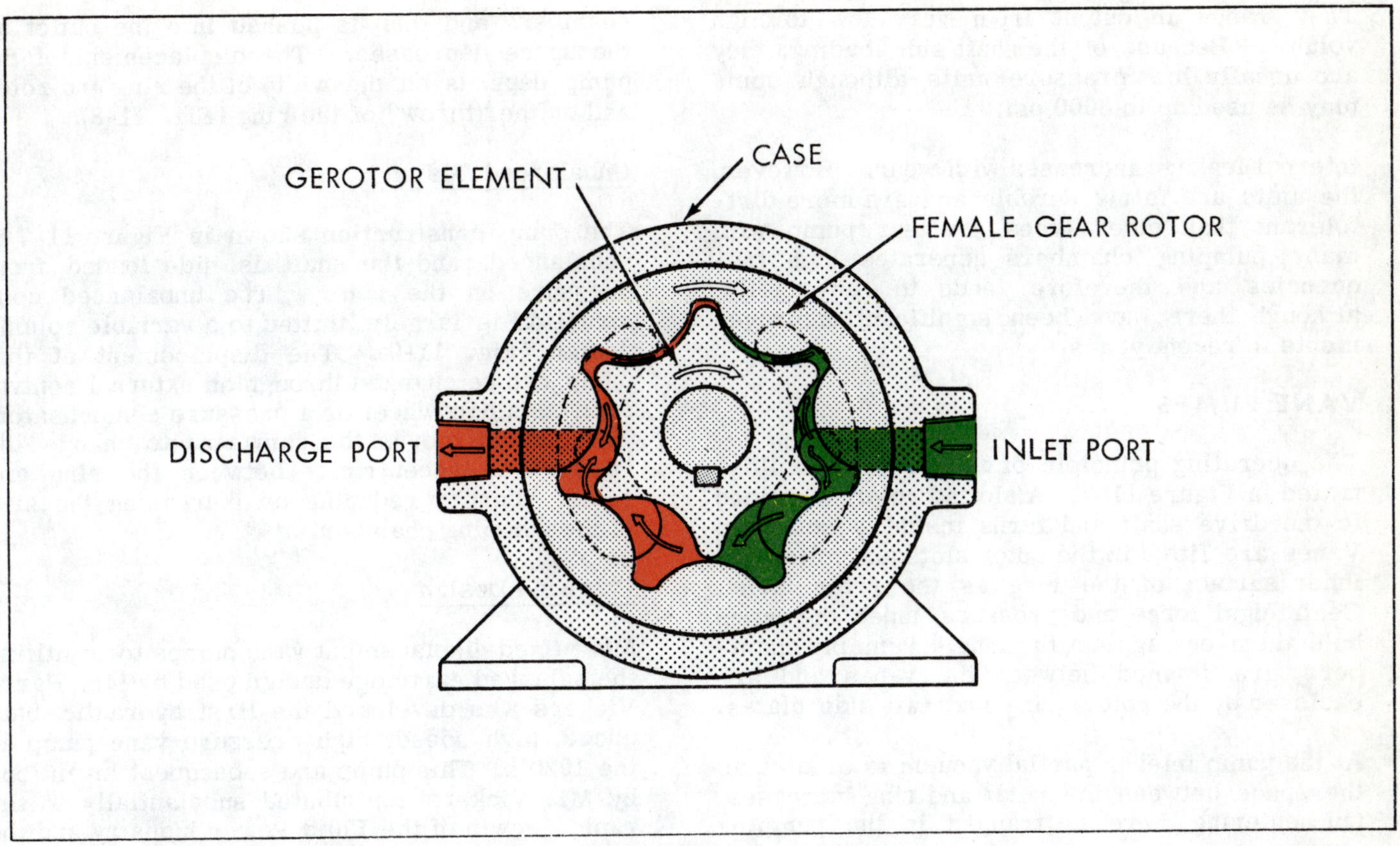

CASE
GERORTOR ELEMENT
FEMALE GEAR ROTOR
DISCHARGE PORT
INLET PORT

Fig. 11-6. Gerotor-Type Pump

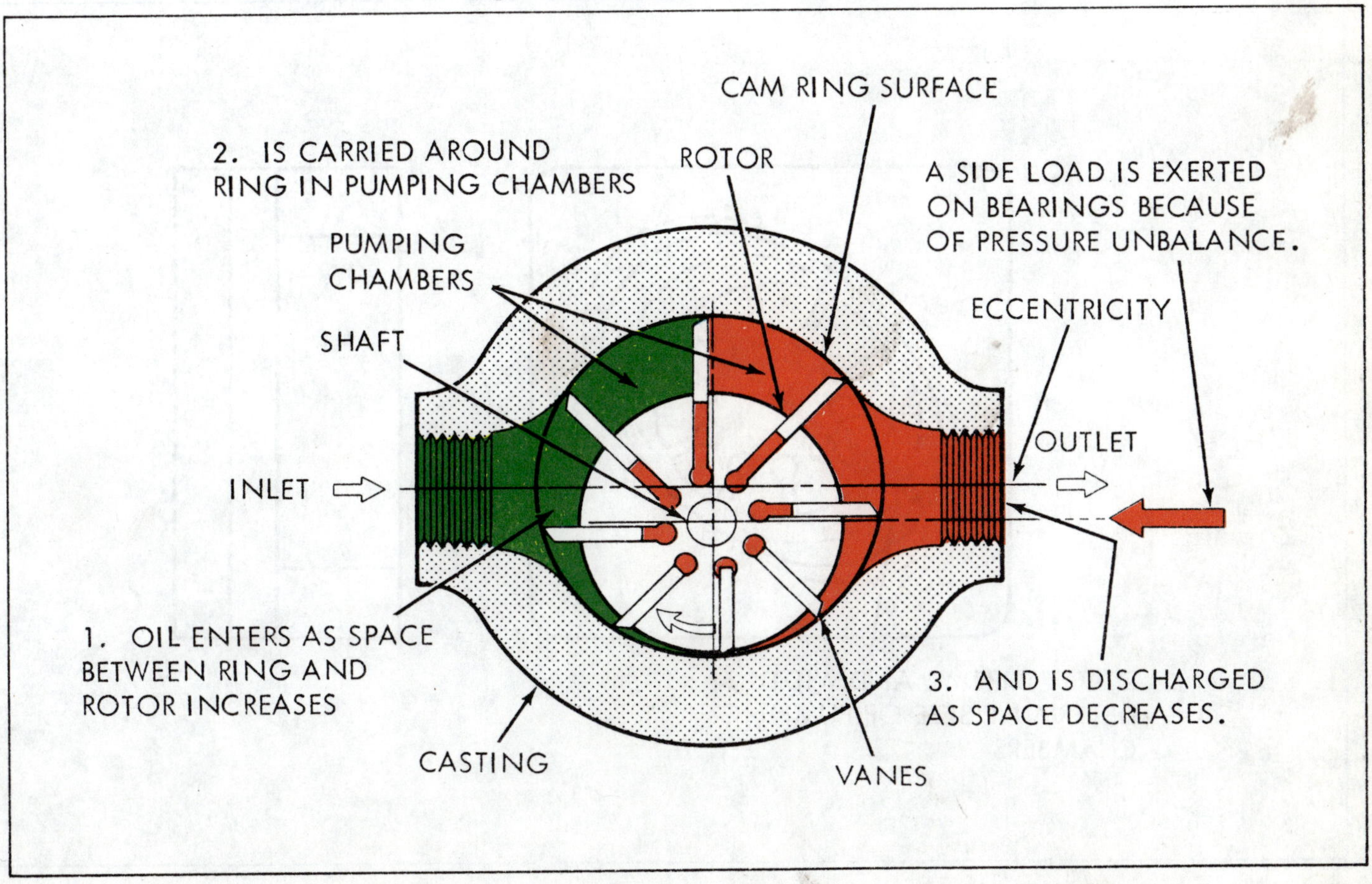

CAM RING SURFACE
2. IS CARRIED AROUND RING IN PUMPING CHAMBERS
ROTOR
A SIDE LOAD IS EXERTED ON BEARINGS BECAUSE OF PRESSURE UNBALANCE.
PUMPING CHAMBERS
ECCENTRICITY
SHAFT
OUTLET
INLET
1. OIL ENTERS AS SPACE BETWEEN RING AND ROTOR INCREASES
3. AND IS DISCHARGED AS SPACE DECREASES.
CASTING
VANES

Fig. 11-7. Unbalanced Vane Pump Operation

They range in output from very low to high volume. Because of the shaft side loading, they are usually low pressure units although some may be used up to 3000 psi.

Internal leakage increases with wear. However, the units are fairly durable and are more dirt-tolerant than other types. A gear pump with many pumping chambers generates high frequencies and therefore tends to run noisily, although there have been significant improvements in recent years.

VANE PUMPS

The operating principle of a vane pump is illustrated in Figure 11-7. A slotted rotor is splined to the drive shaft and turns inside a cam ring. Vanes are fitted to the rotor slots and follow the inner surface of the ring as the rotor turns. Centrifugal force and pressure under the vanes hold them out against the ring. Pumping chambers are formed between the vanes and are enclosed by the rotor, ring and two side plates.

At the pump inlet, a partial vacuum is created as the space between the rotor and ring increases. Oil entering here is trapped in the pumping chambers and then is pushed into the outlet as the space decreases. The displacement of the pump depends on the width of the ring and rotor and on the "throw" of the ring (Fig. 11-8).

Unbalanced Design

The pump construction shown in Figure 11-7 is unbalanced, and the shaft is side-loaded from pressure on the rotor. The unbalanced construction is largely limited to a variable volume design (Fig. 11-9). The displacement of this pump can be changed through an external control such as a handwheel or a pressure compensator. The control moves the cam ring to change the "throw" or eccentricity between the ring and rotor, thereby reducing or increasing the size of the pumping chamber.

Balanced Design

Most fixed displacement vane pumps today utilize the balanced cartridge design used by Mr. Harry Vickers who developed the first hydraulic balanced, high speed, high pressure vane pump in the 1920's. This pump and subsequent inventions by Mr. Vickers contributed substantially to the rapid growth of the Fluid Power industry and the

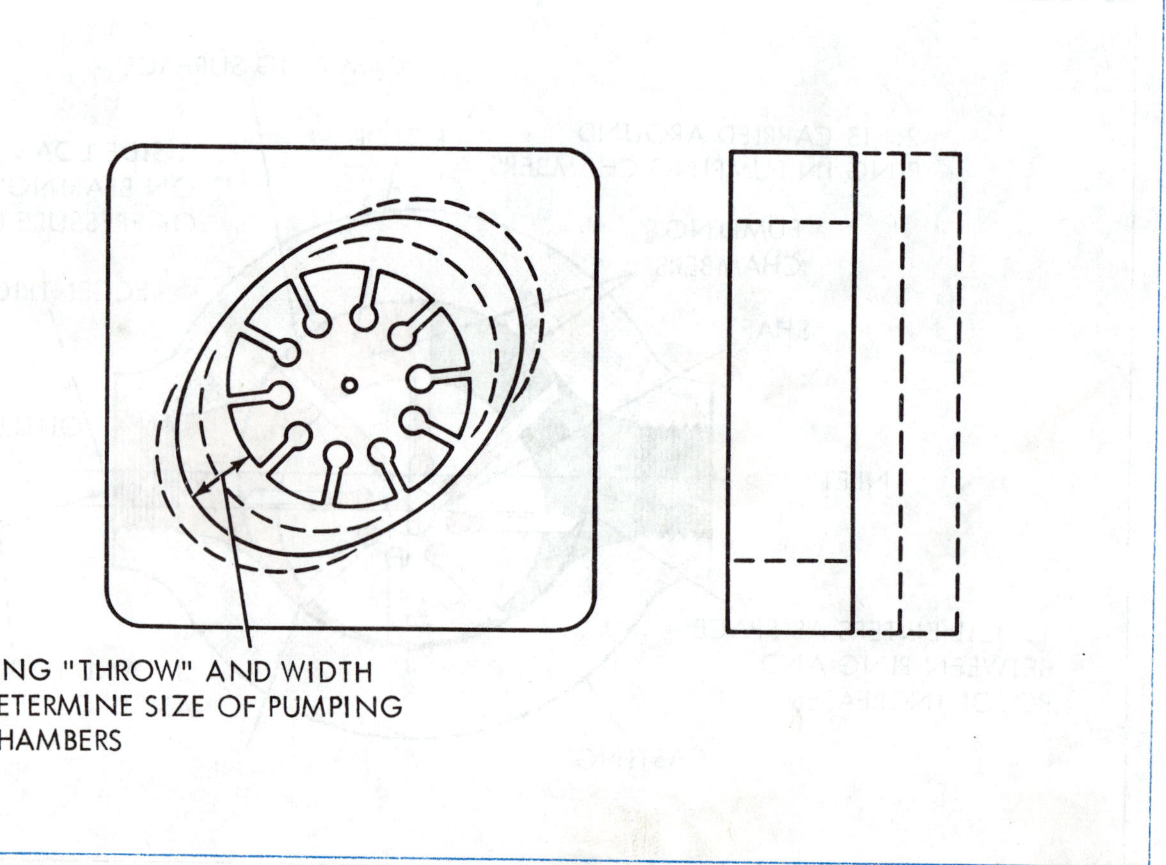

Fig. 11-8. Variations in Vane Pump Displacement

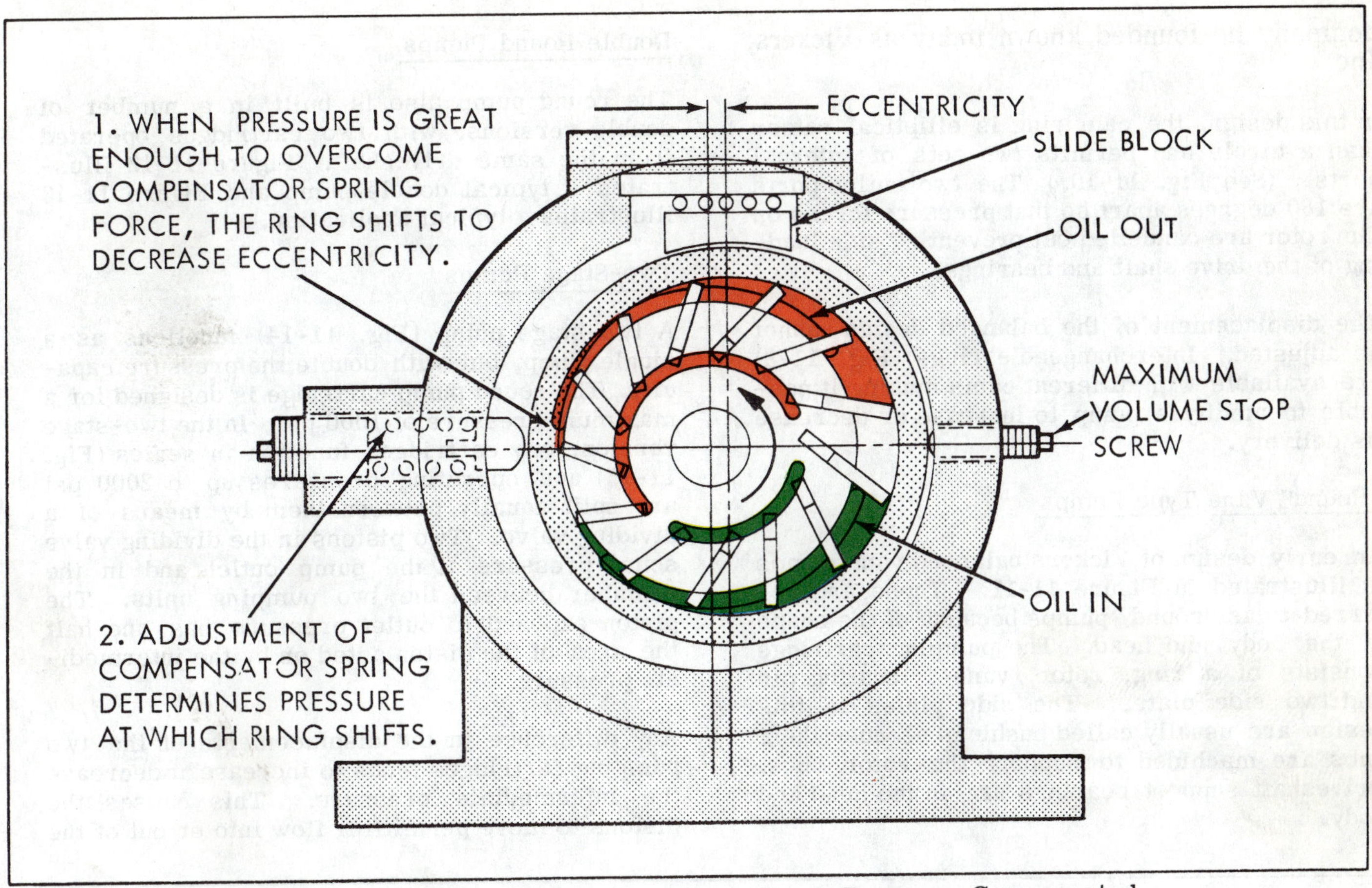

Fig. 11-9. Variable Displacement Vane Pump Pressure Compensated

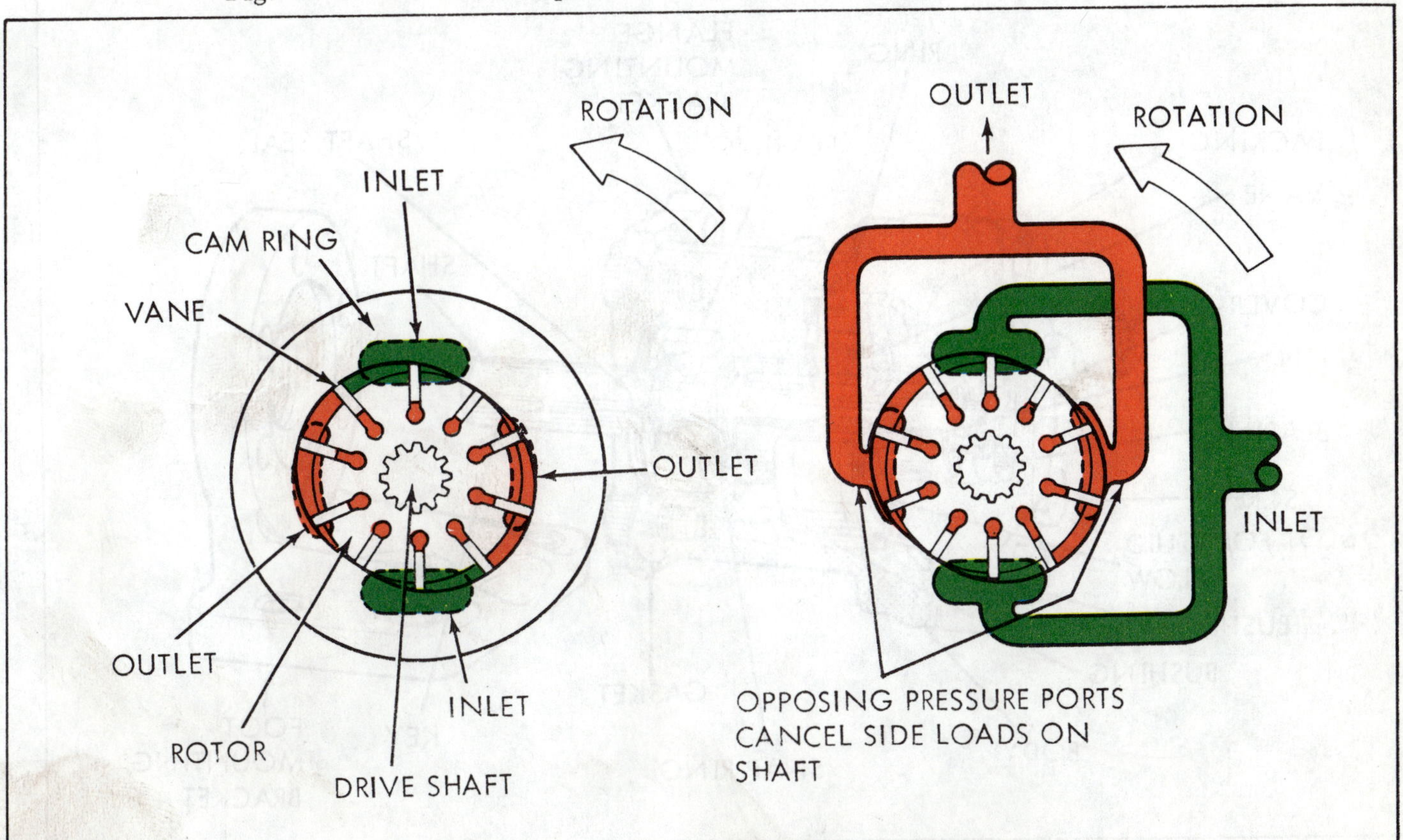

Fig. 11-10. Balanced Vane Pump Principle

company he founded known today as Vickers, Inc.

In this design, the cam ring is elliptical rather than a circle and permits two sets of internal ports. (See Fig. 11-10.) The two outlet ports are 180 degrees apart so that pressure forces on the rotor are cancelled out preventing side loading of the drive shaft and bearings.

The displacement of the balanced design cannot be adjusted. Interchangeable rings (Fig. 11-8) are available with different cams making it possible to modify a pump to increase or decrease its delivery.

"Round" Vane Type Pumps

An early design of Vickers balanced vane pumps is illustrated in Figure 11-11. These are referred to as "round" pumps because of the shape of the body and head. The pumping cartridge consists of a ring, rotor, vanes, locating pin and two side plates. The side plates in this design are usually called bushings because their hubs are machined to support the rotor hubs. Driveshaft support bearings are in the head and body.

Double Round Pumps

The round pump also is built in a number of double versions, with two cartridges operated from the same driveshaft. Figure 11-12 illustrates a typical double pump and Figure 11-13 illustrates a typical application.

Two-Stage Pumps

A two-stage pump (Fig. 11-14) functions as a single pump, but with double the pressure capacity. The round pump cartridge is designed for a maximum pressure of 1000 psi. In the two-stage version, two cartridges function in series (Fig. 11-15) and operating pressures up to 2000 psi are split equally between them by means of a dividing valve. Two pistons in the dividing valve sense pressure at the pump outlet and in the chamber between the two pumping units. The piston exposed to outlet pressure has one-half the area of the piston acted on by the intermediate pressure.

Any difference in the displacements of the two pumping cartridges tends to increase or decrease the intermediate pressure. This causes the pistons to move permitting flow into or out of the

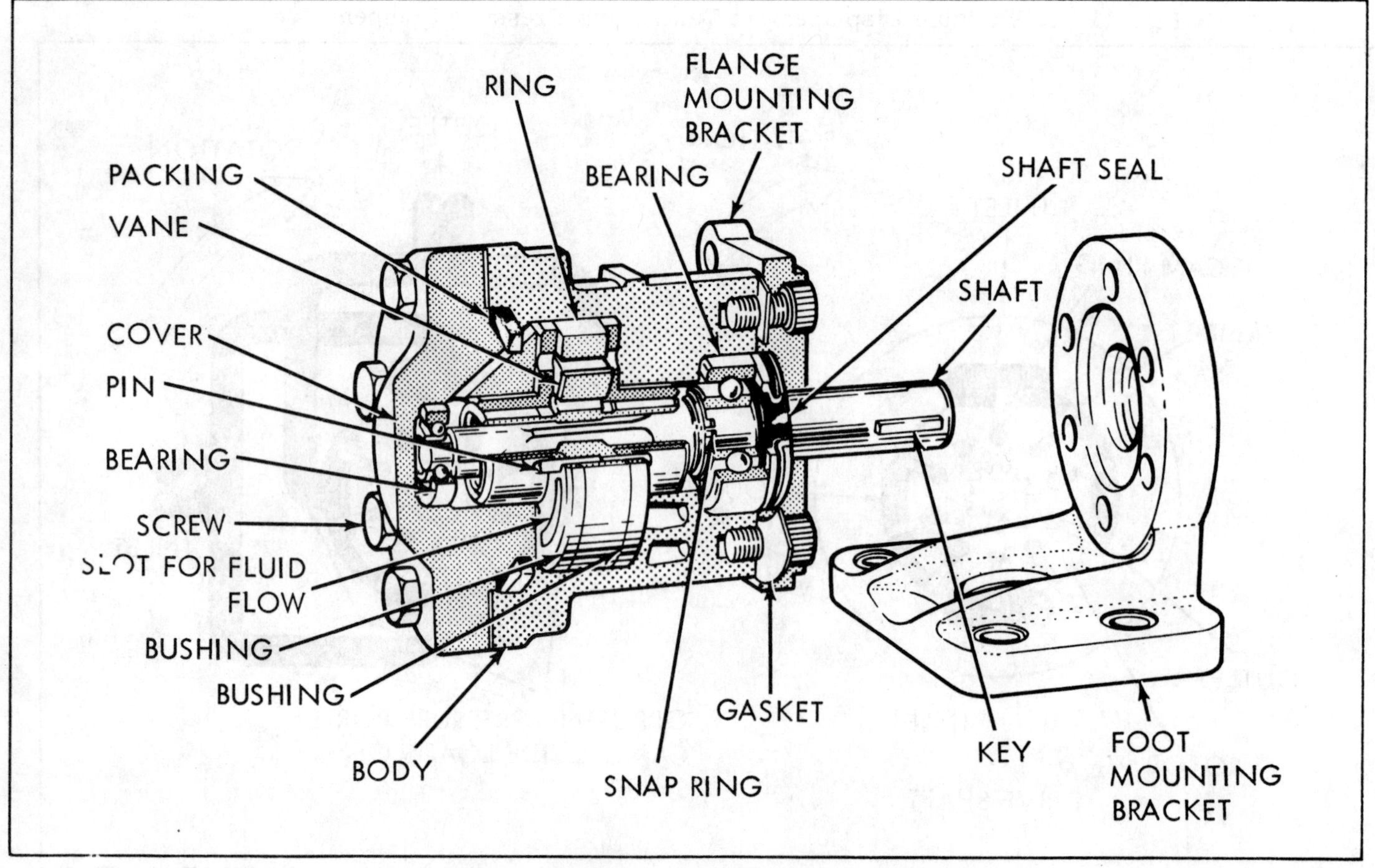

Fig. 11-11. Construction of "Round" Type Pump

chamber as required to assure an equal pressure drop across each stage.

COMBINATION PUMPS

Combination pumps (Fig. 11-16) are double pumps which contain integral valving for relief and unloading functions. A typical unit (Fig. 11-17) consists of two pumping cartridges in a housing with a single inlet port and separate outlets. The relief valve built into the housing is the compound, balanced piston type and the unloading valve is the "R" type. A check valve is installed in the line between the two pump outlets in some combinations. In others, the check valve is replaced by a plug. With the internal valving, the pump is, in effect, a circuit in itself except for the tank, directional valve, and actuator.

Following is a brief description of the most frequently used combinations.

Combination 3 -- Single Outlet Unloading System

To conserve horsepower, the type 3 combination pump may be used instead of a single pump in applications which require a large volume of oil for rapid approach while a cylinder is at low pressure, and reduced volume at high pressure for clamping, pressing or feeding. The outlet port for the high volume unit is plugged and the deliveries of the two cartridges are interconnected through the check valve (Fig. 11-18). A relief valve is connected to the pressure line of the low volume pump and an unloading valve to the high volume outlet.

Low Pressure Advance -- When the load is initially advancing, pressure is below the settings of both valves, and they remain closed. Flow from the high volume unit passes through the check valve and joins the output of the low volume cartridge going to the actuator.

High Pressure Low Volume

When pressure sensed at the pump outlet exceeds the setting of the unloading valve, flow from the large cartridge is bypassed to tank at little or no pressure and the check valve closes. The small volume cartridge continues to deliver oil to the system up to relief valve setting at which time the relief valve opens and returns the oil to the tank. Input horsepower (electric motor size) is determined by the greater of two condi-

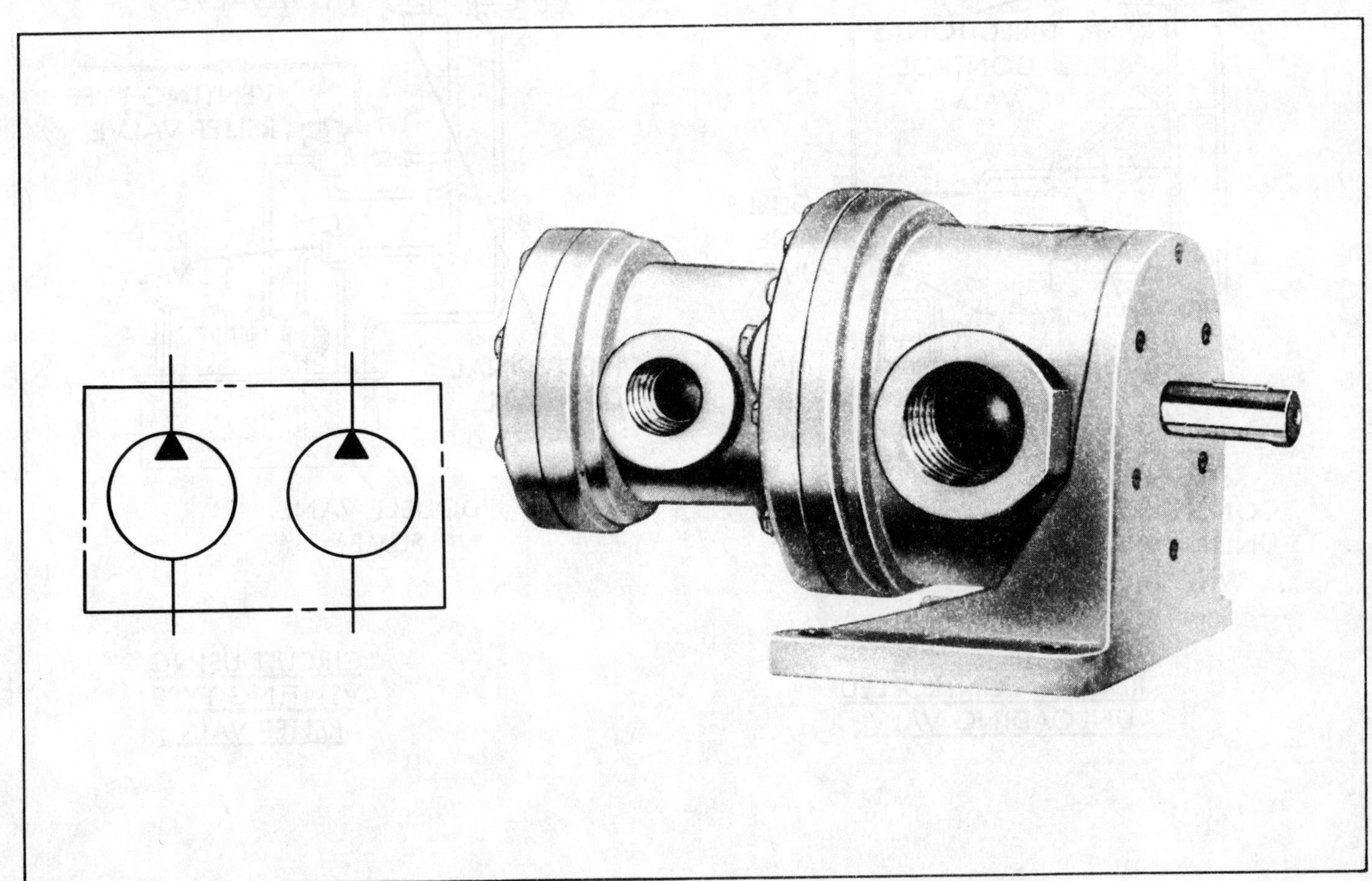

Fig. 11-12. Double "Round" Pump

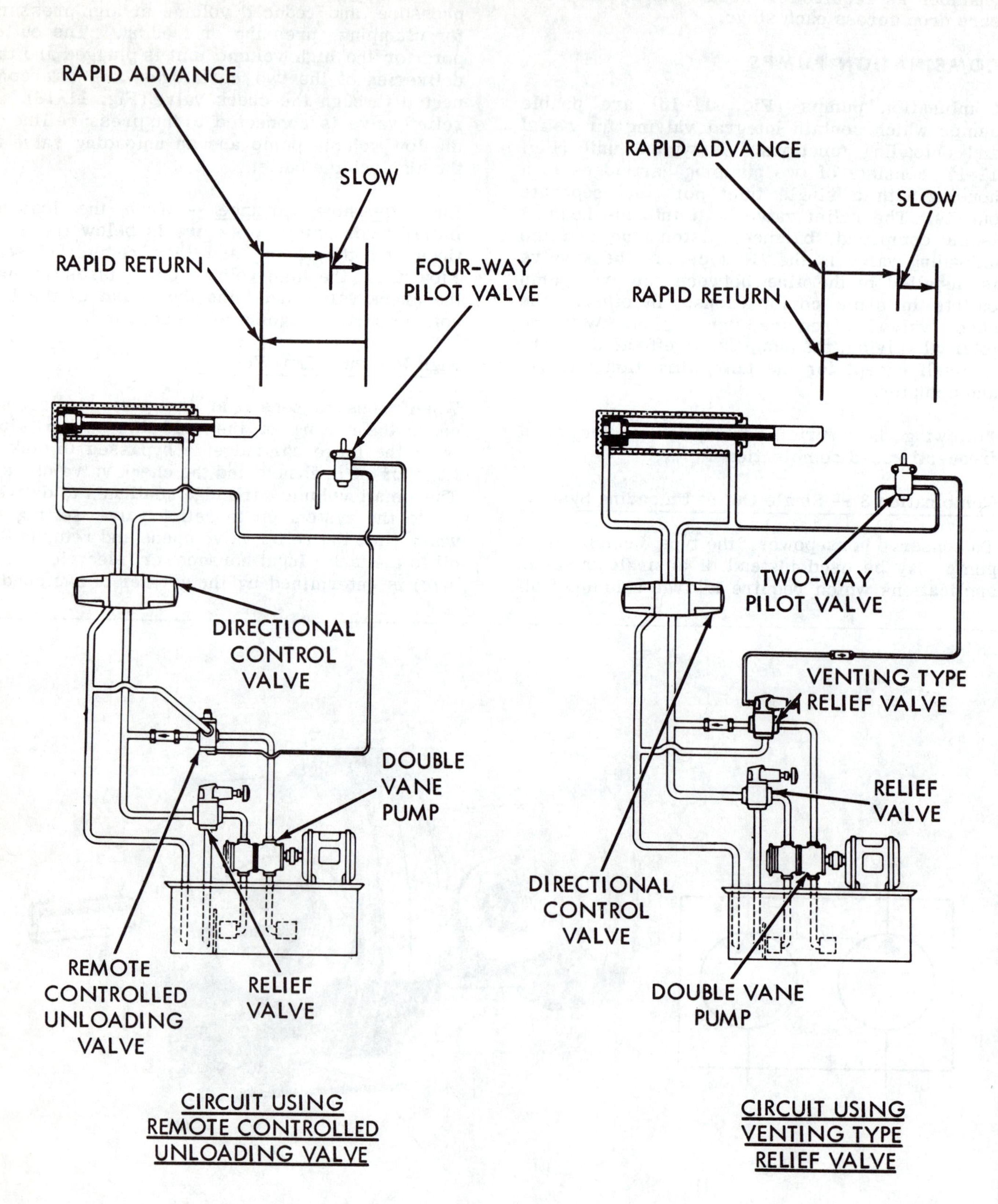

Fig. 11-13. Typical Applications of Double Pump

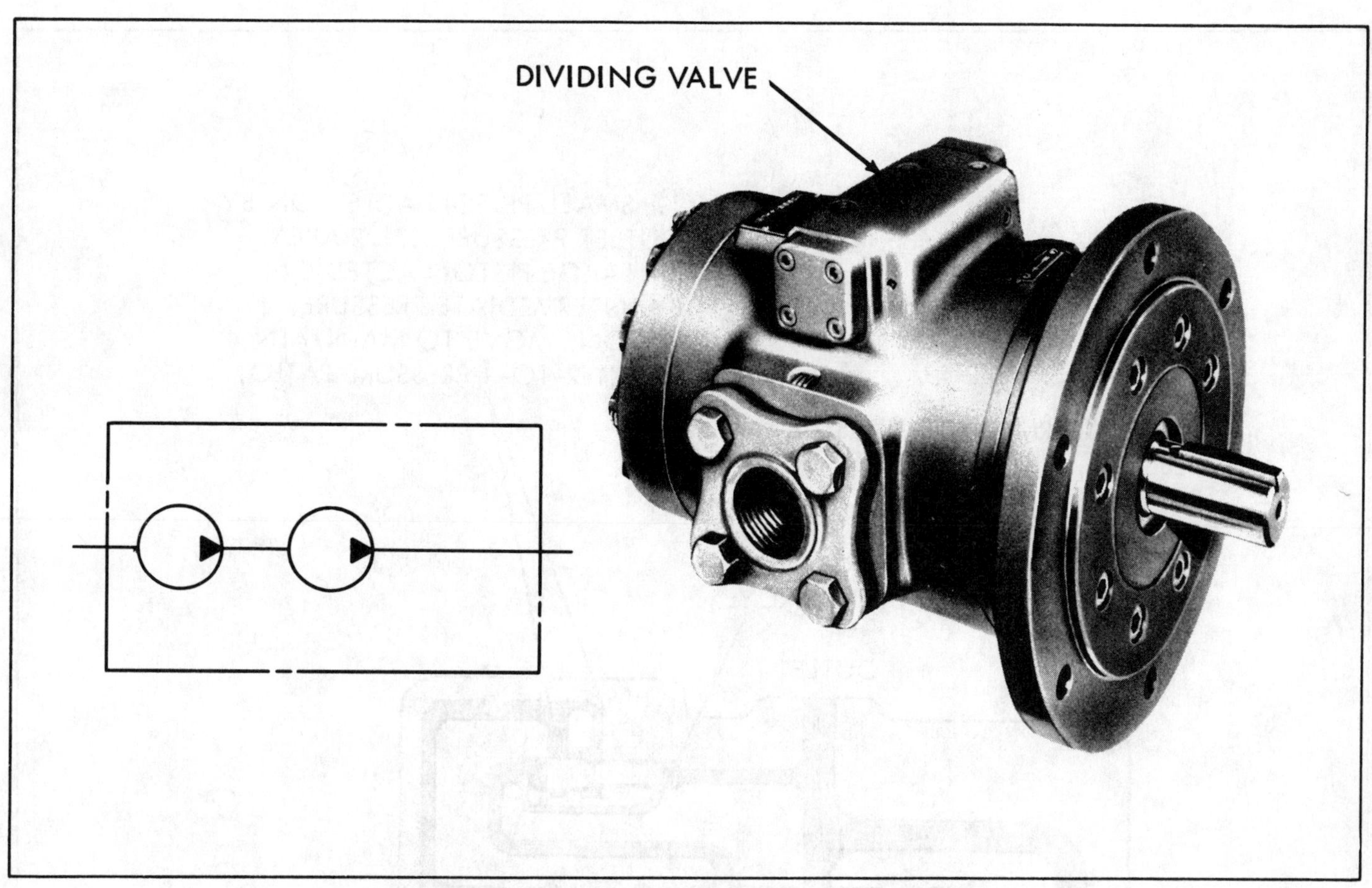

Fig. 11-14. Typical Two Stage Pump

tions: total volume of both cartridges at the unloading valve setting, or volume of the small unit at the relief valve setting plus whatever is required to drive the larger one unloaded. To assure proper operation the unloading valve setting should be adjusted to open at least 150 psi below the relief valve setting.

Combination 33 -- Single Adjustment Unloading System.

This combination is a variation of the combination 3 and simplifies adjustment by providing a single pressure control adjustment. Unloading pressure is automatically set at 125 psi below the relief valve pressure setting. Operation is as follows (see Fig. 11-19):

As in the combination 3 both valves remain closed when outlet pressure is below their settings. Flow from both pumping units is supplied to the system.

When outlet pressure rises to the amount determined by the adjustment, the unloading valve poppet opens and subsequently the spool lifts to unload the large volume pump unit to tank at low pressure. The check valve closes to prevent loss of small cartridge output. During this unloading phase, pressure in the spring chamber at the back side of the relief valve poppet is maintained at the setting of the unloading valve poppet. Maximum outlet pressure, i.e., the relief valve setting, is determined by the pressure in the chamber and the non-adjustable spring behind the poppet. The valve of the latter is such that the relief opens when operating pressure exceeds the unloading valve setting by 125 psi.

Combination 6 -- Two Outlets System

In the combination 6 (Fig. 11-20) a plug is placed in the interconnecting passage so that each unit operates as a separate pump having its own outlet. The R valve is modified to function as a compound relief valve for the large volume pump while the small pump is protected as in the other units.

Other Combinations

The design of these units is such that many other combinations are possible, however, their usage is too infrequent to cover them in detail.

"Square" Vane Type Pumps

The "square" vane type pumps (Fig. 11-21) were

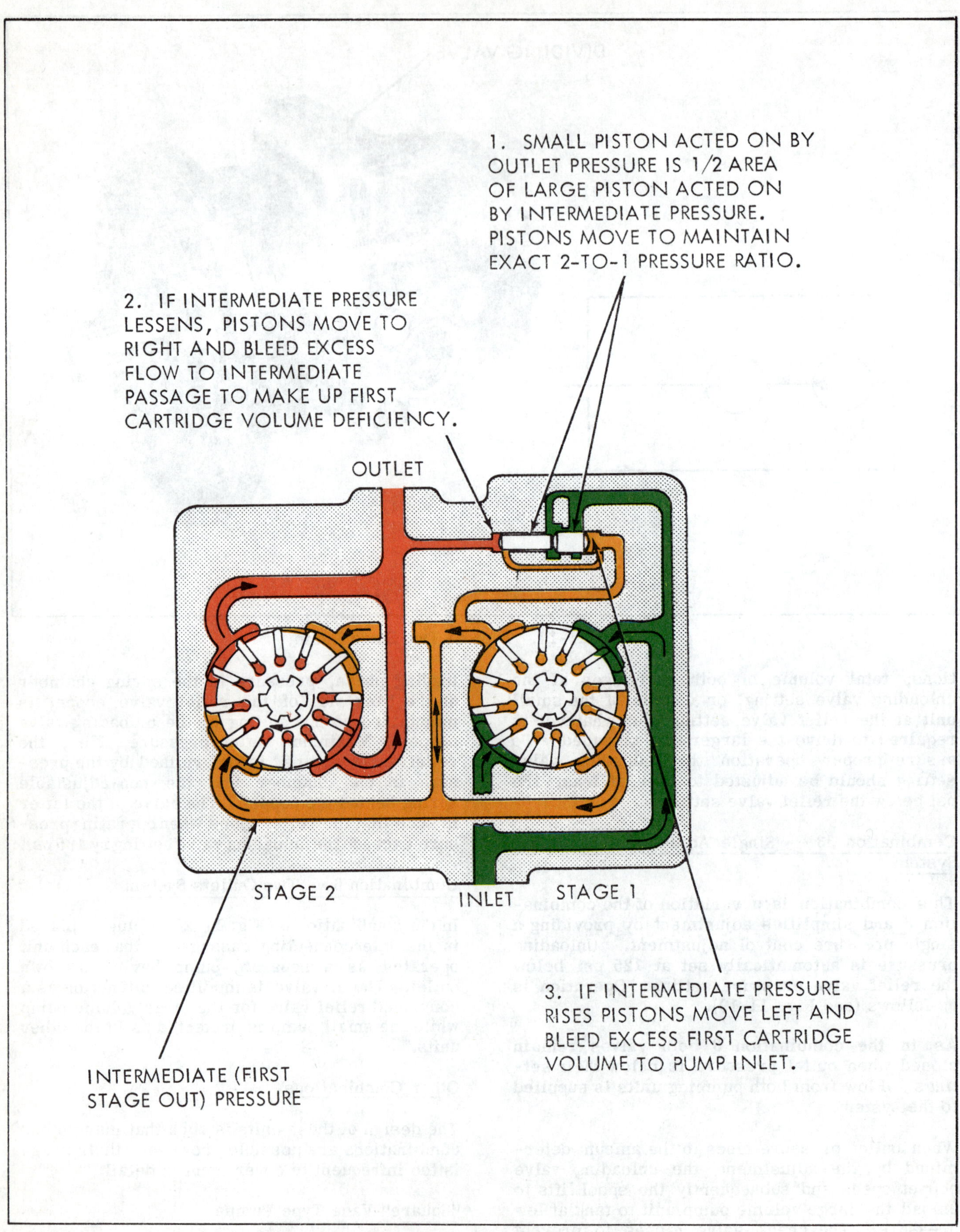

Fig. 11-15. Dividing Valve Splits Pressure Between Two Stages

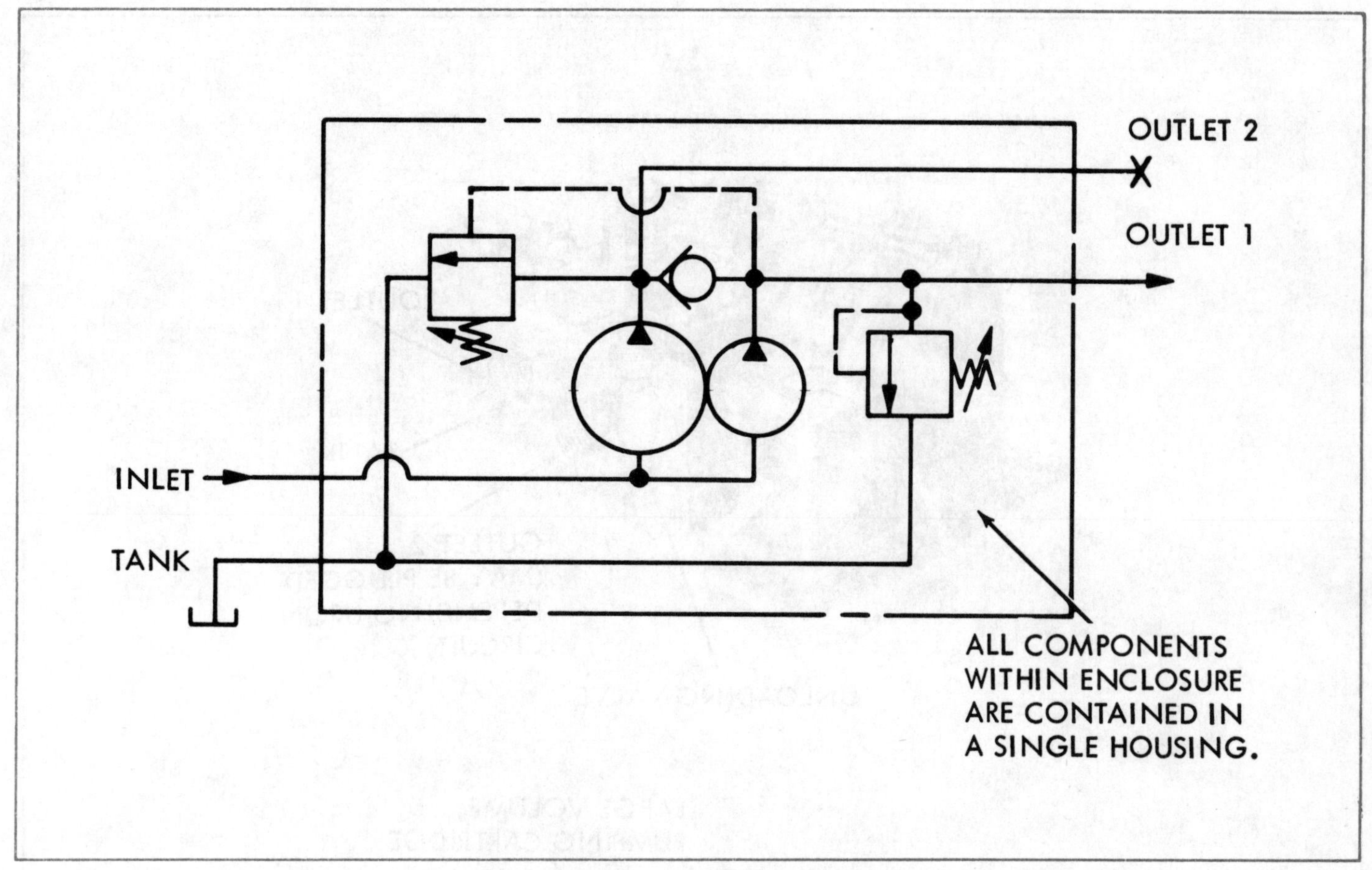

Fig. 11-16. Combination Pump

designed originally for mobile applications. They are also hydraulically balanced, but their construction is much simpler than the round pumps. The cartridge consists of a ring sandwiched between the pump body and cover, a rotor, twelve vanes and a spring loaded pressure plate. The inlet port is in the body and the outlet in the cover which may be assembled in any of four positions for convenience in piping.

Operation

The spring (Fig. 11-22) holds the pressure plate in position against the ring at all times. As outlet pressure builds up it acts with the spring to offset pressures within the cartridge which tend to separate it. Proper running clearance is determined by the (relative) ring and rotor widths.

Initial starting is accomplished by spinning the rotor and shaft fast enough (approximately 600 rpm) for centrifugal force to throw the vanes out against the ring generating the pumping action. An interrupted (annular) groove in the pressure plate permits free flow of pressurized fluid into chambers under the vanes as they move out under the vanes as they move out of the rotor slots. Return flow is restricted as the vanes move back holding them firmly against the ring.

If it is necessary to reverse the drive shaft rotation the ring must be removed and reassembled with the opposite side facing the pump body. Directional arrows cast on the outer edge of the ring facilitate this procedure.

These pumps are manufactured in a variety of sizes. Cartridges with different displacements are available for each of them.

Double pumps have a common inlet in a center housing. (See Fig. 11-23.) The outlet for one, usually the larger unit is in the shaft end body and the other in the cover.

Cartridge construction is essentially the same as in single units making numerous combinations of sizes and displacements possible.

"High Performance" Vane Pumps

The latest design of balanced vane pumps is the high performance series which is capable of higher pressure and speed. A typical single pump of this design is shown in Figure 11-24 and double pump in Figure 11-25. Operation is

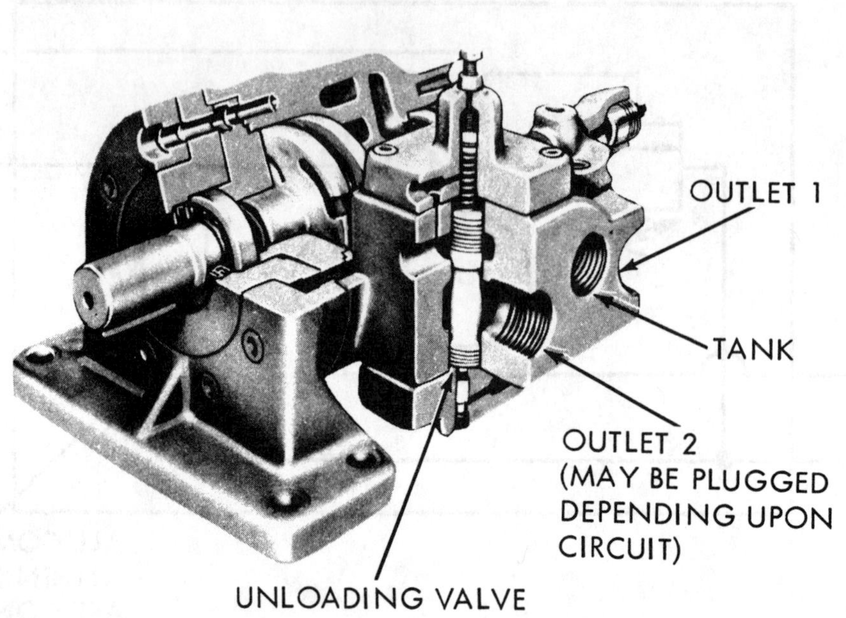

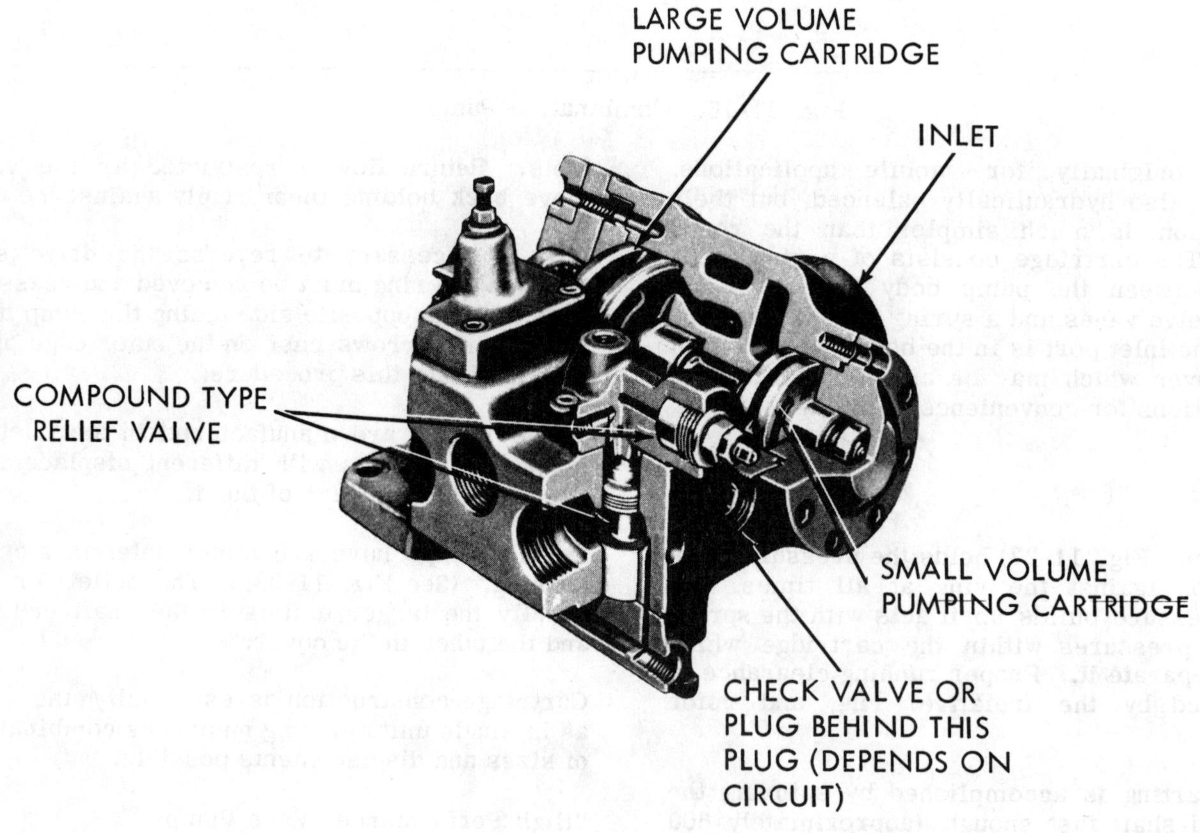

Fig. 11-17. Construction of Typical Combination Pump

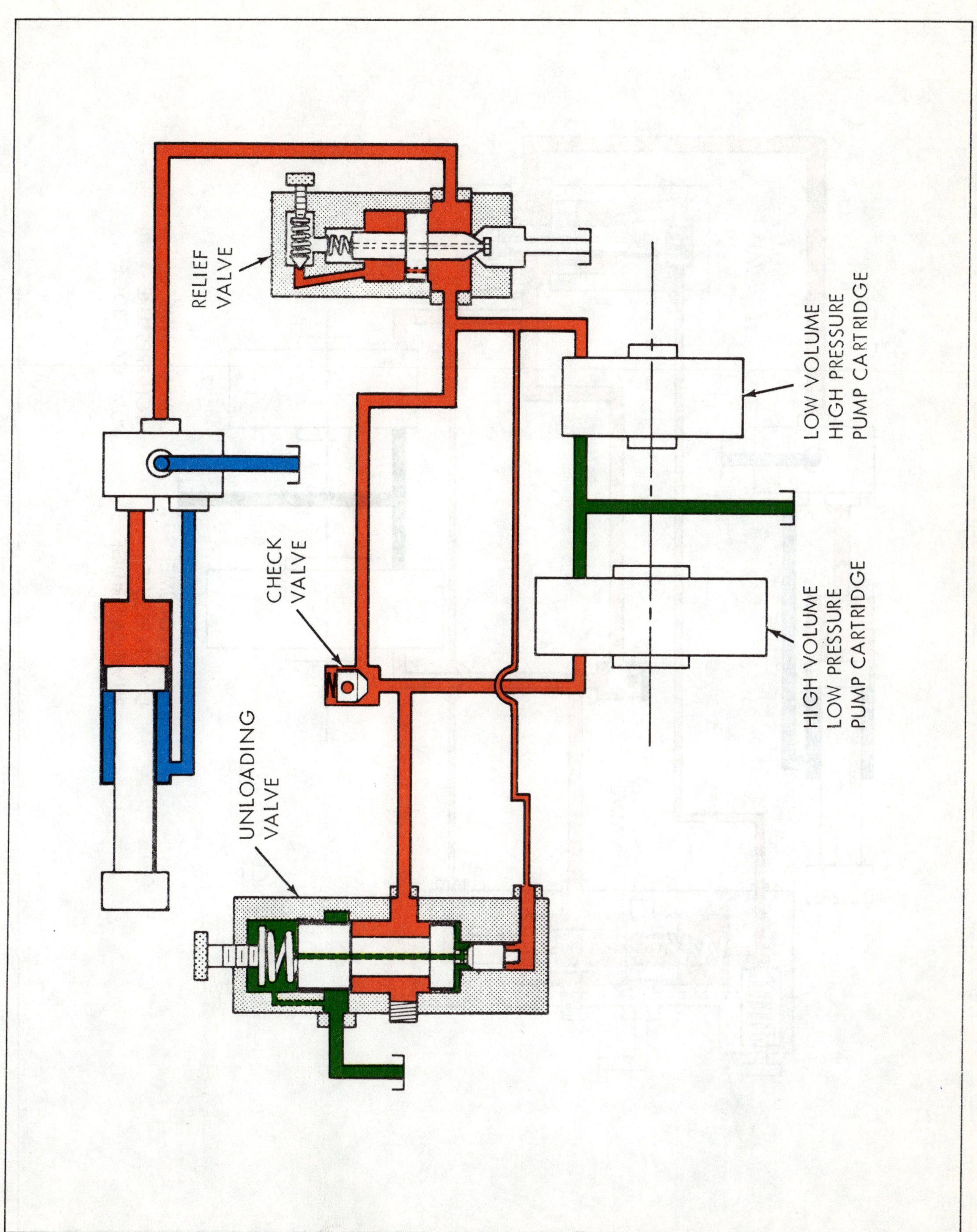

Fig. 11-18. Single Outlet Unloading System

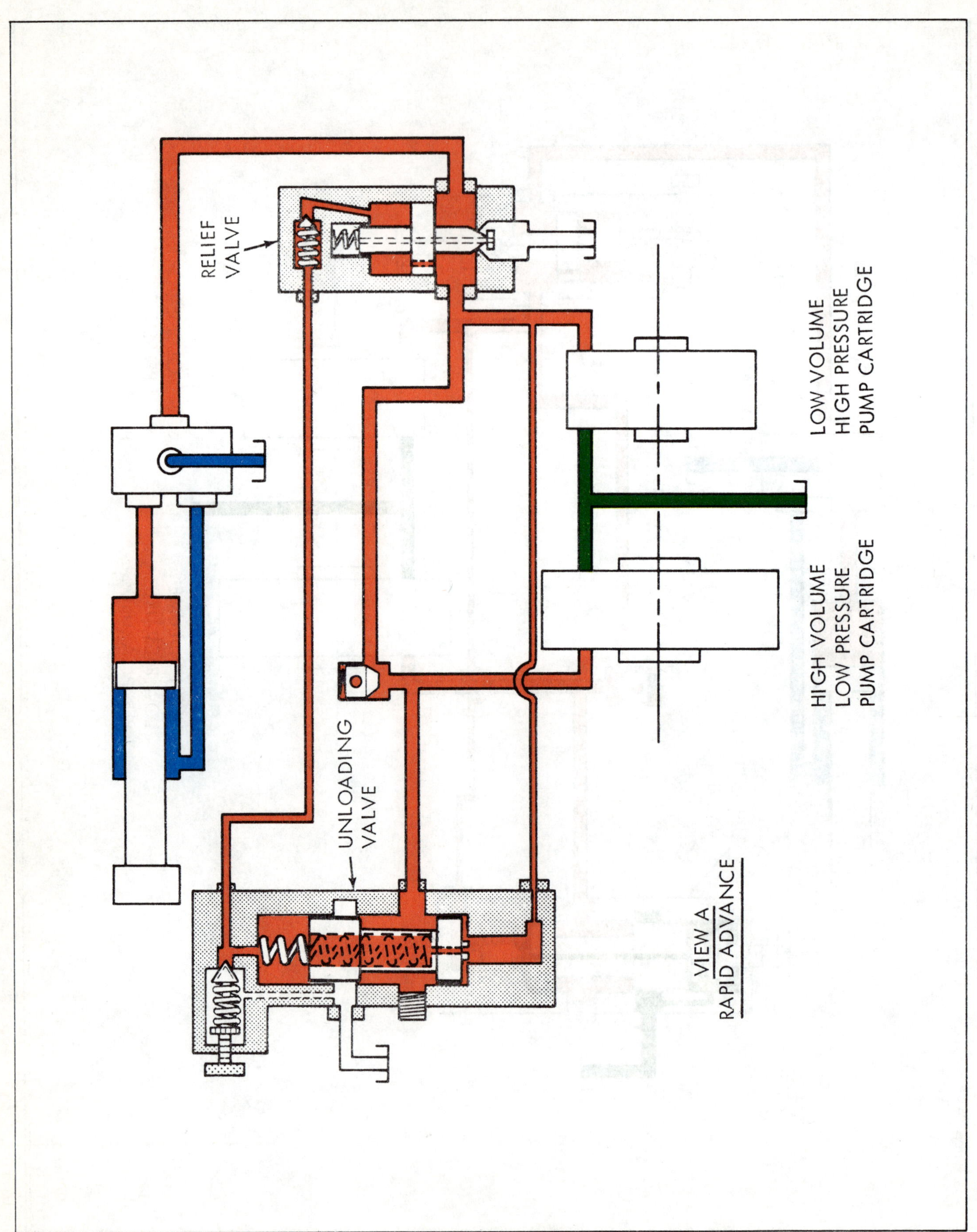

Fig. 11-19. Single Adjustment Unloading System Combination 33

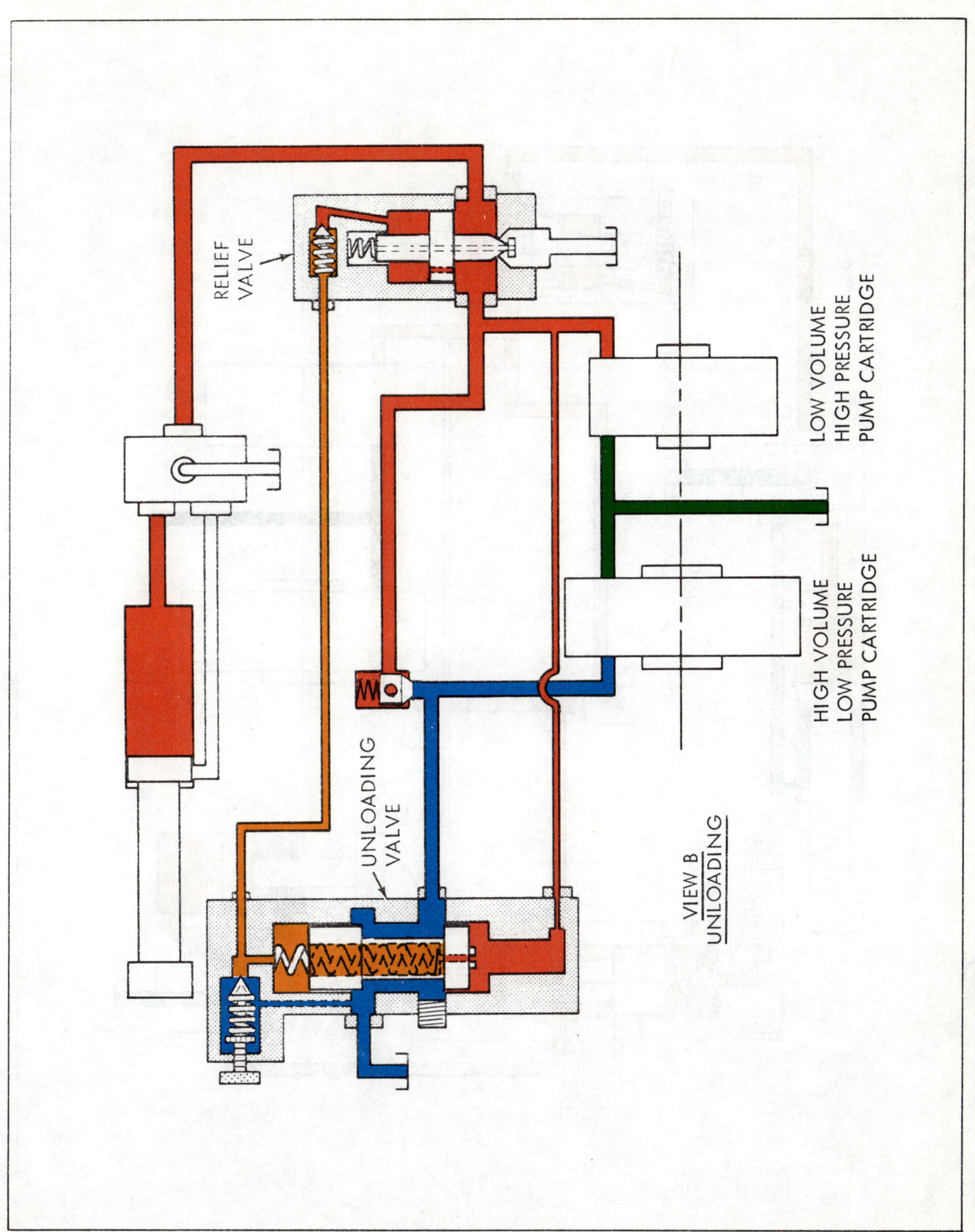

Fig. 11-19. Single Adjustment Unloading System Combination 33

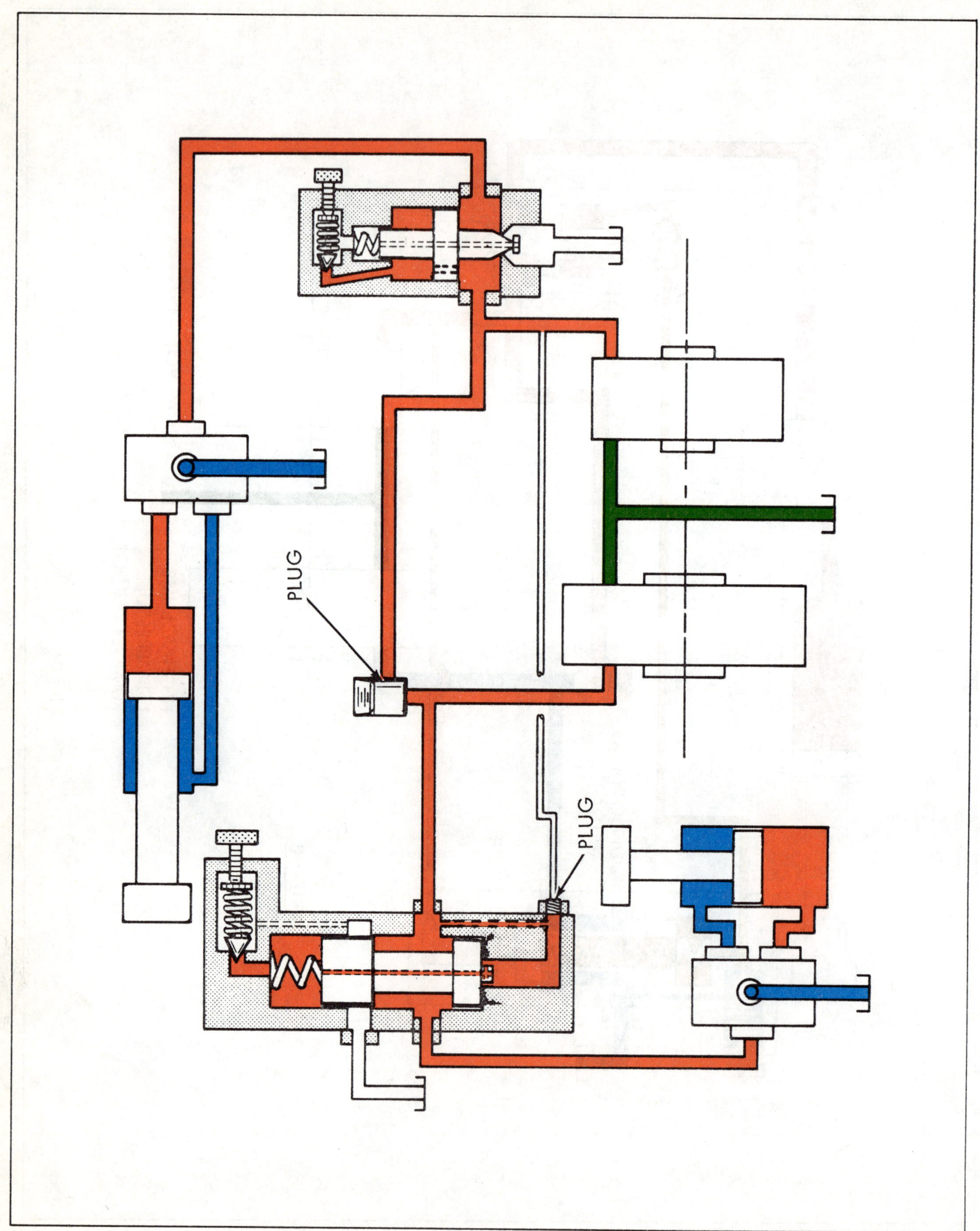

Fig. 11-20. Combination 6

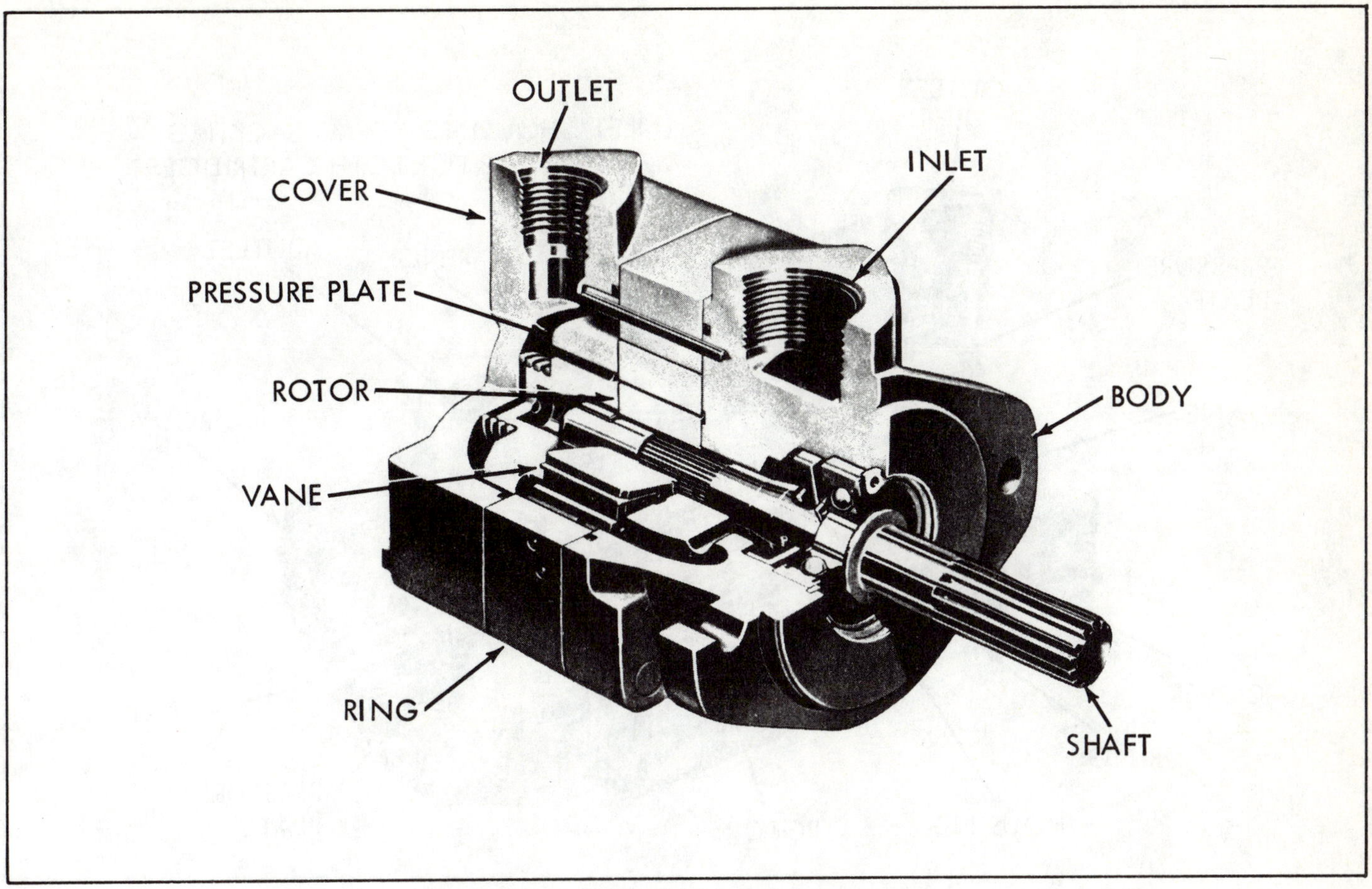

Fig. 11-21. "Square" Design Vane Pumps

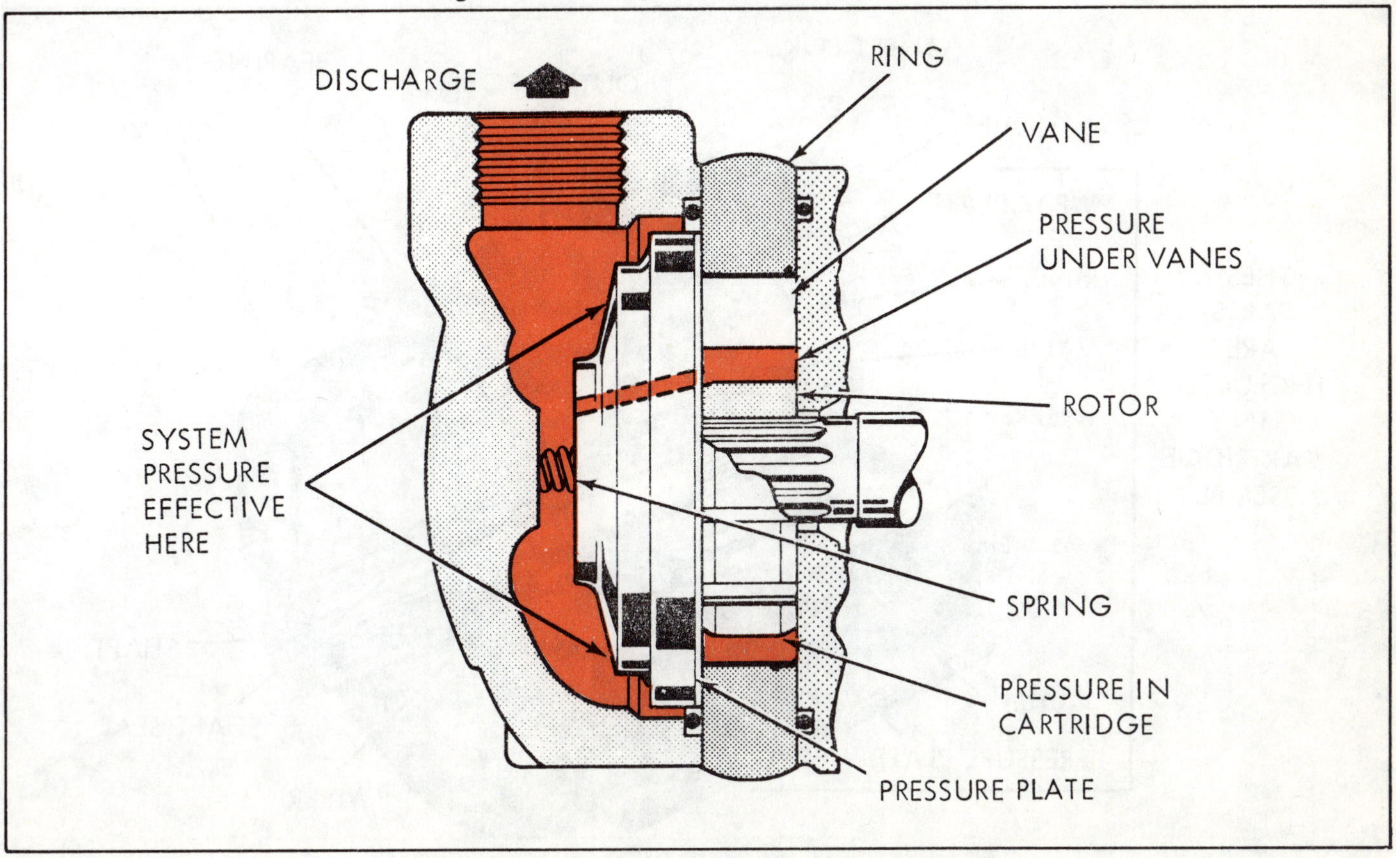

Fig. 11-22. Pressure Plate Seals Cartridge

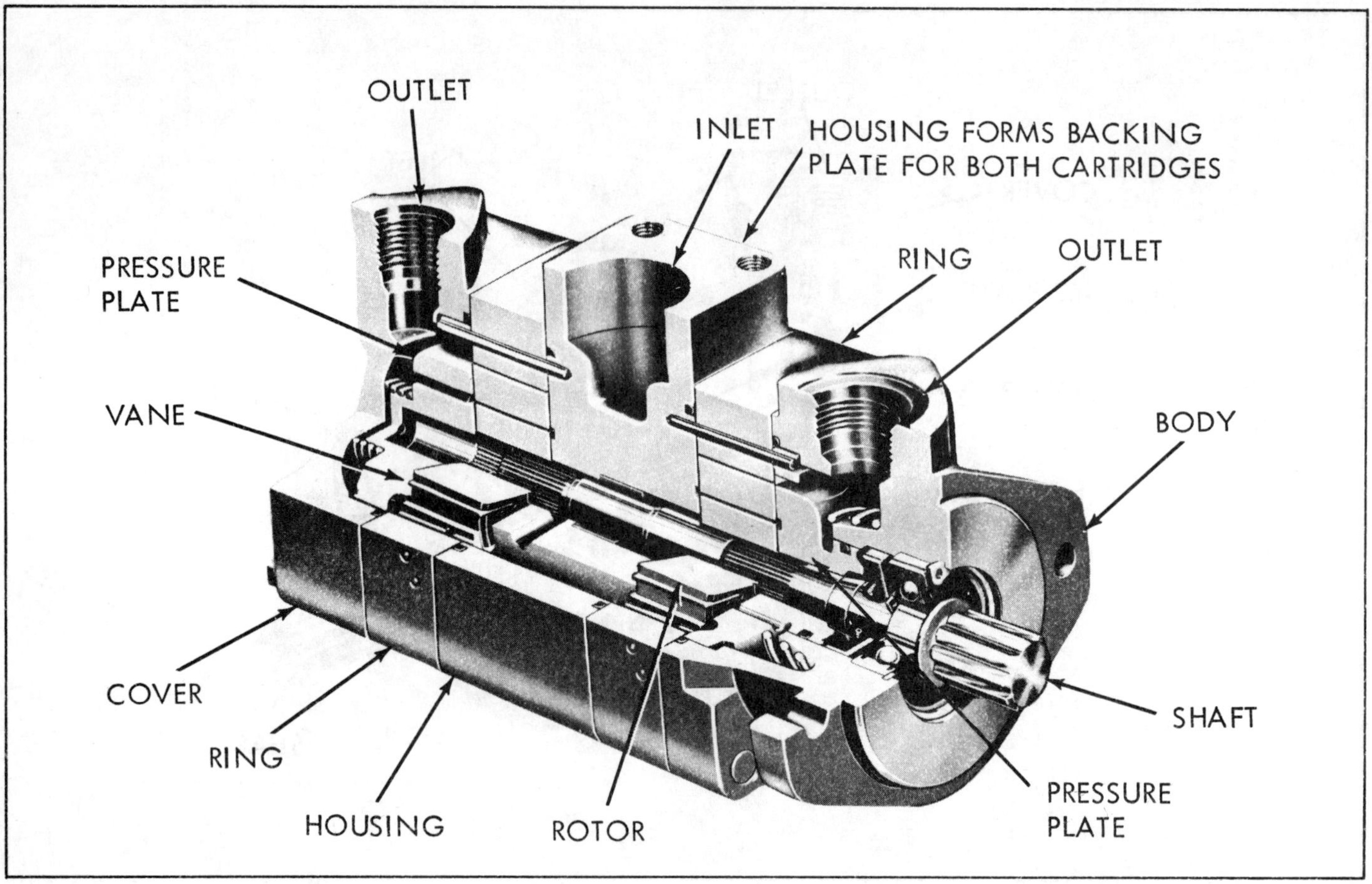

Fig. 11-23. Double "Square" Pump

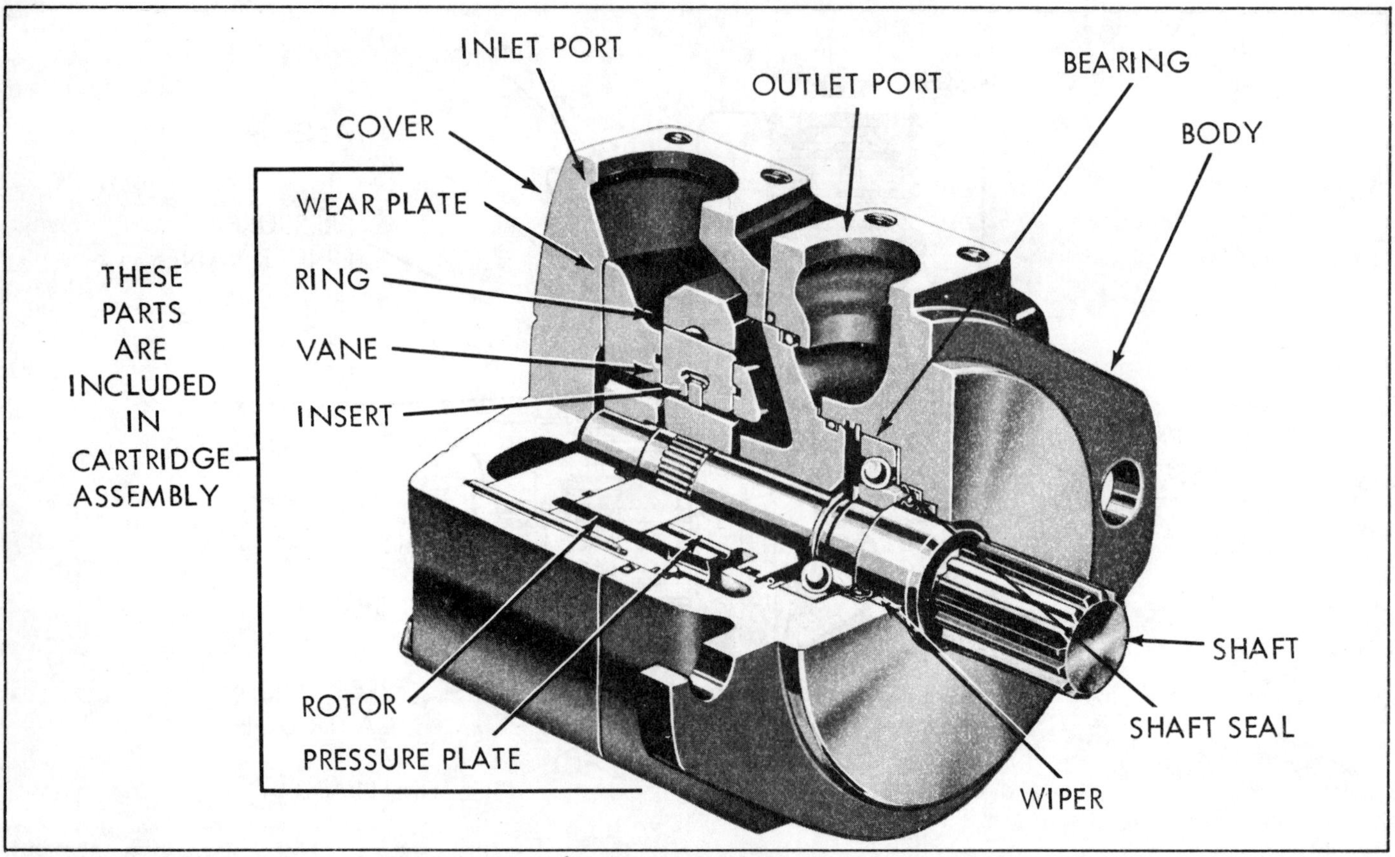

Fig. 11-24. High Performance Pump Construction

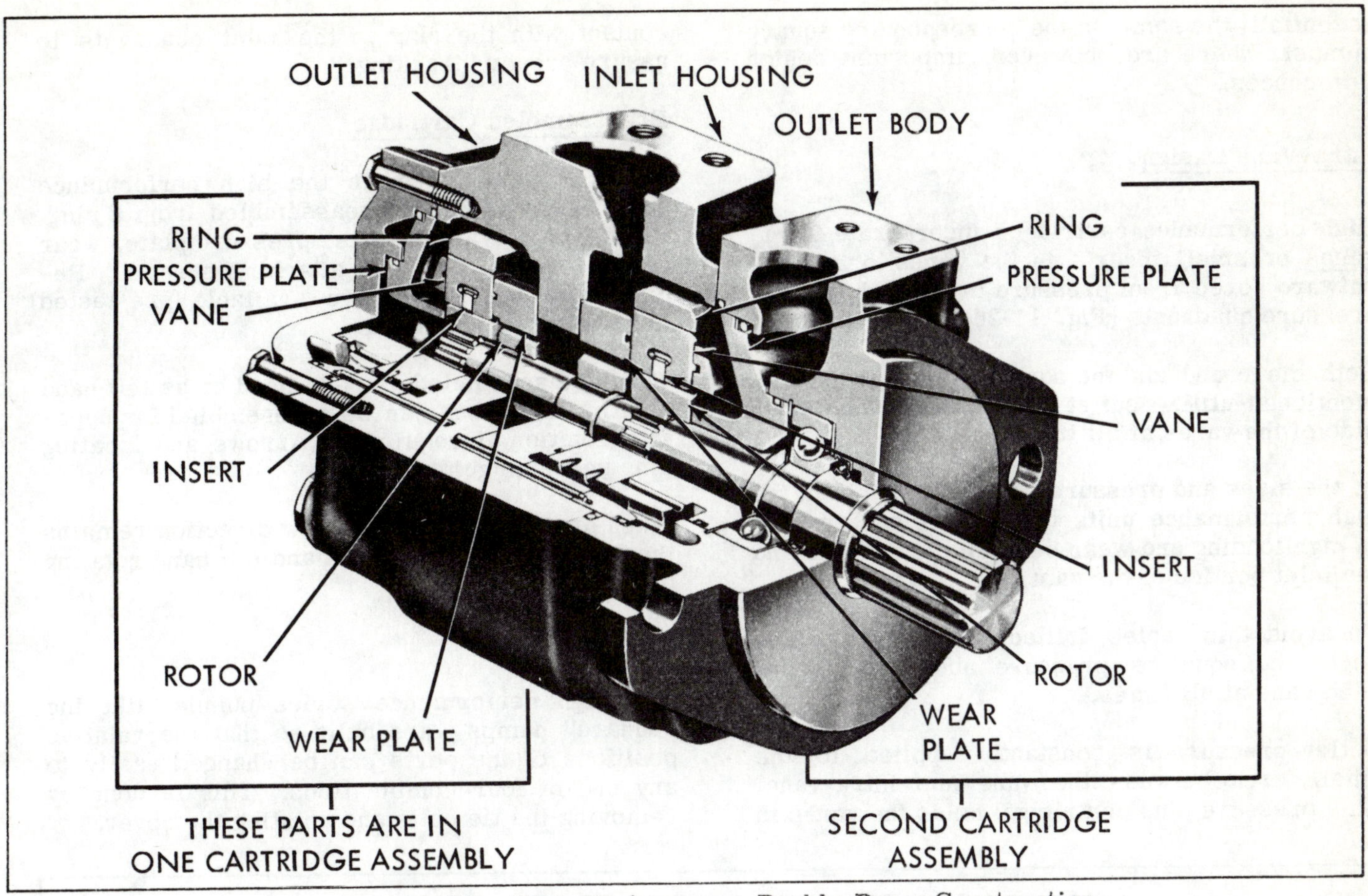

Fig. 11-25. High Performance Double Pump Construction

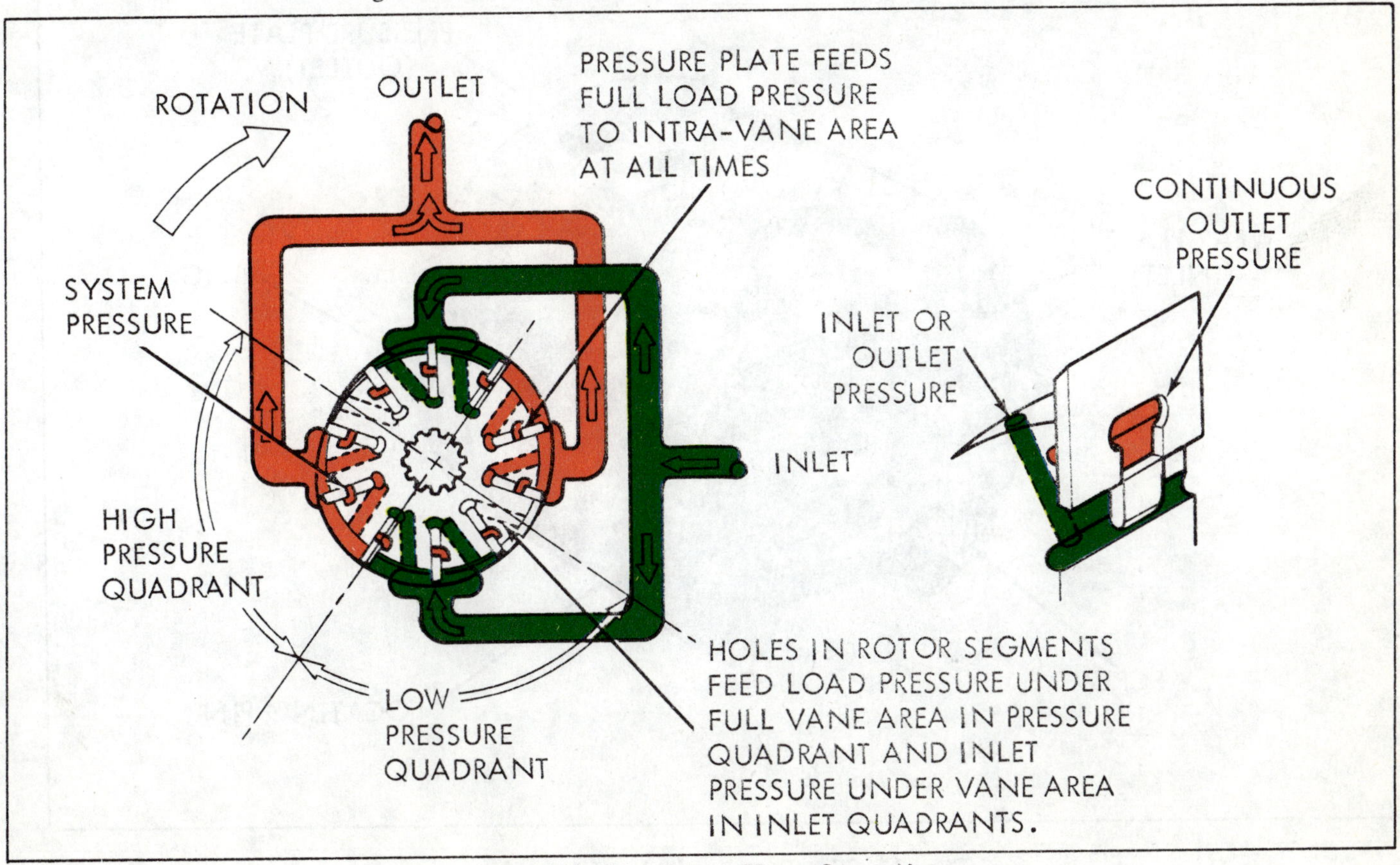

Fig. 11-26. Intra-Vane Operation

essentially the same as the corresponding square pumps. There are, however, important design differences.

Intra-Vane Design

High performance cartridges incorporate intra-vanes or small inserts in the vanes to vary the outward force from pressure in the high and low pressure quadrants (Fig. 11-26).

Both the round and the square pumps previously mentioned utilize outlet pressure on the underside of the vanes at all times.

In the sizes and pressure ranges available in the high performance units this feature could result in high loading and wear between the vane tip and the inlet portion of the cam ring.

To avoid this, holes drilled through the rotor segments equalize pressure above and below each vane at all times.

Outlet pressure is constantly applied to the small area between the vane and intra-vane. This pressure plus centrifugal holds the vanes in contact with the ring in the inlet quadrants to assure proper "tracking."

Preassembled Cartridge

The cartridge used in the high performance pump (Fig. 11-27) is preassembled from a ring, rotor, vanes, vane inserts, pressure plate, wear plate, locating pins and attaching screws. Replacement cartridges are available (pre-tested) for fast replacement.

They are assembled as right hand or as left hand rotating units but can be reassembled for opposite rotation if required. Arrows and locating pins serve as guides.

When properly assembled flow direction remains the same in both the right and left hand rotating units.

Port Positions

The high performance series pumps, like the "square" pumps, are built so that the relative positions of the ports can be changed easily to any one of four combinations. This is done by removing the tie bolts and rotating the cover.

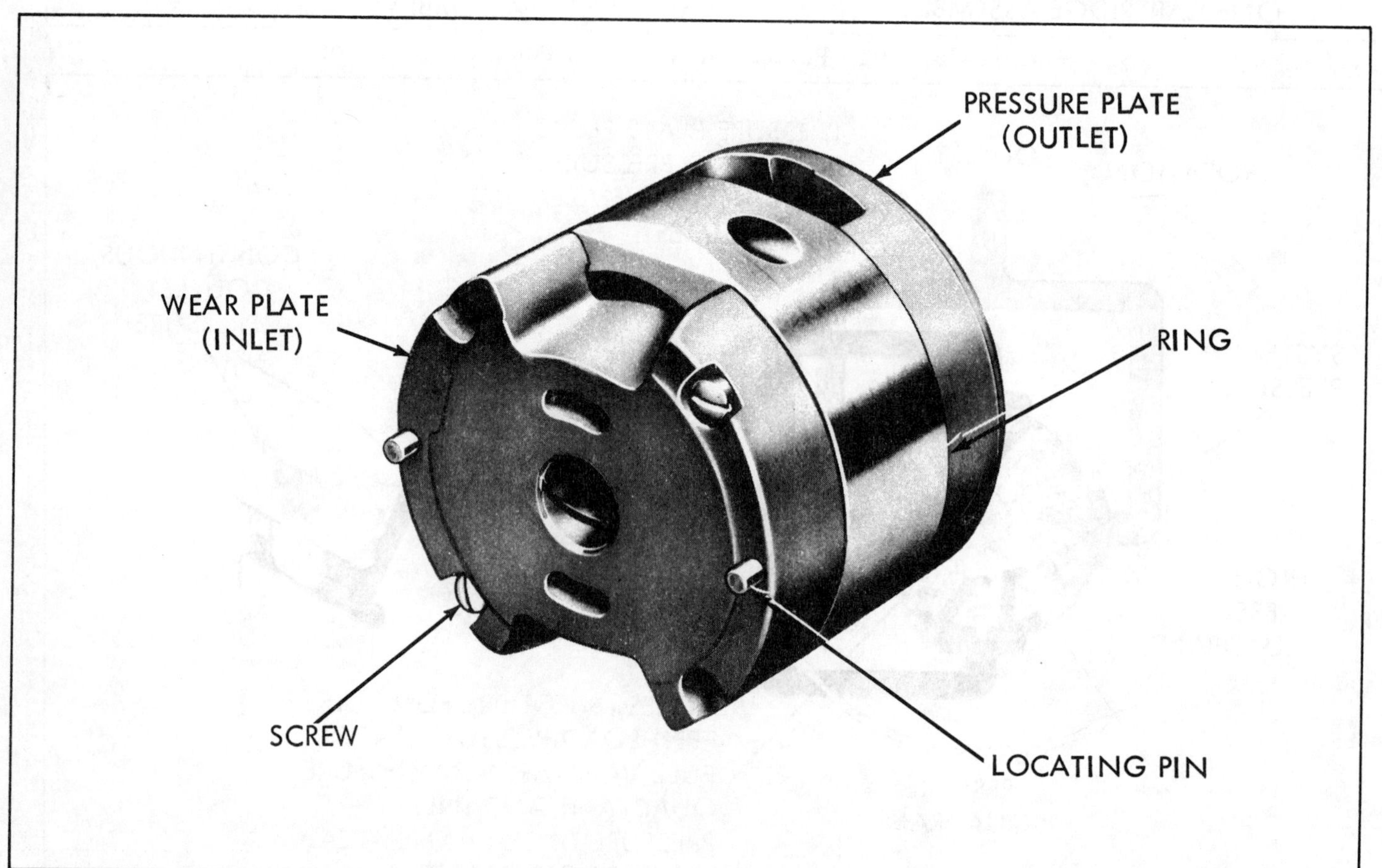

Fig. 11-27. Preassembled Cartridge

Dowel pins locate the cartridge within the cover and it, too, must be rotated, otherwise the inlet port may be restricted.

Vane Pump Operating Characteristics

Vane pumps cover the low to medium high volume ranges with operating pressures up to 3000 psi. They are reliable, efficient, and easy to maintain. The ring surface and vane tips are the points of greatest wear, which is compensated for by the vanes moving farther out of their slots.

Cleanliness and the proper fluid are essential to long life. Petroleum oil having adequate anti-wear qualities is recommended. However, many vane pumps are operating successfully with synthetic fluid.

PISTON PUMPS

All piston pumps operate on the principle that a piston reciprocating in a bore will draw in fluid as it is retracted and expel it on the forward stroke.

Two basic designs are radial and axial, both are available as fixed or variable displacement models. A radial pump has the pistons arranged radially in a cylinder block (Fig. 11-28) while in the axial units the pistons are parallel to each other and to the axis of the cylinder block (Fig. 11-32). The latter may be further divided into in-line (swash plate or wobble plate) and bent axis types.

Radial Piston Pumps

In a radial pump the cylinder block rotates on a stationary pintle and inside a circular reaction ring or rotor. As the block rotates centrifugal force, charging pressure or some form of mechanical action causes the pistons to follow the inner surface of the ring which is offset from the centerline of the cylinder block. As the pistons reciprocate in their bores, porting in the pintle permits them to take in fluid as they move outward and discharge it as they move in.

The size and number of pistons (there may be more than one bank in a single cylinder block) and, of course the length of their stroke determines pump displacement. In some models the displacement can be varied by moving the reaction ring to increase or decrease piston travel.

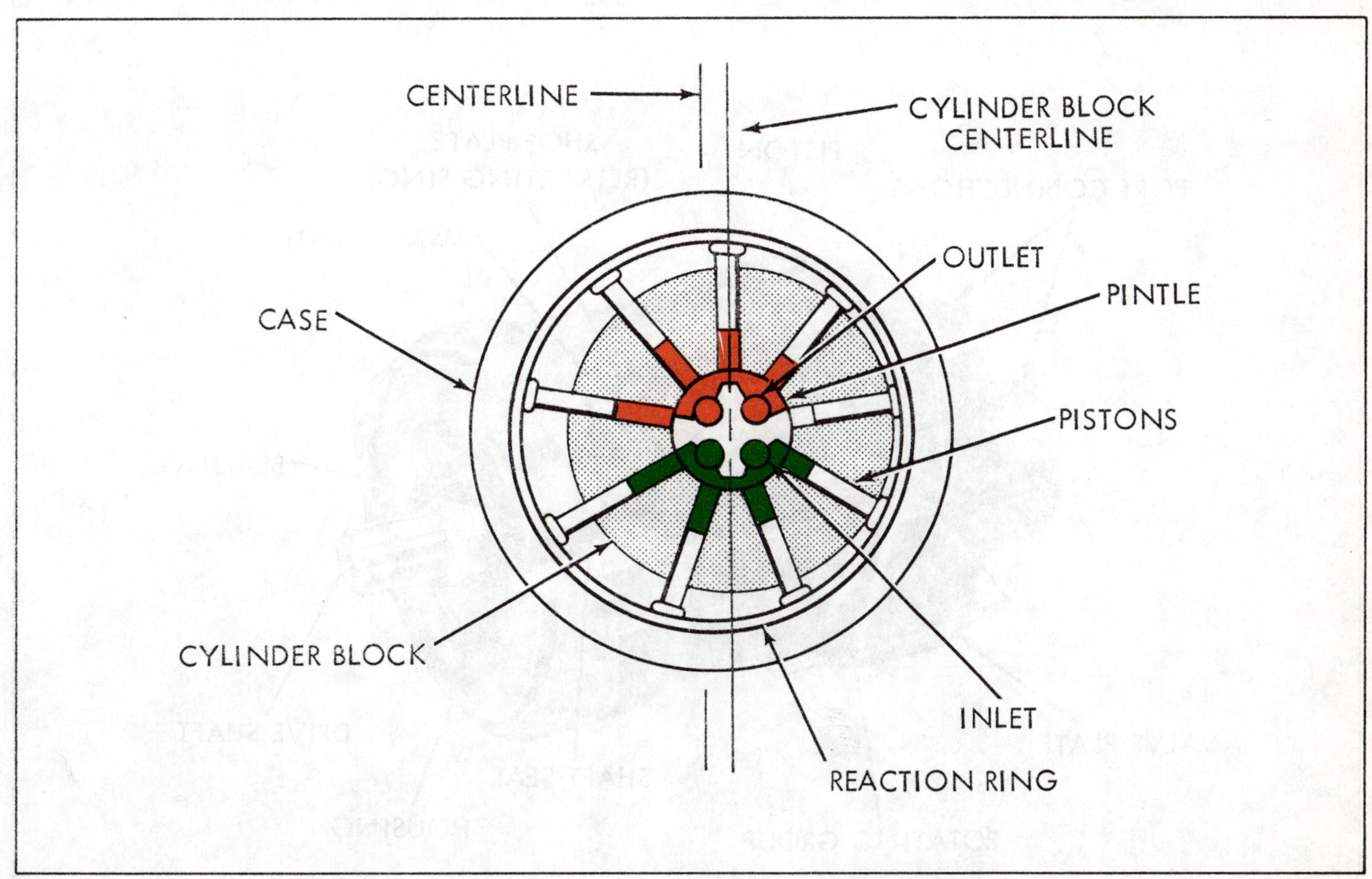

Fig. 11-28. Operation of Radial Piston Pump

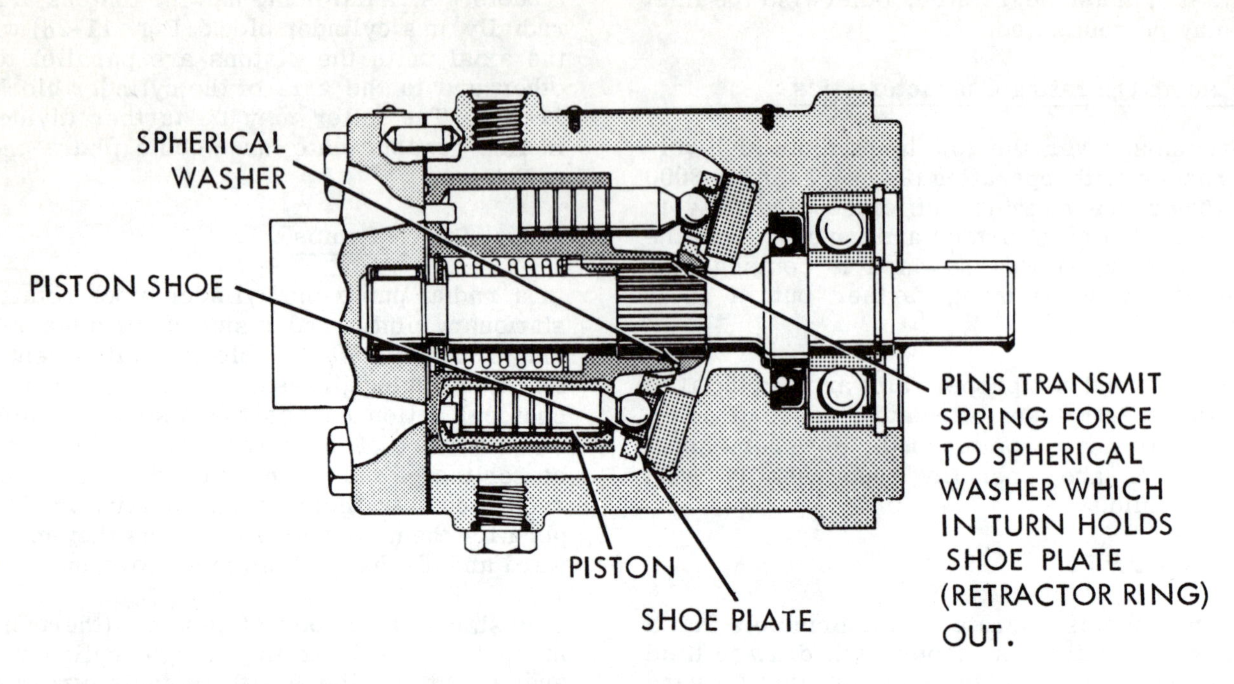

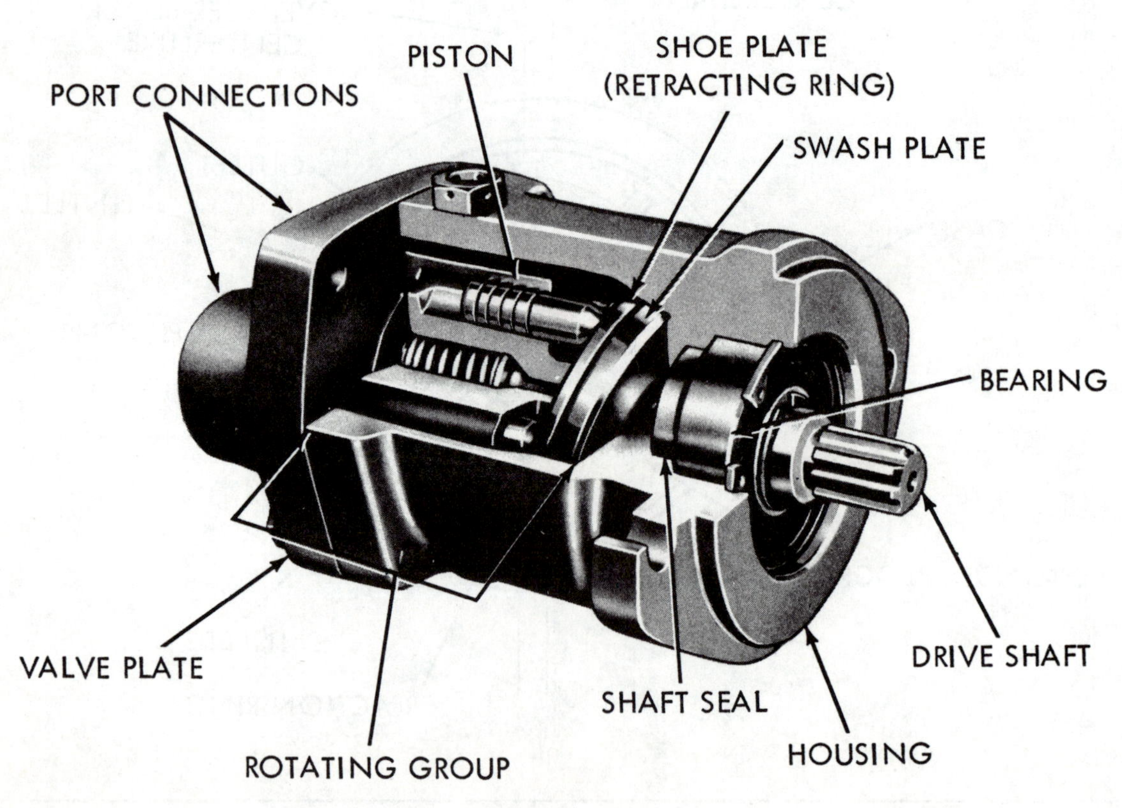

Fig. 11-29. Inline Design Piston Pump

External controls of several types are available for this purpose.

Swash Plate Design Inline Piston Pumps

In axial piston pumps, the cylinder block and drive shaft are on the same centerline and the pistons reciprocate parallel to the drive shaft. The simplest type of axial piston pump is the swash plate inline design (Fig. 11-29).

The cylinder block in this pump is turned by the drive shaft. Pistons fitted to bores in the cylinder are connected through piston shoes and a retracting ring, so that the shoes bear against an angled swash plate.

As the block turns (Fig. 11-30), the piston shoes follow the swash plate, causing the pistons to reciprocate. The ports are arranged in the valve plate so that the pistons pass the inlet as they are being pulled out and pass the outlet as they are being forced back in.

Displacement

In these pumps the displacement is also determined by the size and number of pistons as well as their stroke length, the latter being a function of the swash plate angle.

In variable displacement models of the Inline pump, the swash plate is installed in a movable yoke (Fig. 11-31). "Pivoting" the yoke on pintles changes the swash plate angle to increase or decrease the piston stroke (Fig. 11-32). The yoke can be positioned manually, with a servo control, with a compensator control, or by any of several other means. Figure 11-31 shows a compensator control. Maximum angle on the units shown is limited to $17\frac{1}{2}$ degrees by construction.

Compensator Operation

Operation of the Inline pump compensator control is shown schematically in Figure 11-33. The control consists of a compensator valve balanced between load pressure and the force of a spring, a piston controlled by the valve to move the yoke, and a yoke return spring.

With no outlet pressure, the yoke return spring moves the yoke to the full delivery position. As pressure builds up, it acts against the end of the valve spool. When the pressure is high enough

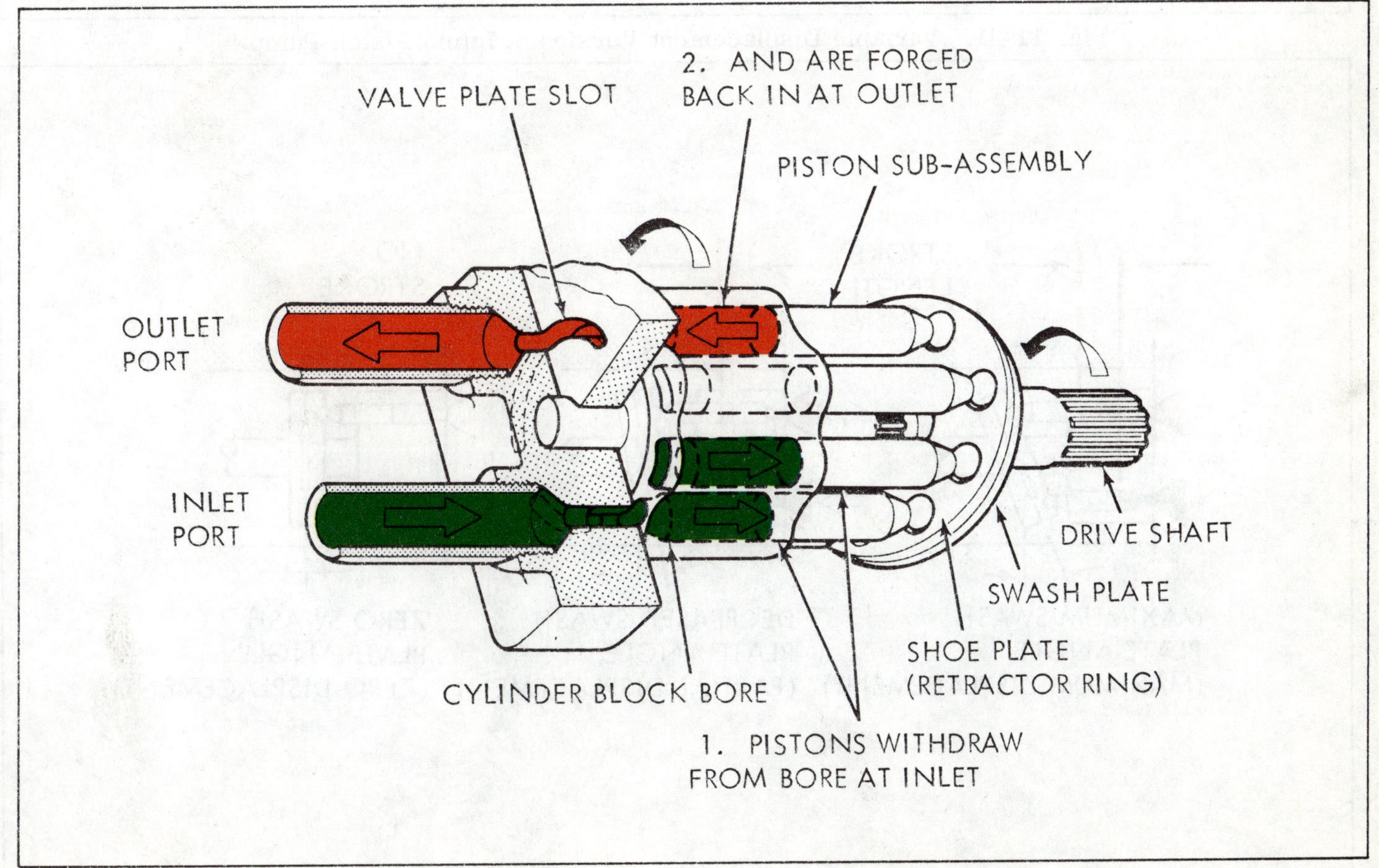

Fig. 11-30. Swash Plate Causes Pistons to Reciprocate

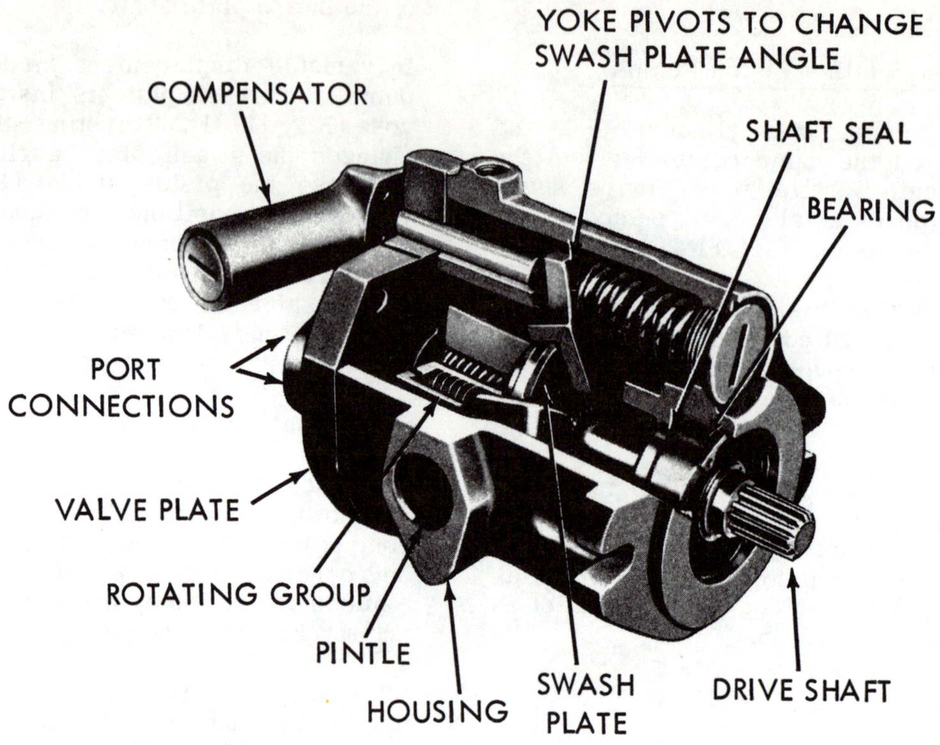

Fig. 11-31. Variable Displacement Version of Inline Piston Pump

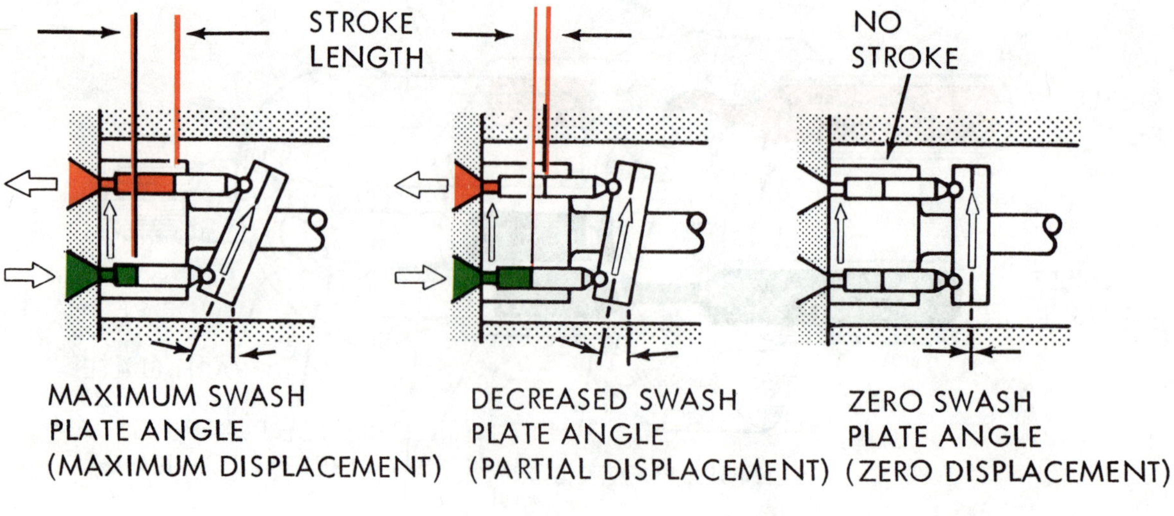

Fig. 11-32. Variation in Pump Displacement

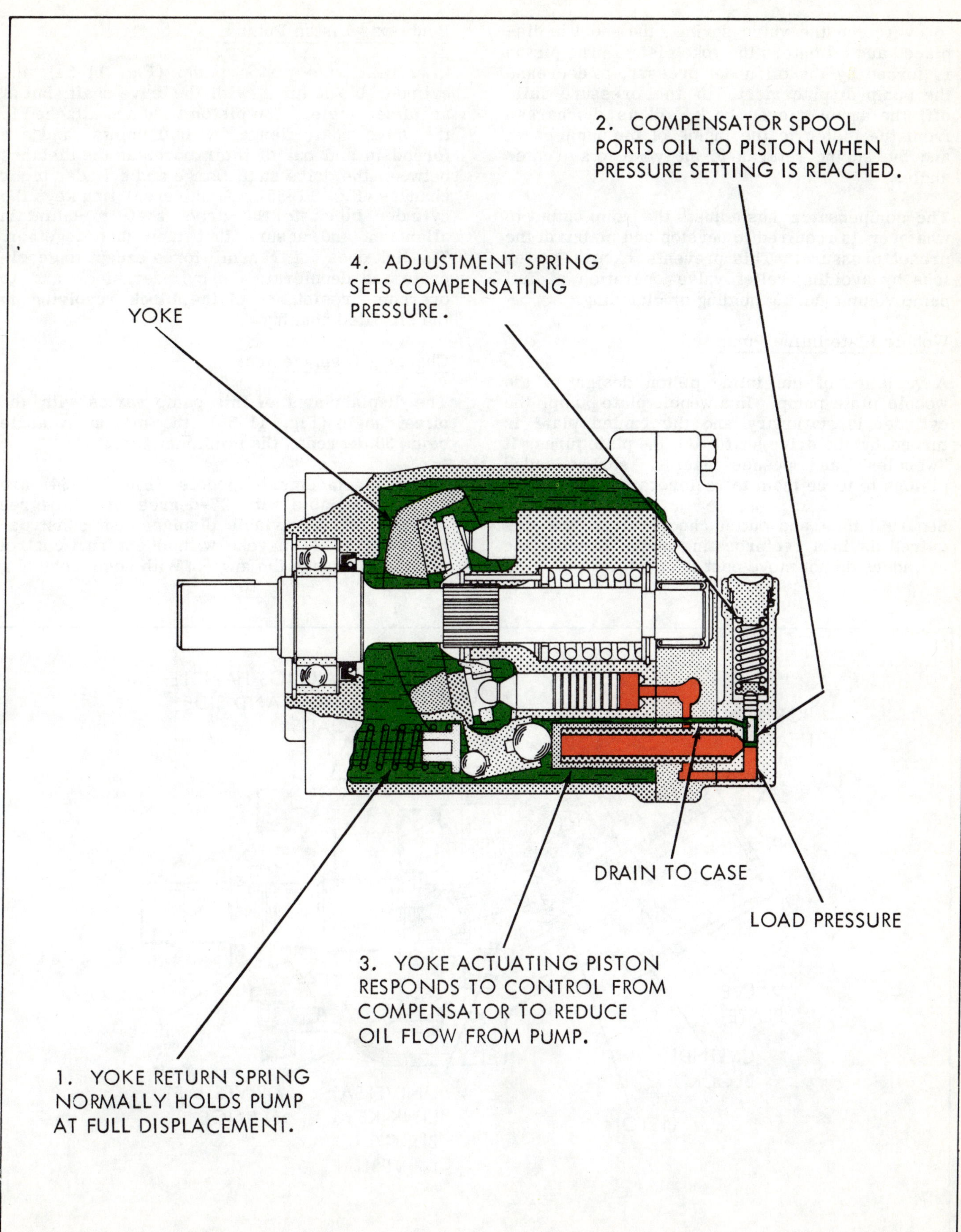

Fig. 11-33. Pressure Compensator Operation

to overcome the valve spring, the spool is displaced and oil enters the yoke piston. The piston is forced by the oil under pressure to decrease the pump displacement. If the pressure falls off, the spool moves back, oil is discharged from the piston to the inside of the pump case, and the spring returns to the yoke to a greater angle.

The compensator thus adjusts the pump output to whatever is required to develop and maintain the preset pressure. This prevents excess power loss by avoiding relief valve operation at full pump volume during holding or clamping.

Wobble Plate Inline Pump

A variation of the inline piston design is the wobble plate pump. In a wobble plate pump, the cylinder is stationary and the canted plate is turned by the drive shaft. As the plate turns, it "wobbles" and pushes against spring-loaded pistons to force them to reciprocate.

Separate inlet and outlet check valves are required as in a reciprocating pump, because the cylinders do not move past the ports.

Bent-Axis Piston Pumps

In a bent axis piston pump (Fig. 11-34), the cylinder block turns with the drive shaft, but at an offset angle. The piston rods are attached to the drive shaft flange by ball joints, and are forced in and out of their bores as the distance between the drive shaft flange and cylinder block changes (Fig. 11-35). A universal link keys the cylinder block to the drive shaft to maintain alignment and assure that they turn together. The link does not transmit force except to accelerate and decelerate the cylinder block and to overcome resistance of the block revolving in the oil filled housing.

Changing Displacement

The displacement of this pump varies with the offset angle (Fig. 11-36), the maximum angle being 30 degrees, the minimum zero.

Fixed displacement models (Fig. 11-34) are usually available with 23-degree or 30-degree angles. In the variable displacement construction (Fig. 11-37) a yoke with an external control is used to change the angle. With some controls,

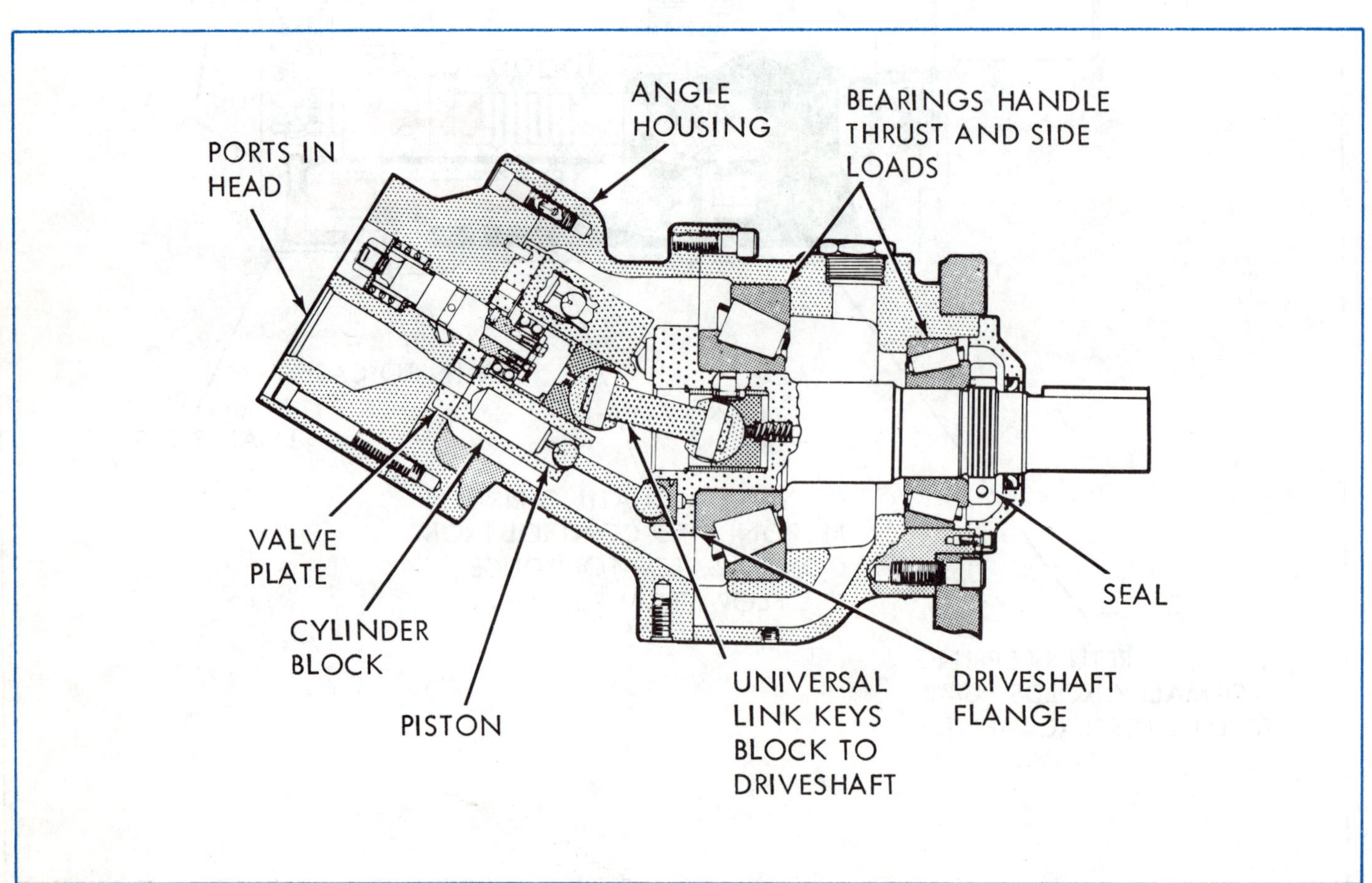

Fig. 11-34. Bent Axis Piston Pump

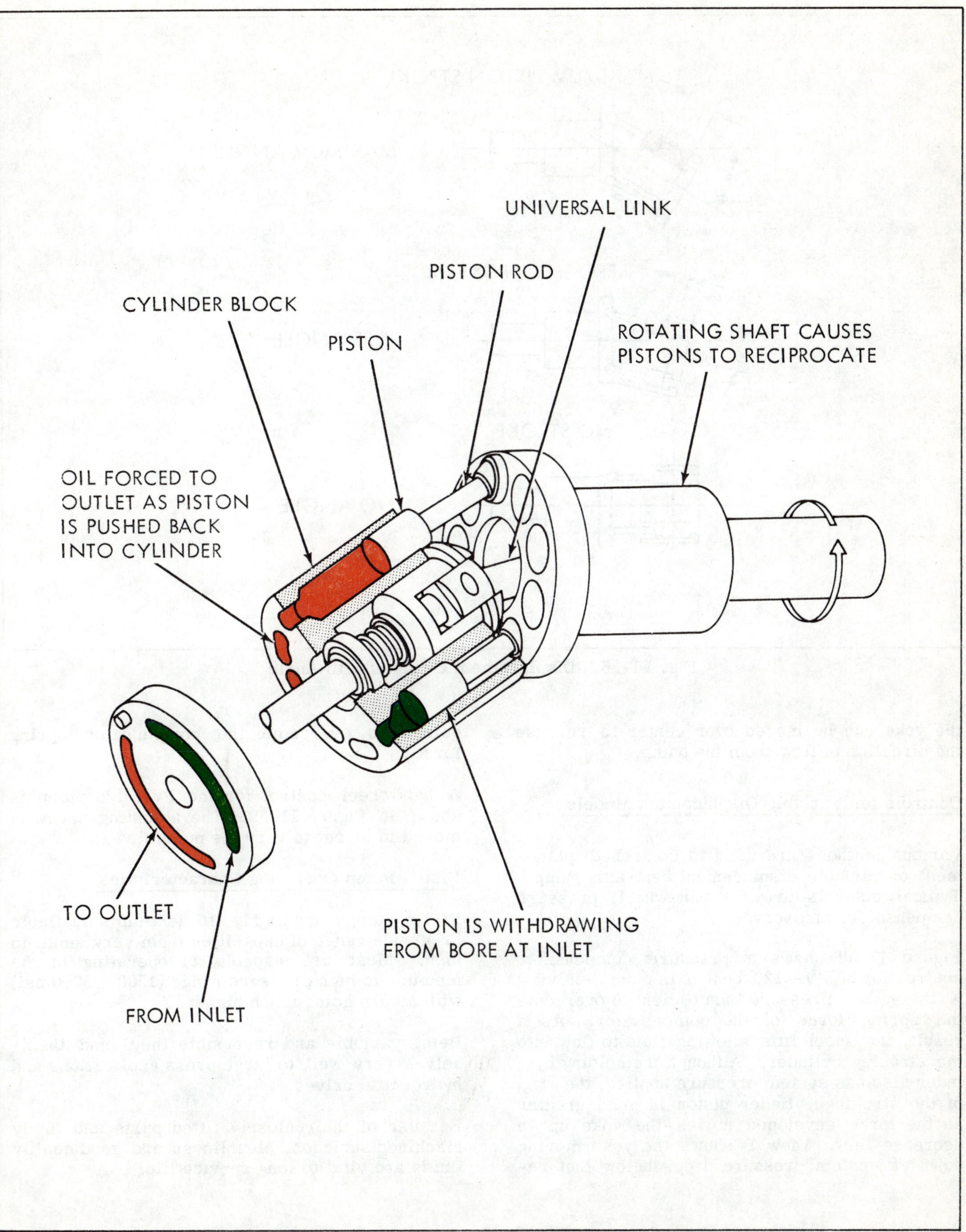

Fig. 11-35. Pumping Action in Bent-Axis Pump

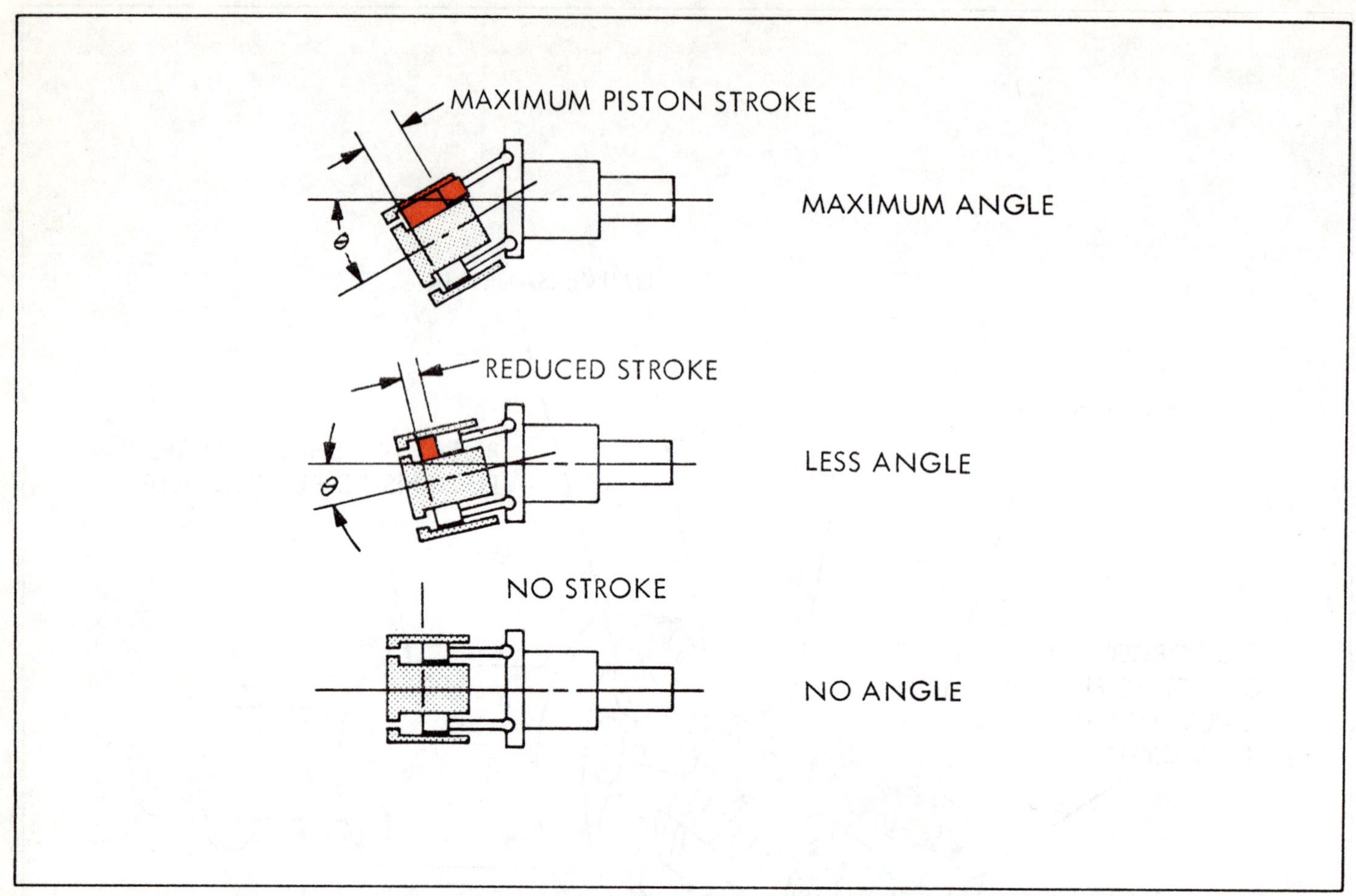

Fig. 11-36. Displacement Changes with Angle

the yoke can be moved over center to reverse the direction of flow from the pump.

Controls for Variable Displacement Models

Various methods are used to control displacement of variable displacement bent-axis pumps. Typical controls are the handwheel, pressure compensator, and servo.

Figure 11-38 shows a pressure compensator control for a PVA 120 bent-axis pump. In view A, the system pressure is sufficient to overcome the spring force of the compensator. As a result, the spool lifts allowing fluid to flow into the stroking cylinder. Although the holding cylinder also has system pressure applied, the area of the stroking cylinder piston is much greater so the force developed moves the yoke up to decrease flow. View B shows the yoke moving down as system pressure drops below that required to overcome the compensator spring force.

A handwheel control for the PVA 120 pump is shown in Figure 11-39. The adjusting screw is moved in or out to vary the pump flow.

Piston Pump Operating Characteristics

Piston pumps are highly efficient units available in a wide range of capacities from very small to high. Most are capable of operating in the medium to high pressure range (1500 - 3000 psi) with others going much higher.

Being variable and reversible they lend themselves very well to large press applications and hydrostatic drives.

Because of their closely fitted parts and finely machined surfaces, cleanliness and good quality fluids are vital to long service life.

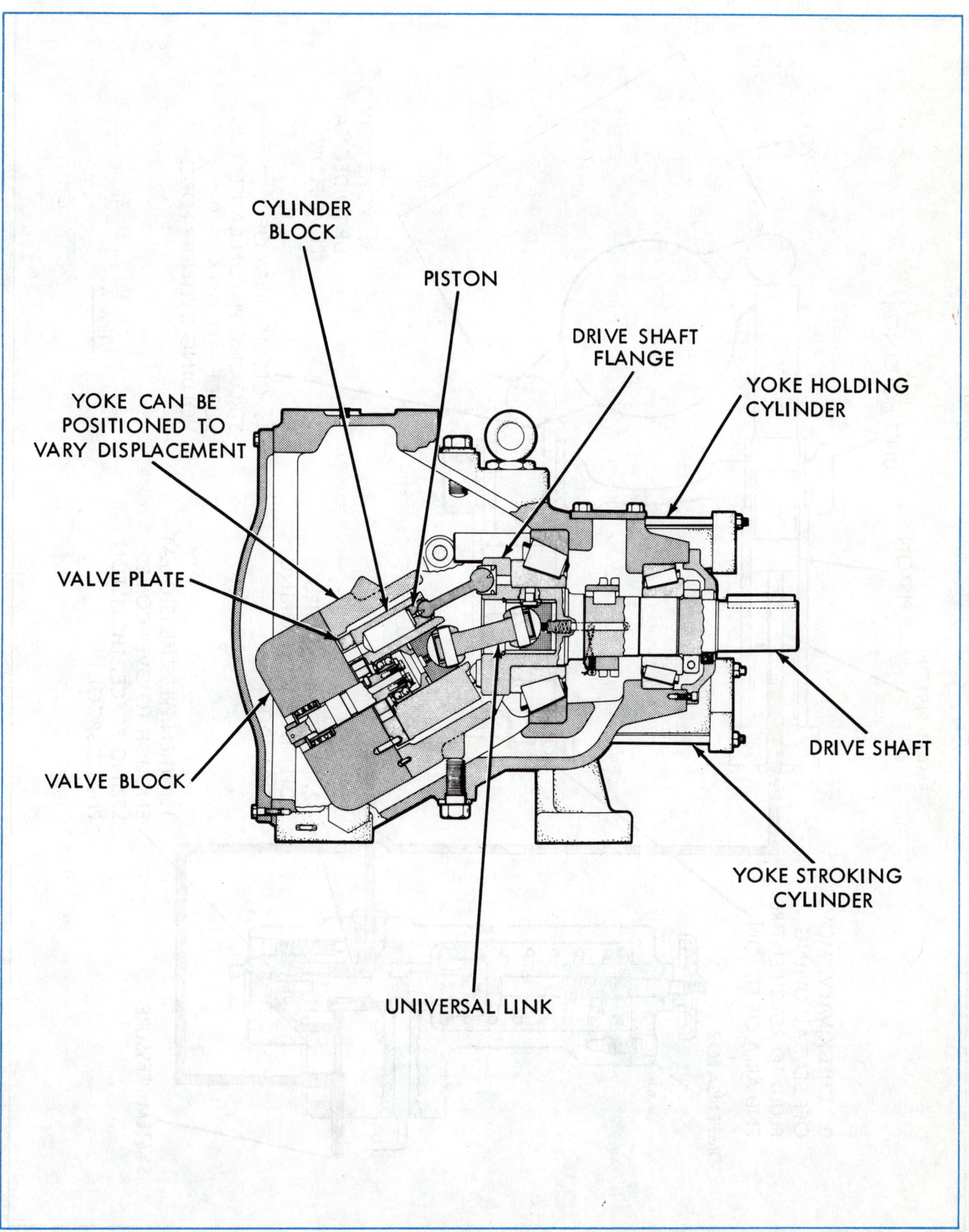

Fig. 11-37. Variable Displacement Bent-Axis Piston Pump

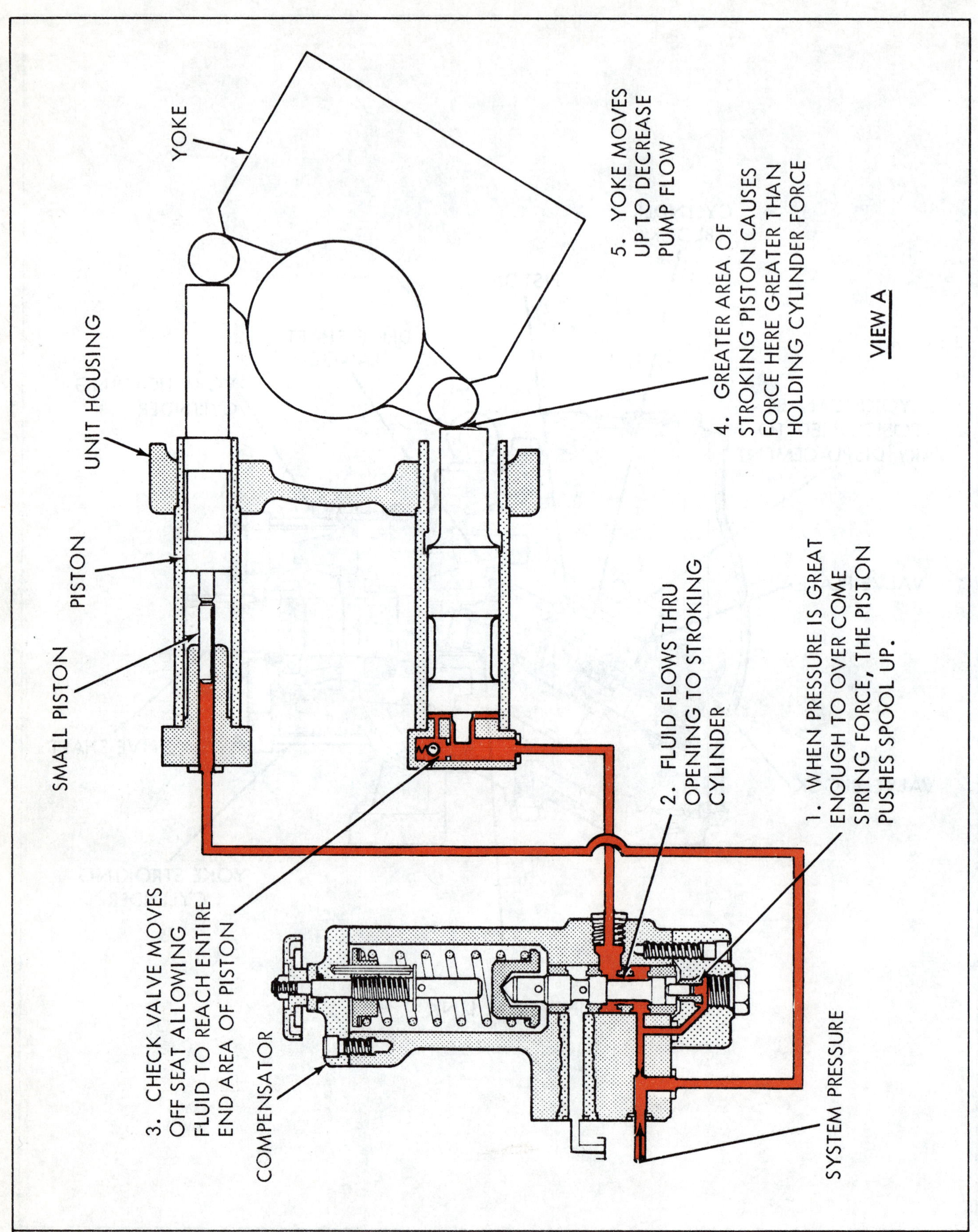

Fig. 11-38. PVA 120 Compensator Control

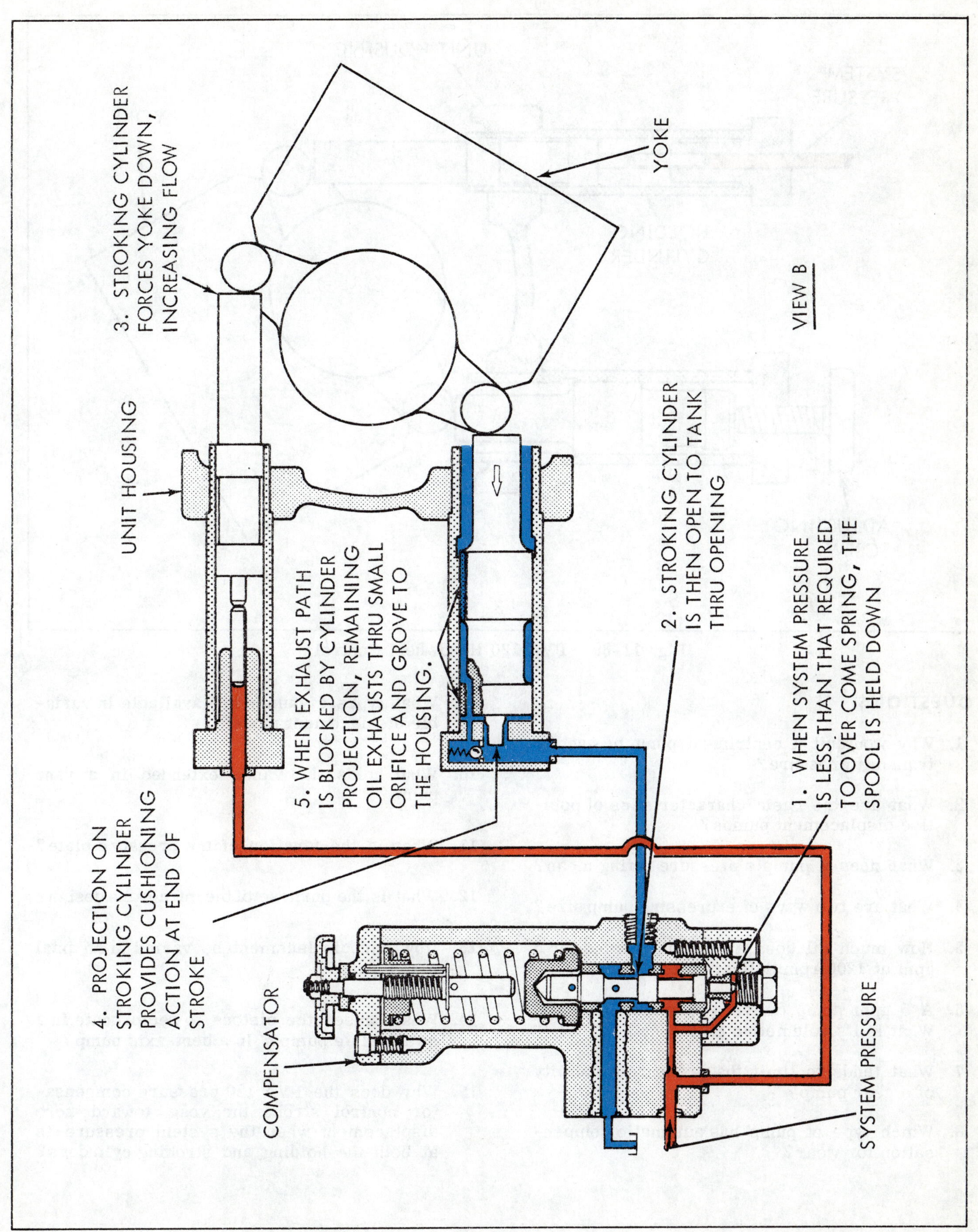

Fig. 11-38. PVA 120 Compensator Control

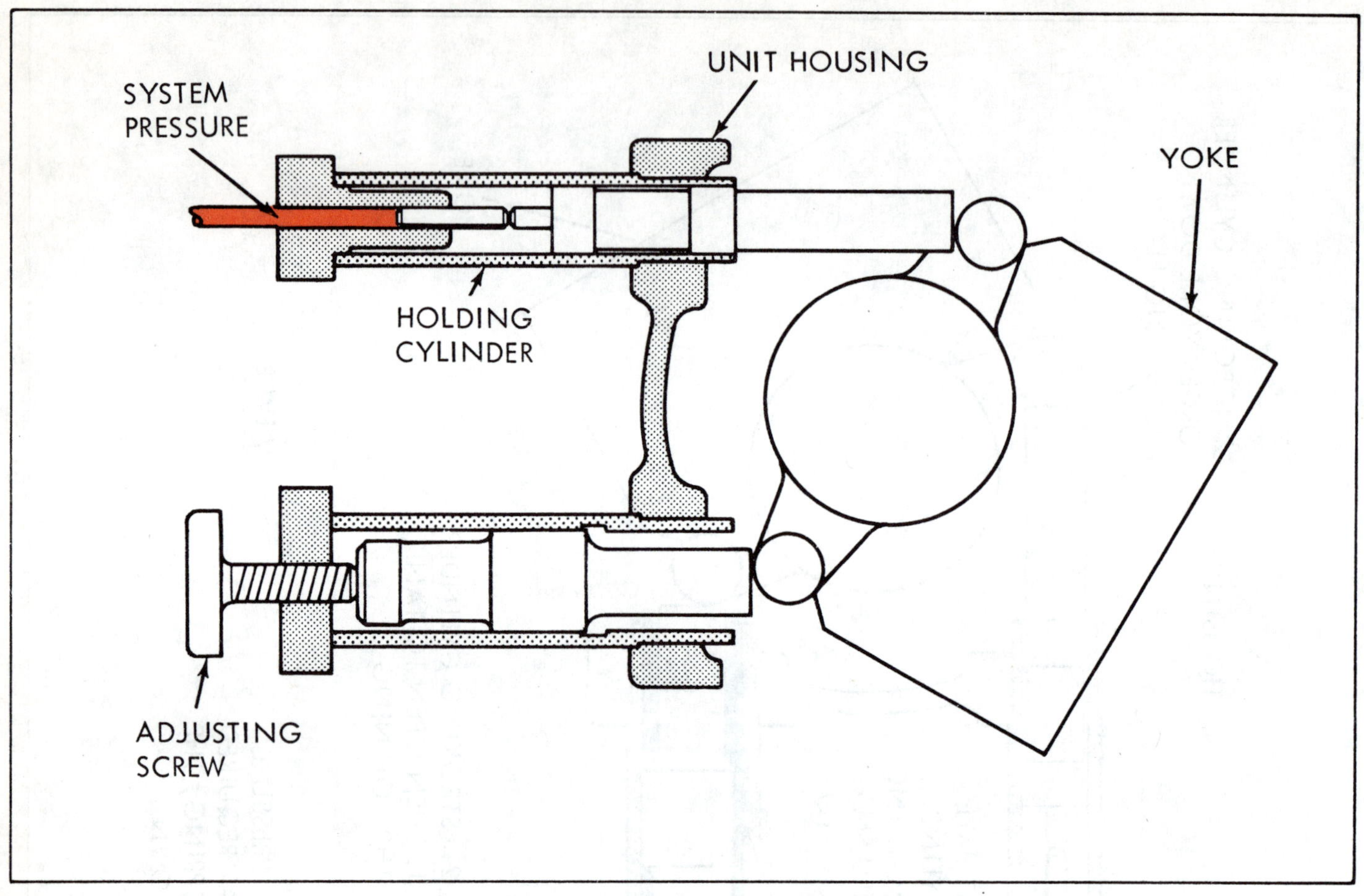

Fig. 11-39. PVA 120 Handwheel Control

QUESTIONS

1. Why wouldn't a centrifugal pump be used to transmit pressure?
2. What are the basic characteristics of positive displacement pumps?
3. What does a pump's pressure rating mean?
4. What are two ways of expressing pump size?
5. How much oil does a vane pump rated for 5 gpm at 1200 rpm deliver at 1800 rpm?
6. A 5 gpm pump delivers $3\frac{1}{2}$ gpm at 3000 psi. What is its volumetric efficiency?
7. What tends to limit the pressure capability of a gear pump?
8. Which type of pump has automatic compensation for wear?
9. What types of pumps are available in variable displacement models?
10. What holds the vanes extended in a vane pump?
11. What is the function of the pressure plate?
12. What is the purpose of the intra-vane design?
13. How can displacement be varied in an axial piston pump?
14. What causes the pistons to reciprocate in a swash plate pump? In a bent-axis pump?
15. Why does the PVA 120 pressure compensator control stroke the yoke toward zero displacement when the system pressure is at both the holding and stroking cylinders?

CHAPTER 12

ACCESSORIES

This chapter deals with various accessories used to perform special functions in hydraulic systems. The subjects covered are accumulators, intensifiers, pressure switches and instruments.

ACCUMULATORS

Unlike gases the fluids used in hydraulic systems cannot be compressed and stored for usage at a different time or place. Where it can be used to advantage an accumulator provides a means of storing these incompressible fluids under pressure. It does so because of the fact that as the hydraulic fluid under pressure enters the accumulator chamber it does one of three things: it compresses a spring, it compresses a gas, or it raises a weight. Any tendency for pressure to drop at the inlet causes the element to react and force the fluid back out.

WEIGHT-LOADED ACCUMULATOR

The earliest type of accumulator built was the weight-loaded design (Fig. 12-1). A vertical ram or piston has provision for adding or removing weights to vary the pressure. Pressure is always equal to the weight imposed divided by the piston or ram area exposed to the hydraulic fluid. This is the only type of accumulator where pressure is constant, whether the chamber is full or nearly empty. Weight-loaded accumulators, however, are heavy and bulky and their use is limited. They may be found on some

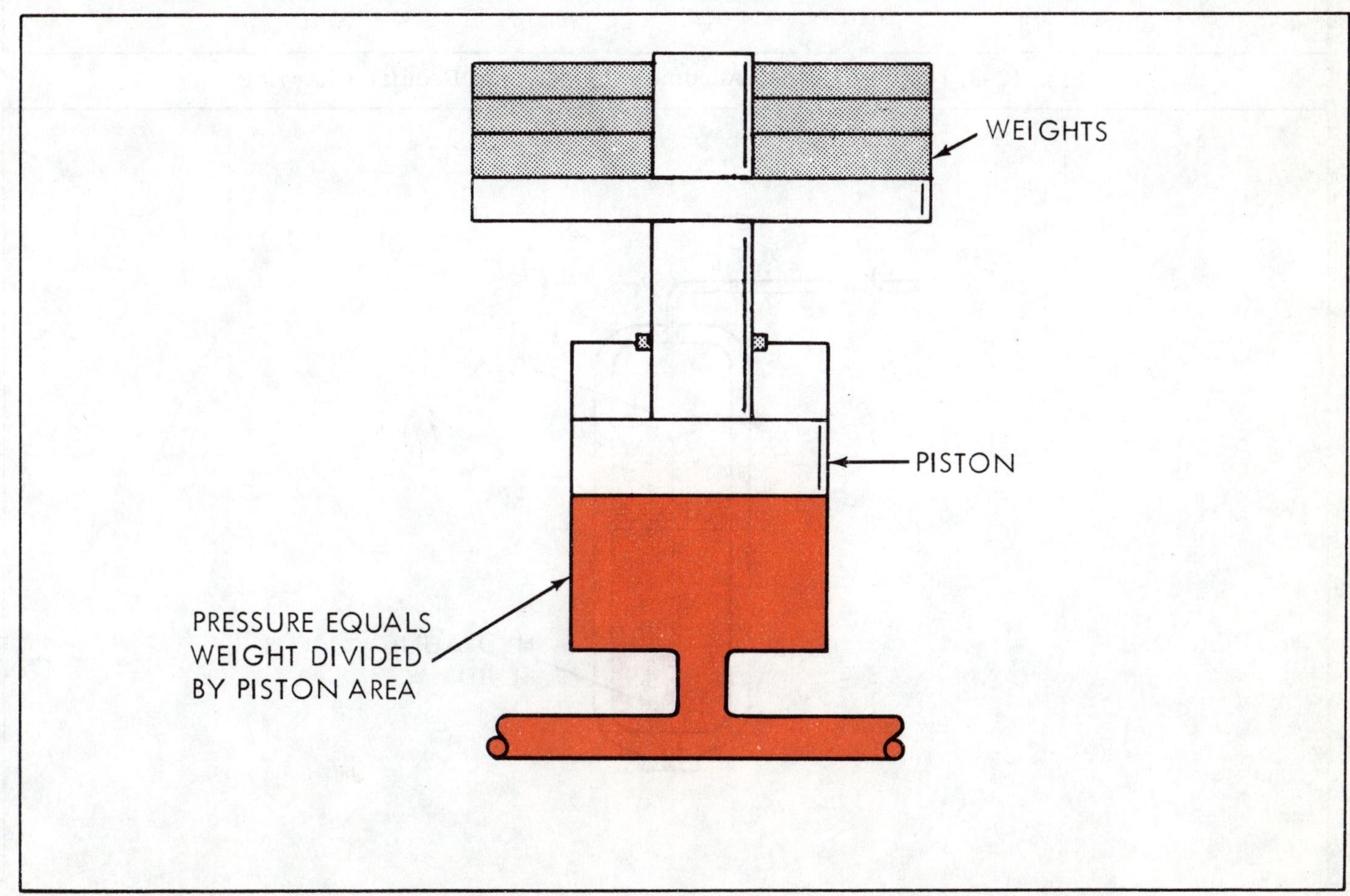

Fig. 12-1. Weighted Accumulator Produces Constant Pressure

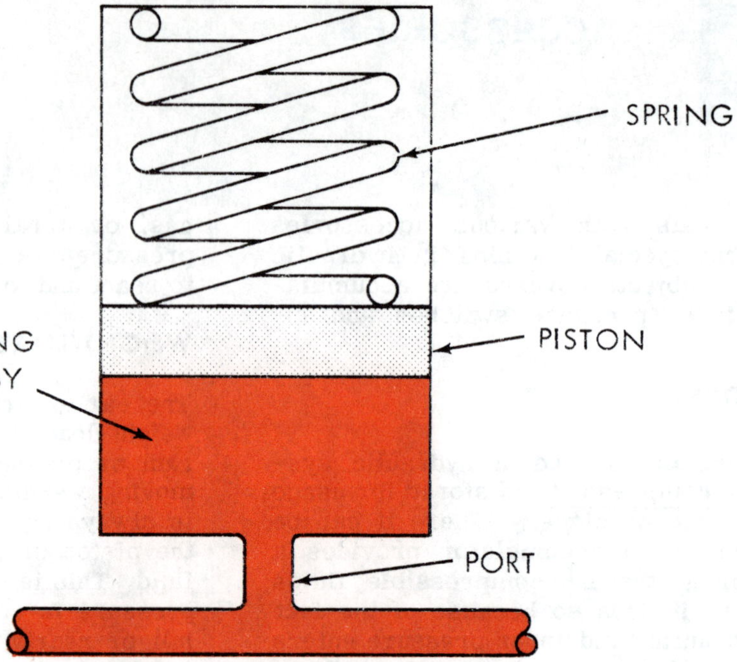

Fig. 12-2. Spring-Loaded Accumulator does not Require Charging

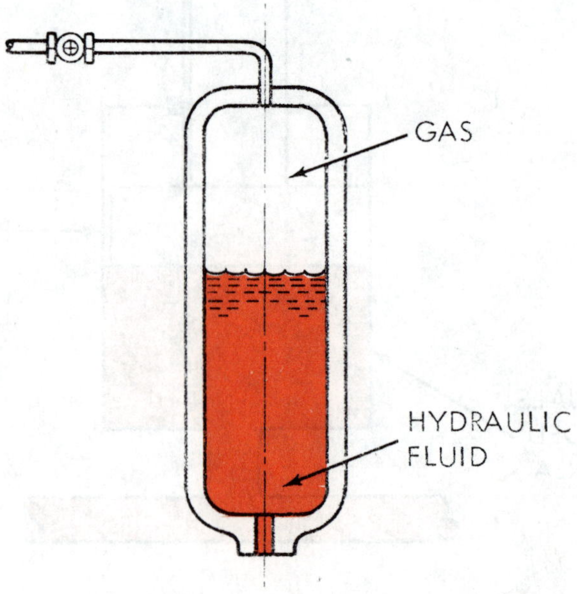

Fig. 12-3. Accumulator with no Separation Between Gas and Liquid

heavy presses where constant pressure is required or in applications where unusually large volumes are necessary.

SPRING-LOADED ACCUMULATORS

In a spring loaded accumulator (Fig. 12-2), pressure is applied to the fluid by compression of a coil spring behind the accumulator piston. The pressure is equal to the instantaneous spring force divided by the piston area.

$$\text{Pressure} = \frac{\text{Spring Force}}{\text{Area}}$$

Where: Spring Force = Spring Constant x (Compression Distance)

The pressure therefore is not constant since the spring force increases as fluid enters the chamber and decreases as it is discharged.

Spring loaded accumulators can be mounted in any position. The spring force, i. e., the pressure range is not easily adjusted, however, and where large quantities of fluid are required the forces involved would make spring sizes impractical.

GAS CHARGED ACCUMULATOR

Probably the most commonly used accumulator is one in which the chamber is pre-charged with an inert gas, usually dry nitrogen. Oxygen should never be used because of its tendency to burn or explode under compression with oil. While air is sometimes used, it is not recommended for the same reason.

A gas charged accumulator should be precharged while empty of hydraulic fluid. Precharge pressures vary with each application and depends upon the working pressure range and volume of fluid required within that range. It should never be less than 1/4 and preferably 1/3 of maximum working pressure. Accumulator pressure varies in proportion to the compression of the gas, increasing as fluid is pumped in and decreasing as it is expelled.

No-Separator Type

Figure 12-3 shows an accumulator with no separator between the hydraulic fluid and the charging gas. Frequently used on diecast machines it must be mounted vertically. It is important to select a pressure volume relationship such that not more than 2/3 of the oil is ever used in operation to avoid accidental discharge of gas into the system.

Diaphragm or Bladder Type

Many accumulators incorporate a synthetic rubber diaphragm or bladder (Fig. 12-4), to contain the gas precharge and separate it from the hydraulic fluid. Since certain fire resistant fluids may not be compatible with conventional diaphragm or bladder materials it is important that proper selection be made.

Available oil can vary between 1/4 and 3/4 of total capacity depending upon operating conditions. Operation outside these limits can cause the separator to stretch or wrinkle and shorten its life.

Piston Type Accumulator

Another method of separating the gas charge from the hydraulic fluid is by means of a free piston (Fig. 12-5). Similar in construction to a hydraulic cylinder the piston under pressure of the gas on one side constantly tries to force the oil out of the opposite side of the chamber. Here too pressure is a function of the compression and varies with the volume of oil in the chamber.

APPLICATIONS

In many hydraulic systems, a large volume of fluid is required to do the work, but the work is done only intermittently in the machine cycle. For example, in die casting, the "shot" cylinder must be moved very rapidly when the piece is being formed, but is idle while the piece is being removed and during the mold closing and opening phases. Rather than use a very high volume pump intermittently, such a system stores fluid from a relatively small volume pump in an accumulator and discharges it during the "shot" portion of the cycle.

Another application is where it is necessary to maintain pressure for extended periods of time. Instead of allowing the pump to run constantly at the relief valve setting, it is used to charge an accumulator. The pump can then be unloaded freely to tank while the accumulator maintains pressure. Pressure switches or unloading valves are used to recycle the pump periodically to replace fluid lost through leakage or valve actuation.

Accumulators may also be installed in a system to absorb shock or pressure surges due to the sudden stopping or reversing of oil flow. In such cases the pre-charge is close to or slightly above maximum operating pressure allowing it to "pick off" pressure peaks without constant or extended flexing of the diaphragm or bladder.

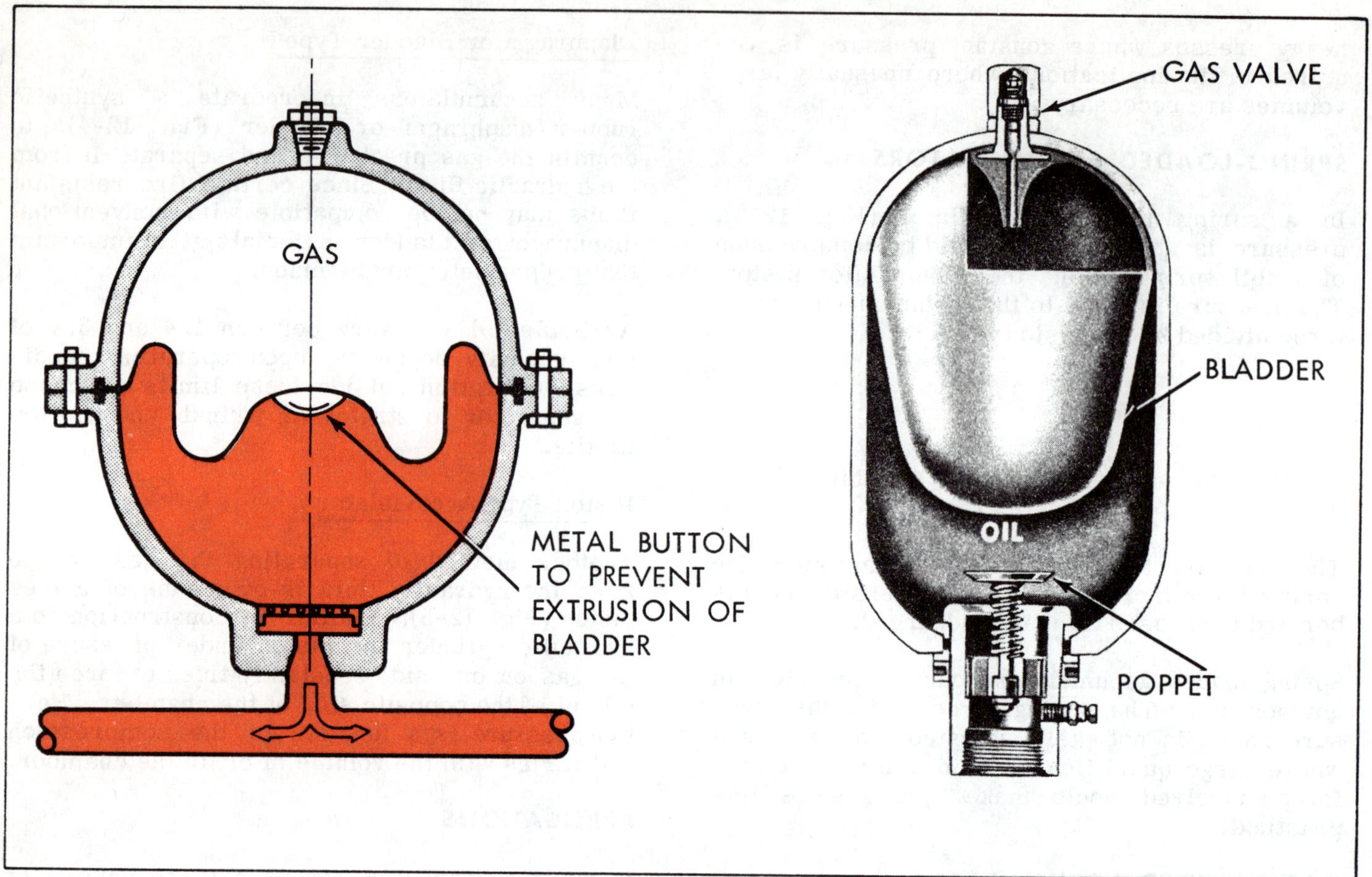

Fig. 12-4. Diaphragm Accumulator Uses Rubber Separator Between Gas and Liquid

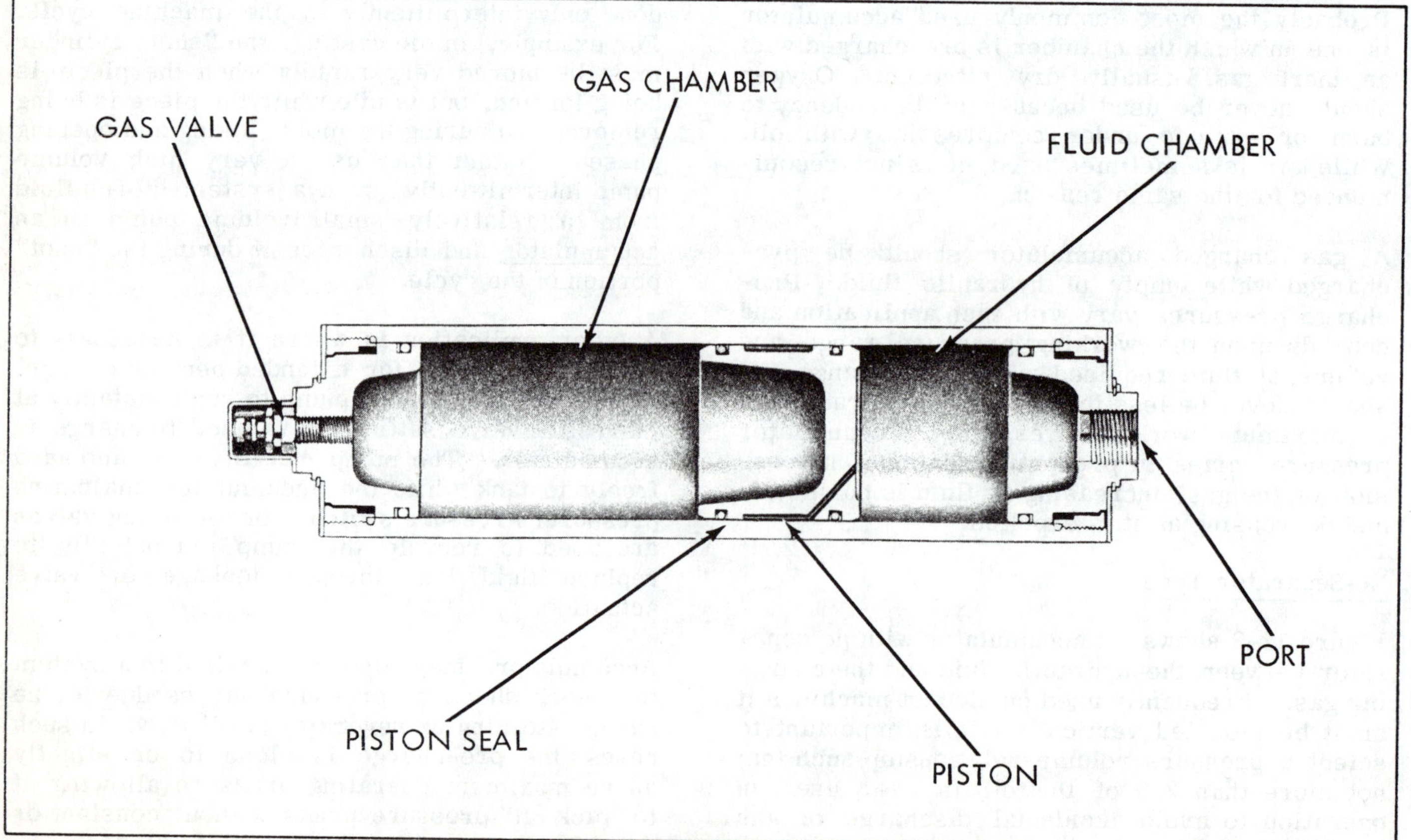

Fig. 12-5. Piston Accumulator is Gas-Charged

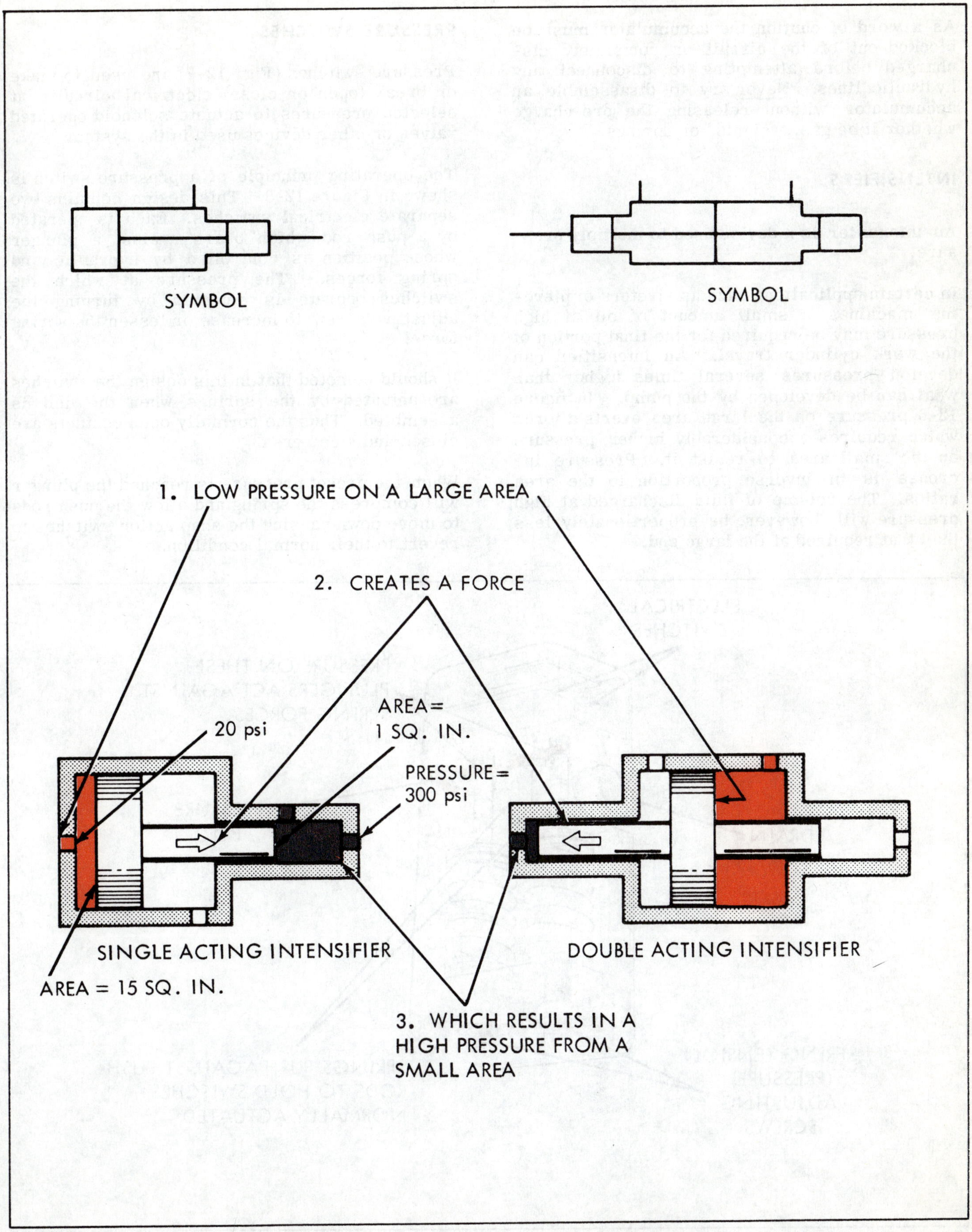

SYMBOL

SYMBOL

1. LOW PRESSURE ON A LARGE AREA

2. CREATES A FORCE

AREA = 1 SQ. IN.

20 psi

PRESSURE = 300 psi

SINGLE ACTING INTENSIFIER

DOUBLE ACTING INTENSIFIER

AREA = 15 SQ. IN.

3. WHICH RESULTS IN A HIGH PRESSURE FROM A SMALL AREA

Fig. 12-6. Intensifier "Boosts" Pressure

As a word of caution the accumulator must be blocked out of the circuit or completely discharged before attempting to disconnect any hydraulic lines. Never try to disassemble an accumulator without releasing the pre-charge whether it be gas, weights, or springs.

INTENSIFIERS

An intensifier is a device used to multiply pressure.

In certain applications such as riveters or piercing machines a small amount of oil at high pressure may be required for the final portion of the work cylinder travel. An intensifier can develop pressures several times higher than what can be developed by the pump. In figure 12-6 pressure on the large area exerts a force which requires a considerably higher pressure on the small area to resist it. Pressure increase is in inverse proportion to the area ratios. The volume of fluid discharged at high pressure will, however, be proportionately less than that required at the large end.

PRESSURE SWITCHES

Pressure switches (Fig. 12-7) are used to make or break (open or close) electrical circuits at selected pressures to actuate solenoid operated valves or other devices used in the system.

The operating principle of a pressure switch is shown in Figure 12-8. This design contains two separate electrical switches. Each is operated by a push rod which bears against a plunger whose position is controlled by hydraulic and spring forces. The pressure at which the switches operate is selected by turning the adjusting screw to increase or lessen the spring force.

It should be noted that in this design the switches are actuated by the springs when the unit is assembled. Thus the normally open contacts are closed and vice versa.

When the preset pressure is reached the plunger will compress the spring and allow the push rods to move down causing the snap action switches to revert to their normal condition.

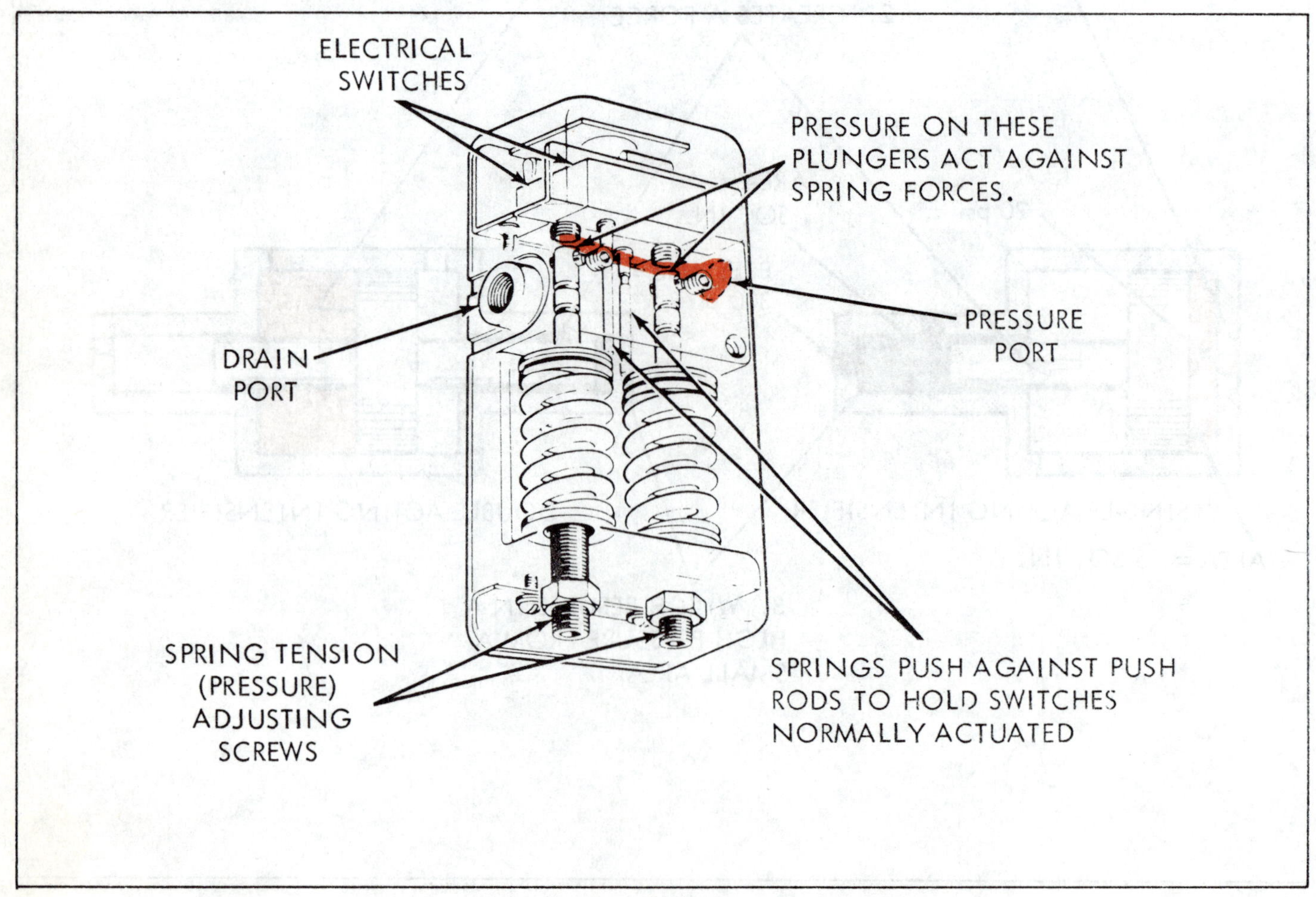

Fig. 12-7. Typical Pressure Switch

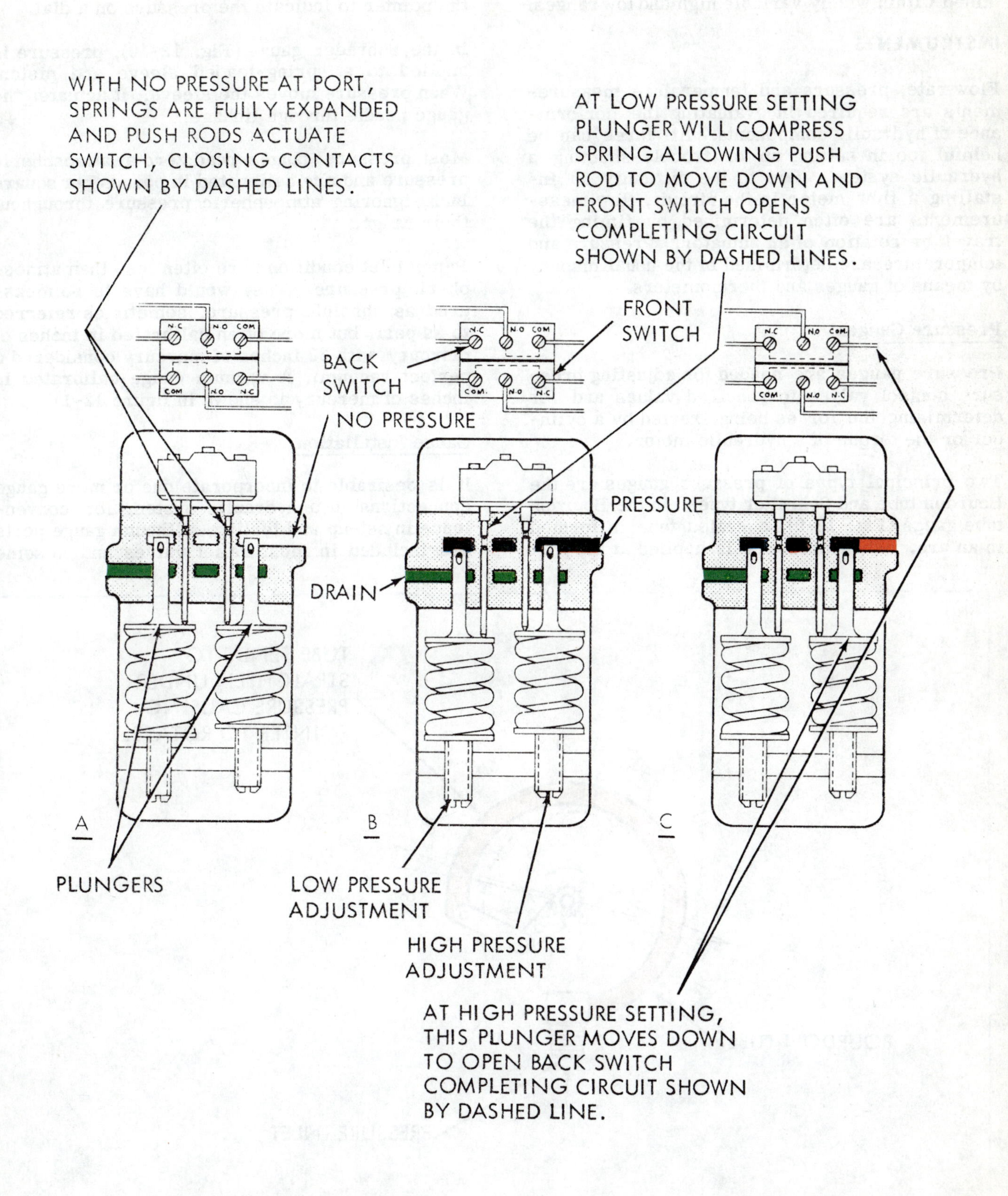

Fig. 12-8. Pressure Switch Operation

By using both switches in conjunction with an electrical relay, system pressures may be maintained within widely variable high and low ranges.

INSTRUMENTS

Flow rate, pressure and temperature measurements are required in evaluating the performance of hydraulic components. All three can be helpful too in setting up or trouble-shooting a hydraulic system. Due to the difficulty in installing a flow meter in the circuit, flow measurements are often determined by timing the travel or rotation of an actuator. Pressure and temperature are determined in the usual manner by means of gauges and thermometers.

Pressure Gauges

Pressure gauges are needed for adjusting pressure control valves to required values and for determining the forces being exerted by a cylinder or the torque of a hydraulic motor.

Two principal types of pressure gauges are the Bourdon tube and Schrader types. In the Bourdon tube gauge (Fig. 12-9), a sealed tube is formed in an arc. When pressure is applied at the port opening, the tube tends to straighten. This actuates linkage to the pointer gear and moves the pointer to indicate the pressure on a dial.

In the Schrader gauge (Fig. 12-10), pressure is applied to a spring-loaded sleeve and piston. When pressure moves the sleeve, it actuates the gauge needle through linkage.

Most pressure gauges read zero at atmospheric pressure and are calibrated in pounds per square inch, ignoring atmospheric pressure throughout their range.

Pump inlet conditions are often less than atmospheric pressure. They would have to be measured as absolute pressure, sometimes referred to as psia, but more often calibrated in inches of mercury with 30 inches of mercury considered a perfect vacuum. A vacuum gauge calibrated in inches of mercury is shown in figure 12-11.

Gauge Installation

It is desirable to incorporate one or more gauge connections in a hydraulic system for convenience in set-up and testing, although gauge ports are included in most relief valves and in some

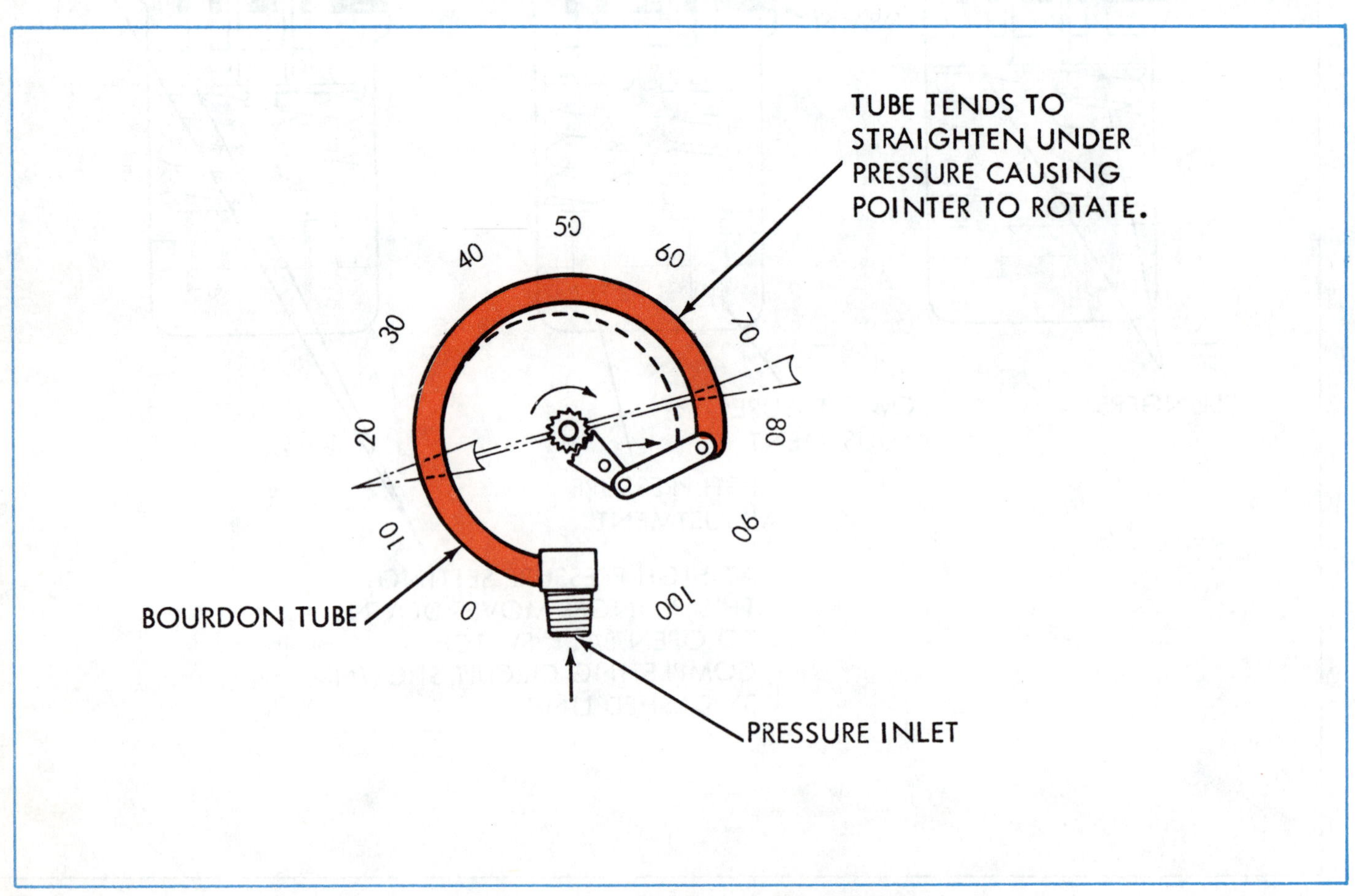

Fig. 12-9. Bourdon Tube Gauge

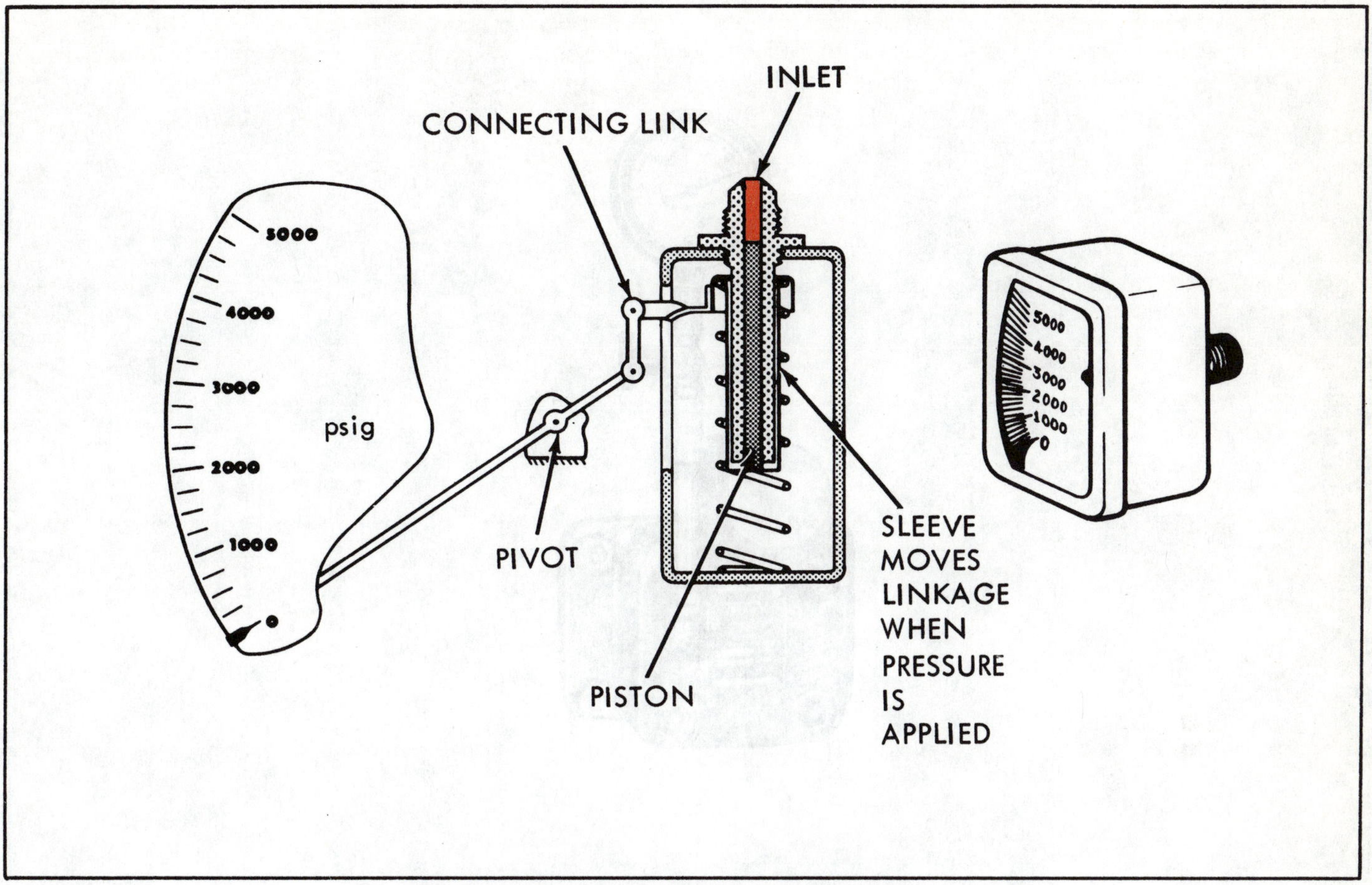

INLET
CONNECTING LINK
psig
PIVOT
PISTON
SLEEVE MOVES LINKAGE WHEN PRESSURE IS APPLIED

Fig. 12-10. - Schrader Gauge Operation

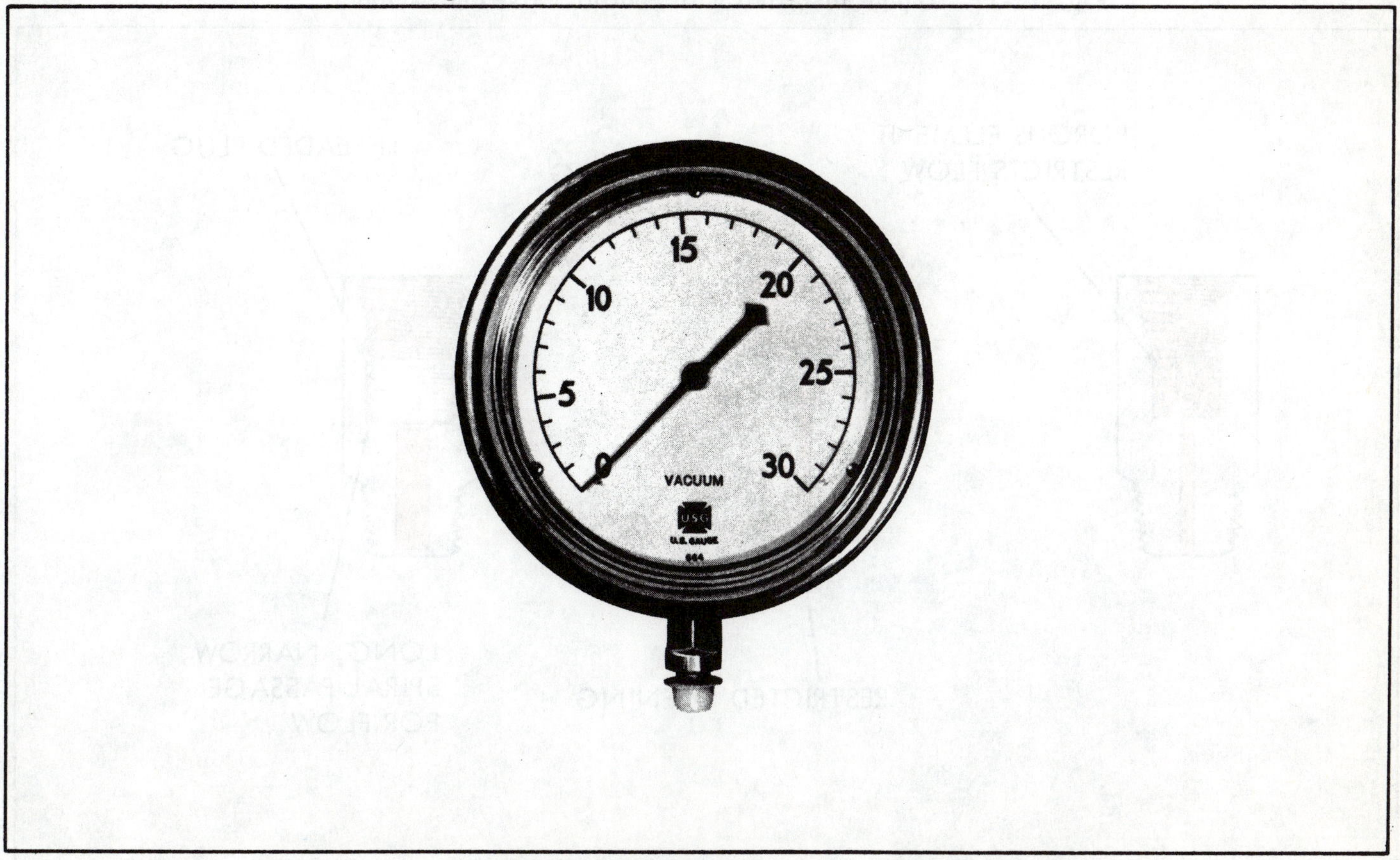

Fig. 12-11. Vacuum Gauge Calibrated in Inches of Mercury

Fig. 12-12. Gauge Installed with Shutoff Valve and Snubber

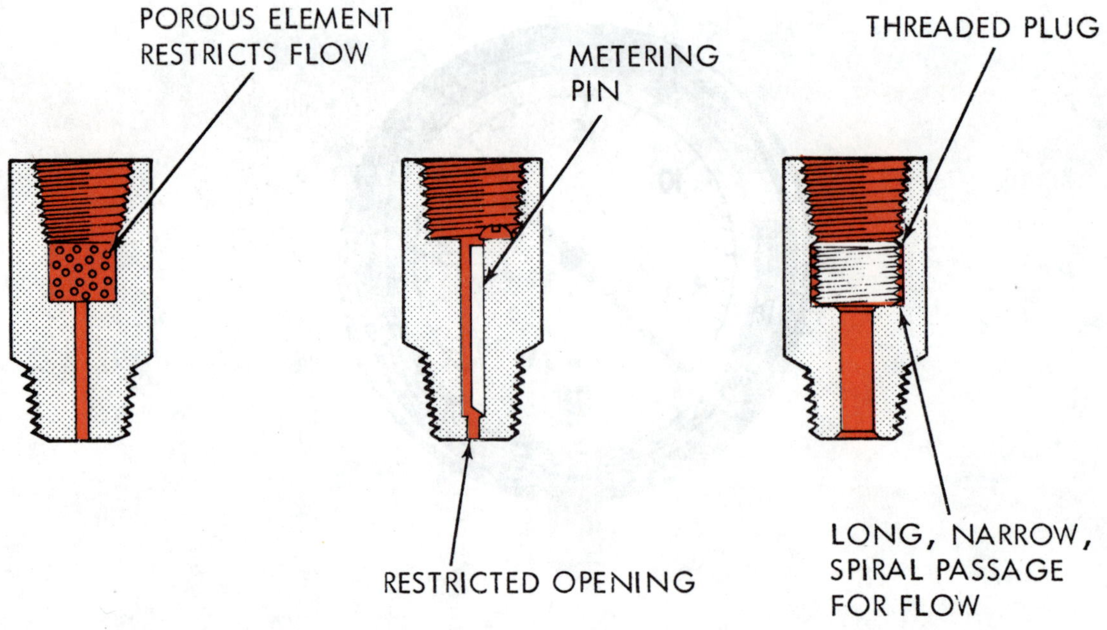

POROUS ELEMENT RESTRICTS FLOW

METERING PIN

THREADED PLUG

RESTRICTED OPENING

LONG, NARROW, SPIRAL PASSAGE FOR FLOW

Fig. 12-13. Gauge Snubbers Guard Against Pressure Surges

other hydraulic components. When a gauge is installed permanently on a machine, a shutoff valve and snubber (Fig. 12-12) are usually installed along with it. The shutoff valve prolongs gauge life by isolating it from the system except when it is desired to make a reading. The snubber (Fig. 12-13) prevents the gauge from oscillating and protects it from pressure surges. A small coil (approx. 2" diam.) of 1/8 inch tubing makes an excellent gauge damping device when commercially made units are not available.

Flow Meters

Flow meters are usually found on test stands, but portable units are available. Some include the flow meter, a pressure gauge and thermometer in a single unit (Fig. 12-14). They are seldom, if ever, connected permanently on a machine. However, coupled into the hydraulic piping they are useful in checking the volumetric efficiency of a pump and determining leakage paths within the circuit.

A typical flow meter (Fig. 12-15) consists of a weight in a calibrated vertical tube. Oil is pumped into the bottom and out the top and raises the weight to a height proportional to the flow. For more accurate measurement, a fluid motor of known displacement can be used to drive a tachometer. The gpm flow is:

$$\text{gpm} = \frac{\text{rpm x displacement (cu in./rev)}}{231}$$

Of course, the tachometer can be calibrated directly in gpm as well as rpm.

Another type of flow meter (Fig. 12-16) incorporates what is called a disk piston which when driven by the fluid passing through the measuring chamber develops a rotary motion which can be transmitted through gearing to indicator hands on a dial.

More sophisticated measuring devices are turbine type flow meters which generate an electrical impulse as they rotate and pressure sensing transducers which may be located at strategic points within the system where they send out electrical signals proportional to the pressures encountered. These signals can be calibrated and observed on an oscilloscope or other readout devices. See figure 12-17.

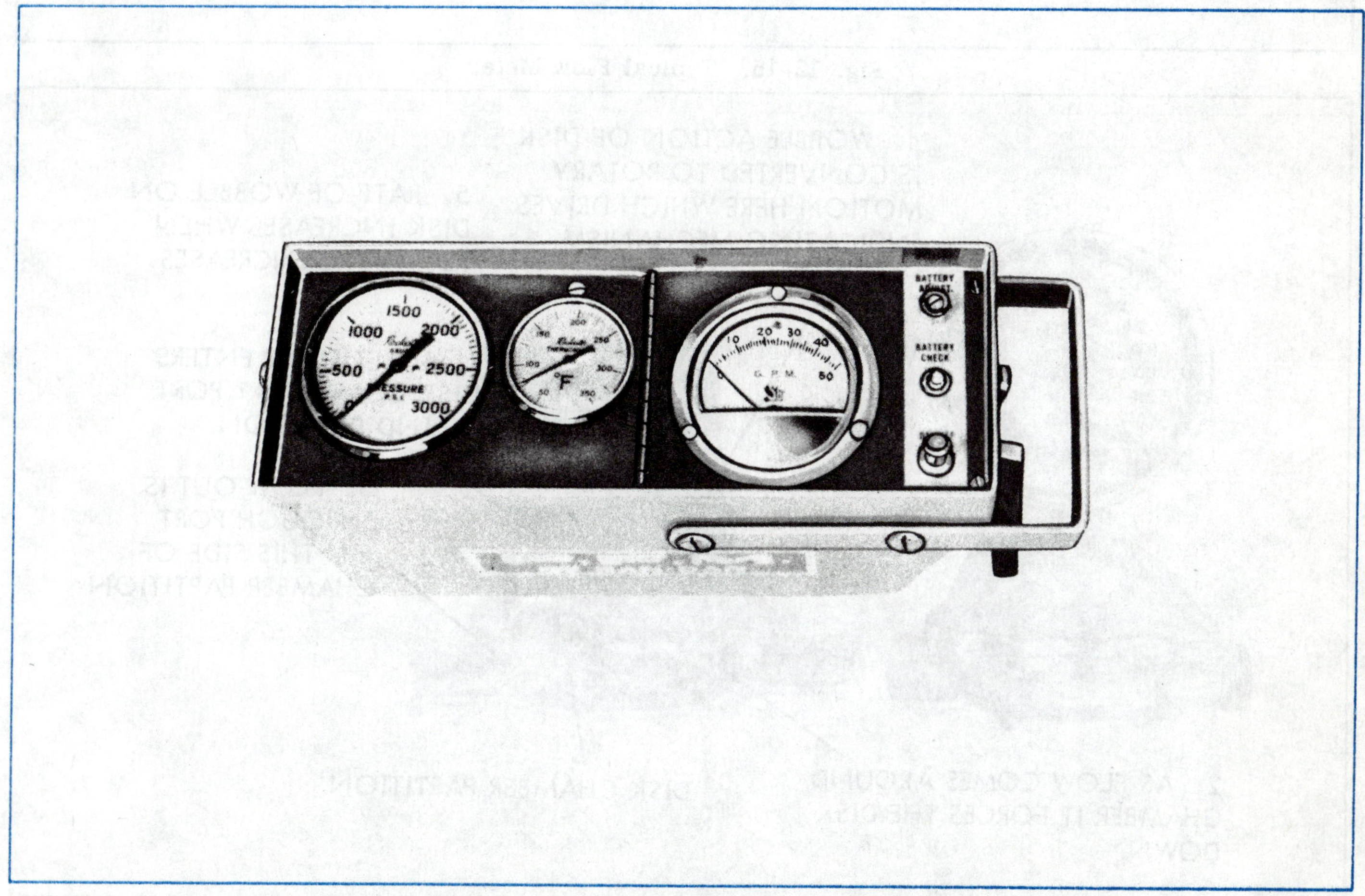

Fig. 12-14. Flow Meter with Pressure Gauge and Thermometer in One Unit

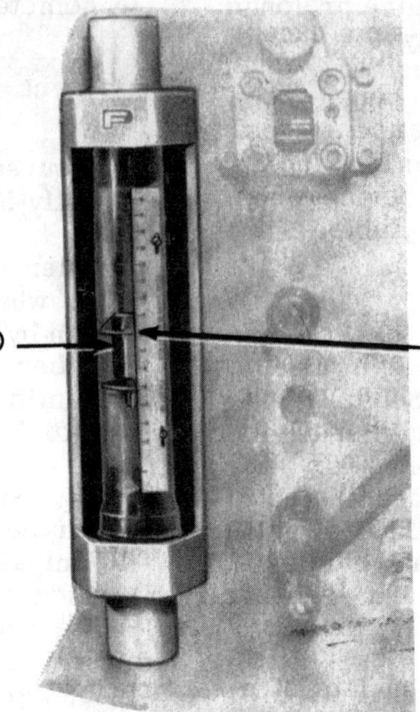

Fig. 12-15. Typical Flow Meter

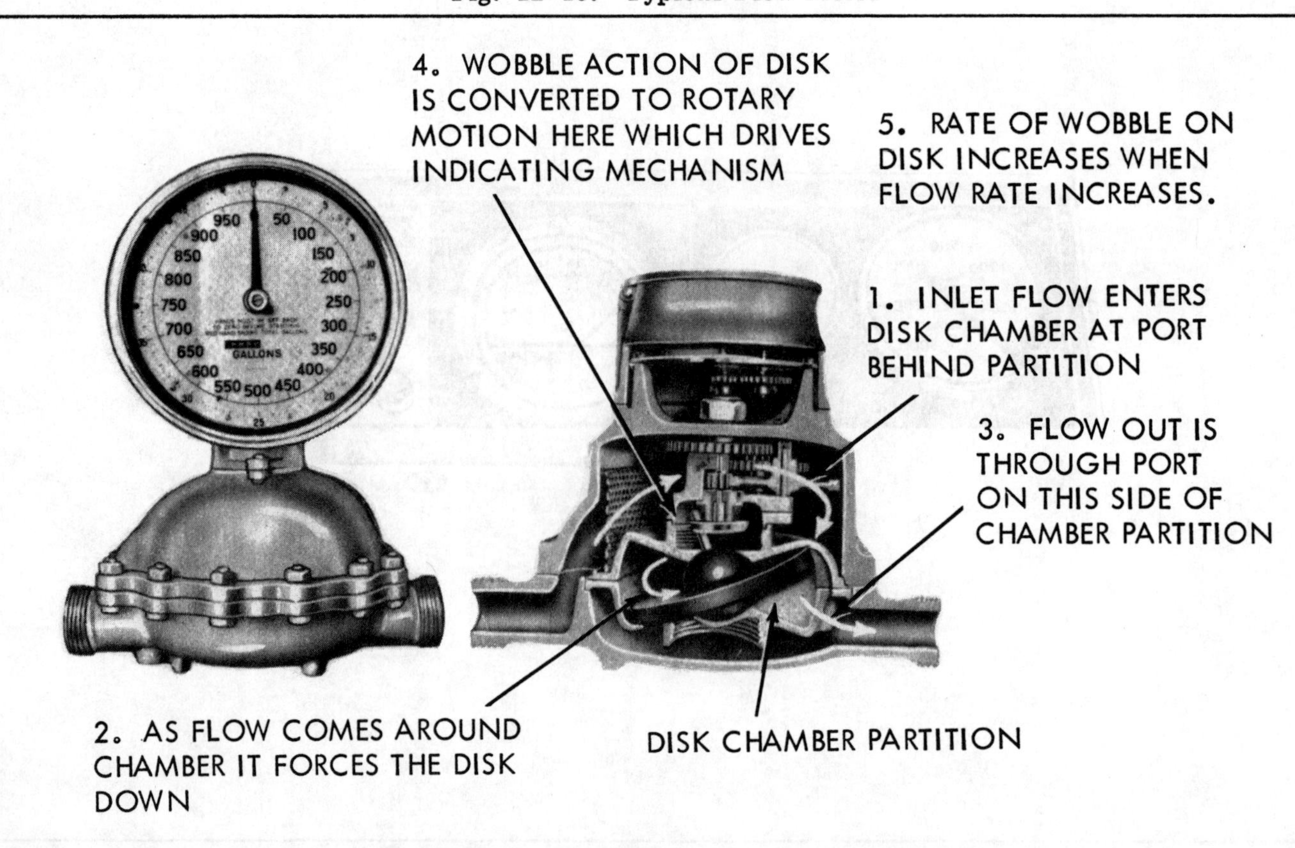

Fig. 12-16. Flow Meter with Disk Piston

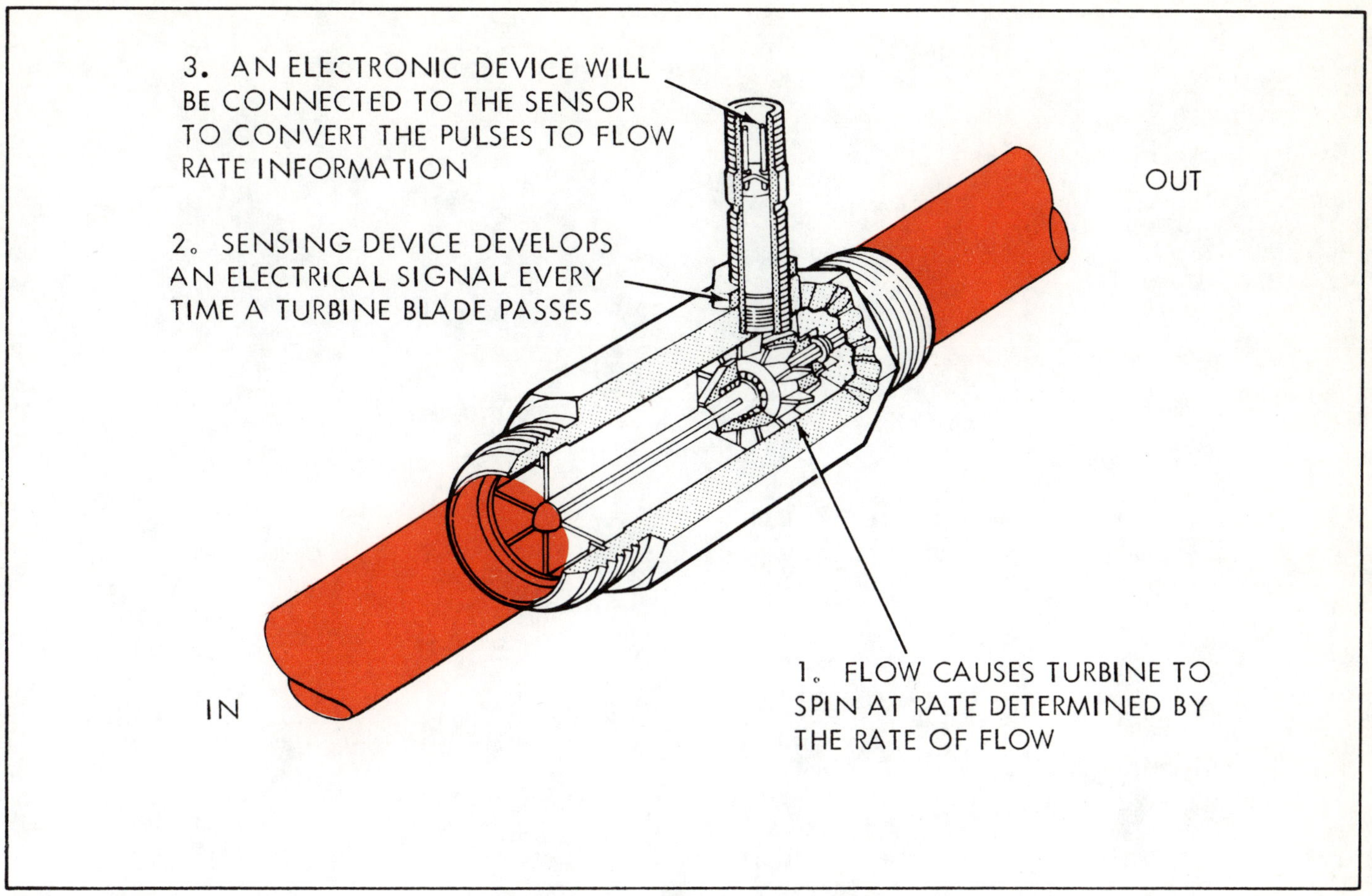

Fig. 12-17. Turbine Flow Meter

Such units are found more often in the laboratory, although they, too, are becoming a part of the equipment used by fluid power technicians in setting up and maintaining equipment.

QUESTIONS

1. Name two functions of an accumulator.
2. Which type of accumulator operates at a constant pressure? How can the pressure be changed?
3. How is pressure created in the free-piston accumulator?
4. What type of gas is preferred for gas-charged accumulators?
5. What prevents the bladder from extruding in a bladder type accumulator?
6. What is the purpose of an intensifier?
7. How is a pressure switch operated?
8. Give three situations where a pressure gauge might be required.
9. How are vacuum gauges calibrated?

CHAPTER 13

INDUSTRIAL HYDRAULIC CIRCUITS

The applications of the principles and components described in this manual are innumerable; as are the possible combinations of components into systems. It would be impossible to illustrate here more than a sampling of hydraulic circuits.

Those described in this chapter are typical of systems used in industrial machinery, and illustrate the basic principles of applying hydraulics to various kinds of work.

Many of the circuits are presented in cutaway or pictorial diagrams for ease in following oil flow. Graphical diagrams are shown for all circuits presented to aid in understanding the use of symbols.

UNLOADING CIRCUITS

An "unloading" circuit is a system where a pump's outlet is diverted to tank at low pressure during part of the cycle. The pump may be unloaded because load conditions at times would exceed the available input power, or simply to avoid wasting power and generating heat during idle periods.

Two-Pump Unloading System

It is often desirable to combine the delivery of two pumps for more speed while a cylinder is advancing at low pressure. When the high speed is no longer required or the pressure rises to the point where the combined volume would exceed the input horsepower, the larger of the two pumps is unloaded.

Low Pressure Operation

Figure 13-1, view A, shows the arrangement of components in such a system and the flow condition at low pressure. Oil from the large volume pump passes through the unloading valve and over the check valve to combine with the low volume pump output. This condition continues so long as system pressure is lower than the setting of the unloading valve.

High Pressure Operation

In view B, system pressure exceeds the setting of the unloading valve which opens permitting the large volume pump to discharge to the tank at little or no pressure. The check valve closes preventing flow from the pressure line through the unloading valve.

In this condition, much less power is used than if both pumps had to be driven at high pressure. However, the final advance is slower because of the smaller volume output to the system.

When motion stops, the small volume pump discharges over the relief valve at its setting.

Two Maximum Pressures Plus Venting

The network shown in Figure 13-2 can be incorporated in a hydraulic system to allow selection of two maximum pressures, plus venting. The highest maximum pressure will be set at the pilot stage of the main relief valve. A lesser pressure can be set by the remote control relief valve. The solenoid-operated four-way valve switches between the controls.

Venting

In view A, both solenoids of the directional valve are de-energized. The open-center spool is centered by the valve springs, and the vent port on the relief valve is opened to tank. The balanced piston opens pump flow to tank at the pressure equivalent of the light spring--about 20 psi.

Intermediate Maximum Pressure

In view B, the left-hand solenoid of the directional valve is energized. The valve spool is shifted to connect the relief valve vent port to the remote control valve. This valve now operates as the pilot stage for the balanced piston. Pump flow is diverted to tank when the remote valve setting is reached.

High Maximum Pressure

In view C, the opposite solenoid of the directional

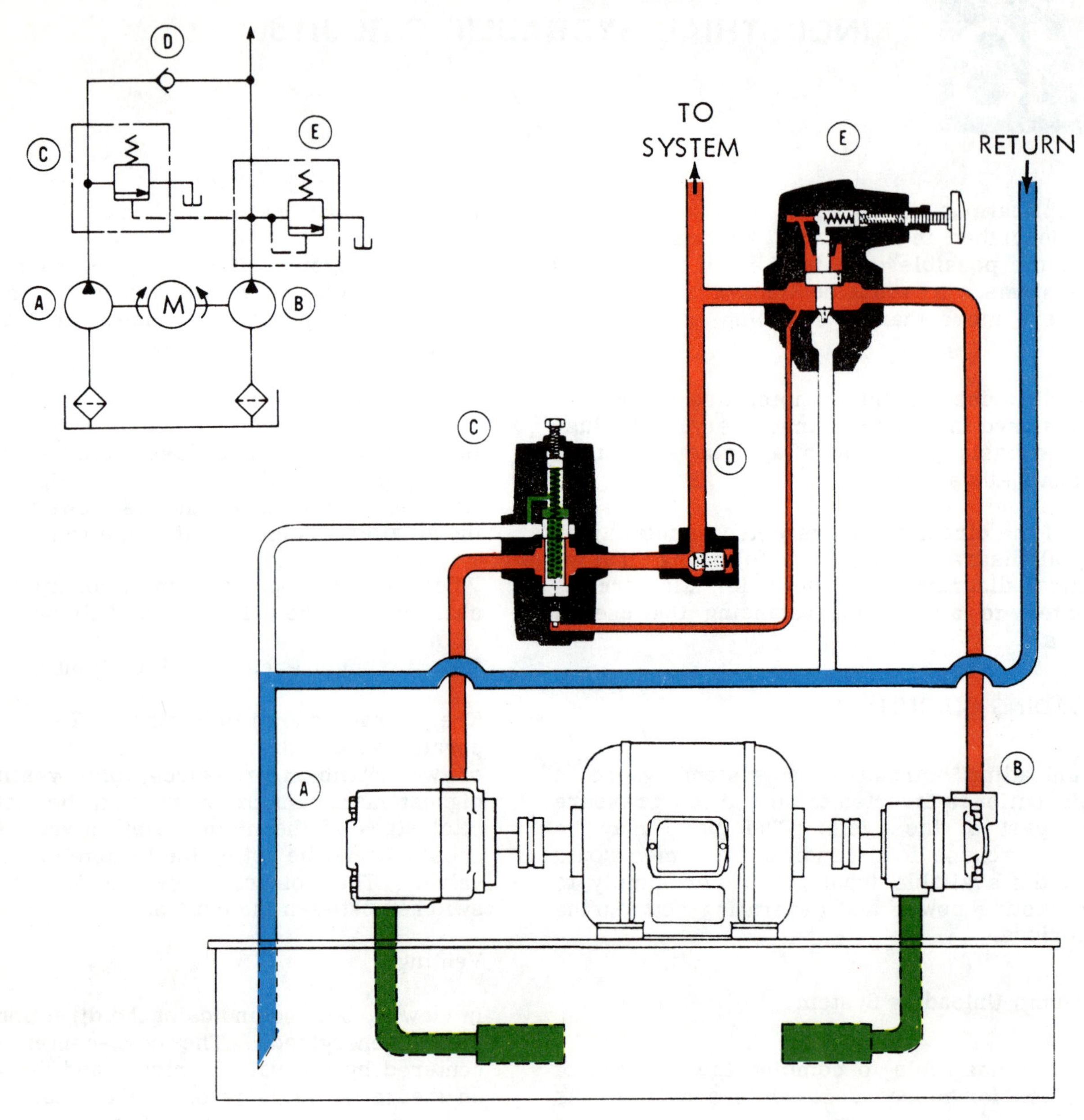

LOW PRESSURE OPERATION

System pressure is less than the adjusted settings of pressure control valves (C) and (E). Therefore, both (C) and (E) are in their normally closed positions. Delivery of pump (B) is directed into the system through (E). Delivery of pump (A) is directed through (C) and check valve (D) and combines with delivery of (B) to also be directed into the system.

Fig. 13-1, View A. - Unloading Circuit - Low Pressure Operation

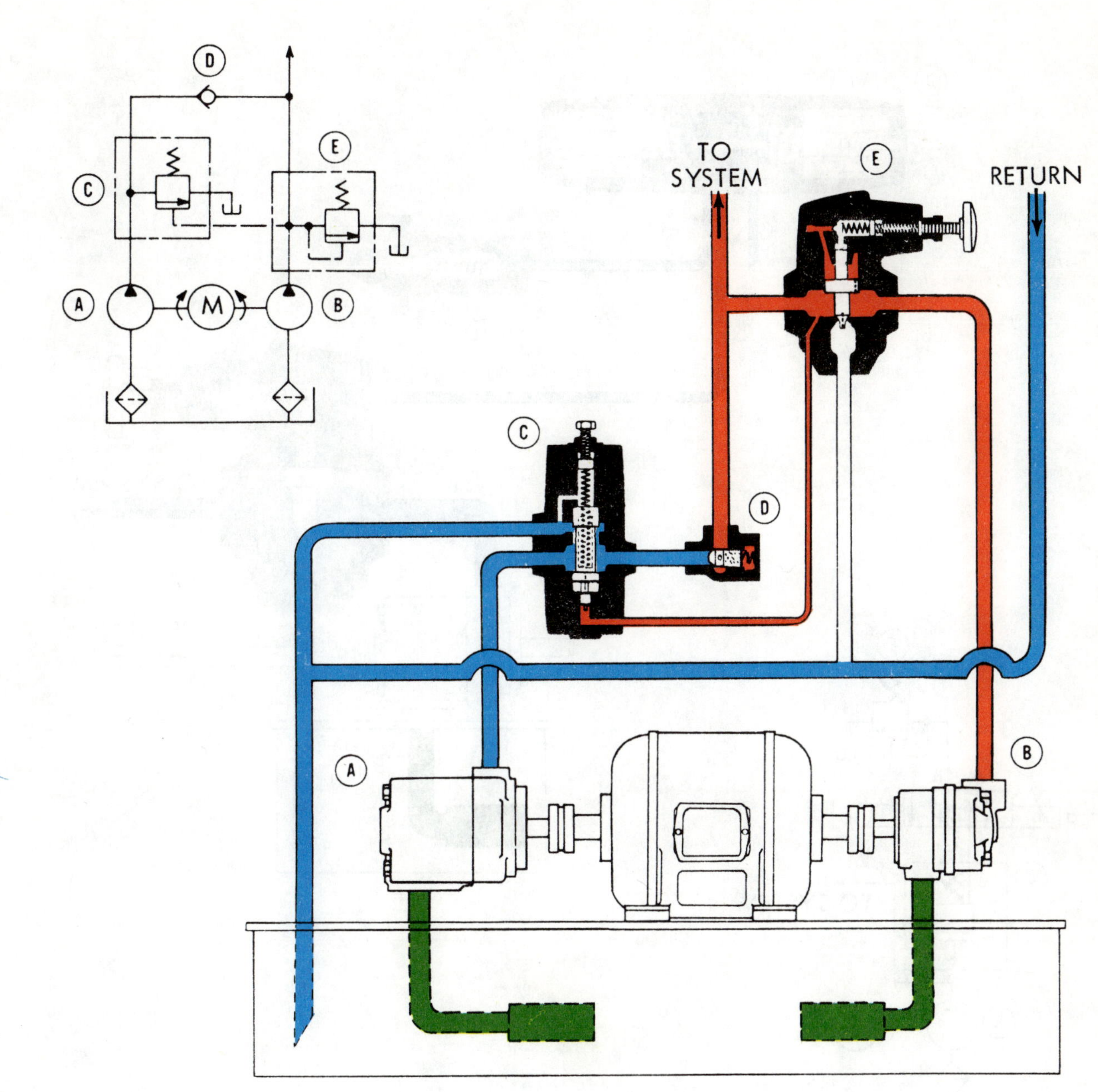

HIGH PRESSURE OPERATION

System pressure is less than the adjusted setting of relief valve (E) and higher than setting of unloading valve (C). Valve (E) is in its normally closed position and valve (C) is held open by system pressure. Delivery of pump (B) is directed into the system through (E). Check valve (D) is closed and delivery of pump (A) returns freely to tank through (C).

Fig. 13-1, View B. - Unloading Circuit - High Pressure Operation

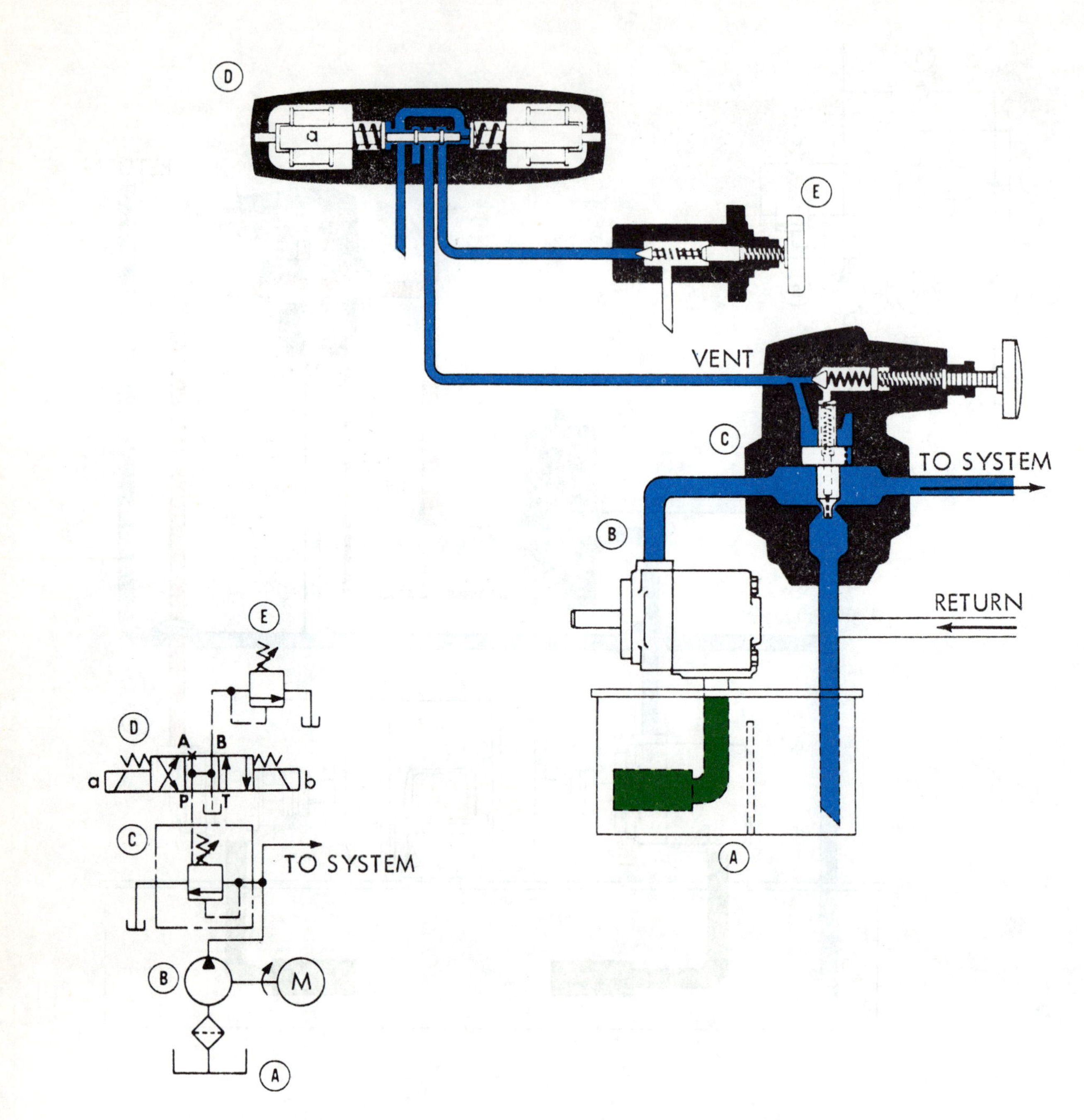

VENTING

Both solenoids of directional valve (D) are de-energized. Spool of (D) is spring centered and connects all ports of (D) to tank. Therefore, vent connection of relief valve (C) is open to tank through (D) and delivery of pump (B) unloads to tank through (C).

Fig. 13-2, View A. - Two Maximum Pressures plus Venting - Venting Operation

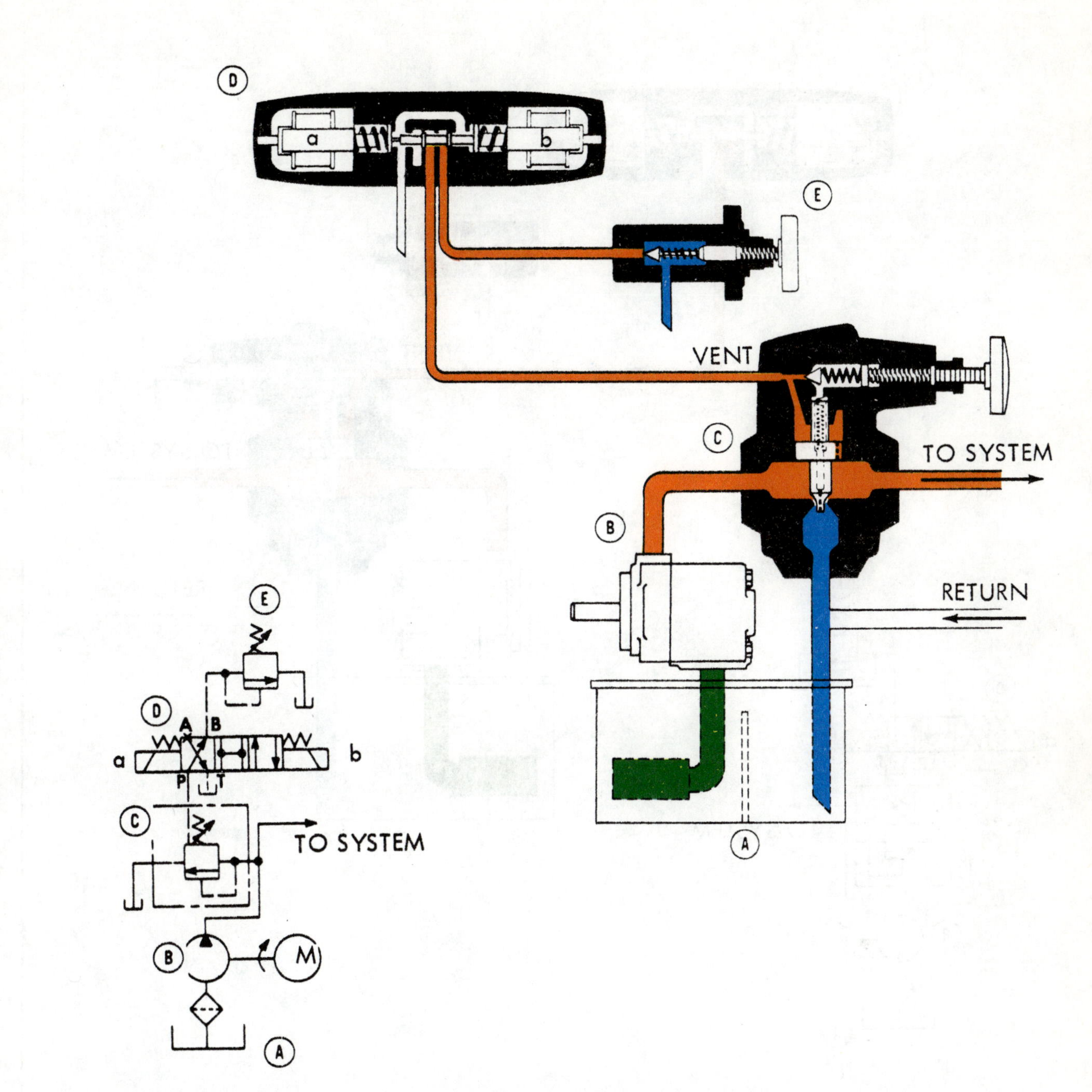

INTERMEDIATE PRESSURE

Solenoid "a" of directional valve (D) is held energized. Spool of (D) is shifted to connect the vent of relief valve (C) to pressure port of remote control relief valve (E). System pressure is limited by (E) which remotely controls (C).

Fig. 13-2, View B. - Two Maximum Pressures plus Venting - Intermediate Pressure

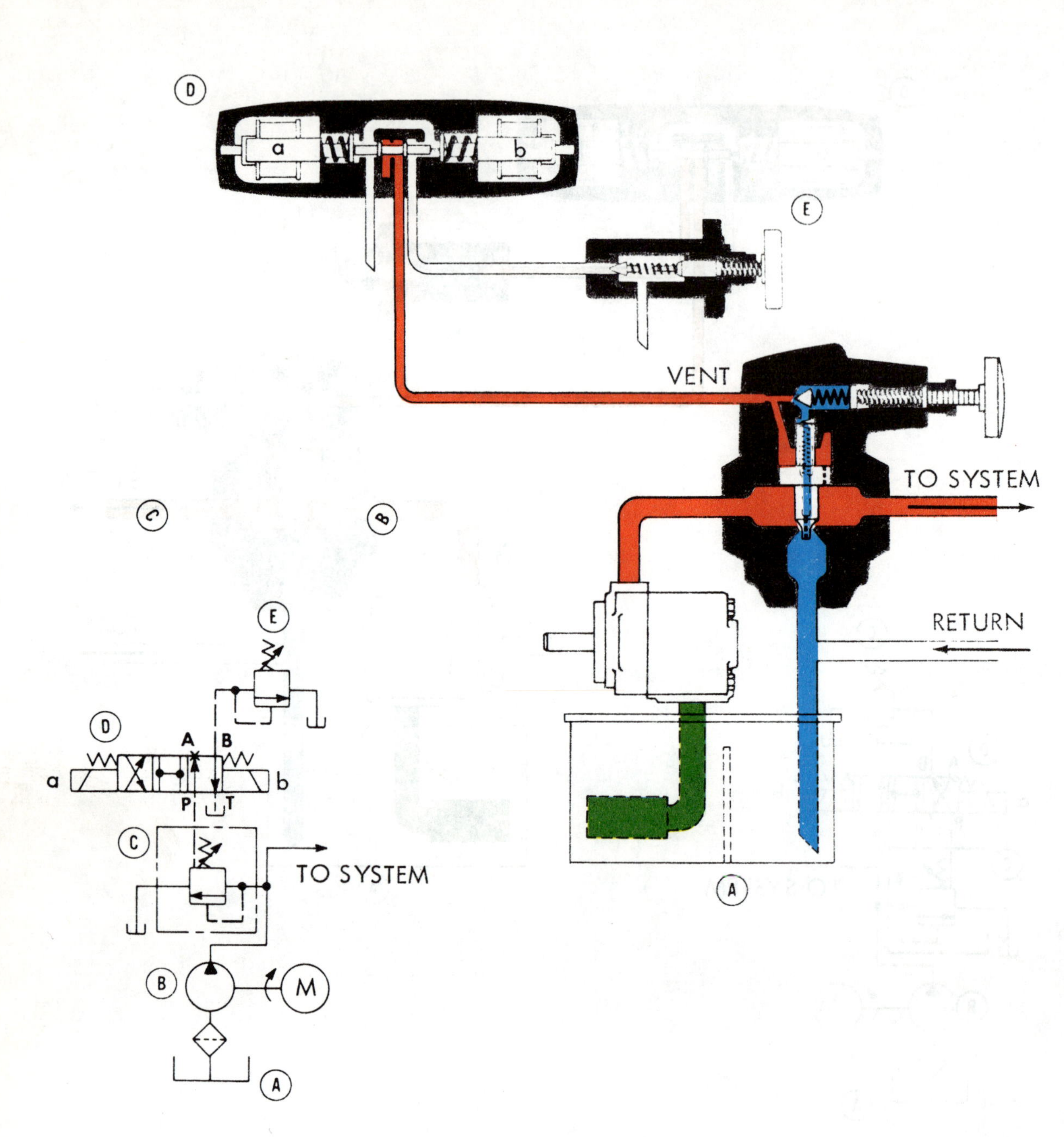

HIGH MAXIMUM PRESSURE

Solenoid "b" of directional valve (D) is held energized. Spool of (D) is shifted to connect vent of relief valve (C) to a plugged port in (D). System pressure is limited by (C).

Fig. 13-2, View C. - Two Maximum Pressures plus Venting - High Pressure valve is energized. The spool has shifted to connect the relief valve vent port to a "dead end" against a plugged port in the directional valve. The relief valve now functions at the setting of its integral pilot stage.

AUTOMATIC VENTING AT END OF CYCLE

In systems where it is not necessary to hold pressure at the end of a cycle, it is possible to unload the pump by automatically venting the relief valve. Figure 13-3 shows such a system using a cam operated pilot valve to vent the relief valve.

Mid-Stroke Extending (View A)

The machine cycle begins when the solenoid of the spring offset directional valve is energized. Pump output is to the cap end of the cylinder. The vent line from the directional valve is blocked at the cam-operated pilot valve. (Note that pilot valve has only two flow paths instead of the usual four.)

Mid-Stroke Retracting (View B)

At the end of the extension stroke, the limit switch was contacted by the cam on the cylinder rod to break the solenoid circuit. The directional valve has shifted to retract the cylinder. The relief valve vent connection is still blocked.

Automatic Stop (View C)

At the end of the retraction stroke, the cam on the cylinder opens the venting pilot valve. The relief valve vent port is thus connected to the line from the cap end of the cylinder, and the valve is vented through the inline check valve, the directional valve and the right angle check valve. Pilot pressure for the directional valve is maintained at a value determined by the spring loads in the balanced piston of the relief valve, vent line check valve, and the tank line check valve. (In this circuit a high vent spring in the relief valve would have eliminated the need for the right angle check valve.)

Push Button Start (View D)

When the start button is depressed it energizes the solenoid, the directional valve shifts to direct pump output into the cap end of the cylinder. This causes the check valve in the vent line to close deventing the relief valve. Pressure again builds up and the cycle is repeated.

ACCUMULATOR PUMP UNLOADING-- ELECTRIC CONTROL

In an accumulator charging circuit, the pump is unloaded when a preset pressure is reached and cuts back in to recharge the accumulator when the pressure drops to a predetermined minimum.

A spring offset solenoid operated directional valve (Fig. 13-4) actuated by a pressure switch is used to vent and devent the relief valve as required.

Charging (View A)

The two micro switches of the pressure switch are inter-connected to an electric relay in such a manner that at the low pressure setting the solenoid is energized and the relief valve vent connection is blocked. The pump output flows through the relief valve and check valve into the system where it charges the accumulator.

Unloading (View B)

When pressure reaches the maximum setting of the pressure switch the solenoid is de-energized and the relief valve is vented to unload the pump to tank. The check valve closes to prevent back flow from the accumulator and maintain pressure in the system.

Accumulator - Pump Unloading - Hydraulic Control

Another means of pump unloading in accumulator circuits is through the use of a direct acting unloading valve, illustrated in Figure 13-5.

Charging (View A)

In the "cut in" or "charging" condition the unloading section of the unloading valve is closed and the integral check valve permits pump flow to be directed into the system. When system demand is less than pump delivery rate, flow is directed into the accumulator and system pressure increases.

Unloading (View B)

When increasing pressure reaches the adjusted setting of the unloading valve it is caused to snap open and the integral check valve immediately closes. Pump delivery returns freely to tank through the unloading section of the unloading valve and the integral check valve permits system pressure to be sustained by the accumulator.

When some oil is taken from the accumulator, either to perform work or due to system leakage, the pressure decreases. When the decreasing pressure reaches a pre-determined percentage of the unloading valve's setting, the unloading section snaps closed and the system reverts to the condition shown in view A.

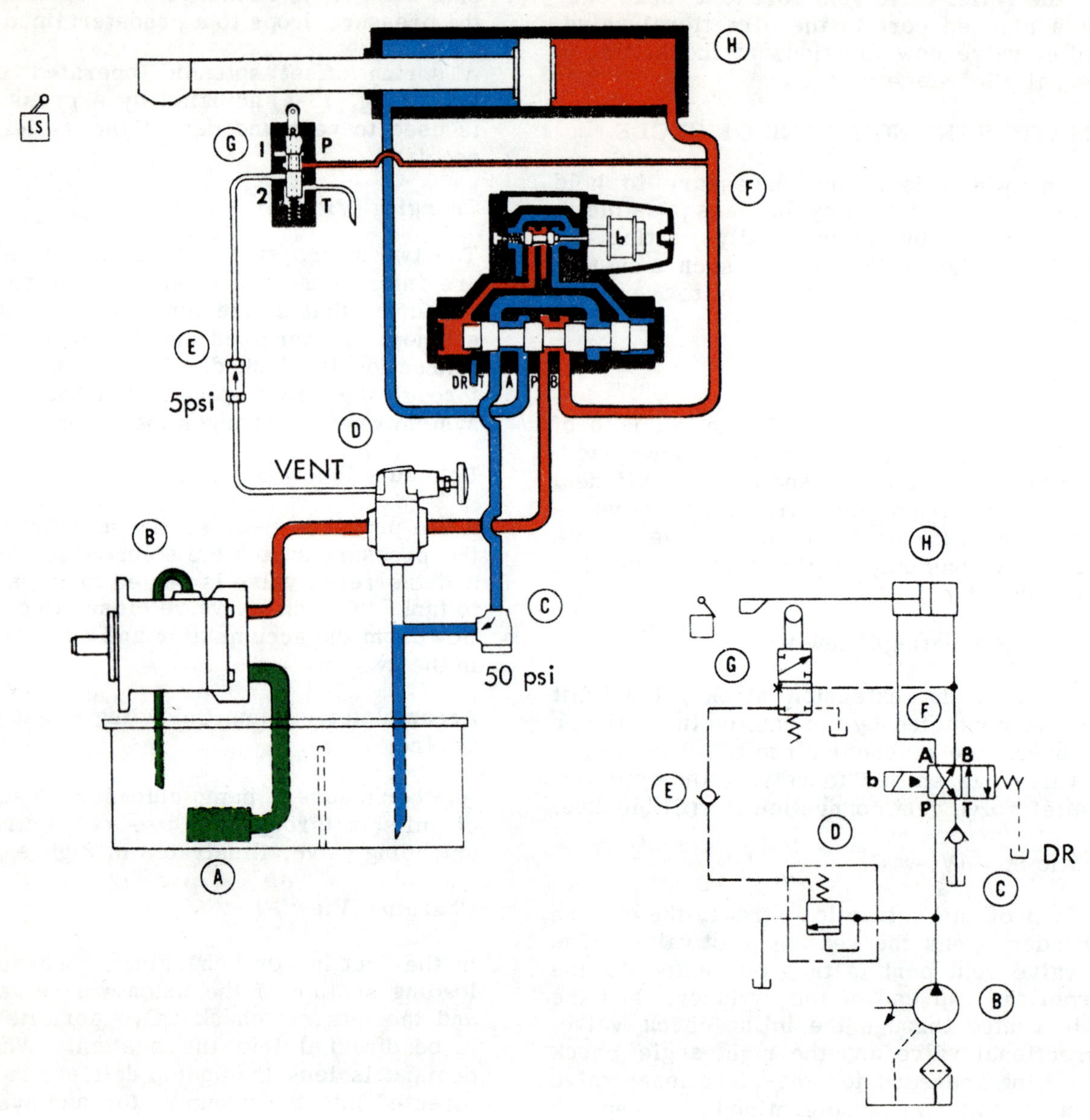

MID STROKE-EXTENDING

Solenoid "B" of valve (F) is held energized during the extending stroke. Vent line from valve (D) is blocked at valve (G). Delivery of pump (B) is directed through (F) into head end of cylinder (H). Discharge from rod end of (H) flows to tank through valves (F) and (C).

Fig. 13-3, View A. - Automatic Venting at Cycle End - Extend Operation

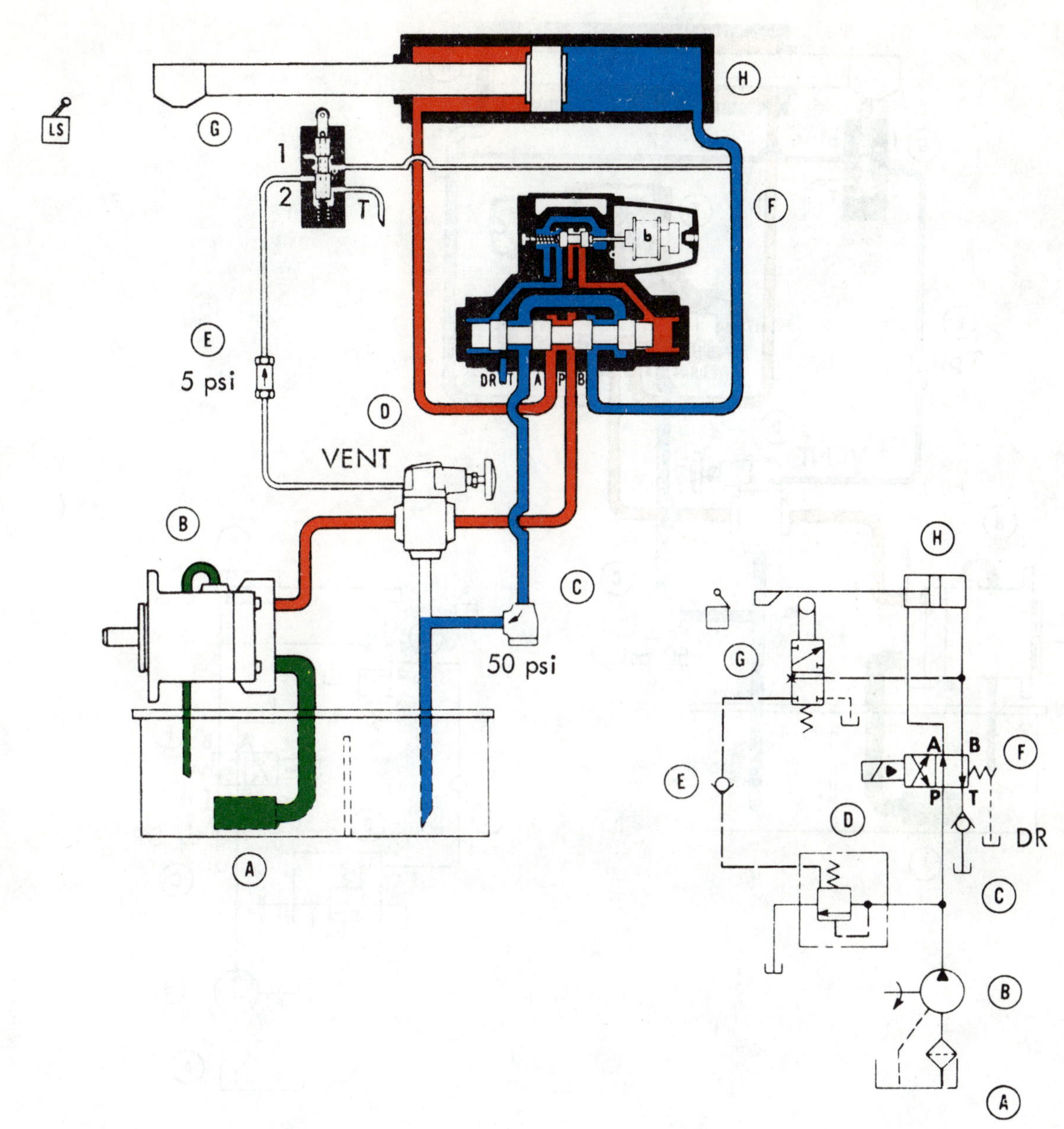

MID STROKE-RETRACTING

At end of extension stroke, cam on cylinder (H) contacts limit switch LS. This causes solenoid "B" of valve (F) to be de-energized. (F) shifts to the spring offset position and directs delivery of pump (B) into rod end of (H). Discharge from head end of (H) flows to tank through valves (F) and (C).

Fig. 13-3, View B. - Automatic Venting at Cycle End - Retract Operation

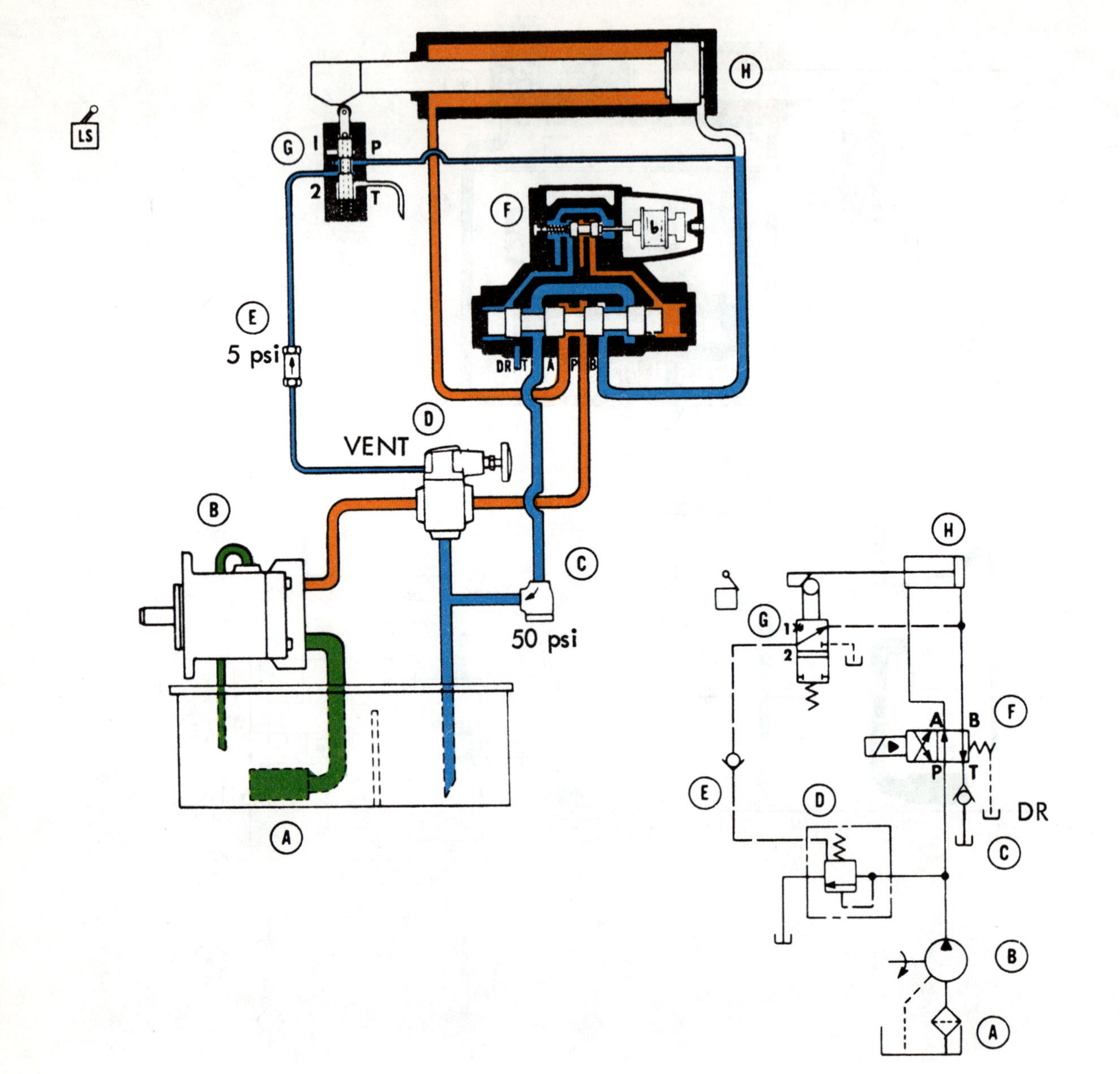

AUTOMATIC STOP

At end of retraction stroke, cam on cylinder (H) depresses valve (G). Valve (D) is now vented through valves (E), (G), (F), and (C). Delivery of pump (B) returns to tank over valve (D) at low pressure. Pressure drop through (C) assures pilot pressure for operation of (F).

Fig. 13-3, View C. - Automatic Venting at Cycle End - Automatic Stop

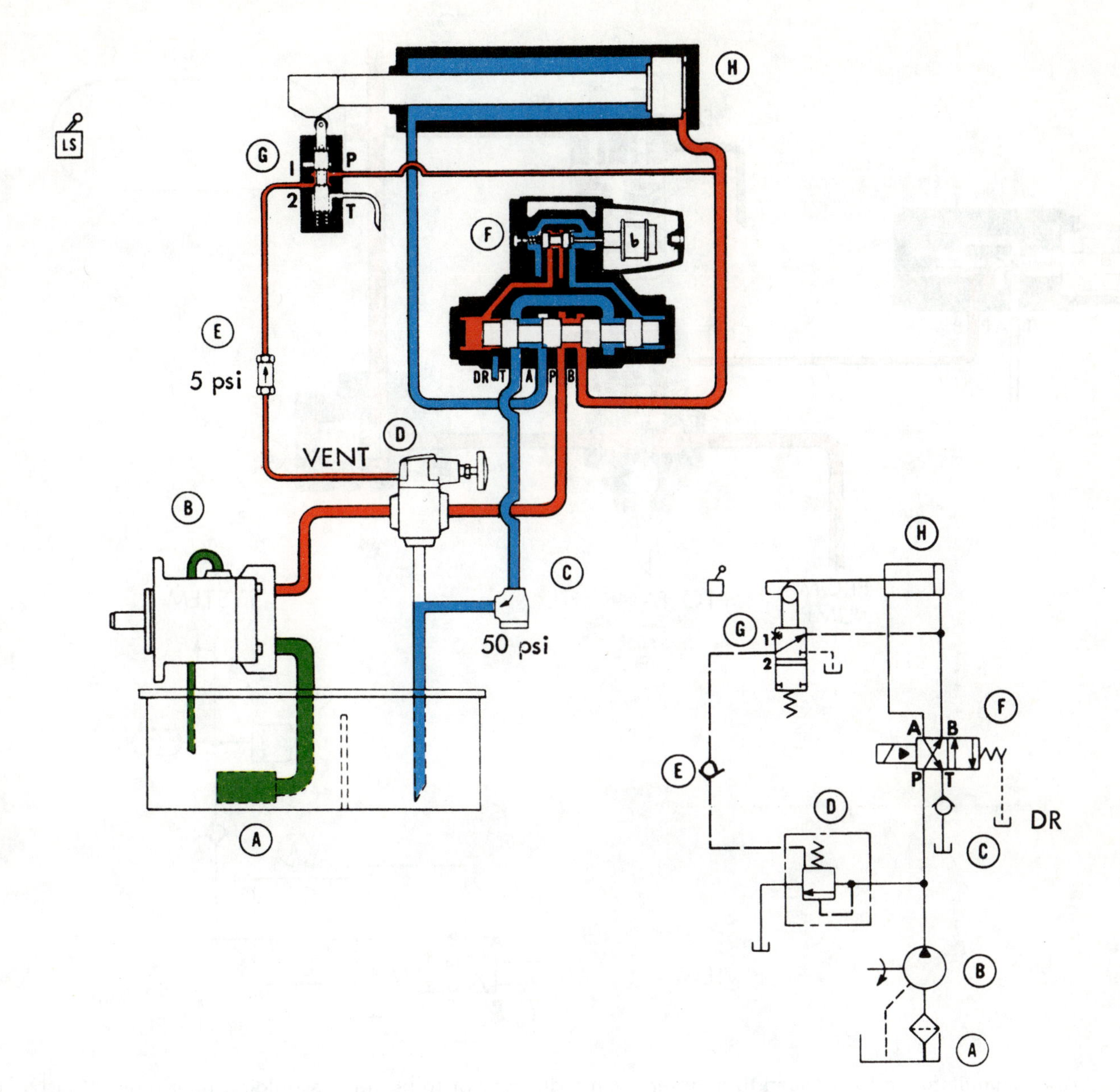

PUSH BUTTON START

Depressing a push button causes solenoid "B" of valve (F) to be held energized. (F) shifts to connect head end of cylinder (H) to pump (B), and rod end of (H) to tank. Pilot flow from vent of (D) stops and check valve (E) closes. Pressures equalize through balance hole in hydrostat of (D) causing it to start to close. Acceleration of (H) takes place during the closing of the hydrostat of (D).

Fig. 13-3, View D. - Automatic Venting at Cycle End - Push Button Start

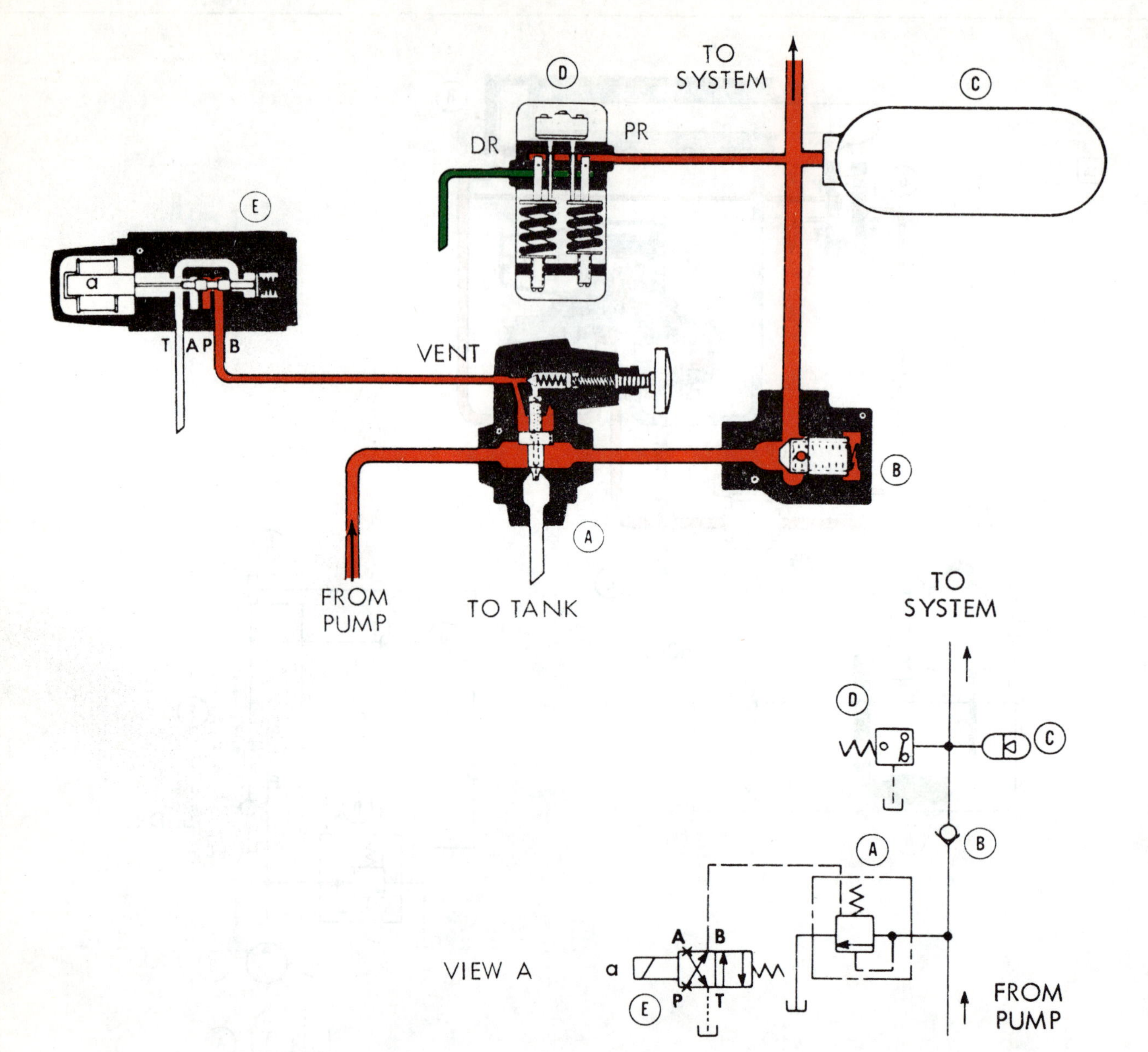

An accumulator may be used to augment pump delivery or to perform a holding operation. This circuit shows a method of unloading the pump when the accumulator is fully charged. It consists of relief valve (A), check valve (B), accumulator (C), dual pressure switch (D) and directional valve (E). Pressure setting of (A) is higher than the high setting of (D).

The electric control circuit performs the following operations: 1) energizes solenoid (Ea) when pump motor is started; 2) de-energizes (Ea) when system pressure reaches the high setting of switch (D); 3) energizes (Ea) when system pressure reduces to the low setting of switch (D); 4) de-energizes (Ea) when pump motor is stopped.

View A shows circuit condition when system pressure is below the low setting of switch (D). Solenoid (Ea) is energized to shift valve (E) and block the vent connection of valve (A). Valve (A) is de-vented and pump delivery is directed through valve (B) into the system. Accumulator (C) is charged with fluid if system volumetric demand is less than delivery rate of the pump.

Fig. 13-4, View A. - Accumulator Pump Unloading (Electric Control) - Charging

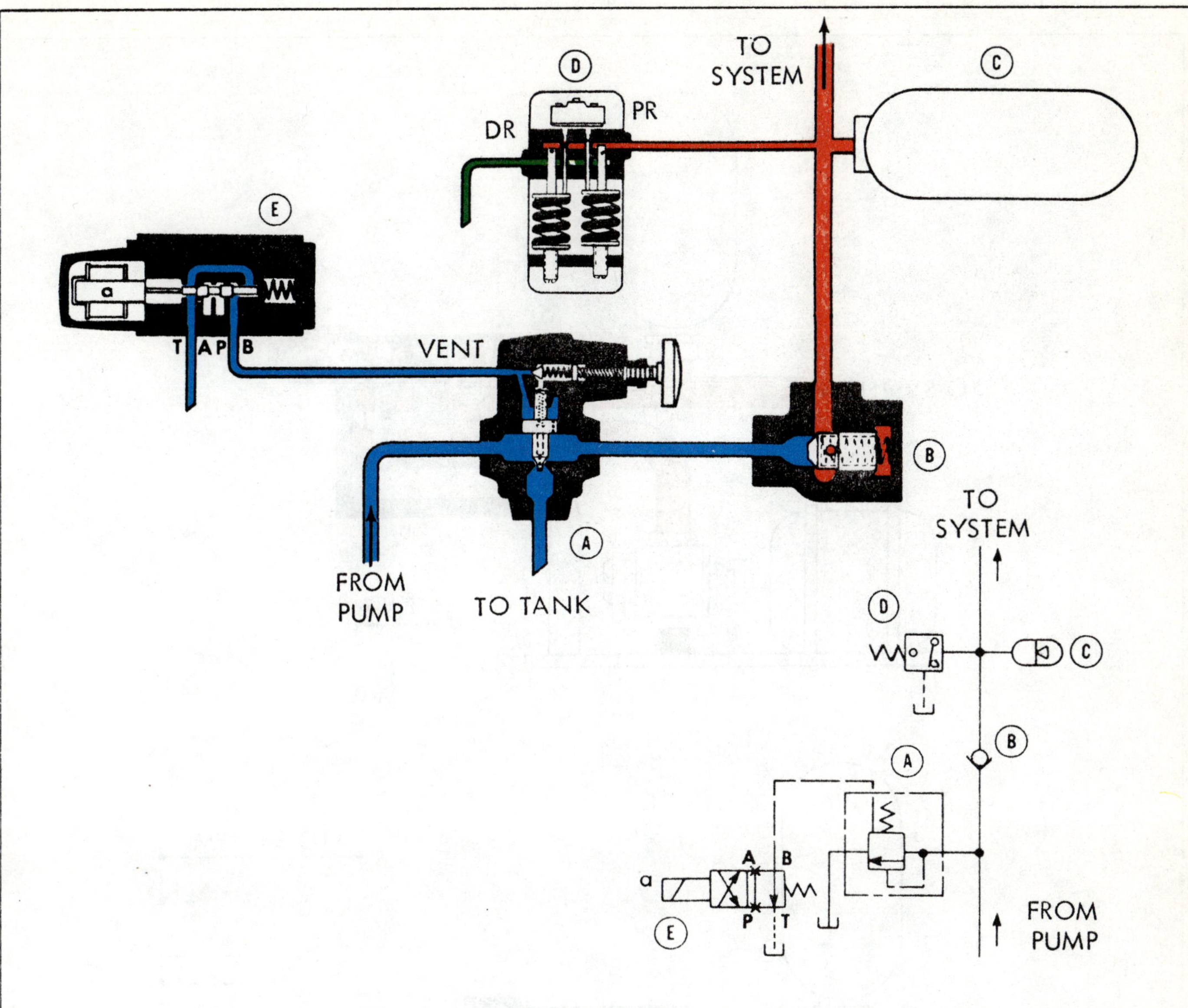

View B shows circuit condition when accumulator (C) is charged and system pressure has reached the high setting of switch (D). Solenoid (Ea) is de-energized to vent valve (A). The pump is unloaded, its delivery being returned freely to tank through valve (A). Check valve (B) closes to permit accumulator (C) to hold pressure and maintain a volume supply in the system.

Charging and unloading continue automatically until pump motor is stopped. The dual pressure switch provides means to adjust the pressure difference between pump "cut-in" and pump "cut-out". The high setting of switch (D) is the maximum pressure control for the system with overload protection provided by valve (A).

Fig. 13-4, View B. - Accumulator Pump Unloading (Electric Control) - Unloading

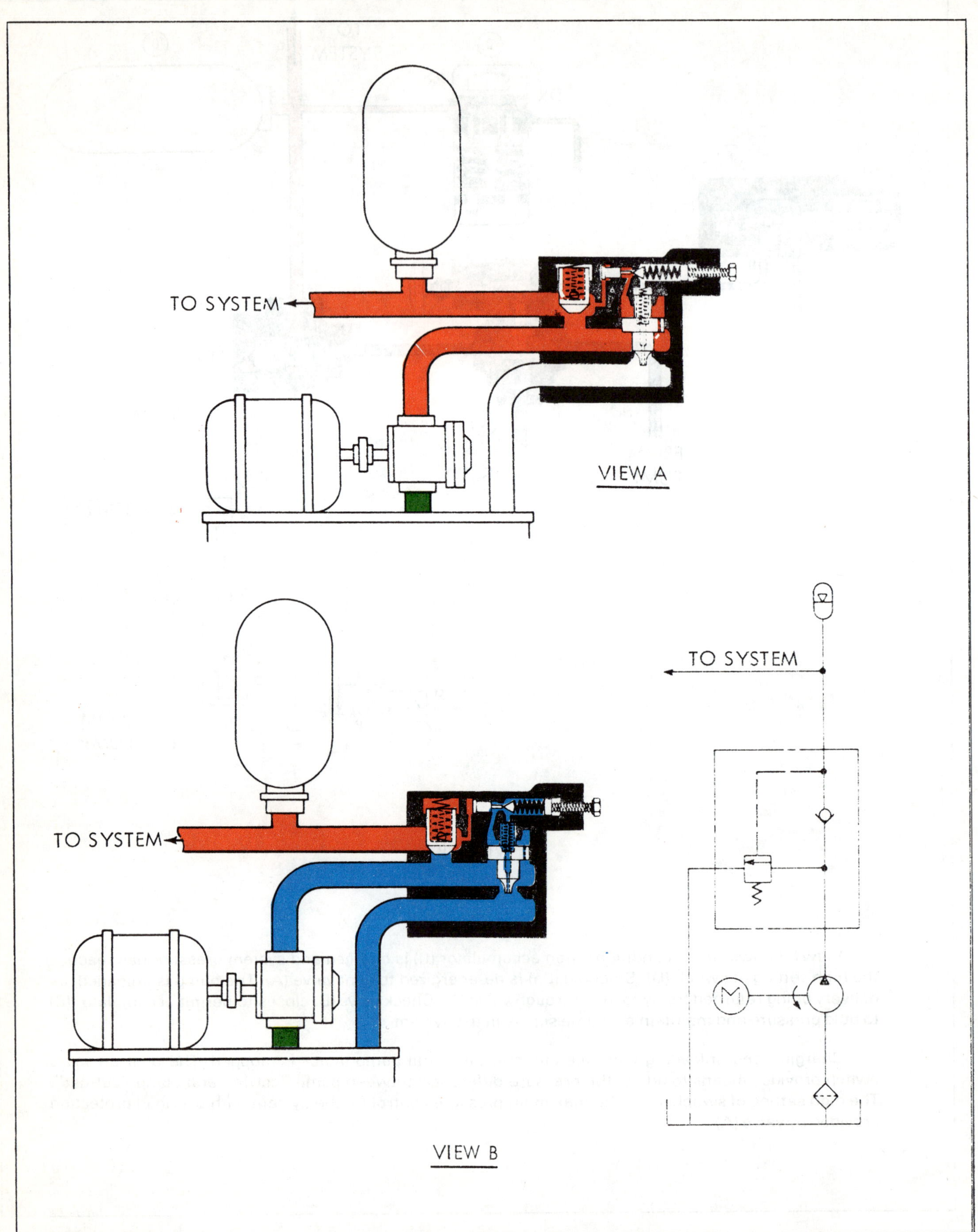

Fig. 13-5. - Accumulator Pump Unloading (Hydraulic Control)

ACCUMULATOR SAFETY CIRCUITS

Accumulator Safety--Bleed-Off

The circuit in Figure 13-6 is used to automatically bleed-off a charged accumulator when the pump is shut down to prevent accidental operation of an actuator or to make it safe to open the system for service. The bleed-off is accomplished through a spring offset directional valve and a fixed restriction.

The directional valve solenoid is actuated by the prime mover switch, so that the solenoid energizes whenever the pump is started (View A). This blocks the bleed passage during normal operation.

When the pump is shut down (View B), the spool spring shifts the directional valve and opens the accumulator to tank through the restriction.

The manual valve shown is used to control accumulator discharge rate to the system. The auxiliary relief is set slightly higher than the system relief valve and limits pressure rise from heat expansion of the gas charge.

The accumulator must have a separator, i. e., diaphragm, bladder or piston to prevent loss of gas preload each time the machine is shut down.

Blocked Accumulator Safety Circuit

It is also possible to block a charged accumulator to permit service on the balance of the system thus not losing the stored energy. This method (Fig. 13-7) uses a spring offset directional valve to control a pilot-operated check valve. Again, the directional valve solenoid is tied into the prime mover control.

When the pump starts (view A), the solenoid is energized and the directional valve sends pilot pressure to open the check valve, permitting flow into and out of the accumulator.

Stopping the pump (view B) de-energizes the solenoid. The directional valve vents the pilot line of the check valve. The check valve closes to block the accumulator from the system and allows it to maintain its oil under pressure.

The small needle valve is used only to drain oil from the accumulator prior to replenishing the gas preload. The large manual valve and auxiliary relief valve perform the same functions as in the bleed-off system.

RECIPROCATING CIRCUITS

Conventional reciprocating circuits use a four-way directional valve piped directly to a cylinder or motor to provide reversal. Retracting speed is faster than extending speed because of rod volume when a differential cylinder is used.

A non-conventional reciprocating hook-up is the regenerative circuit, where oil from the rod end of the cylinder is directed into the cap end to increase speed.

Regenerative Advance

The principle of the regenerative circuit is shown in Figure 13-8. Note that the "B" port on the directional valve, which would conventionally connect to the cylinder is plugged and the rod end of the cylinder is connected directly to the pressure line. With the valve shifted to connect the "P" port to the cap end (view A), flow out of the rod end joins pump delivery to increase the cylinder speed. In the reverse condition (view B), flow from the pump is directly to the rod end.

Exhaust flow from the cap end returns to the tank conventionally through the directional valve.

If the ratio of cap end area to rod end annular area in the cylinder is 2:1, the cylinder will advance and retract at the same speed. However, the pressure during advance will be double the pressure required for a conventional hook-up. This is because the same pressure in the rod end, effective over half the cap end area, opposes the cylinder's advance. With a higher ratio of areas extending speed will increase proportionally.

Regenerative Advance with Pressure Changeover to Conventional Advance

The regenerative principle also can be used to increase advance speed with a changeover to conventional advance to double the final force (Fig. 13-9). In this system, a normally closed "R" type pressure control valve in effect plugs the "B" port of the directional during regenerative advance. When the pressure setting of the "R" valve is reached, it opens to route oil from the rod end conventionally to tank through the directional valve.

The 5 psi check valve permits oil from the rod end of the cylinder to join pump delivery during regenerative advance, but prevents pump delivery from taking this route to tank during conventional advance.

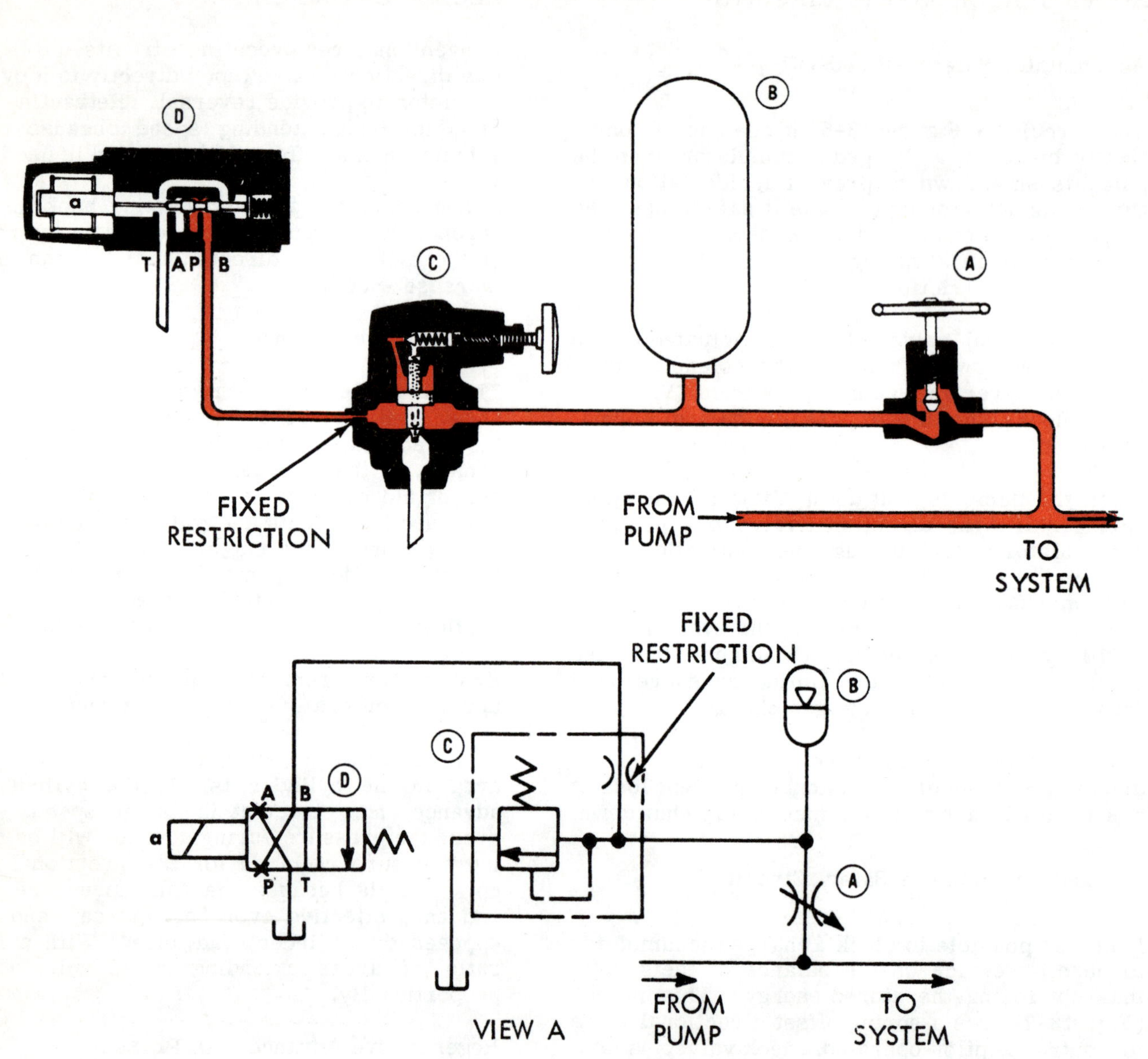

The charge in accumulator (B) is automatically bled-off to permit safe servicing of the system when pump motor is stopped. The circuit consists of needle valve (A), accumulator (B), relief valve (C) and directional valve (D). An electrical control circuit holds solenoid (Da) energized when the pump motor is running and de-energizes it when the motor is stopped.

View A shows circuit condition during normal operation of the system when the pump motor is running. Solenoid (Da) is energized to shift valve (D) and block flow to tank from accumulator (B). Accumulator is charged or discharged through valve (A) as dictated by requirements of the system. Needle valve (A) is often used to control rate of accumulator discharge to the system.

Fig. 13-6, View A. - Accumulator Bleed-Off Circuit - Normal Operation

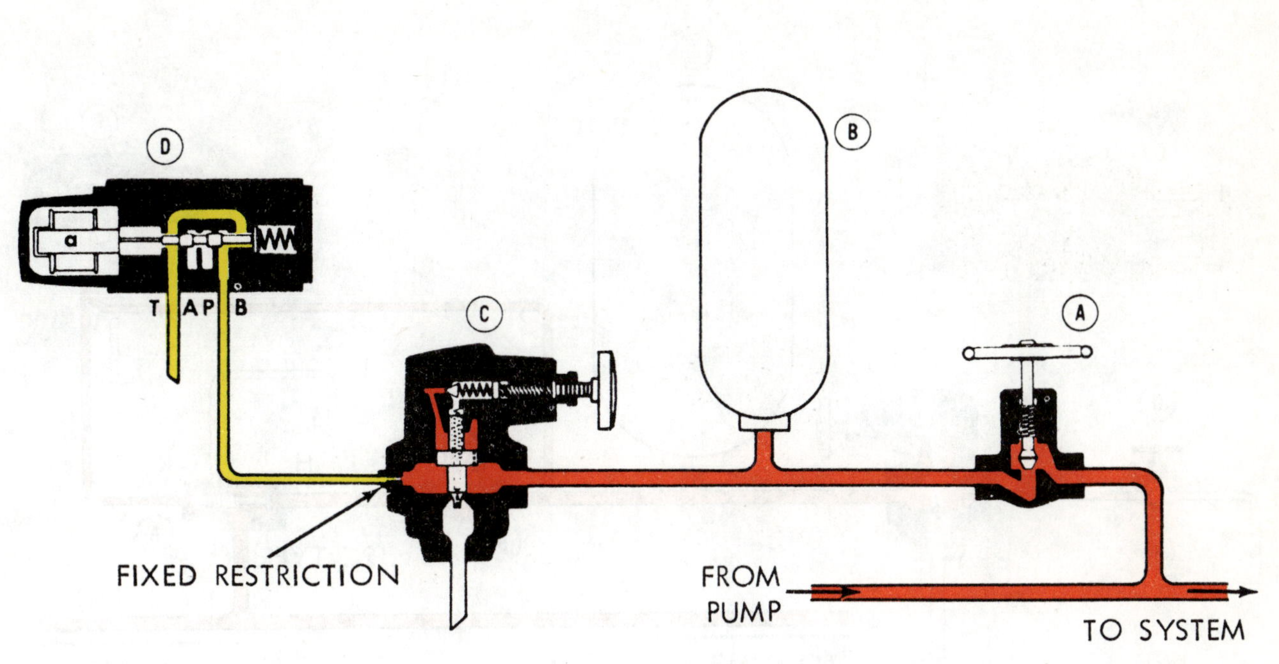

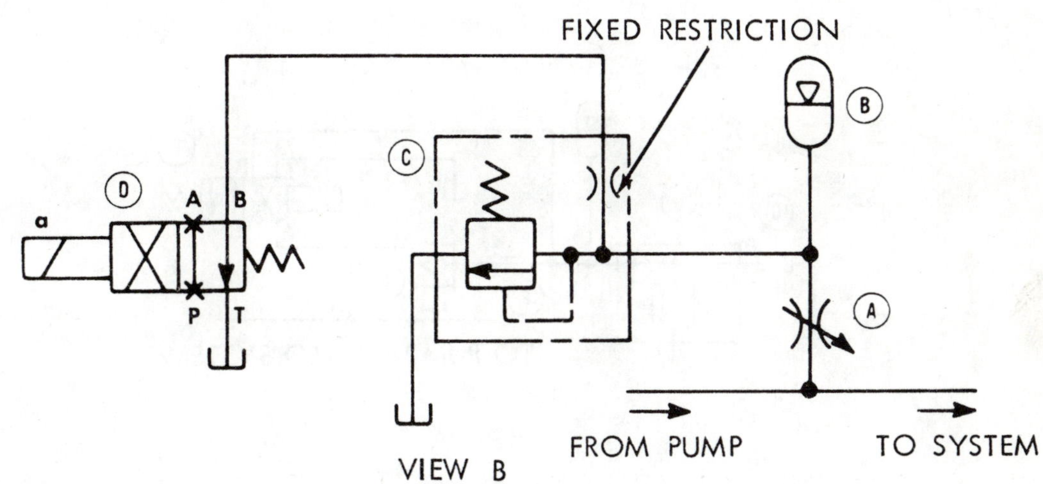

View B shows circuit condition when the pump motor is stopped. Solenoid (Da) is de-energized and the charge in accumulator (B) is bled-off to tank through valve (D). Rate of bleed-off is controlled by a fixed restriction at valve (C).

Valve (C) is set slightly higher than the maximum pressure control and provides protection against excessive pressures due to thermal expansion.

Fig. 13-6, View B. - Accumulator Bleed-Off Circuit - Bleed-Off

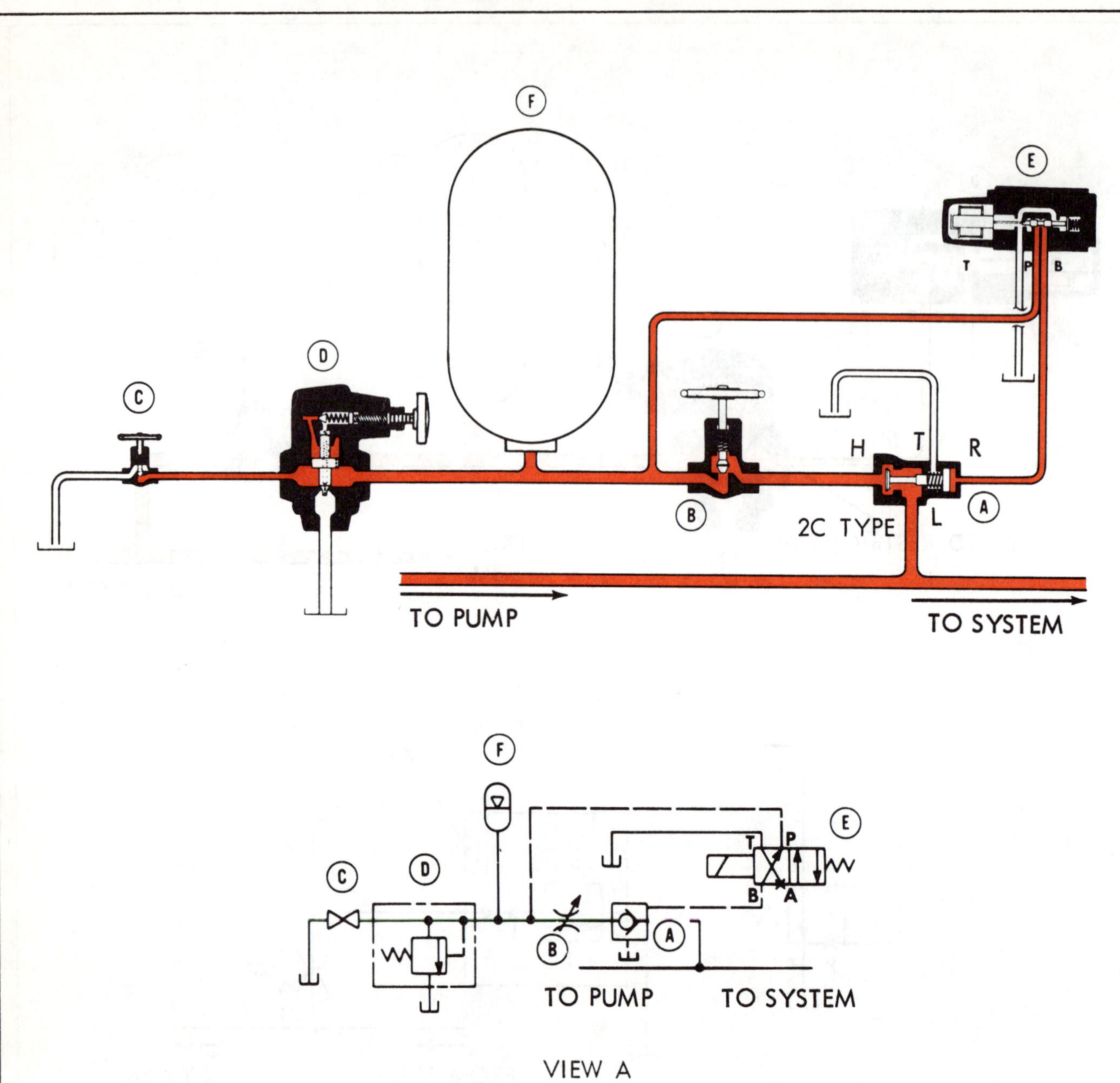

VIEW A

The accumulator is automatically isolated from the hydraulic system when pump motor is stopped to permit safe servicing of remaining hydraulic components. The circuit consists of pilot operated check valve (A), needle valve (B), needle valve (C), relief valve (D), directional valve (E) and accumulator (F). An electrical control circuit holds solenoid (Ea) energized when the pump motor is running and de-energizes it when the motor is stopped.

View A shows circuit condition during normal operation of the system. Solenoid (Ea) is energized to shift valve (E). System pressure acts on remote control connection of valve (A) holding it open. Accumulator (C) is charged or discharged thru valves (A) and (B) as dictated by system requirements. Valve (B) is often used to control rate of accumulator discharge to the system. Valve (D) is slightly higher than the maximum pressure control and provides protection against excessive pressures due to thermal expansion.

Fig. 13-7, View A. - Accumulator Blocking Circuit - Normal Operation

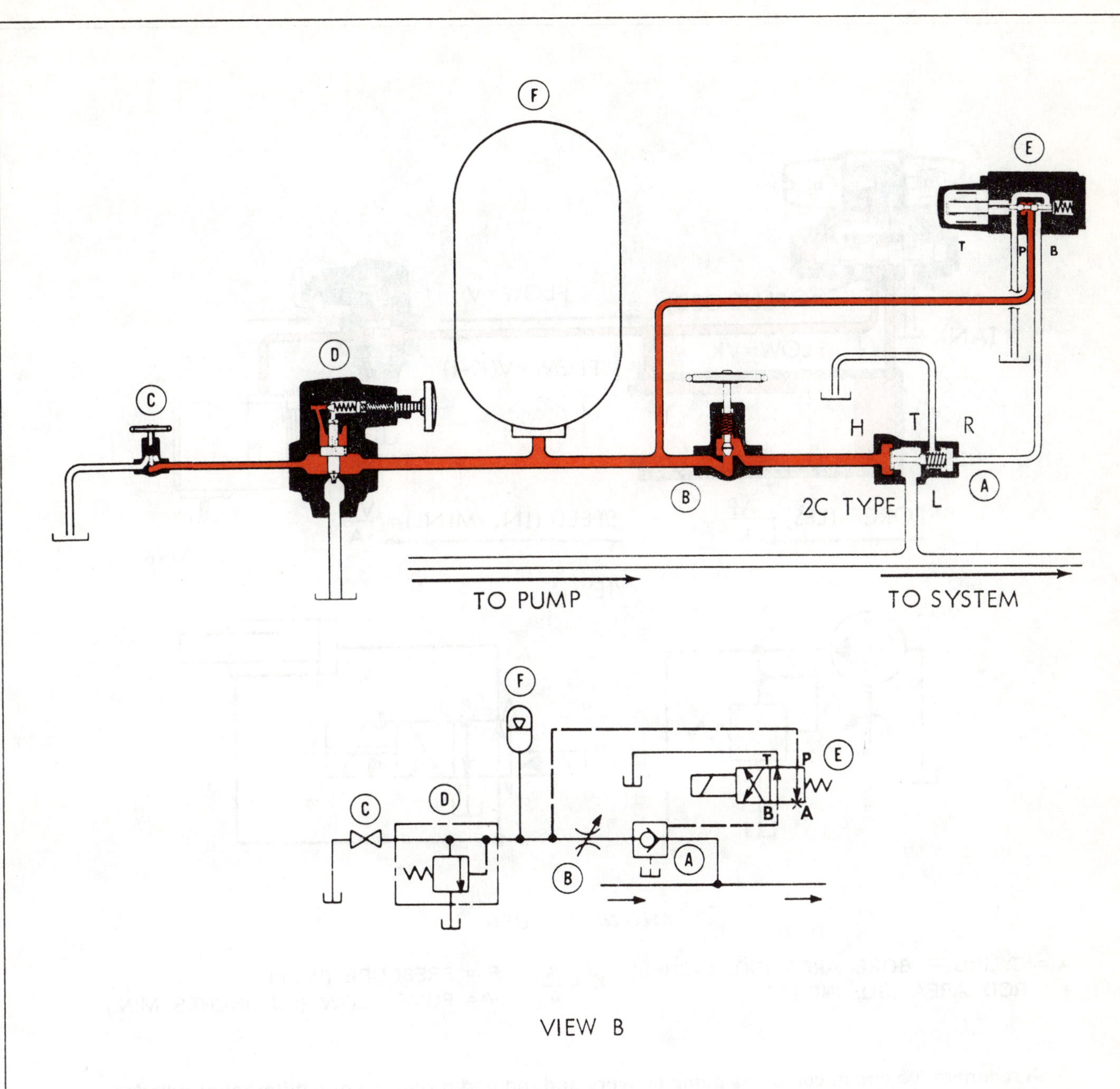

VIEW B

View B shows circuit condition when pump motor is stopped. Solenoid (Ea) is de-energized and remote control connection of valve (A) is connected to tank. Accumulator pressure acts on "H" connection of valve (A) holding it closed. Flow from accumulator (F) to the system is blocked. The charge in accumulator (F) may be bled-off to tank thru valve (C).

Fig. 13-7, View B. - Accumulator Blocking Circuit - Accumulator Blocked

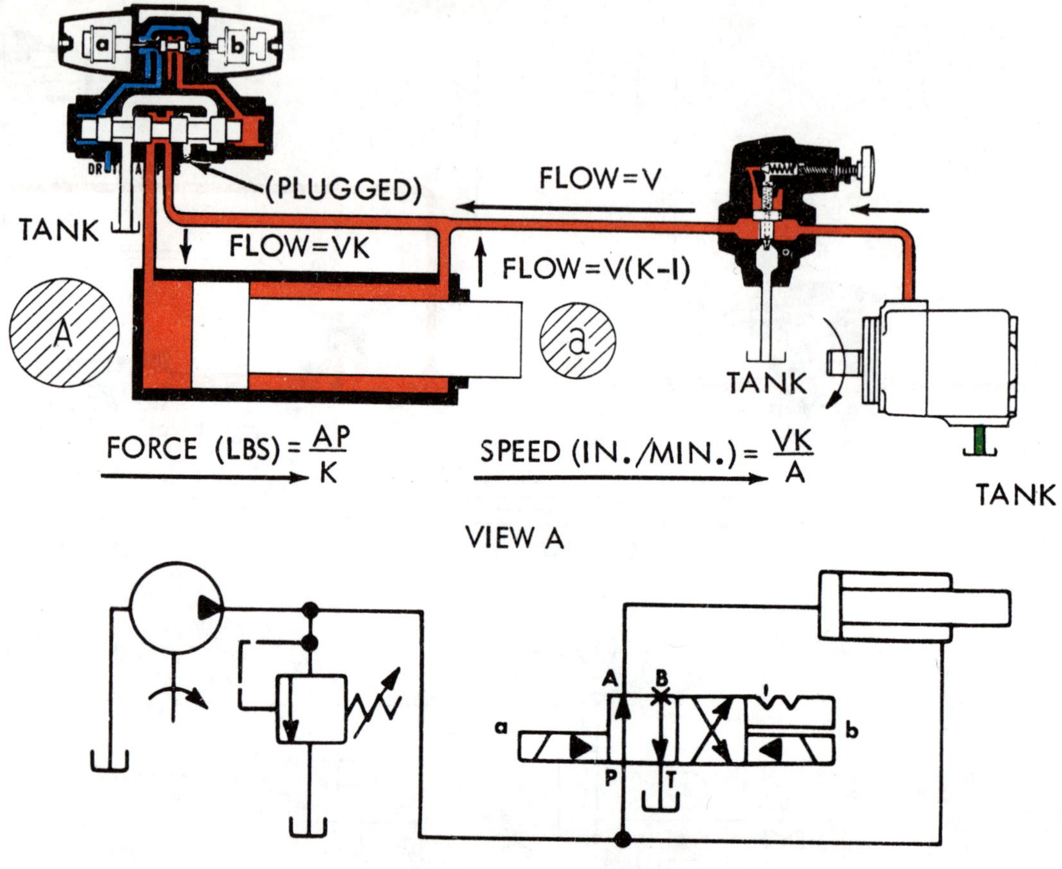

KNOWN VALUES

A = CYLINDER BORE AREA (SQ. INCHES)
a = ROD AREA (SQ. INCHES)

$K = \frac{A}{a}$

P = PRESSURE (P.S.I.)
V = PUMP FLOW (CU. INCHES/MIN.)

A regenerative circuit combines pump delivery and rod end discharge of a differential cylinder to obtain rapid speed when extending. Pressure is equal at both the head and rod ends during regenerative movement.

View A shows flow condition during regenerative advance. Pump delivery and rod end discharge are directed to the cylinder head end thru the directional valve. Equal pressure acting on the difference in areas creates a larger force at the head end to extend the cylinder.

Formulas shown are used to calculate speeds, forces and flow rates. They also show that, during regenerative advance, speed increases and force decreases proportionately as the ratio of areas increase.

Fig. 13-8, View A. - Regenerative Circuit - Advancing

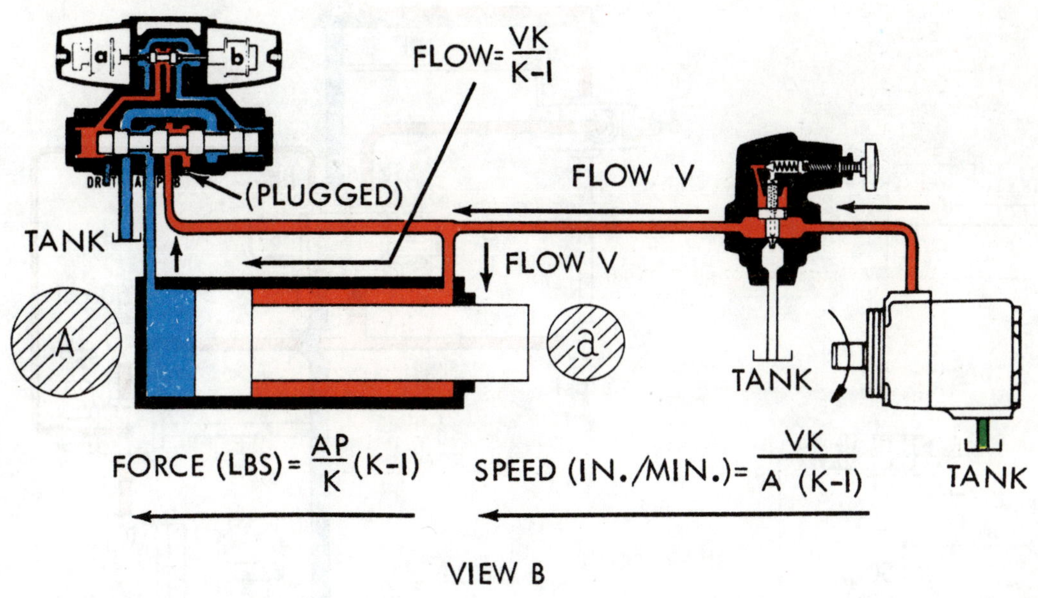

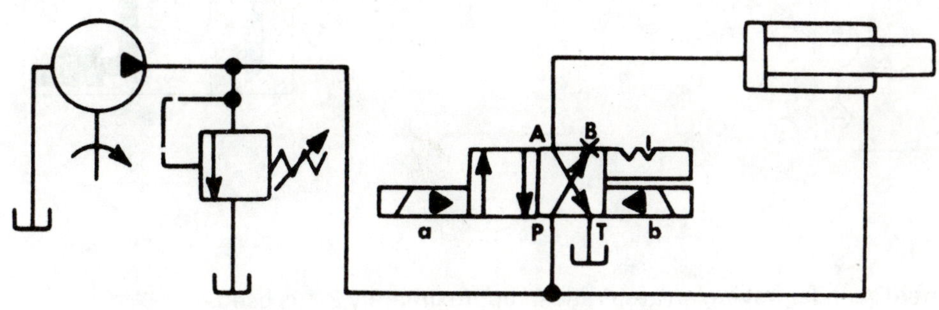

View B shows conventional flow to and from the cylinder during return movement. Pump delivery is to rod end only with head end open to tank thru the directional valve. Pressure acts only on the rod end.

Fig. 13-8, View B. - Regenerative Circuit - Retracting

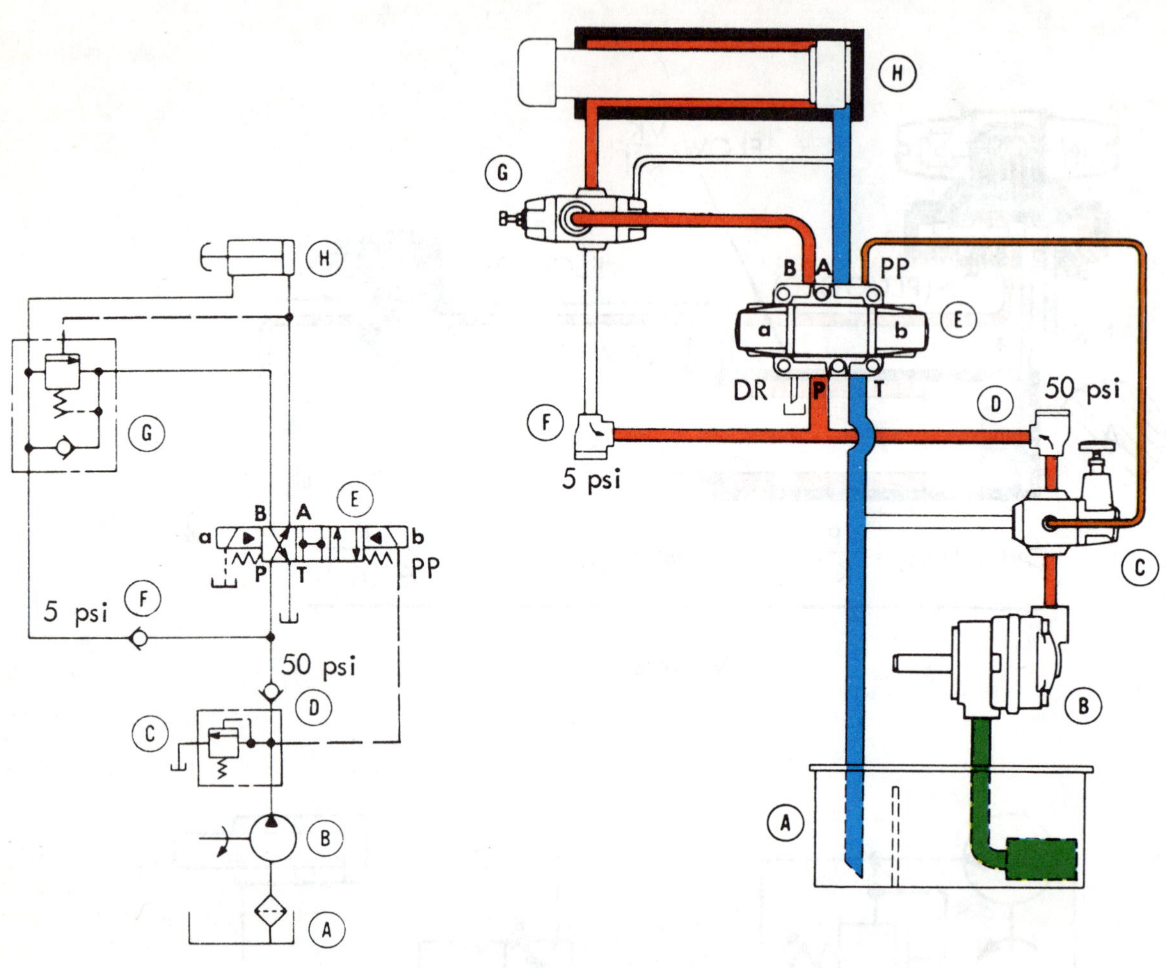

A differential cylinder having an area ratio of approximately 2:1 is used.

Rapid approach is obtained when, with solenoid "A" energized and operating pressure lower than setting of valve (G), pump flow is directed into head end of (H) with discharge from (H) directed through valves (G) and (F) to combine with pump flow. Piston speed is determined by pump flow and crossectional area of rod of (H).

When work resistance is encountered, pressure increase causes (G) to open permitting discharge from rod end of (H) to flow freely to tank through (G) and (E). Piston of (H) slows to half speed, but potential thrust is now a function of full piston area and maximum operating pressure.

Rapid return is obtained when, with solenoid "B" energized, pump flow is directed through (E) and integral check valve in (G) into rod end of (H). Discharge from head end of (H) is freely to tank through (E). Piston speed is determined by pump flow and annular area of (H) and is same as advance speed.

Valve (C) limits maximum pressure and provides overload protection. Valve (D) assures pilot pressure for operation of (E).

Fig. 13-9. - Regenerative Advance with Changeover to Conventional

When the directional valve shifts to retract the cylinder, pump output is through the check valve in the "R" valve to the rod end.

CLAMPING AND SEQUENCE CIRCUITS

In many applications, such as clamping a workpiece and then machining it, it is necessary to have operations occur in a definite order, and to hold pressure at the first operation while the second occurs. Following are two of several such circuits.

Sequencing Circuit

Figure 13-10 shows a method of having machine motions occur in a definite sequence, using one directional valve and two sequence valves. (The counterbalance valve shown is used to control the descent of the vertical cylinder.)

The sequence is:

View A--Extend Cylinder H
View B--Extend Cylinder J while holding pressure on Cylinder H
View C--Retract Cylinder J
View D--Retract Cylinder H

This system can be used for clamping only if it is not necessary to hold the work piece (Cylinder H) while the work cylinder retracts. If the work must be held until the work cylinder retracts, a second directional valve is used as in the following circuit.

Controlled Pressure Clamping Circuit

The circuit shown in Figure 13-11 provides sequencing plus a controlled clamping pressure, which can be held while the work cylinder is feeding and retracting. The sequence of operations is:

Pressing start button shifts directional valve, and clamp cylinder extends.

Upon contact with work piece a limit switch actuates solenoid of directional valve 2 to initiate work stroke. Sequence valve assures clamp pressure is maintained at predetermined minimum during work stroke. Pressure reducing valve limits clamp pressure to safe maximum when higher pressure is required for work stroke. Additional electric controls can reverse work cylinder directional valve while pressure is maintained on clamp.

Clamp opens after work cylinder is fully retracted.

COUNTERBALANCE CIRCUIT

A typical "RC" type counterbalance circuit (Figure 13-12) is used to operate a vertical cylinder with the rate of descent controlled by delivery from the pump. The counterbalance valve prevents the load from falling freely on the downward stroke.

In view A, the cylinder is being raised. Flow from the pump to the head end passes freely over the integral check valve.

View B shows the hold position where pressure generated by the load alone is not sufficient to overcome setting of counterbalance valve spring.

View C shows the load being lowered with pressure on the head of the piston providing the additional force required to cause the counterbalance valve to open.

BRAKE CIRCUIT

Figure 13-13 shows an application of the "RC" type valve to hold a back pressure in a rotary motor when needed and to brake the motor when the open center directional valve is shifted to neutral.

View A shows the motor accelerating with the brake valve held wide open by load pressure in the auxiliary remote control connection. View B shows the operation when the motor tries to overrun the pump creating a lower pressure in the drive line. Neutral braking through backpressure is shown in view C.

FEED CIRCUITS

Meter-In Flow Control

Figure 13-14 shows the operation of a pressure compensated flow control to control speed of the extending stroke. In view A, the directional valve is shifted to extend the cylinder, in view B to retract it. Since the flow control valve is placed in the line to the cap end of the cylinder the control is meter-in. The flow control is bypassed by a check valve to provide a rapid return stroke. Any tendency of the load to move in the forward direction could cause it to run away. Pump delivery in excess of flow control setting is diverted to tank over the relief valve.

Meter-Out Flow Control

A meter-out circuit is illustrated in Figure 13-15. The difference is that the flow control is placed downstream from the cylinder. Since exhaust oil is regulated by the flow control,

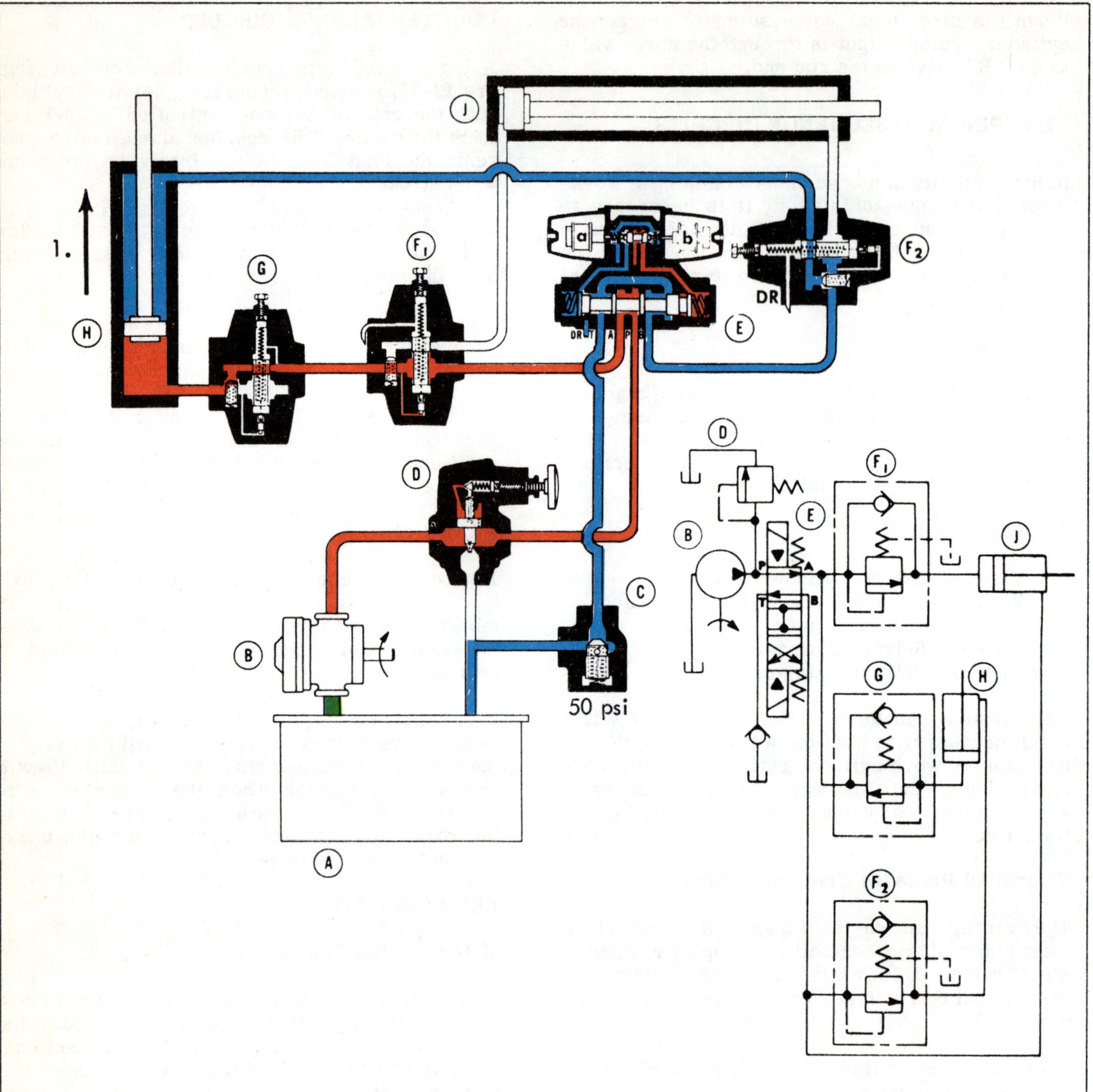

Phase #1 - Solenoid (Ea) Energized

Delivery of (B) is directed through valves (D), (E), (F1), and integral check valve of (G) into the head end of (H). Discharge from rod end of (H) flows freely to tank through integral check valve of (F2), and valves (E) and (C).

Fig. 13-10, View A. - Sequence Circuit - Extend Cylinder H

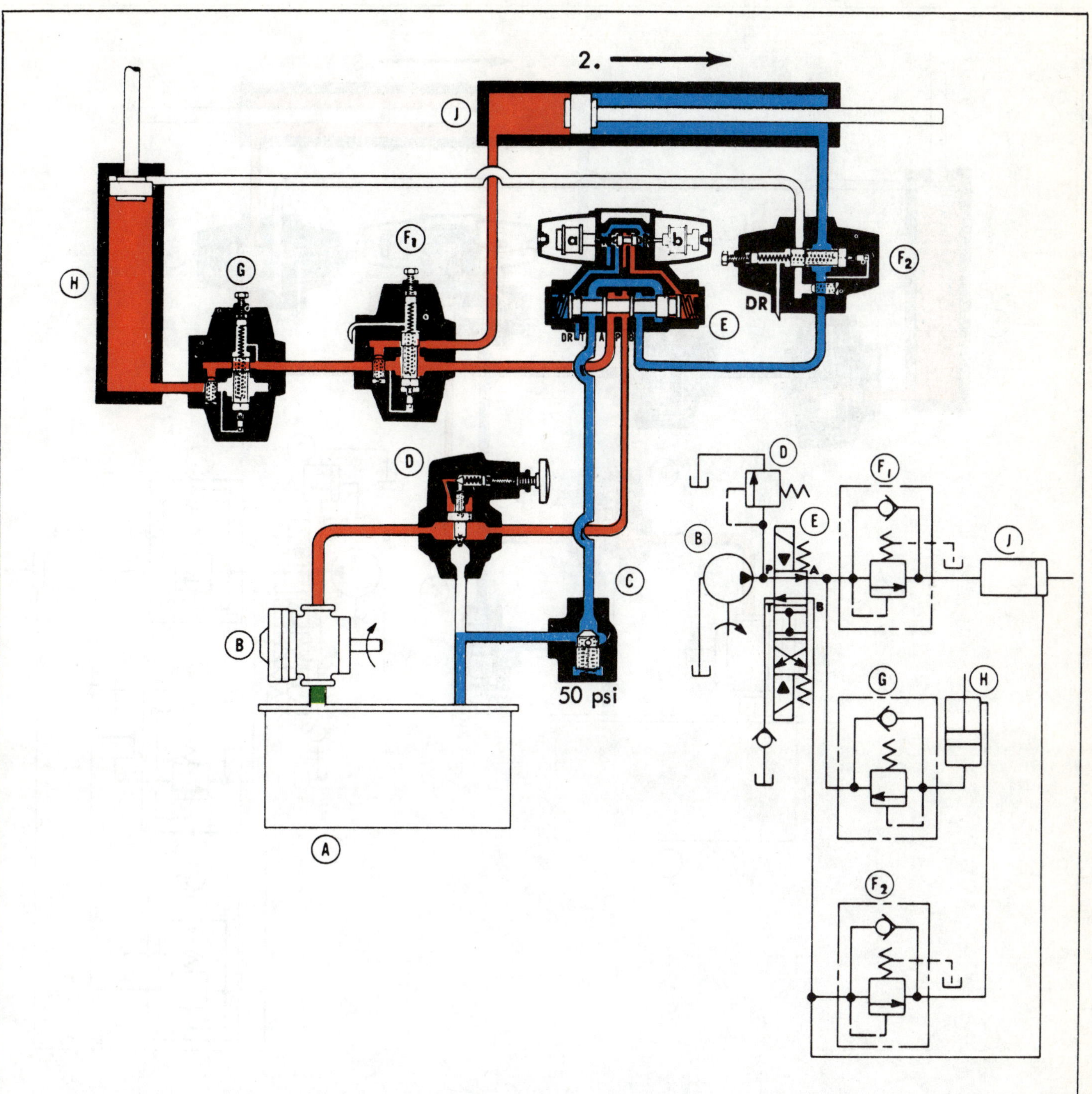

Phase #2 - Solenoid (Ea) Energized

Pressure increase, on completion of phase #1, causes flow to sequence through (F1) into head end of (J). Discharge from rod end of (J) flows freely to tank through valves (F2), (E), and (C). Valve (F1) assures minimum pressure equal to its setting in (H) during extension stroke of (J). When (J) is fully extended, pressure increases to setting of valve (D) which provides overload protection for (B).

Fig. 13-10, View B. - Sequence Circuit - Extend Cylinder J While Holding H

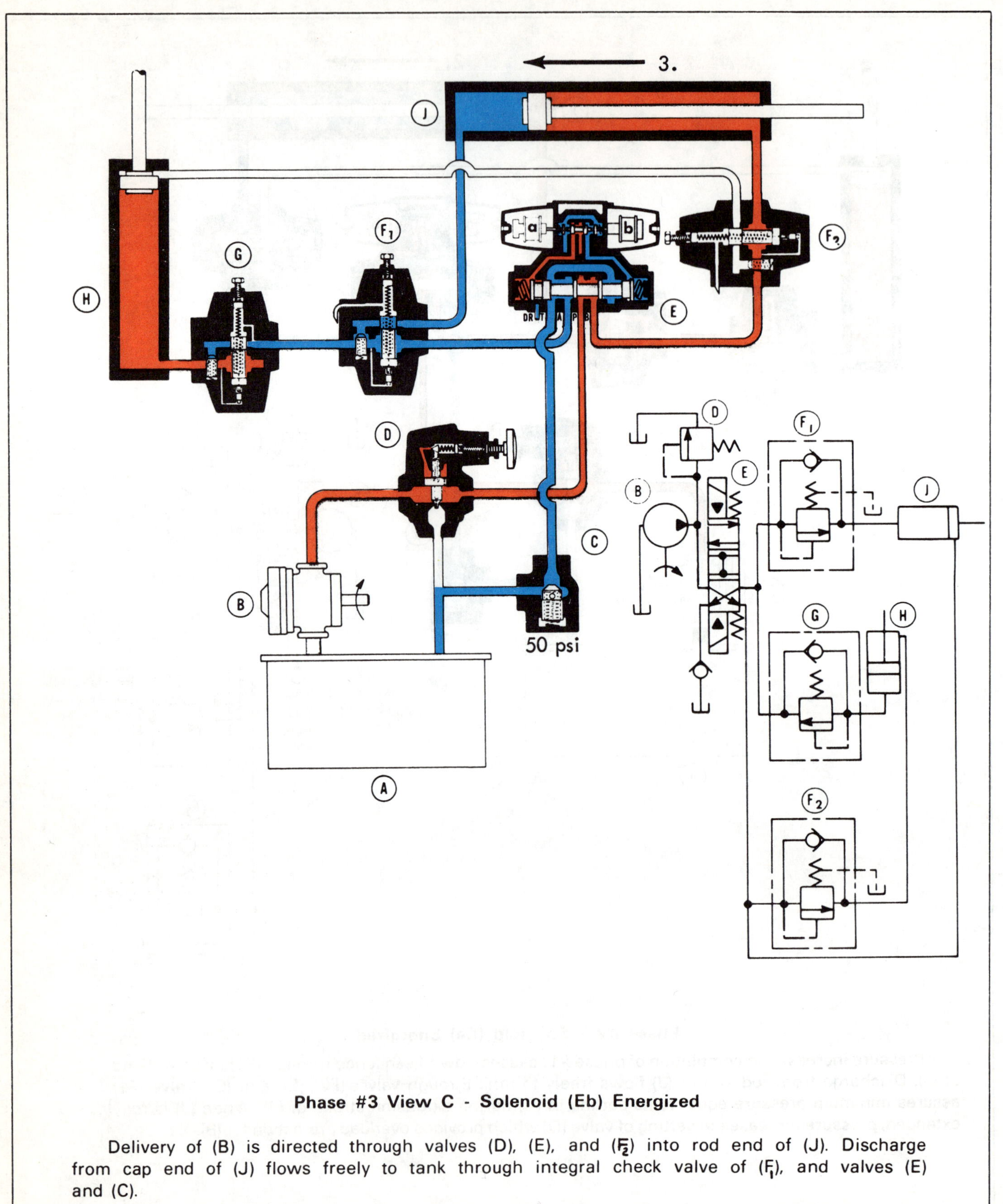

Phase #3 View C - Solenoid (Eb) Energized

Delivery of (B) is directed through valves (D), (E), and (F_2) into rod end of (J). Discharge from cap end of (J) flows freely to tank through integral check valve of (F_1), and valves (E) and (C).

Fig. 13-10, View C. - Sequence Circuit - Retract Cylinder J

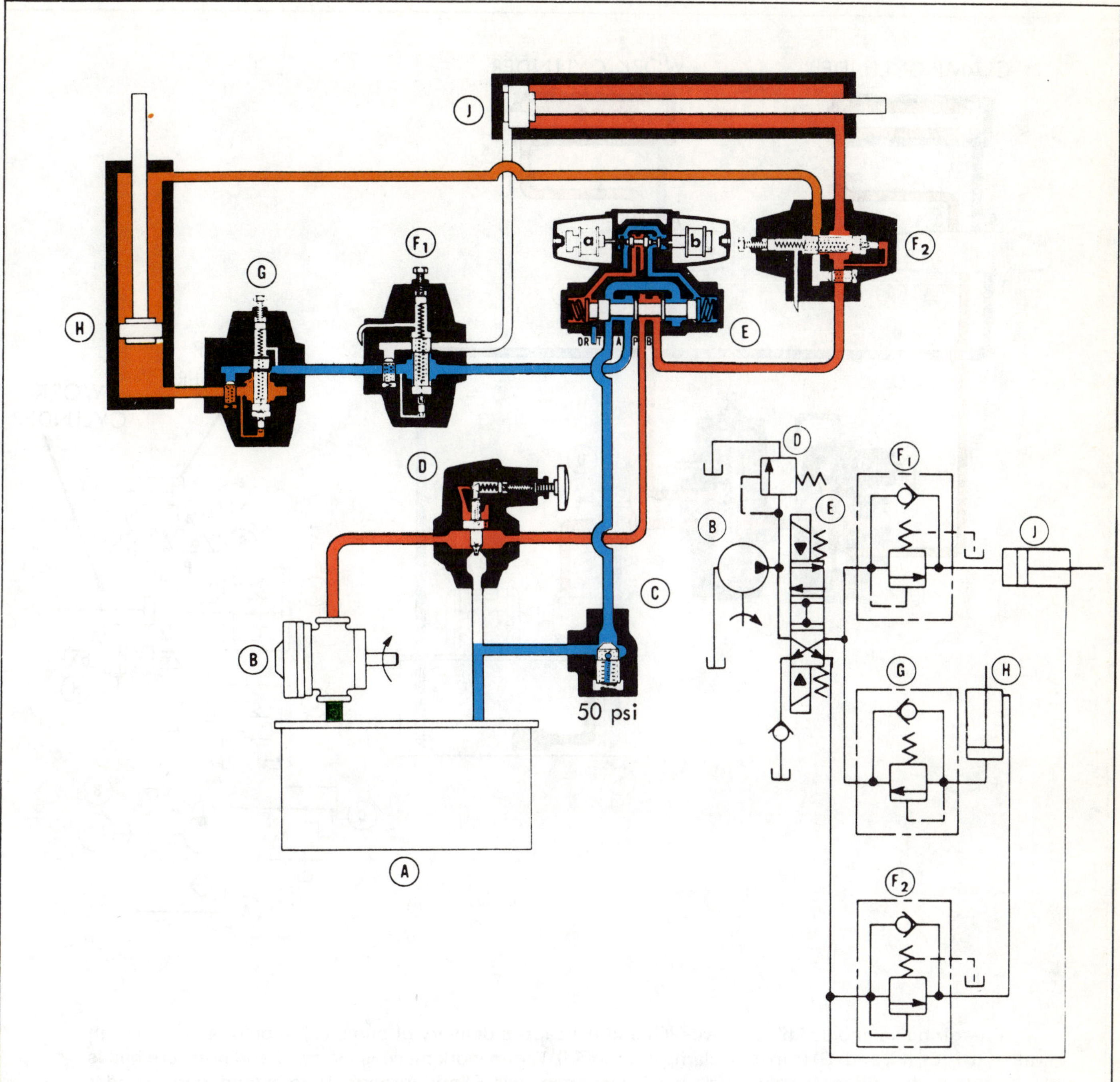

Phase #4 - Solenoid (Eb) Energized

Pressure increase, on completion of phase #3, causes flow to sequence through (F2) into rod end of (H). Discharge from head end of (H) flows through (G) at its pressure setting and then freely to tank through valves (F1), (E), and (C). Valve (F2) assures minimum pressure equal to its setting in rod end of (J) during retraction of (H). Valve (G) provides back pressure to prevent (H) from falling out of control in lowering.

Fig. 13-10, View D. - Sequence Circuit - Retract Cylinder H

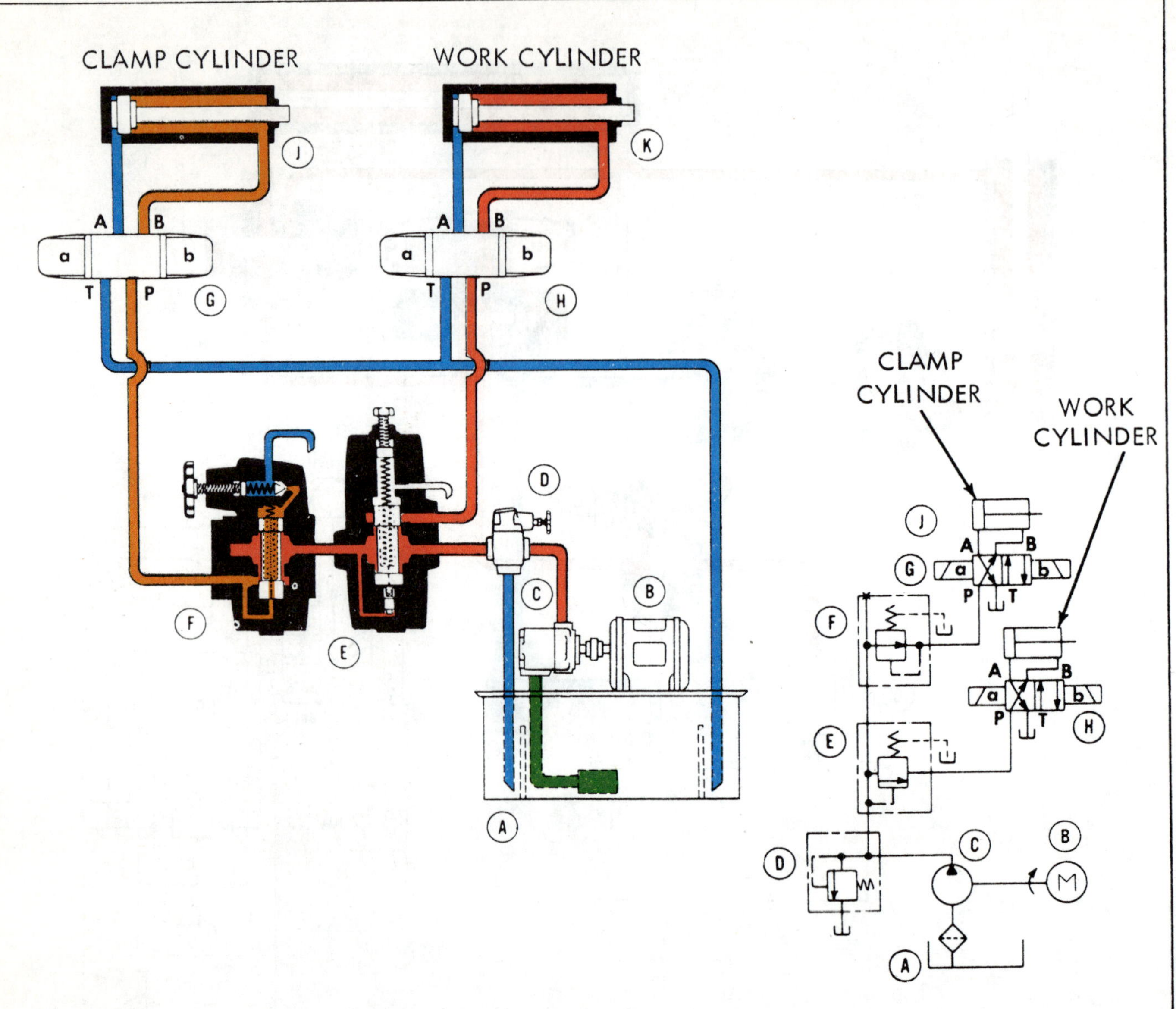

Energizing solenoids "b" of valves (G) and (H) causes delivery of pump (C) to be directed through valves (D), (E), (F), and (G) to extend clamp cylinder (J). When work piece is clamped and pressure builds up to setting of sequence valve (E), flow will sequence over (E) and through (H) to extend work cylinder (K). Valve (E) assures minimum pressure, equal to its setting, during operation of (K). Reducing valve (F) limits the maximum pressure in (J).

De-energizing solenoid "b" of (H) and energizing "a" of (H) causes delivery of (C) to be sequenced over (E) and through (H) to retract (K). When (K) is fully retracted, solenoid "b" of (G) is de-energized and "a" energized. Delivery of (C) is directed through valves (E), (F), and (G) to retract (J).

Fig. 13-11. - Controlled Pressure Clamping Circuit

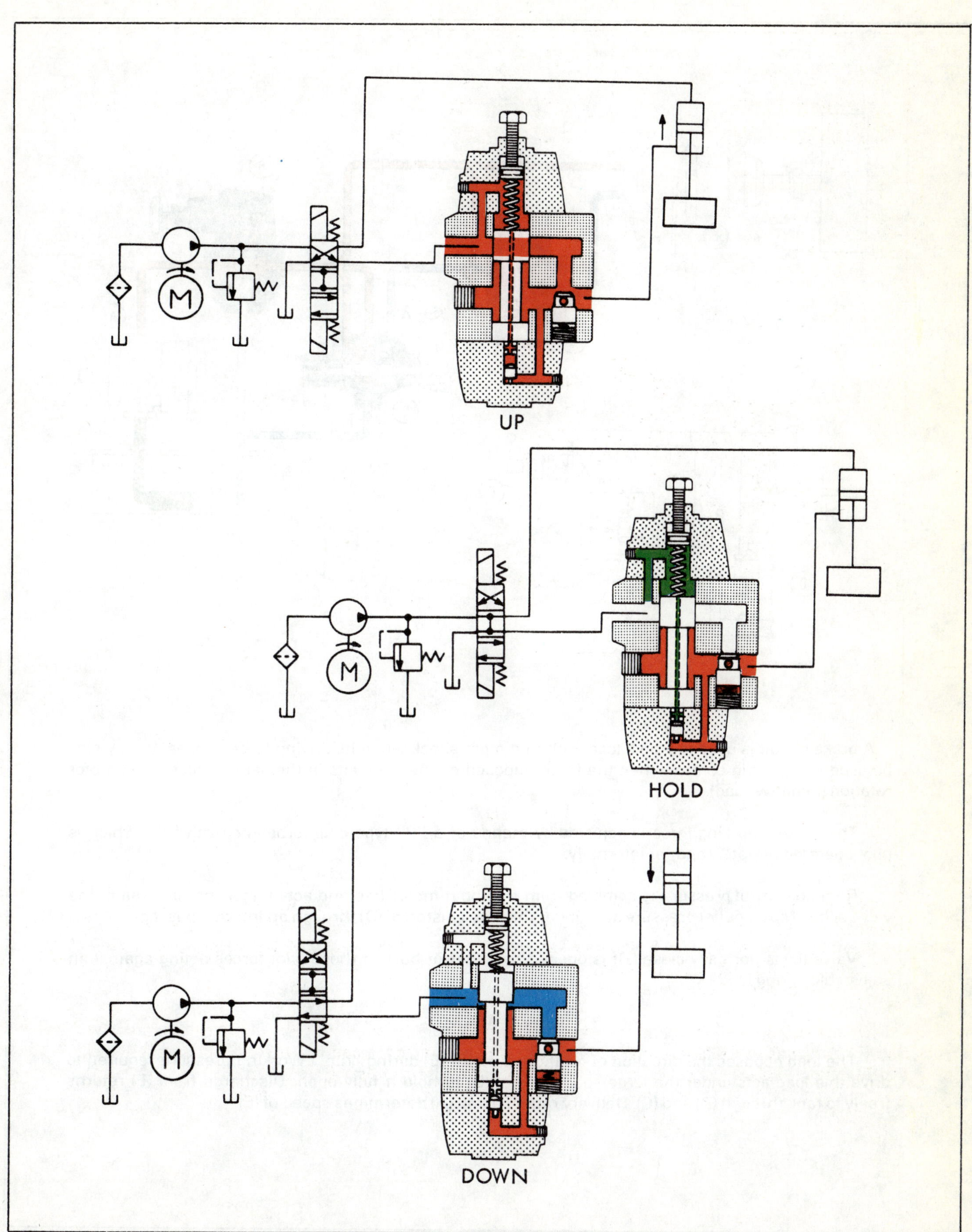

Fig. 13-12. - Counterbalance Circuit

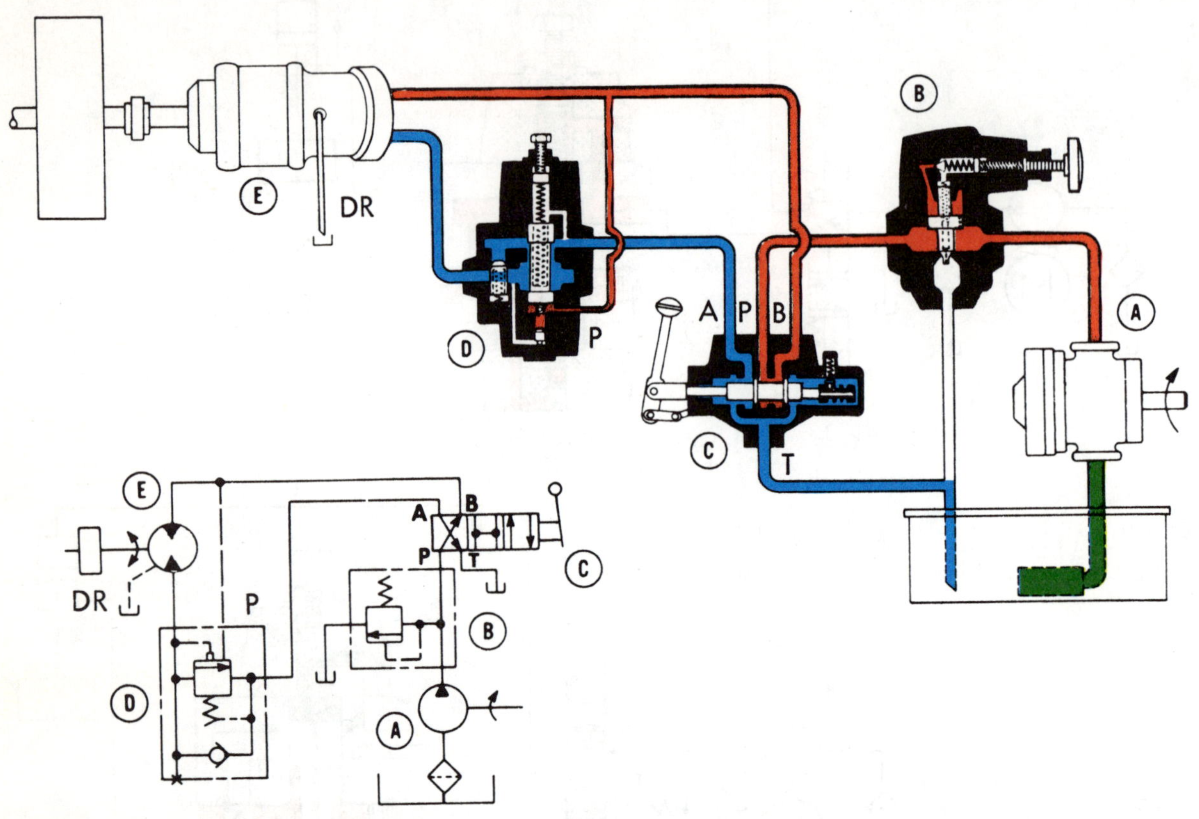

A brake circuit is used to stop a load with minimum shock when its driving force ceases. It may also be used to maintain control when the force imposed by the load acts in the same direction as motor rotation (negative load).

The desired braking force is adjusted by means of a "P" type counterbalance valve (D) which is pilot operated remotely and/or internally.

Remote control pressure is sampled from the input motor line and acts under the full area of the valve spool. Motor outlet pressure acts under the small piston of (D) through an internal passage.

Valve (D) is normally closed. It is opened by either or both of these pilot forces acting against an adjustable spring load.

RUN

The load opposes the direction of rotation of motor (E) during "run". Working pressure required to drive this load acts under the large spool area of (D) to hold it fully open. Discharge from (E) returns freely to tank through (D) and (C). Delivery rate of pump (A) determines speed of (E).

Fig. 13-13, View A. - Brake Circuit -

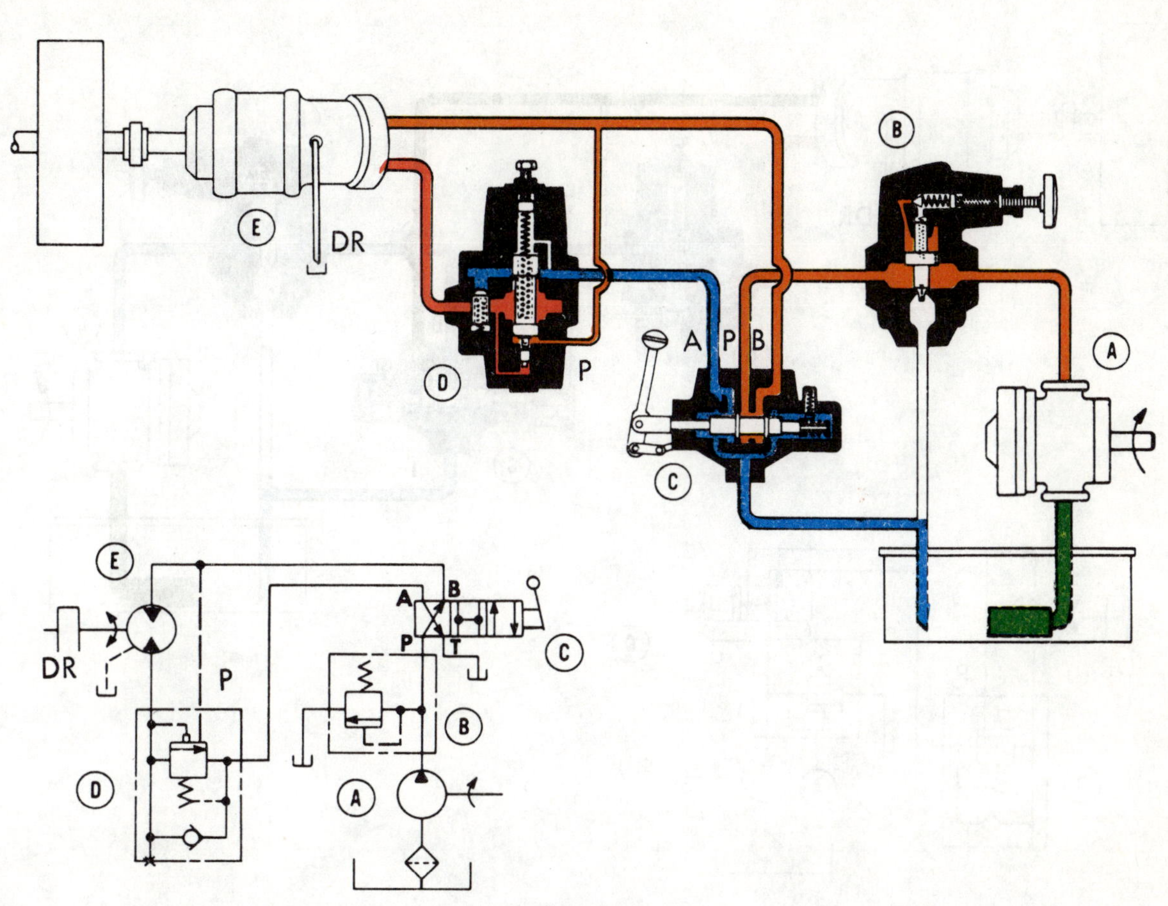

NEGATIVE LOAD

The load may act in the same direction as rotation of motor (E) in certain applications. This "negative load" assumes a portion of the driving force on motor (E) which reduces pressure at the motor inlet.

Reduced pressure at motor inlet, effective under the valve spool of (D), permits the spool to move toward its closed position, thus restricting the discharge from (E).

Restricted flow through (D) creates back pressure in the outlet of (E). This back pressure acts under the small piston of (D).

The sum of the pressures acting under the valve spool and small piston of (D) holds the valve spool at the restricting position required for sufficient back pressure to maintain control of the load on (E).

The extent of negative loading determines the amount of back pressure on (E).

Fig. 13-13, View B. - Brake Circuit - Overrunning Load

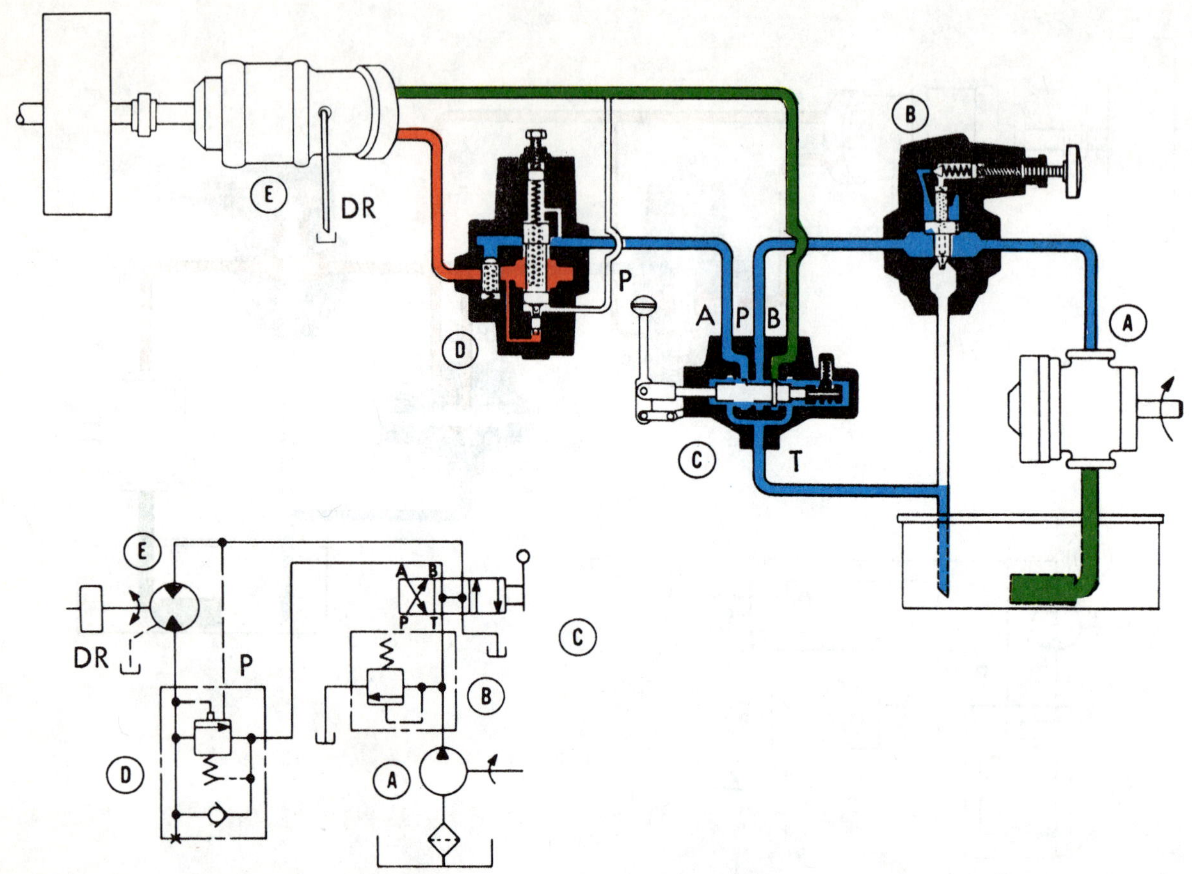

BRAKING

Valve (C) is shifted to the "neutral" position to brake the load on motor (E). Pump (A) delivery is open to tank through valve (C).

Load inertia continues to drive (E) causing it to act as a pump. Inlet fluid to (E) is supplied through (C).

With the inlet of (E) open to tank, pilot pressure under the valve spool of (D) becomes zero permitting it to move toward the closed position. This restricts discharge from (E) creating back pressure at its outlet.

Back pressure at outlet of (E) acts under the small piston of (D) opposing the spring force. These two opposing forces hold the valve spool at the restricting position. Adjusted setting of (D) therefore determines braking pressure and rate of decerlation.

Fig. 13-13, View C. - Brake Circuit - Braking

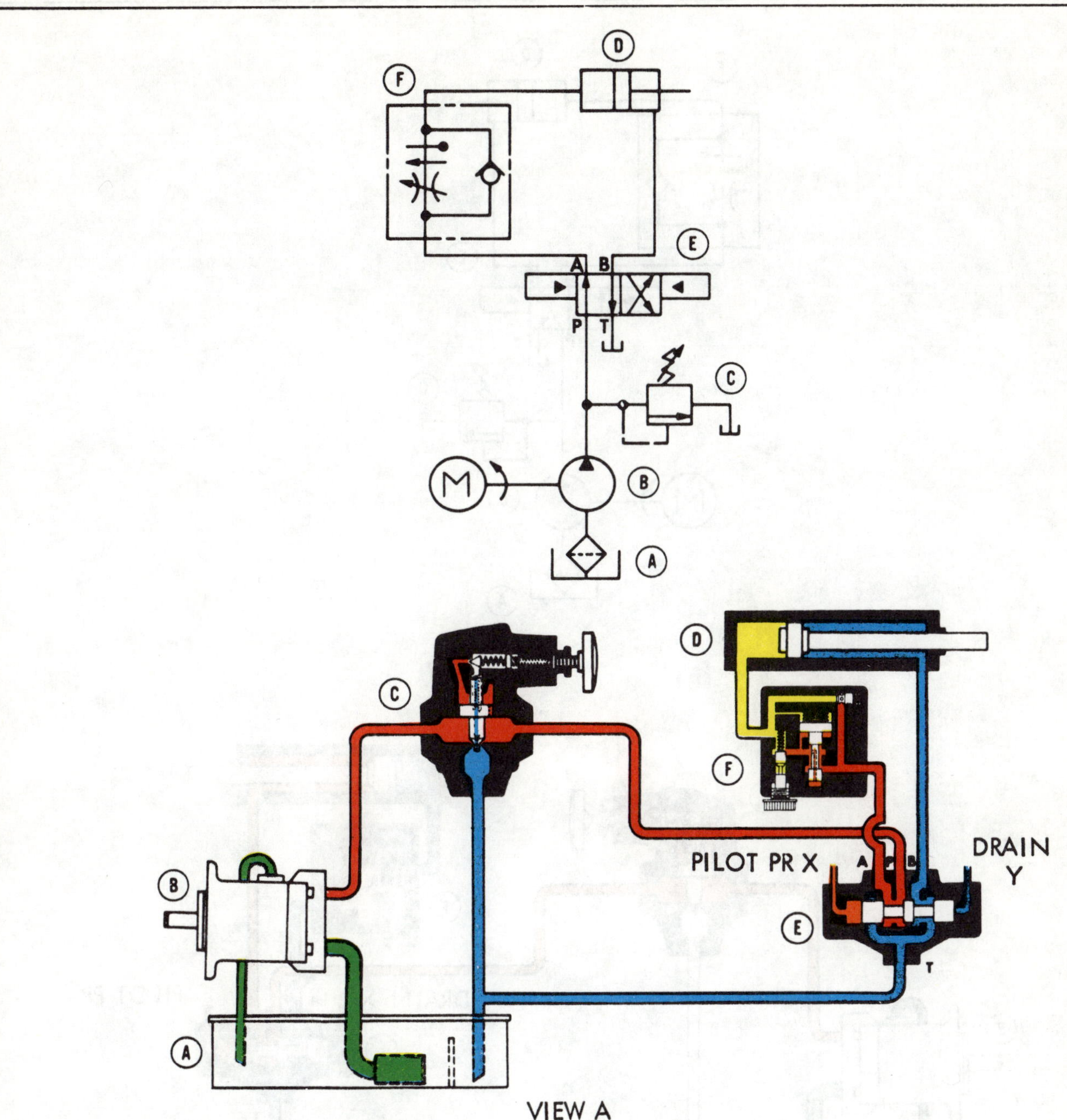

Flow path is from reservoir (A) through pump (B), relief valve (C), directional valve (E), and flow control (F), to cylinder (D). Flow path from (D) is through (E) to (A).

Valve (F) meters less flow than (B) delivers. Excess must return through (C) to (A). Valve (C) determines pressure imposed on (B). Input power is a function of delivery of (B) and pressure setting of (C) regardless of work load or piston speed. Valve (C) should be set only high enough to assure the recommended minimum pressure drop across (F) when maximum work load is encountered.

Piston speed is a function of piston area and flow metered-in by (F). Since piston area is constant, piston speed can be affected only by a fluctuation of flow through (F). Speed is independent of variation in pump discharge.

View A shows piston extending control of cylinder is maintained only when work load opposes direction of piston movement because discharge from (D) returns freely to (A).

Fig. 13-14, View A. Meter-In Flow Control - Metering

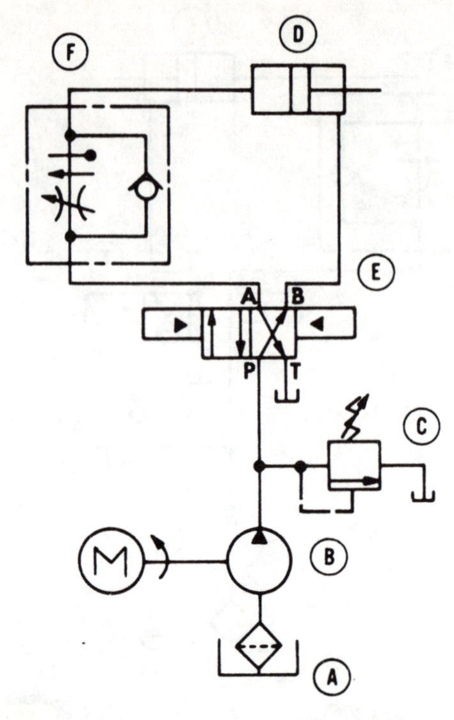

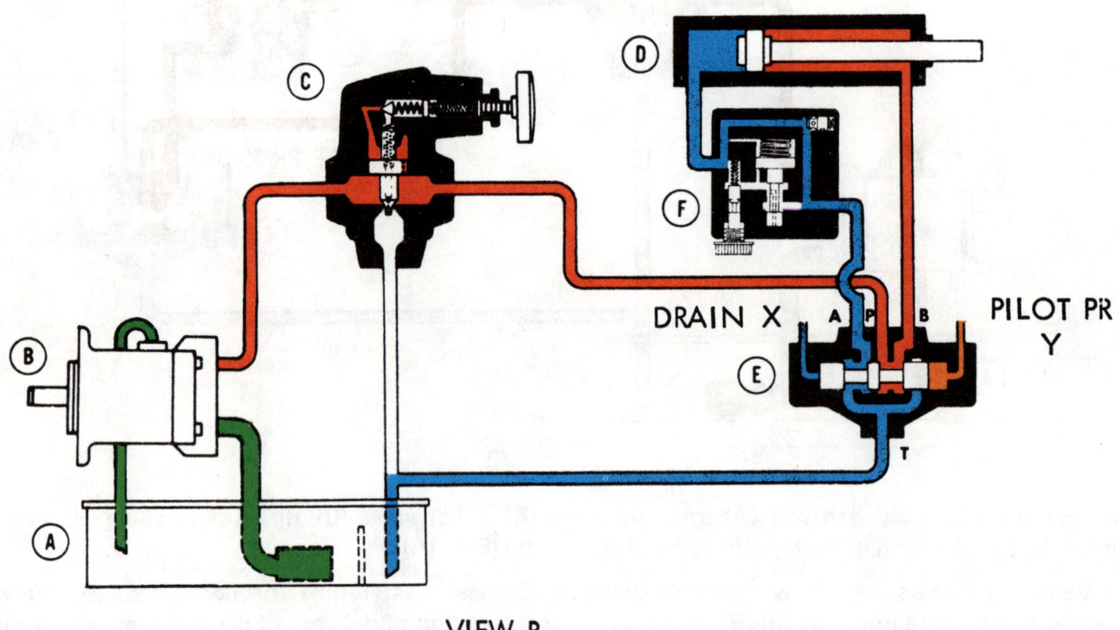

View B shows rapid return stroke with check valve used to bypass flow control.

Fig. 13-14, View B. - Meter-In Flow Control - Free Flow

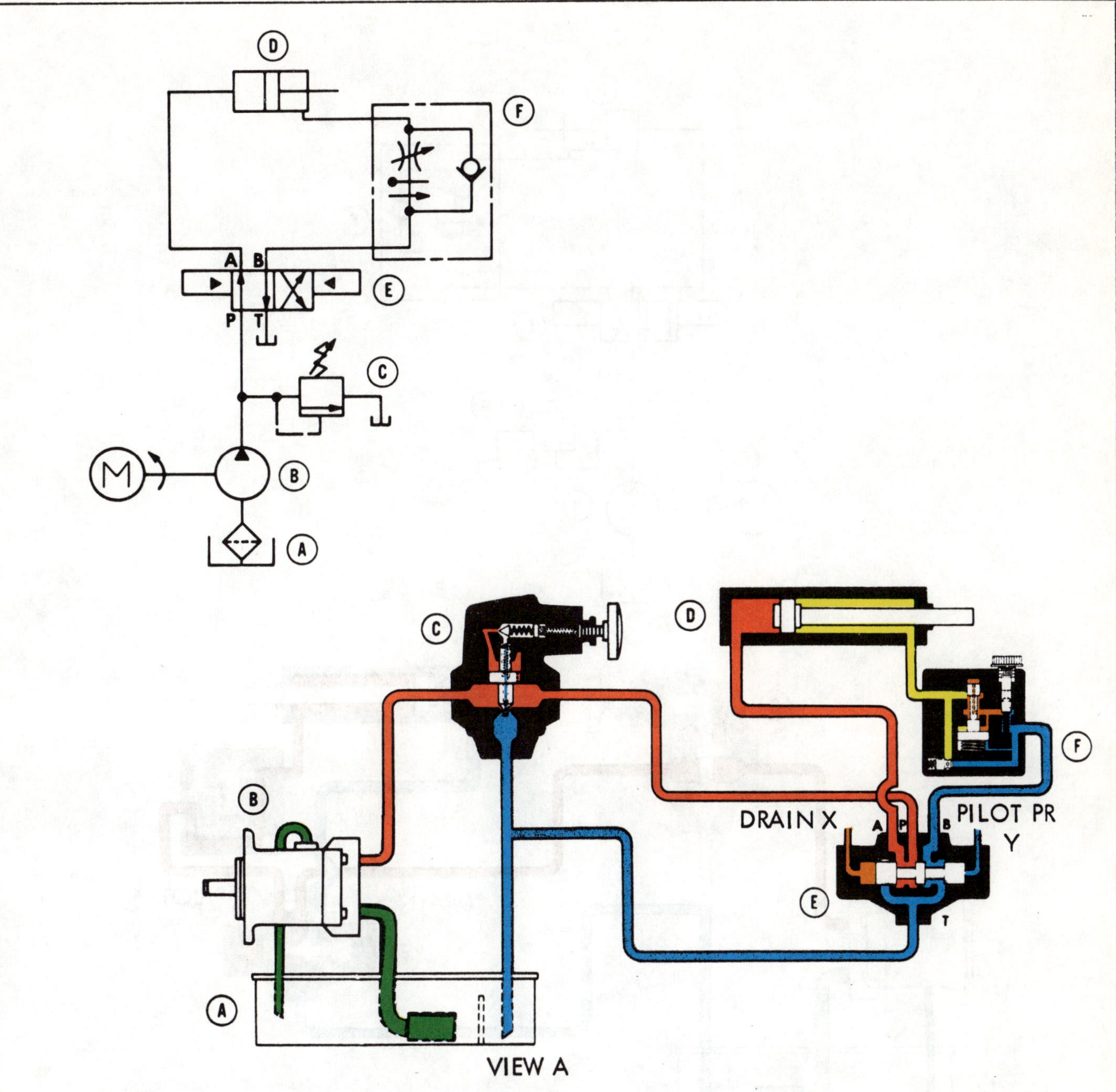

Flow path is from reservoir (A) through pump (B), relief valve (C), and direction valve (E) to cylinder (D). Flow path from (D) is through flow control (F) and (E) to (A).

Valve (F) meters less flow than that which would be discharged from rod end of (D) if all of the delivery of (B) were directed into (D). Pump flow in excess of that to (D) must return through (C) to (A). Valve (C) determines pressure imposed on (B). Input power is a function of the delivery of (B) and pressure setting of (C) regardless of work load or piston speed. Valve (C) should be set only high enough to assure the recommended minimum pressure drop across (F) when maximum work load is encountered.

Piston speed is a function of piston area and flow metered-out of (D) by (F). Since piston area is constant, piston speed can be affected only by a fluctuation of flow through (F). Speed is independent of variation in pump discharge.

Fig. 13-15, View A. Meter-Out Flow Control - Metering

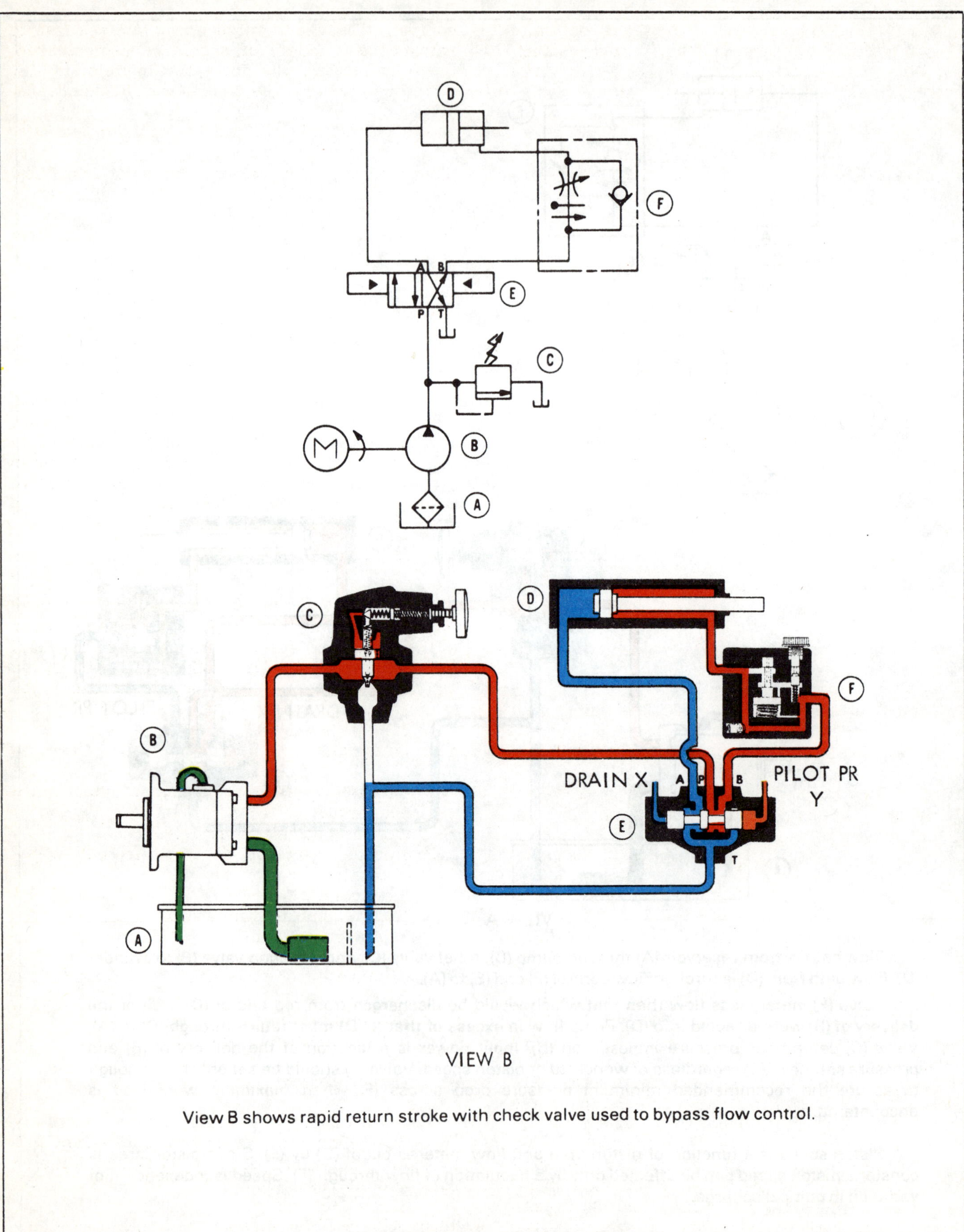

View B shows rapid return stroke with check valve used to bypass flow control.

Fig. 13-15, View B. - Meter-Out Flow Control - Free Flow

speed is constant regardless of the direction of forces imposed by the work load. In this circuit, too, the pump must operate at the relief valve setting during the feed stroke.

Bleed-Off Flow Control

In Figure 13-16, the flow control valve meters oil from the pressure line to tank rather than to the system, providing speed control in both directions. While less precise than the two previous methods, the bleed-off circuit permits some savings in horsepower since operating pressure is only that required to move the cylinder. Excess pump flow returns to tank through the flow control.

RAPID ADVANCE TO FEED CIRCUITS

Three methods of making a transition from rapid advance to a slower feed speed in a meter-out circuit are shown in Figure 13-17.

In view A, a deceleration valve is piped in parallel with the meter-out flow control. During rapid advance, exhaust flow from the cylinder rod end passes freely through the deceleration valve. When the cam closes the deceleration valve, the oil must take the other path through the flow control. To retract the cylinder, oil into the rod end flows freely over the check valve in the deceleration valve.

In view B, the parallel flow path around the flow control valve is through a spring offset, solenoid operated directional valve. This valve allows free return flow so long as its solenoid is energized. When the feed position is reached, a limit switch is tripped and breaks the solenoid circuit. The directional valve shifts to block exhaust flow, which must then go through the flow control. A separate check valve is provided for free flow in to the cylinder for a rapid return stroke.

Leakage past the sliding spools of the deceleration valve shown in view A and the directional valve in view B would affect the feed rate.

View C shows the transition made with a pilot-operated check valve which because of its low leakage characteristics is used where extremely accurate feed rates are required. Pilot pressure to this check valve is supplied through an offset directional valve. With the directional valve solenoid energized, the check valve is opened to allow free flow from the cylinder rod end. Tripping the limit switch at the feed position breaks the solenoid circuit and vents the check valve pilot line allowing the poppet to seat. Exhaust flow is then forced to pass through the flow control. The check valve opens for free flow to retract the cylinder. It should be remembered, too, that leakage past the cylinder piston is also a factor which must be considered in such cases.

Flow Control and Overload Relief Valve

Where meter in control can be used, the flow control and relief valve (Fig. 13-18) will control feed speed. An additional directional valve is incorporated to by-pass the flow control for rapid advance and rapid return.

The circuit is shown in the "feed" part of the cycle. The three-position, spring-centered directional valve is directing flow to the cylinder. The offset directional valve is blocking the parallel path around the flow control. Its solenoid can be energized (during rapid advance and return) to divert pump delivery around the flow control valve. The relief valve incorporated in the flow control provides overload protection in all operating conditions. The 50 psi check valve in the tank line assures that pilot pressure is always available to shift the directional valves.

ROTARY HYDROSTATIC DRIVES

The purpose of any transmission or "drive" is to match the torque and speed of the prime mover to the torque and speed requirements of the load. Hydraulic or hydrostatic drives utilize positive displacement pumps, motors and various controls for this purpose.

The advantages of a hydrostatic drive include:

* Infinitely variable regulation of output speed and torque.
* Ease and accuracy of control.
* Smooth, stepless acceleration and speed changes.
* Low inertia.
* Low weight-to-power ratio.
* Increased reliability.
* Flexibility of component location.
* Elimination of drive shafts and complicated gearing.
* Dynamic braking
* Built in overload protection.

Open Circuit Drives

In an open circuit the pump draws its supply from the reservoir. Its output is directed to a hydraulic motor and discharged from the motor back into the reservoir.

Figure 13-19 illustrates an open circuit containing the basic units required for a reversible hydrostatic drive.

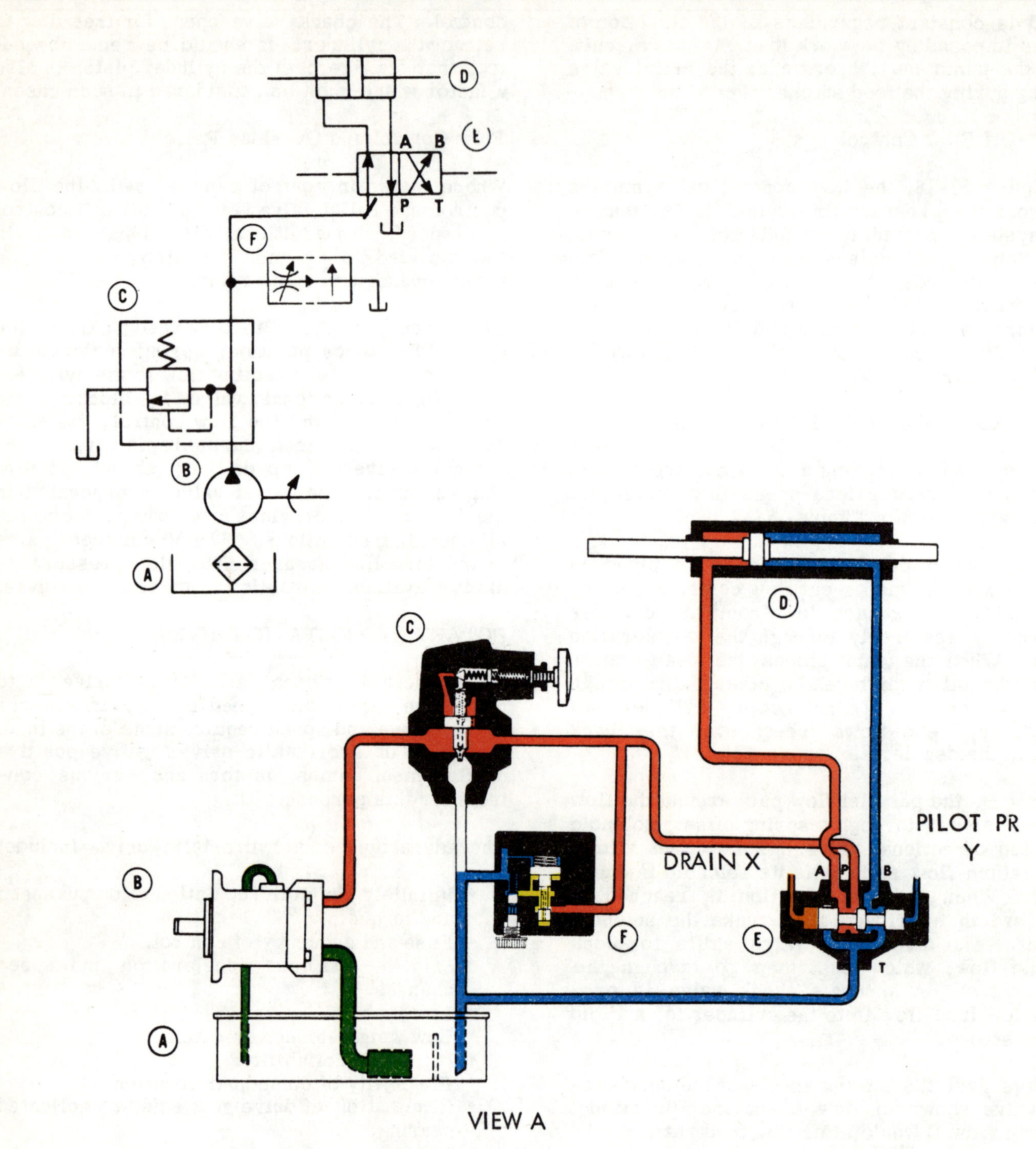

Flow path is from reservoir (A) through pump (B) and relief valve (C) to "tee" into a flow path to cylinder (D) through directional valve (E) and a flow path to (A) through flow control (F). Flow path from (D) is through (E) to (A).

Figure 13 - 16, View A. Bleed-Off Flow Control - Cylinder Extending

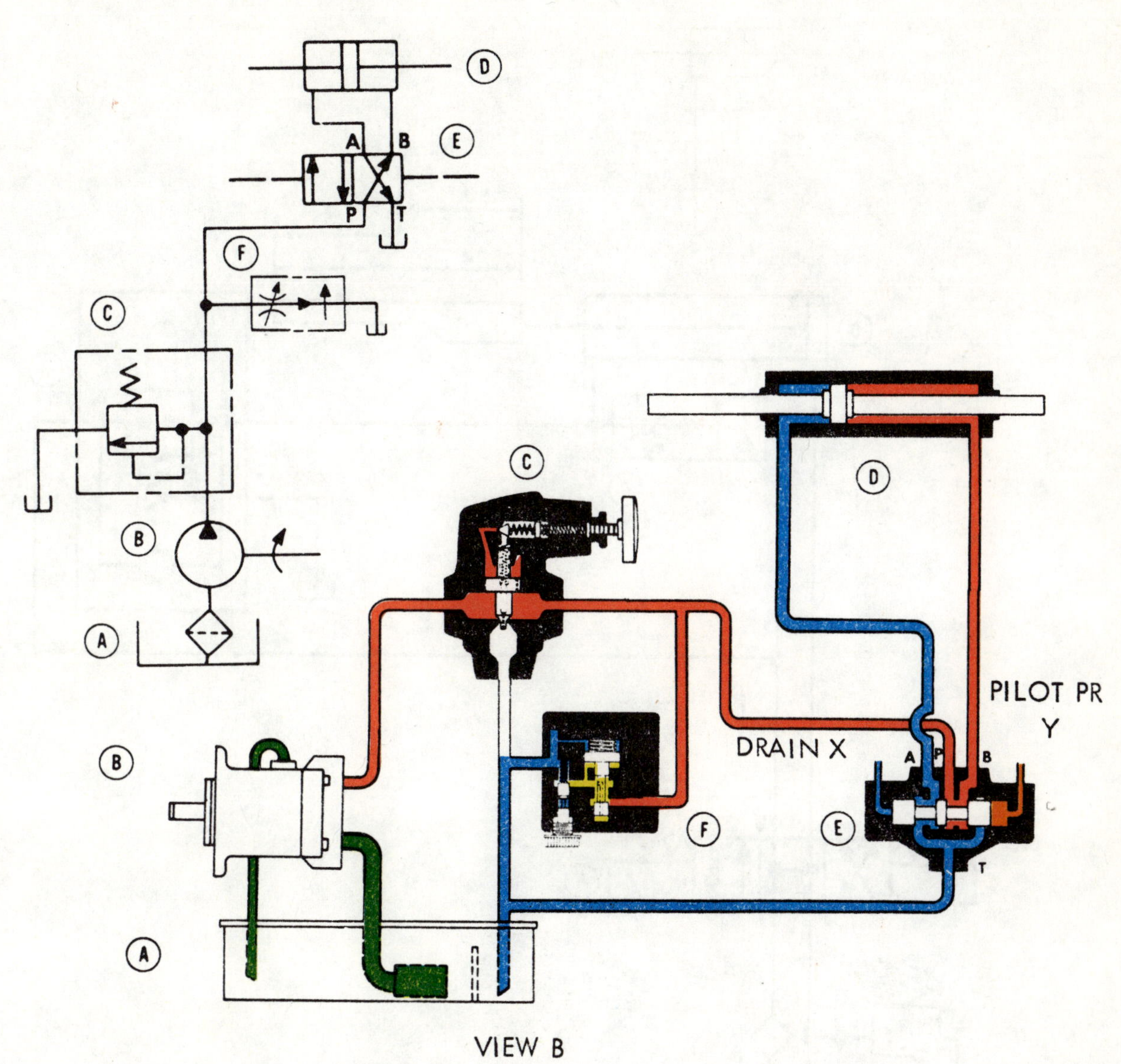

Pressure imposed on (B) is created by the work load. Input power is a function of delivery of (B) and working pressure and varies directly with the work load. Valve (C) limits maximum operating pressure and protects the hydraulic system against pressure overload.

Piston speed is a function of piston area and delivery of (B) less flow metered to tank through (F). Since area of piston is constant, piston speed is affected only by flow variation. Variations may be caused by fluctuations of either flow through (F), or delivery or (B), or both.

With (F) teed-off the pressure line, piston speed is controlled in both directions. View A shows piston extending and View B shows piston retracting. In either case, control of (D) is maintained only when the work load opposes direction of piston movement because discharge from (D) returns freely to tank.

Figure 13 - 16, View B. Bleed-Off Flow Control - Cylinder Retracting

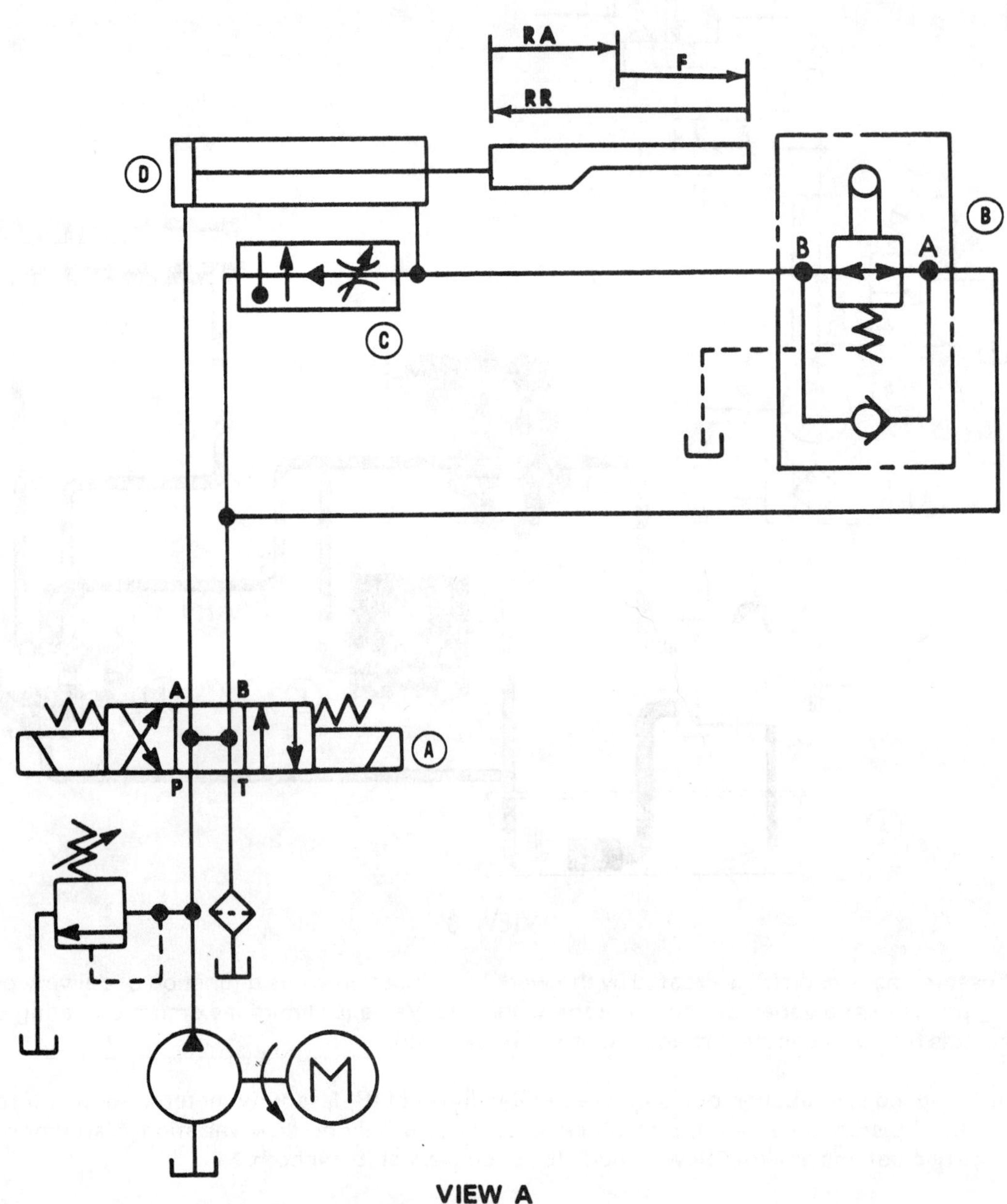

VIEW A

Directional valve (A) is shifted to direct flow into head end of cylinder (D). Discharge from rod end of (D) flows through normally open deceleration valve (B) and freely to tank through (A) for rapid advance. At end of rapid advance, cam on (D) depresses spool of (B) to the closed position. Discharge from rod end of (D) is then metered by flow control (C) for the feed stroke. Valve (A) is reversed to direct flow through integral check valve of (B) freely into rod end of (D) for rapid return.

Fig. 13-17, View A. - Rapid Advance to Feed - Using Deceleration Valve

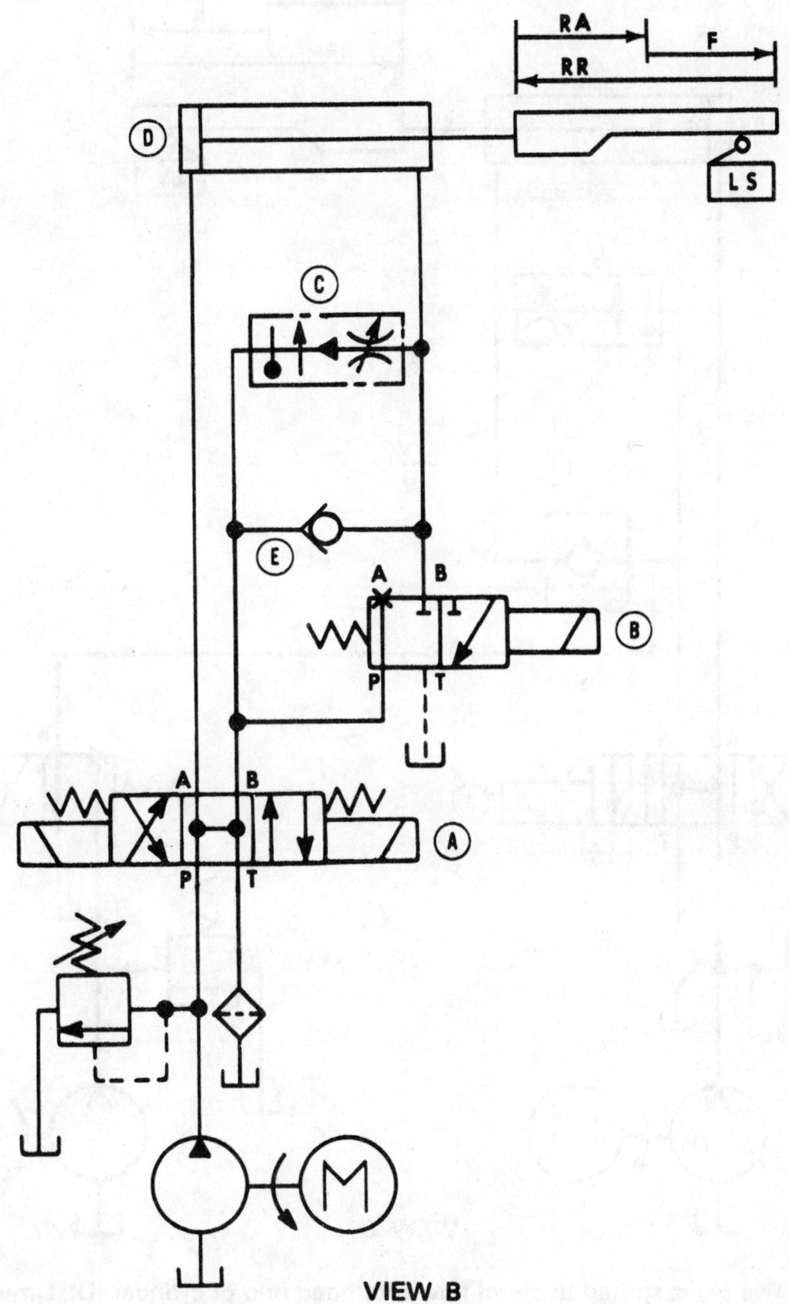

VIEW B

Directional valve (A) is shifted to direct flow into head end of cylinder (D). Directional valve (B) is shifted from its normal spring offset position to permit discharge from rod end of (D) to flow freely to tank through (B) and (A) for rapid advance. At end of rapid advance, cam on (D) contacts a limit switch to allow (B) to spring return to its normal position. Discharge from rod end of (D) is then metered through flow control (C) for the feed stroke. Valve (A) is reversed to direct flow through check valve (E) freely into rod end of (D) for rapid return.

Fig. 13-17, View B. - Rapid Advance to Feed - Using Directional Valve

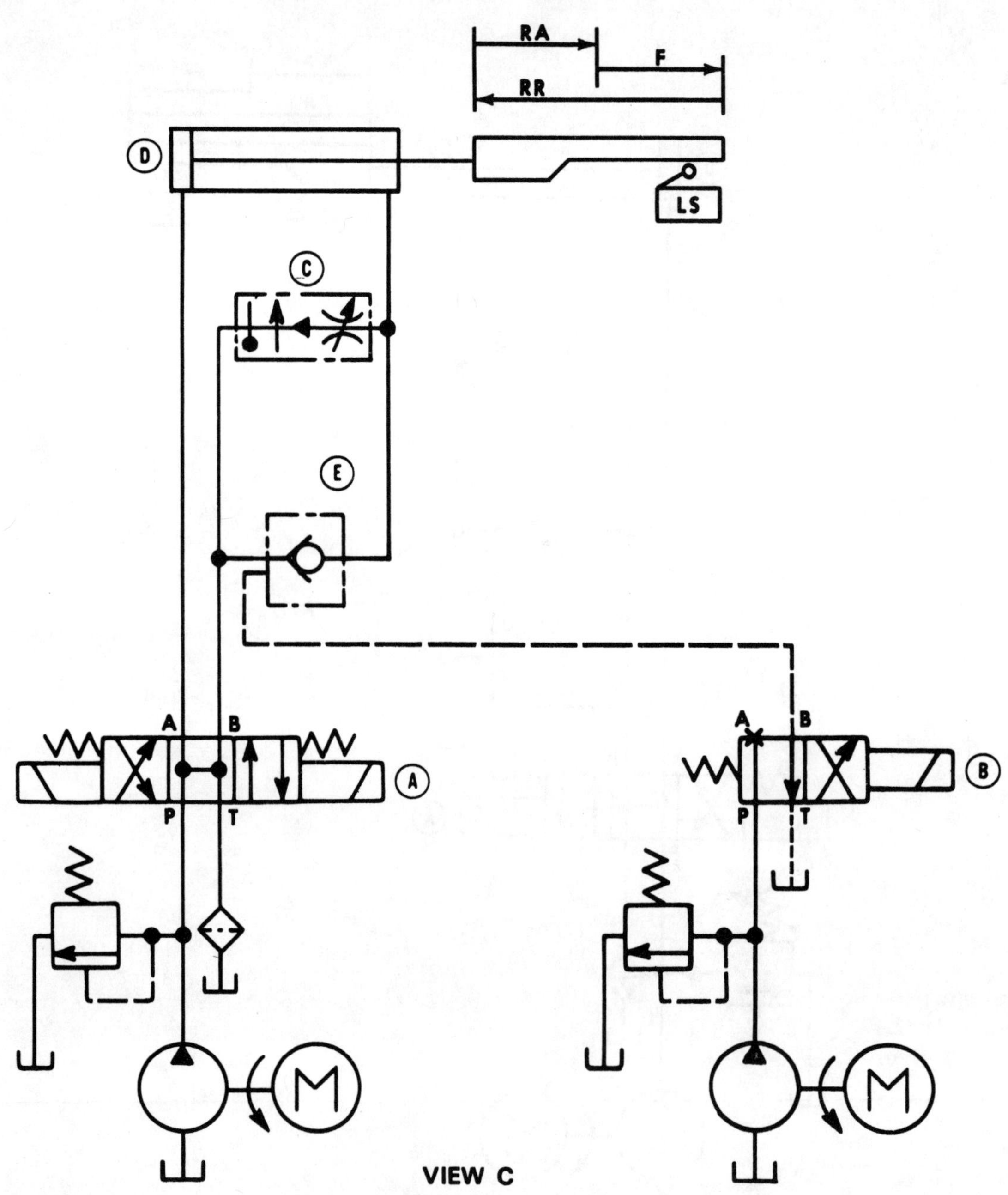

Directional valve (A) is shifted to direct flow into head end of cylinder (D). Directional valve (B) is shifted to direct pilot flow to remotely open pilot operated check valve (E). Discharge from rod end of (D) flows freely to tank through (E) and (A). At end of rapid advance, cam on (D) contacts a limit switch to allow (B) to spring return to its normal position. Valve (E) closes and discharge from rod end of (D) is then metered by flow control (C) for the feed stroke. Valve (A) is reversed to direct flow freely through (E) into rod end of (D) for rapid return.

Fig. 13-17, View C. - Rapid Advance to Feed - Using Pilot Operated Check Valve

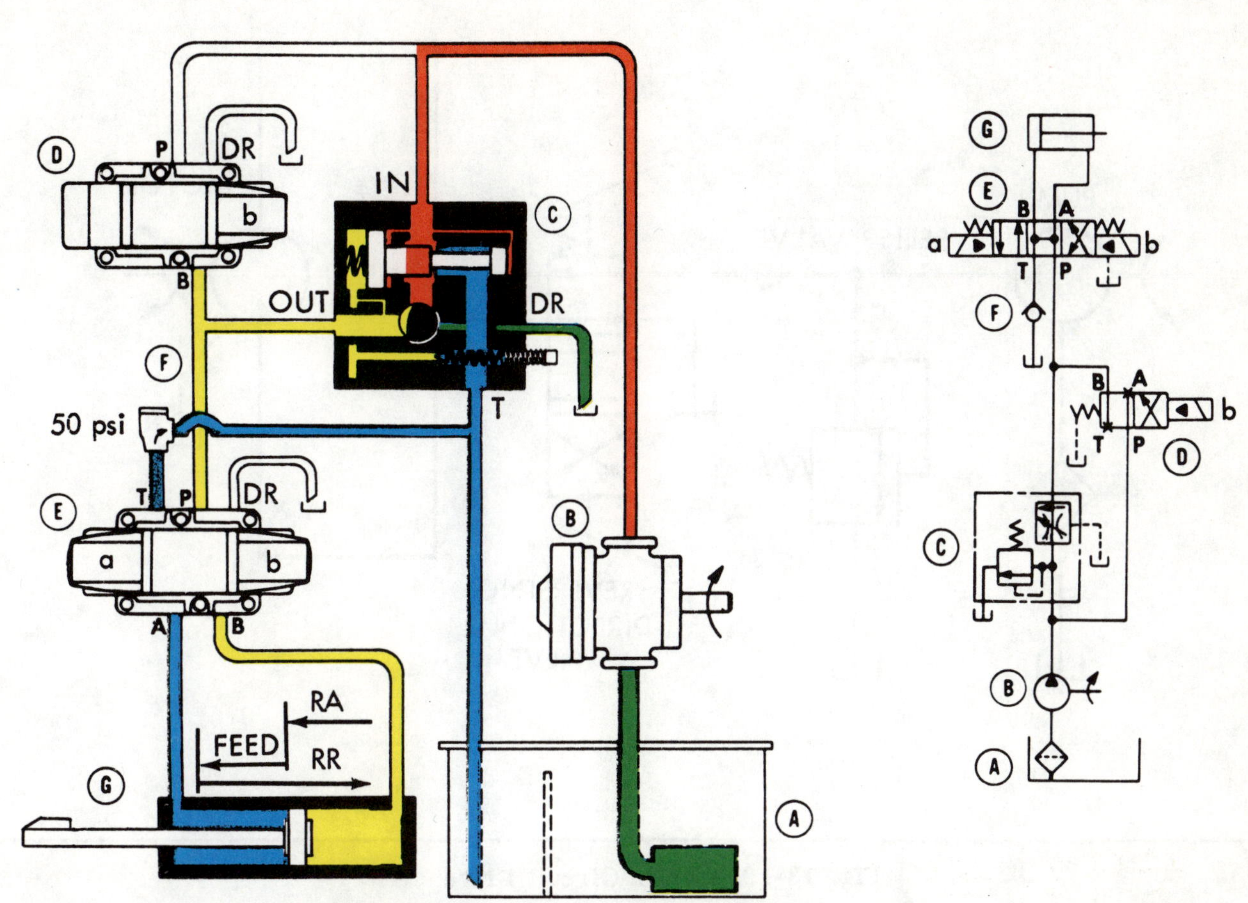

Throttle adjustment of valve (C) determines feed speed and pressure adjustment of (C) determines maximum operating pressure and provides overload protection.

Solenoids of valves (D) and (E) are de-energized for "idle" permitting (D) to spring-offset and (E) to spring-center. Volume passed through throttle of (C) is returned to tank through the open center of (E) and through valve (F) which assures minimum pilot pressure for operation of (D) and (E). Excess pump delivery is returned to tank through (C) at a pressure equivalent to the low spring values in valve (F) and on spool of (C). Pump is considered to be "unloaded".

Valve (E) is the main directional control determining direction of motion of piston of cylinder (G). Solenoid "A" is held energized for advance, solenoid "B" for return. Valve (D) is a bypass valve which, with its solenoid "B" energized, permits pump flow to bypass the throttle of (C) for rapid motion of (G). Its solenoid is de-energized to block the bypass for "feed".

Operating pressure of pump (B) during rapid motion of (G) is equal to the back-pressure imposed by (F) plus the equivalent in work resistance. During "feed", operating pressure of pump is equal to back-pressure of (F), work resistance, plus spring load on spool of (C). Pump pressure is at all times proportional to and only slightly higher than equivalent in work load. During "feed" excess pump delivery not permitted to pass through throttle of (C) is returned to tank through (C).

Fig. 13-18. - Flow Control and Relief Valve

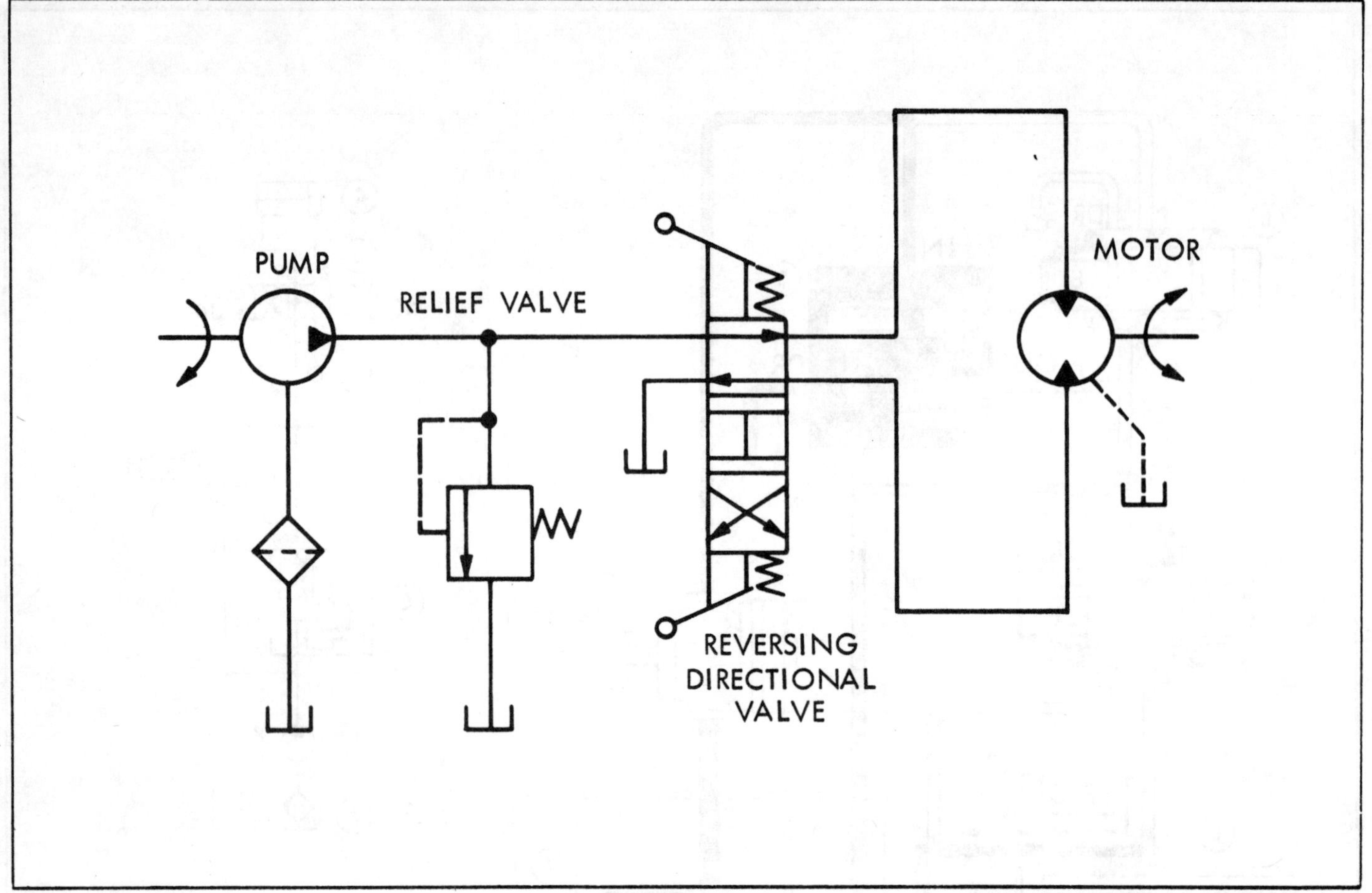

Fig. 13-19. - Open Circuit Drive

If the pump and motor both have the same displacement, theoretical output speed and torque will be equal to input speed and torque. The drive thus functions simply as a liquid drive shaft. Should the motor be twice the pump displacement, output speed would be half input, but output torque would be double input. Other combinations of displacement would produce output speed proportional to pump motor displacement ratio and torque proportional to motor-pump displacement ratio.

This type of drive, using a fixed displacement pump, can incorporate a speed control in the form of a flow control valve.

Maximum torque, of course, is limited by the setting of the relief valve.

Closed Circuit Drives

In a closed circuit drive, exhaust oil from the motor is returned directly to the pump inlet. Figure 13-20 shows a single direction drive of this type. Motor speed is adjustable by varying the pump displacement. Torque is a function of motor displacement and the relief valve setting. Make up oil to replenish leakage from the "closed loop" flows into the low pressure side of the circuit through a line from the reservoir.

Reversible Closed Circuit Drives

Many closed circuit drives include a variable displacement, reversible pump with one outlet connected to a motor port and the other motor port connected to the opposite pump outlet (Fig. 13-21).

This permits the motor to be driven in either direction at infinitely variable speeds each determined by the position of the pump displacement control. In the circuit shown, losses from internal leakage are replaced by a replenishing pump which maintains a positive pressure on the low pressure side of the system. Overload protection is provided by cross-line relief valves.

Characteristics of Closed Circuit Drives

Closes circuit drives may be designed with fixed or variable pumps and motors in any combination. Following are their characteristics.

Fixed Displacement Pump and Motor - Output speed and torque equal input speed and torque if

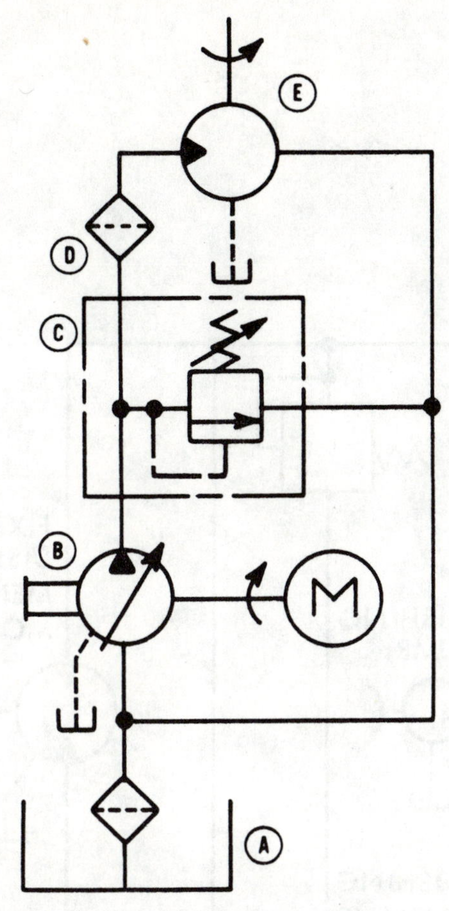

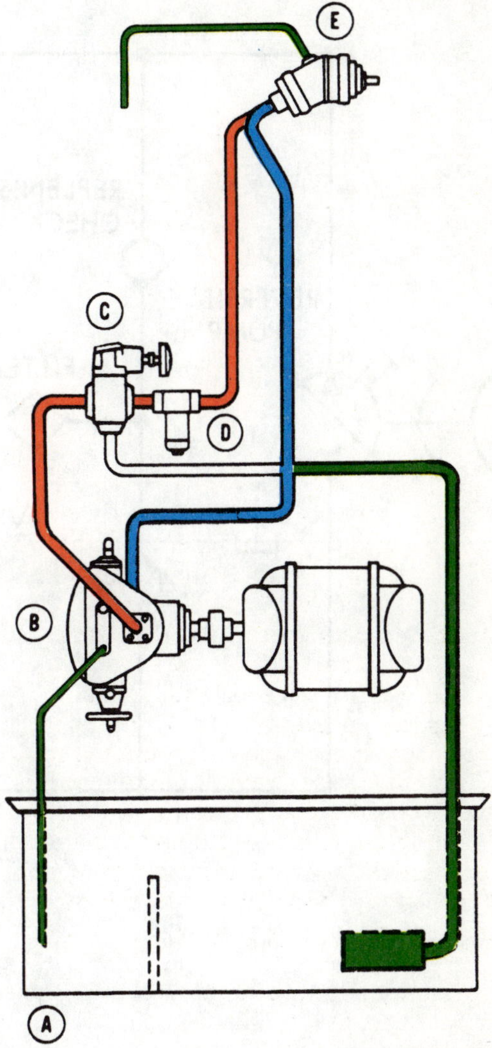

Variable delivery of pump (B) is directed through valve (C) and filter (D) to drive motor (E). Discharge from (E) returns to inlet of (B) along with make-up oil from reservoir (A). Valve (C) limits maximum torque of motor (E) and provides overload protection for system.

Fig. 13-20. - Closed Circuit Drive - One Direction

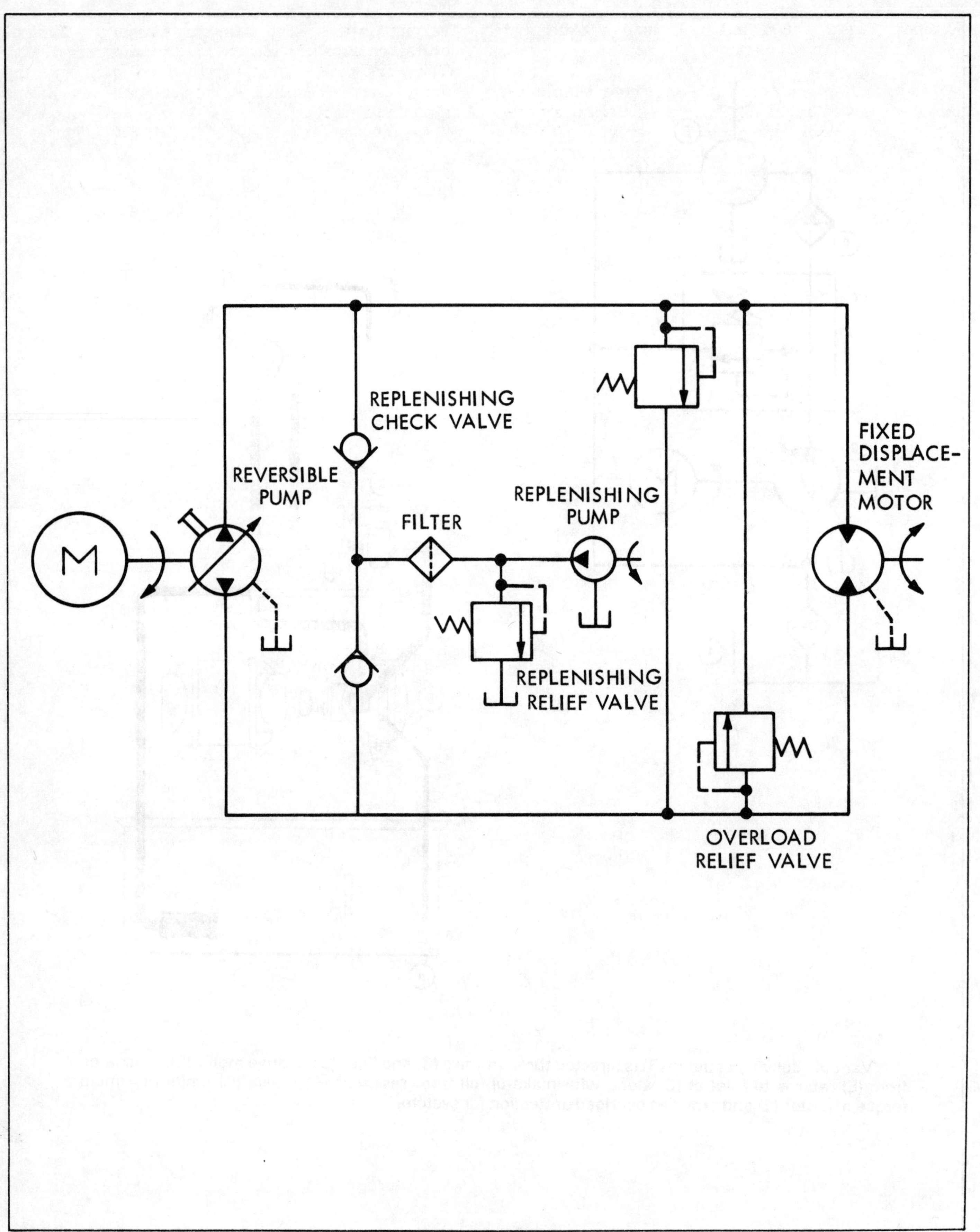

Fig. 13-21. Closed Circuit Drive - Reversible

displacements are equal. If not, torque and speed change in proportion.

Variable Displacement Pump, Fixed Displacement Motor - This combination is called a constant torque variable horsepower drive. Torque and pressure are always proportional, regardless of speed. The speed depends on the output of the pump, which is variable. An over-center control on the pump permits reversing the output rotation.

Fixed Displacement Pump, Variable Displacement Motor - When the motor displacement can be changed, but the pump's cannot, the power is always proportional to pressure. This combination is called a constant horsepower, variable torque drive. If the motor is pressure compensated, any increase in load (torque) results in a proportional decrease in speed.

Variable Displacement Pump and Motor - Some drive applications require various combinations of torque and power vs. speed. A variable displacement pump and motor then permit a very wide speed range, plus the operating characteristics of either the constant torque or constant horsepower drive.

"Packaged" Drives

Closed circuit drives for many applications are available as integral units with all controls and valving in a single, compact housing. Typical units are shown in Figure 13-22 and Figure 13-23. They are built with the motor integral with the drive, or in "split" versions with the motor mounted remotely.

QUESTIONS

1. Why is a high volume pump often unloaded in high pressure operation?
2. When a remote relief valve is used to provide a second maximum pressure, which valve must be set higher?
3. What are two ways of unloading the pump when an accumulator becomes charged?
4. Name two methods of transition from rapid advance to feed.
5. How can pressure be held on a clamp while the work cylinder is withdrawn?
6. To what degree does the brake valve restrict flow when a motor is accelerating?
7. Explain the difference between an open-circuit and a closed-circuit drive.
8. What combination of pump and motor will produce a constant torque drive?

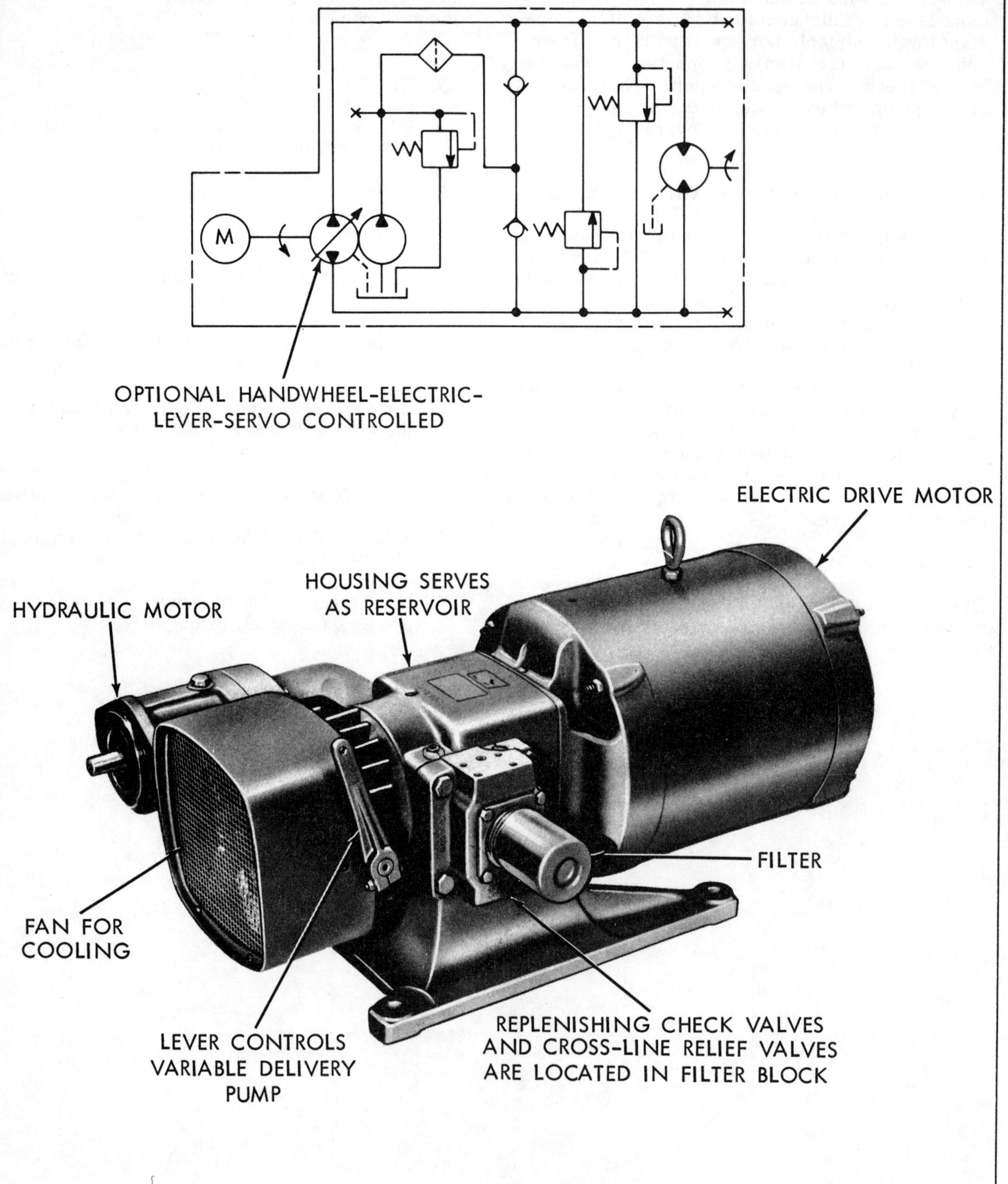

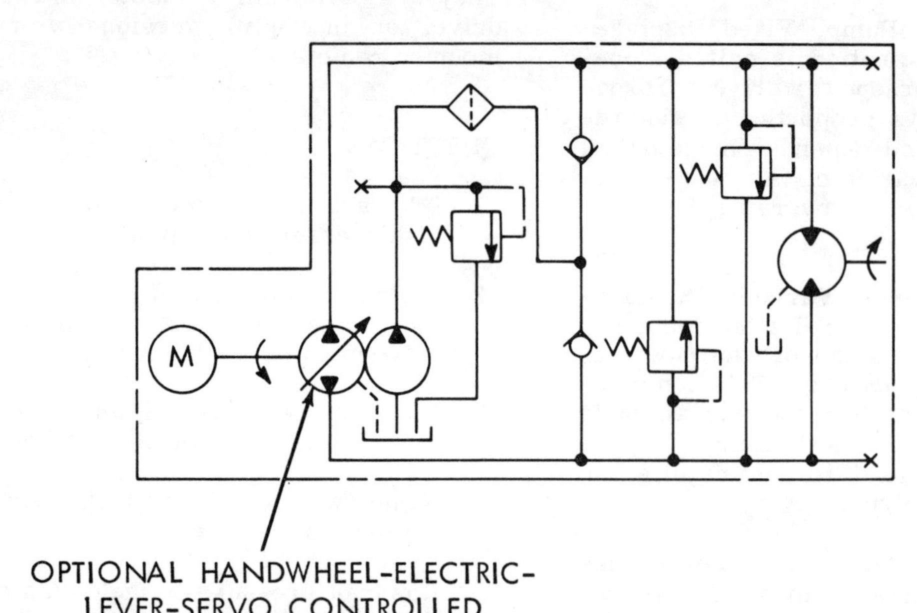

ELECTRIC DRIVE MOTOR

HOUSING SERVES
AS RESERVOIR

HYDRAULIC MOTOR

FILTER

FAN FOR
COOLING

LEVER CONTROLS
VARIABLE DELIVERY
PUMP

REPLENISHING CHECK VALVES
AND CROSS-LINE RELIEF VALVES
ARE LOCATED IN FILTER BLOCK

Fig. 13-22. - HAS Packaged Drive

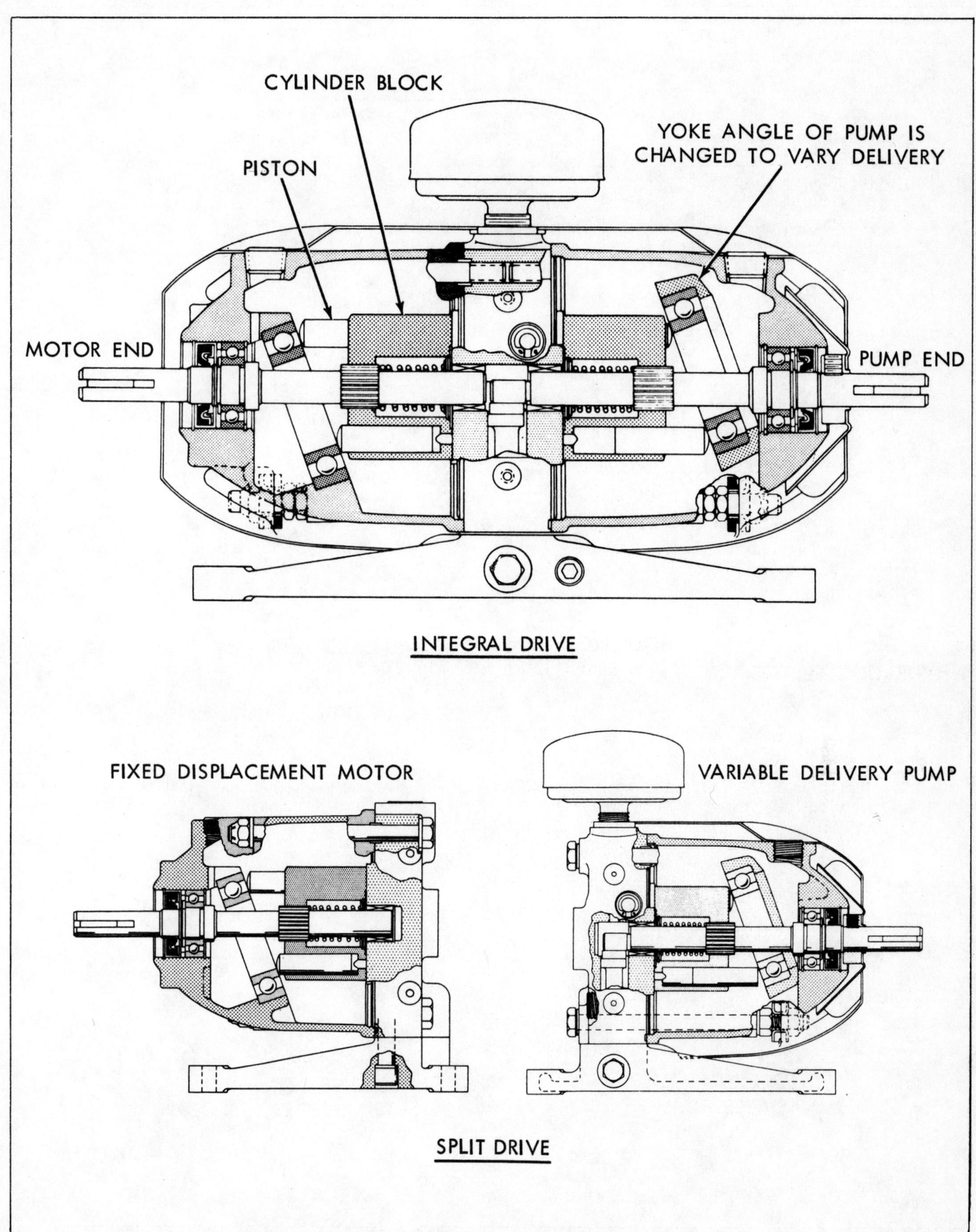

Fig. 13-23. - TR3 Packaged Drive

DEFINITIONS OF TECHNICAL TERMS

NOTE

Definitions listed relate to the context in which these terms are used in this manual. A more general definition of terms is given in the "Glossary of Terms for Fluid Power NFPA Recommended Standard T2.70.1".

ABSOLUTE - A measure having as its zero point or base the complete absence of the entity being measured.

ABSOLUTE PRESSURE - The pressure above zero absolute, i.e., the sum of atmospheric and gage pressure. In vacuum related work it is usually expressed in millimeters of mercury (mm Hg).

ACCUMULATOR - A container in which fluid is stored under pressure as a source of fluid power.

ACTUATOR - A device for converting hydraulic energy into mechanical energy. A motor or cylinder.

AERATION - Air in the hydraulic fluid. Excessive aeration causes the fluid to appear milky and components to operate erratically because of the compressibility of the air trapped in the fluid.

AMPLIFIER - A device for amplifying the error signal sufficiently to cause actuation of the stroke control. Several types of servo amplifiers are used at the present time: electronic (DC, AC, phase sensitive, and magnetic) and mechanical.

AMPLITUDE OF SOUND - The loudness of a sound.

ANNULAR AREA - A ring shaped area--often refers to the net effective area of the rod side of a cylinder piston, i.e., the piston area minus the cross-sectional area of the rod.

ATMOSPHERE (ONE) - A pressure measure equal to 14.7 psi.

ATMOSPHERIC PRESSURE - Pressure exerted by the atmosphere at any specific location. (Sea level pressure is approximately 14.7 pounds per square inch absolute.)

BACK CONNECTED - A condition where pipe connections are on normally unexposed surfaces of hydraulic equipment. (Gasket mounted units are back connected.)

BACK PRESSURE - A pressure in series. Usually refers to pressure existing on the discharge side of a load. It adds to the pressure required to move the load.

BAFFLE - A device, usually a plate, installed in a reservoir to separate the pump inlet from return lines.

BLEED-OFF - To divert a specific controllable portion of pump delivery directly to reservoir.

BREATHER - A device which permits air to move in and out of a container or component to maintain atmospheric pressure.

BY-PASS - A secondary passage for fluid flow.

CARTRIDGE

1. The replaceable element of a fluid filter.
2. The pumping unit from a vane pump, composed of the rotor, ring, vanes and one or both side plates.

CAVITATION - A localized gaseous condition within a liquid stream which occurs where the pressure is reduced to the vapor pressure.

CHAMBER - A compartment within a hydraulic unit. May contain elements to aid in operation or control of a unit. Examples: spring chamber, drain chamber, etc.

CHANNEL - A fluid passage, the length of which is large with respect to its cross-sectional dimension.

CHARGE (supercharge)

1. To replenish a hydraulic system above atmospheric pressure.

2. To fill an accumulator with fluid under pressure (see precharge pressure).

CHARGE PRESSURE - The pressure at which replenishing fluid is forced into the hydraulic system (above atmospheric pressure).

CHECK VALVE - A valve which permits flow of fluid in one direction only.

CHOKE - A restriction, the length of which is large with respect to its cross-sectional dimension.

CIRCUIT - An arrangement of components interconnected to perform a specific function within a system.

CLOSED CENTER VALVE - One in which all ports are blocked in the center or neutral position.

CLOSED CENTER CIRCUIT - One in which flow through the system is blocked in neutral and pressure is maintained at the maximum pressure control setting.

CLOSED LOOP - A system in which the output of one or more elements is compared to some other signal to provide an actuating signal to control the output of the loop.

COMMAND SIGNAL (or input signal) - An external signal to which the servo must respond.

COMPENSATOR CONTROL - A displacement control for variable pumps and motors which alters displacement in response to pressure changes in the system as related to its adjusted pressure setting.

COMPONENT - A single hydraulic unit.

COMPRESSIBILITY - The change in volume of a unit volume of a fluid when it is subjected to a unit change in pressure.

CONTROL - A device used to regulate the function of a unit (see Hydraulic Control, Manual Control, Mechanical Control, and Compensator Control).

COOLER - A heat exchanger used to remove heat from the hydraulic fluid.

COUNTERBALANCE VALVE - A pressure control valve which maintains back pressure to prevent a load from falling.

CRACKING PRESSURE - The pressure at which a pressure actuated valve begins to pass fluid.

CUSHION - A device sometimes built into the ends of a hydraulic cylinder which restricts the flow of fluid at the outlet port, thereby arresting the motion of the piston rod.

CYLINDER - A device which converts fluid power into linear mechanical force and motion. It usually consists of a movable element such as a piston and piston rod, plunger rod, plunger or ram, operating within a cylindrical bore.

DEADBAND - The region or band of no response where an error signal will not cause a corresponding actuation of the controlled variable.

DECOMPRESSION - The slow release of confined fluid to gradually reduce pressure on the fluid.

DELIVERY - The volume of fluid discharged by a pump in a given time, usually expressed in gallons per minute (gpm).

DE-VENT - To close the vent connection of a pressure control valve permitting the valve to function at its adjusted pressure setting.

DIFFERENTIAL CURRENT - The algebraic summation of the current in the torque motor; measured in MA (milliamperes).

DIFFERENTIAL CYLINDER - Any cylinder in which the two opposed piston areas are not equal.

DIRECTIONAL VALVE - A valve which selectively directs or prevents fluid flow to desired channels.

DISPLACEMENT - The quantity of fluid which can pass through a pump, motor or cylinder in a single revolution or stroke.

DITHER - A low amplitude, relatively high frequency periodic electrical signal, sometimes superimposed on the servo valve input to improve system resolution. Dither is expressed by the dither frequency (Hz) and the peak-to-peak dither current amplitude (ma).

DOUBLE ACTING CYLINDER - A cylinder in which fluid force can be applied to the movable element in either direction.

DRAIN - A passage in, or a line from, a hydraulic component which returns leakage fluid independently to reservoir or to a vented manifold.

EFFICIENCY - The ratio of output to input. Volumetric efficiency of a pump is the actual output in gpm divided by the theoretical or design output. The overall efficiency of a hydraulic system is the output power divided by the input power. Efficiency is usually expressed as a percent.

ELECTRO-HYDRAULIC SERVO VALVE - A directional type valve which receives a variable or controlled electrical signal and which controls or meters hydraulic flow.

ENERGY - The ability or capacity to do work. Measured in units of work.

ENCLOSURE - A rectangle drawn around a graphical component or components to indicate the limits of an assembly.

ERROR (signal) - The signal which is the algebraic summation of an input signal and a feedback signal.

FEEDBACK (or feedback signal) - The output signal from a feedback element.

FEEDBACK LOOP - Any closed circuit consisting of one or more forward elements and one or more feedback elements.

FILTER - A device whose primary function is the retention by a porous media of insoluble contaminants from a fluid.

FLOODED - A condition where the pump inlet is charged by placing the reservoir oil level above the pump inlet port.

FLOW CONTROL VALVE - A valve which controls the rate of oil flow.

FLOW RATE - The volume, mass, or weight of a fluid passing through any conductor per unit of time.

FLUID

1. A liquid or gas.
2. A liquid that is specially compounded for use as a power-transmitting medium in a hydraulic system.

FOLLOW VALVE - A control valve which ports oil to an actuator so the resulting output motion is proportional to the input motion to the valve.

FORCE - Any push or pull measured in units of weight. In hydraulics, total force is expressed by the product P (force per unit area) and the area of the surface on which the pressure acts. $F = P \times A$.

FOUR-WAY VALVE - A directional valve having four flow paths.

FREQUENCY - The number of times an action occurs in a unit of time. Frequency is the basis of all sound. A pump or motor's basic frequency is equal to its speed in revolutions per second multiplied by the number of pumping chambers.

FRONT CONNECTED - A condition wherein piping connections are on normally exposed surfaces of hydraulic components.

FULL FLOW - In a filter, the condition where all the fluid must pass through the filter element or medium.

GAUGE PRESSURE - A pressure scale which ignores atmospheric pressure. Its zero point is 14.7 psi absolute.

HEAD - The height of a column or body of fluid above a given point expressed in linear units. Head is often used to indicate gage pressure. Pressure is equal to the height times the density of the fluid.

HEAT - The form of energy that has the capacity to create warmth or to increase the temperature of a substance. Any energy that is wasted or used to overcome friction is converted to heat. Heat is measured in calories or British Thermal Units (BTU's). One BTU is the amount of heat required to raise the temperature of one pound of water one degree Fahrenheit.

HEAT EXCHANGER - A device which transfers heat through a conducting wall from one fluid to another.

HORSEPOWER - (HP) - The power required to lift 550 pounds one foot in one second or 33,000 pounds one foot in one minute. A horsepower is equal to 746 watts or to 42.4 British Thermal Units per minute.

HYDRAULIC BALANCE - A condition of equal opposed hydraulic forces acting on a part in a hydraulic component.

HYDRAULIC CONTROL - A control which is actuated by hydraulically induced forces.

HYDRAULICS - Engineering science pertaining to liquid pressure and flow.

HYDRODYNAMICS - Engineering science pertaining to the energy of liquid flow and pressure.

HYDROSTATICS - Engineering science pertaining to the energy of liquids at rest.

KINETIC ENERGY - Energy that a substance or body has by virtue of its mass (weight) and velocity.

LAMINAR (FLOW) - A condition where the fluid particles move in continuous parallel paths. Streamline flow.

LEVERAGE - A gain in output force over input force by sacrificing the distance moved. Mechanical advantage or force multiplication.

LIFT - The height a body or column of fluid is raised; for instance, from the reservoir to the pump inlet. Lift is sometimes used to express a negative pressure or vacuum. The opposite of head.

LINE - A tube, pipe or hose which acts as a conductor of hydraulic fluid.

LINEAR ACTUATOR - A device for converting hydraulic energy into linear motion--a cylinder or ram.

MANIFOLD - A fluid conductor which provides multiple connection ports.

MANUAL CONTROL - A control actuated by the operator, regardless of the means of actuation. Example: Lever or foot pedal control for directional valves.

MANUAL OVERRIDE - A means of manually actuating an automatically-controlled device.

MAXIMUM PRESSURE VALVE - See relief valve.

MECHANICAL CONTROL - Any control actuated by linkages, gears, screws, cams or other mechanical elements.

METER - To regulate the amount or rate of fluid flow.

METER-IN - To regulate the amount of fluid flow into an actuator or system.

METER-OUT - To regulate the flow of the discharge fluid from an actuator or system.

MICRON - One-millionth of a meter or about .00004 inch.

MICRON RATING - The size of the particles a filter will remove.

MOTOR - A device which converts hydraulic fluid power into mechanical force and motion. It usually provides rotary mechanical motion.

OPEN CENTER CIRCUIT - One in which pump delivery flows freely through the system and back to the reservoir in neutral.

OPEN CENTER VALVE - One in which all ports are interconnected and open to each other in the center or neutral position.

ORIFICE - A restriction, the length of which is small in respect to its cross-sectional dimensions.

PASSAGE - A machined or cored fluid conducting path which lies within or passes through a component.

PILOT PRESSURE - Auxiliary pressure used to actuate or control hydraulic components.

PILOT VALVE - An auxiliary valve used to control the operation of another valve. The controlling stage of a 2-stage valve.

PISTON - A cylindrically shaped part which fits within a cylinder and transmits or receives motion by means of a connecting rod.

PLUNGER - A cylindrically shaped part which has only one diameter and is used to transmit thrust. A ram.

POPPET - That part of certain valves which prevents flow when it closes against a seat.

PORT - An internal or external terminus of a passage in a component.

POSITIVE DISPLACEMENT - A characteristic of a pump or motor which has the inlet positively sealed from the outlet so that fluid cannot recirculate in the component.

POTENTIOMETER - A control element in the servo system which measures and controls electrical potential.

POWER - Work per unit of time. Measured in horsepower (hp) or watts.

POWER PACK - An integral power supply unit usually containing a pump, reservoir, relief valve and directional control.

PRECHARGE PRESSURE - The pressure of compressed gas in an accumulator prior to the admission of liquid.

PRESSURE - Force per unit area; usually expressed in pounds per square inch (psi).

PRESSURE DROP - The difference in pressure between any two points of a system or a component.

PRESSURE LINE - The line carrying the fluid from the pump outlet to the pressurized port of the actuator.

PRESSURE OVERRIDE - The difference between the cracking pressure of a valve and the pressure reached when the valve is passing full flow.

PRESSURE PLATE - A side plate in a vane pump or motor cartridge on the pressure port side.

PRESSURE REDUCING VALVE - A valve which limits the maximum pressure at its outlet regardless of the inlet pressure.

PRESSURE SWITCH - An electric switch operated by fluid pressure.

PROPORTIONAL FLOW - In a filter, the condition where part of the flow passes through the filter element in proportion to pressure drop.

PUMP - A device which converts mechanical force and motion into hydraulic fluid power.

RAM - A single-acting cylinder with a single diameter plunger rather than a piston and rod. The plunger in a ram-type cylinder.

RECIPROCATION - Back-and-forth straight line motion or oscillation.

REGENERATIVE CIRCUIT - A piping arrangement for a differential type cylinder in which discharge fluid from the rod end combines with pump delivery to be directed into the head end.

RELIEF VALVE - A pressure operated valve which by-passes pump delivery to the reservoir, limiting system pressure to a predetermined maximum value.

REPLENISH - To add fluid to maintain a full hydraulic system.

RESERVOIR - A container for storage of liquid in a fluid power system.

RESTRICTION - A reduced cross-sectional area in a line or passage which produces a pressure drop.

RETURN LINE - A line used to carry exhaust fluid from the actuator back to sump.

REVERSING VALVE - A four-way directional valve used to reverse a double-acting cylinder or reversible motor.

ROTARY ACTUATOR - A device for converting hydraulic energy into rotary motion--a hydraulic motor.

SEQUENCE

1. The order of a series of operations or movements.
2. To divert flow to accomplish a subsequent operation or movement.

SEQUENCE VALVE - A pressure operated valve which, at its setting, diverts flow to a secondary line while holding a predetermined minimum pressure in the primary line.

SERVO MECHANISM (servo) - A mechanism subjected to the action of a controlling device which will operate as if it were directly actuated by the controlling device, but capable of supplying power output many times that of the controlling device, this power being derived from an external and independent source.

SERVO VALVE

1. A valve which modulates output as a function of an input command.
2. A follow valve.

SIGNAL - A command or indication of a desired position or velocity.

SINGLE ACTING CYLINDER - A cylinder in which hydraulic energy can produce thrust or motion in only one direction. (May be mechanically or gravity returned.)

SLIP - Internal leakage of hydraulic fluid.

SPOOL - A term loosely applied to almost any moving cylindrically shaped part of a hydraulic component which moves to direct flow through the component.

STRAINER - A coarse filter.

STREAMLINE FLOW - (See laminar flow.)

STROKE

1. The length of travel of a piston or plunger.
2. To change the displacement of a variable displacement pump or motor.

SUB-PLATE - An auxiliary mounting for a hydraulic component providing a means of connecting piping to the component.

SUCTION LINE - The hydraulic line connecting the pump inlet port to the reservoir or sump.

SUMP - A reservoir'

SUPERCHARGE - (See charge.)

SURGE - A transient rise of pressure or flow.

SWASH PLATE - A stationary canted plate in an axial type piston pump which causes the pistons to reciprocate as the cylinder barrel rotates.

SYNCHRO - A rotary electromagnetic device generally used as an AC feedback signal generator which indicates position. It can also be used as a reference signal generator.

TACHOMETER -- (AC) (DC) - A device which generates an AC or DC signal proportional to the speed at which it is rotated and the polarity of which is dependent on the direction of rotation of the rotor.

TANK - The reservoir or sump.

THROTTLE - To permit passing of a restricted flow. May control flow rate or create a deliberate pressure drop.

TORQUE - A rotary thrust. The turning effort of a fluid motor usually expressed in inch pounds.

TORQUE CONVERTER - A rotary fluid coupling that is capable of multiplying torque.

TORQUE MOTOR - A type of electromechanical transducer having rotary motion used in the input stages of servo valves.

TRANSDUCER (or feedback transducer) - An element which measures the results at the load and sends a signal back to the amplifier.

TURBULENT FLOW (TURBULENCE) - A condition where the fluid particles move in random paths rather than in continuous parallel paths.

TURBINE - A rotary device that is actuated by the impact of a moving fluid against blades or vanes.

TWO-WAY VALVE - A directional control valve with two flow paths.

UNLOAD - To release flow (usually directly to the reservoir), to prevent pressure being imposed on the system or portion of the system.

UNLOADING VALVE - A valve which by-passes flow to tank when a set pressure is maintained on its pilot port.

VACUUM - Pressure less than atmospheric pressure. It is usually expressed in inches of mercury (in. Hg) as referred to the existing atmospheric pressure.

VALVE - A device which controls fluid flow direction, pressure, or flow rate.

VELOCITY

1. The speed of flow through a hydraulic line. Expressed in feet per second (fps) or inches per second (ips).

2. The speed of a rotating component measured in revolutions per minute (rpm).

VENT

1. To permit opening of a pressure control valve by opening its pilot port (vent connection) to atmospheric pressure.

2. An air breathing device on a fluid reservoir.

VISCOSITY - A measure of the internal friction or the resistance of a fluid to flow.

VISCOSITY INDEX - A measure of the viscosity-temperature characteristics of a fluid as referred to that of two arbitrary reference fluids.

VOLUME

1. The size of a space or chamber in cubic units.

2. Loosely applied to the output of a pump in gallons per minute (gpm).

WOBBLE PLATE - A rotating canted plate in an axial type piston pump which pushes the pistons into their bores as it "wobbles".

WORK - Exerting a force through a definite distance. Work is measured in units of force multiplied by distance; for example, pound-foot.

STANDARD GRAPHICAL SYMBOLS

THE SYMBOLS SHOWN CONFORM TO THE AMERICAN NATIONAL STANDARDS INSTITUTE (ANSI) SPECIFICATIONS. BASIC SYMBOLS CAN BE COMBINED IN ANY COMBINATION. NO ATTEMPT IS MADE TO SHOW ALL COMBINATIONS.

LINES AND LINE FUNCTIONS	
LINE, WORKING	
LINE, PILOT (L>20W)	
LINE, DRAIN (L<5W)	
CONNECTOR	
LINE, FLEXIBLE	
LINE, JOINING	
LINE, PASSING	
DIRECTION OF FLOW, HYDRAULIC PNEUMATIC	
LINE TO RESERVOIR ABOVE FLUID LEVEL BELOW FLUID LEVEL	
LINE TO VENTED MANIFOLD	
PLUG OR PLUGGED CONNECTION	
RESTRICTION, FIXED	
RESTRICTION, VARIABLE	

PUMPS	
PUMP, SINGLE FIXED DISPLACEMENT	
PUMP, SINGLE VARIABLE DISPLACEMENT	
MOTORS AND CYLINDERS	
MOTOR, ROTARY, FIXED DISPLACEMENT	
MOTOR, ROTARY VARIABLE DISPLACEMENT	
MOTOR, OSCILLATING	
CYLINDER, SINGLE ACTING	
CYLINDER, DOUBLE ACTING	
CYLINDER, DIFFERENTIAL ROD	
CYLINDER, DOUBLE END ROD	
CYLINDER, CUSHIONS BOTH ENDS	

MISCELLANEOUS UNITS	
DIRECTION OF ROTATION (ARROW IN FRONT OF SHAFT)	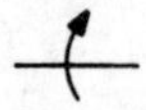
COMPONENT ENCLOSURE	
RESERVOIR, VENTED	
RESERVOIR, PRESSURIZED	
PRESSURE GAGE	
TEMPERATURE GAGE	
FLOW METER (FLOW RATE)	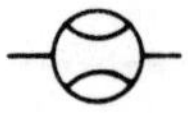
ELECTRIC MOTOR	 M
ACCUMULATOR, SPRING LOADED	
ACCUMULATOR, GAS CHARGED	
FILTER OR STRAINER	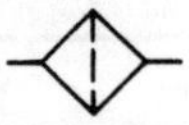
HEATER	
COOLER	
TEMPERATURE CONTROLLER	
INTENSIFIER	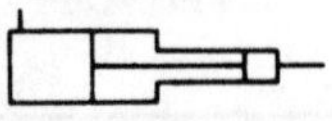
PRESSURE SWITCH	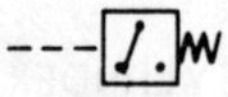 W
BASIC VALVE SYMBOLS	
CHECK VALVE	
MANUAL SHUT OFF VALVE	
BASIC VALVE ENVELOPE	
VALVE, SINGLE FLOW PATH, NORMALLY CLOSED	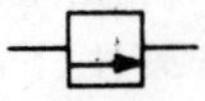

BASIC VALVE SYMBOLS (CONT.)	
VALVE, SINGLE FLOW PATH, NORMALLY OPEN	
VALVE, MAXIMUM PRESSURE (RELIEF)	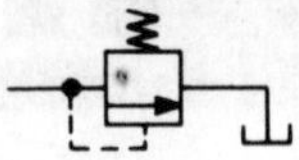
BASIC VALVE SYMBOL, MULTIPLE FLOW PATHS	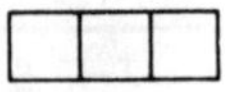
FLOW PATHS BLOCKED IN CENTER POSITION	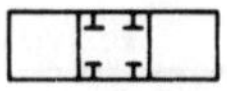
MULTIPLE FLOW PATHS (ARROW SHOWS FLOW DIRECTION)	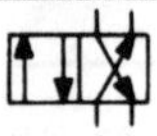
VALVE EXAMPLES	
UNLOADING VALVE, INTERNAL DRAIN, REMOTELY OPERATED	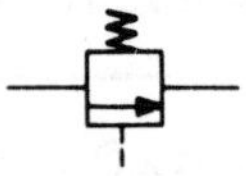
DECELERATION VALVE, NORMALLY OPEN	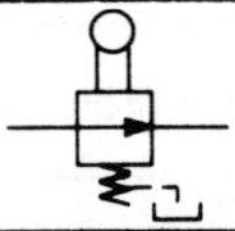
SEQUENCE VALVE, DIRECTLY OPERATED, EXTERNALLY DRAINED	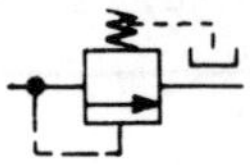
PRESSURE REDUCING VALVE	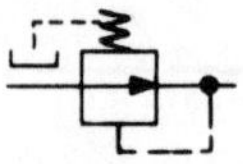
COUNTER BALANCE VALVE WITH INTEGRAL CHECK	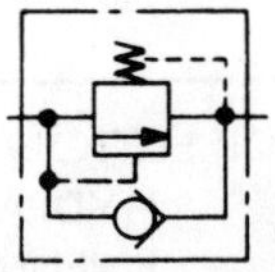
TEMPERATURE AND PRESSURE COMPENSATED FLOW CONTROL WITH INTEGRAL CHECK	
DIRECTIONAL VALVE, TWO POSITION, THREE CONNECTION	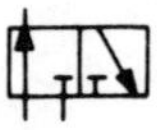
DIRECTIONAL VALVE, THREE POSITION, FOUR CONNECTION	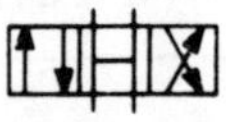
VALVE, INFINITE POSITIONING (INDICATED BY HORIZONTAL BARS)	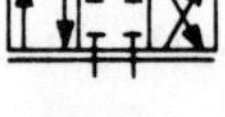

METHODS OF OPERATION	
PRESSURE COMPENSATOR	
DETENT	
MANUAL	
MECHANICAL	
PEDAL OR TREADLE	
PUSH BUTTON	

METHODS OF OPERATION	
LEVER	
PILOT PRESSURE	
SOLENOID	
SOLENOID CONTROLLED, PILOT PRESSURE OPERATED	
SPRING	
SERVO	

INDEX

A

Absolute pressure 1-21, 2-5
Accumulators . . . . 12-1 thru 12-6
Actuators 6-1
Aeration 1-8
Amplifier, servo 8-3
Atmosphere 2-3

B

Baffle plate 5-1
Balanced piston relief valve 9-1
Barometer 2-3
Bent-axis piston pump 11-29
Bernoulli's principle 2-9
Bleed-off circuit 10-2
Brake valves 9-11
Breather 5-1

C

Cartridge, pump 11-14, 11-23
Catalysts 3-7
Cavitation 1-8
Charge (pre-charge) 12-3
Check valves . . . . 7-1 thru 7-7
Closed circuit 13-45
Combination pumps . . . . 11-10 thru 11-12
Compensator control (pressure),
  hydraulic motor 6-24
  piston pump 11-26
Compound relief valve 9-1
Coolers 5-13
Counterbalance valves 9-7
Cylinders
  options 6-6
  ratings 6-4
  types 6-1
Cracking pressure 9-1

D

Deceleration valves 7-23
Demulsibility 3-8
Directional controls . . . . 7-1 thru 7-30
Displacement 6-6, 11-2
Dither 8-4
Double pumps 11-9, 11-14

E

Efficiency, volumetric 11-2
Electro-hydraulic servo valves . . 8-3 thru 8-10
Energy,
  conservation of 1-3, 2-9
  kinetic and potential 2-9

F

Feedback 8-1, 8-2
Filters 5-3
Flapper type servo valve 8-4
Flow
  laminar and turbulent 2-6
  principles 2-5
  velocity in pipes 1-19
Flow controls . . . . 10-1 thru 10-7
Flow meters 12-11
Fluids 3-1
Follow valve 8-1
Four-way valves 7-7

G

Gauges, pressure 12-8
Gear pump 11-3
Graphical symbols 2-9

H

Head (pressure) 2-2
Heat exchangers 5-13
"High Performance" vane pump 11-14
Horsepower 1-21
Hose, hydraulic
  fittings 4-8
  installation 4-11
  safety factor 4-8
  sizing 4-8, 4-9
Hydrodynamic device 2-2, 11-1
Hydrostatic device 2-2, 11-2

I

Instrumentation 12-8
Intensifiers 12-6
Intra-vane design (pump) 11-23

INDEX (Cont.)

J

Jet pipe servo valve 8-4

K

Kinetic energy 2-9

L

Laminar flow 2-6
Leakage 4-11
Line
  preparation 4-10
  sizing 1-19, 4-8
Linear actuators 6-1
Lubrication 3-1, 3-7

M

Maximum pressure valve 9-1
Mechanical servo 8-1
Mesh 5-4
Meter - in circuit 10-1
Meter - out circuit 10-1
Micron 5-4
Motors, hydraulic
  application formulas 6-10
  bent-axis piston 6-24
  gear 6-11
  inline piston 6-17
  low speed, high torque 6-17
  pressure compensated 6-24
  rating 6-6
  vane 6-1

N

Nomograph chart 4-9

O

Oil 3-1
Oxidation resistance 3-7

P

Pascal's law 1-1
Pilot choke 7-23
Pilot valve 7-15
Pipe
  installation 4-10
  preparation 4-10, 4-12
  sealing 4-1
  sizing 4-1
Piston pumps . . . . 11-24 thru 11-31
Piston pump controls . . . . . . 11-26 thru 11-31
Pour point 3-7
Power 1-21
Pre-charge 12-3
Pressure
  absolute 1-8, 2-5
  at pump inlet 1-8
  atmospheric 2-3
  causes . . . . 1-10 thru 1-13, 2-2
  definition 1-3, 2-1
  drop 1-13
  in a column 1-4
Pressure controls . . . . 9-1 thru 9-16
Pressure gauges 12-8
Pressure override 9-1
Pressure reducing valves
  direct operating 9-13
  pilot operated 9-15
Pressure switch 12-6
Proportional flow filter 5-10
Pump
  efficiency 11-2
  ratings 11-2

R

"R" valves . . . . 9-3 thru 9-13
Radial piston pumps 11-24
Regenerative circuit 13-15
Relief valves
  compound 9-1
  "R" type 9-7
  simple 9-1
Remote flow control 10-7
Reservoirs 5-1

S

SAE numbers 3-5
Schematic symbols 2-9

INDEX (Cont.)

Seals . . . . 4-11 thru 4-23
Servo valves
flapper 8-4
jet pipe 8-4
mechanical 8-1
single-stage 8-3
two-stage. 8-4
Sequence valves
compound "X" and "Y" 9-13
"R" type 9-7
Slippage 11-2
Solenoid 7-10
Speed
of a cylinder 1-19
of a hydraulic motor 6-11
"Square" pump 11-12
Strainers 5-3

T

Torque 1-22
Torque generators 6-28
Torque motor 8-3
Torricelli, Evangelista 2-2
Tubing
fitting. 4-3
installation 4-10
size 4-3
Turbulent flow 2-6
Two-stage pump. 11-9
Two-way valves 7-7

U

Unloading relief valve 9-16
Unloading valve 9-7

V

Vacuum
at pump inlet 1-8
measuring 2-3
Vane pumps. . . . . 11-7 thru 11-24
Venting. 9-3
Venturi effect 2-9
Viscosity 3-3
Viscosity index 3-6
Volume control . . . . 10-1 thru 10-7
Volumetric efficiency 11-2

W

Wobble plate pump 11-29
Work 1-21

X

"X" type sequence valve. 9-13

Y

"Y" type sequence valve. 9-13